STUDENT STUDY GUIDE
FOR CAMPBELL'S

BIOLOGY

SECOND EDITION

Martha R. Taylor, Ph.D.

Cornell University

The Benjamin/Cummings Publishing Company, Inc.

Redwood City, California • Fort Collins, Colorado • Menlo Park, California
Reading, Massachusetts • New York • Don Mills, Ontario • Wokingham, U.K.
Amsterdam • Bonn • Sydney • Singapore • Tokyo • Madrid • San Juan

Sponsoring Editor: Robin Williams
Production Coordinator: Eleanor Renner Brown
Copyeditor: Merry Finley
Composition: Merry Finley, Desktop Productions

ISBN 0-8053-1802-X

ABCDEFGHIL-AL-8932109

The Benjamin/Cummings Publishing Company, Inc.
390 Bridge Parkway, Suite 102
Redwood City, California 94065

CONTENTS

When I first thought about writing a student study guide for Neil Campbell's *Biology*, I wanted to call it a structuring guide. The purpose of this book is to help you to structure and organize your developing knowledge of biology and to create your own personal understanding of the topics covered in the text. The four components of each chapter of this study guide are designed to help you in this task. The *Framework* identifies the overall picture; it provides a conceptual framework into which the chapter information fits. The *Chapter Summary* condenses the major concepts of each chapter. The *Structure Your Knowledge* section directs you to organize and relate these concepts by completing tables, answering essay questions, and constructing concept maps. In the *Test Your Knowledge* section, you are provided with a battery of objective questions to test your understanding. Suggested answers to the *Structure Your Knowledge* and *Test Your Knowledge* questions are provided in the *Answer Section* at the end of each unit.

What are the concept maps that often appear in this study guide? A concept map is a diagram that shows how ideas are organized and related. The *structure* of a concept map is a hierarchically organized cluster of concepts, enclosed in boxes and connected with lines that explicitly state the relationships between the concepts. The *function* of a concept map is to help you structure your understanding of a topic and create personal meaning. The *value* of a concept map is in the thinking and evaluating required to create a map.

Developing a concept map for a group of concepts requires that you evaluate the relative importance of the concepts (which are most inclusive and important? which are less important and subordinate to other concepts?); arrange the concepts in a meaningful cluster; and draw connections between them that help you make their meanings explicit. (Additional information about concept mapping may be found in *Learning How to Learn*, Joseph D. Novak and D. Bob Gowin, Cambridge University Press, 1984.)

This book uses concept maps in several ways. A map of a chapter may be presented in the *Framework* section to show the organization of the key concepts in that chapter. More often, you will develop a concept map on a certain subset of ideas from the chapter. The answer section will present *suggested* concept maps. A concept map is an individual picture of your understanding at the time you make the map. Meanings change and grow as you gain more knowledge and experience in an area. As your understanding of an area develops, your concept map will evolve—sometimes becoming more complex and interrelated, sometimes becoming simplified and streamlined. Do not look to the answer section for the "right" concept map. *After* you have organized your own thoughts, look at the suggested map to make sure you have included the key concepts (although you may have added more), to check that the connections you have made are reasonable, and perhaps to see another way to organize the information.

In the multiple choice questions presented in each chapter, you are asked to choose the best answer. Some answers may be partly correct; almost all choices have been written to test your ability to think and discriminate. Make sure you understand why the other choices are incorrect, as well as why the given answer is correct.

Biology is a fascinating and broad subject. Campbell's *Biology* is filled with terminology and facts that are useful for understanding the major themes of modern biology. This *Student Study Guide* is intended to help you learn and recall information and, most importantly, to encourage and guide you as you develop your own understanding of and appreciation for biology.

CHAPTER 1

INTRODUCTION: THEMES IN THE STUDY OF LIFE

FRAMEWORK

This chapter outlines seven themes that unify the study of biology and describes the scientific construction of biological knowledge. A course in biology is neither a vocabulary course nor a classification exercise for the diverse forms of life. Biology is a collection of facts and concepts that are structured within theories and organizing principles. Recognizing the common themes within biology will help you to structure your knowledge of this fascinating and challenging study of life.

CHAPTER SUMMARY

Biology is the scientific study of life, an extension of our innate interest in life in its diverse forms. The scope of biology is immense, spanning from the submicroscopic realm of molecules to the complex web of ecosystems, from the present back through 4 billion years of evolutionary history. Recent advances in research methods, coupled with the large number of contemporary biologists working throughout the many subfields of biology, have led to an explosion of information. A beginning student can make sense of this expanding universe of knowledge by focusing on a few enduring themes that unify the study of biology.

A Hierarchy of Organization

Biological organization is based on a hierarchy of structural levels, ranging from atoms, through biological molecules, organelles, cells, tissues, organs, organ systems, organisms, populations, and communities, to ecosystems. The study of biology includes understanding the various levels of biological organization and the connection between these levels.

Emergent Properties

The interactions among the components of a particular level of organization lead to the emergence of novel properties at the next level: the whole is often greater than the sum of its parts. The emergent properties of life do not support the doctrine of vitalism, which claims that life is driven by forces that defy explanation. The principles of physics and chemistry can be used to unravel and understand the properties of life that arise from the hierarchy of structural organization.

The characteristics of life include the ability to take in energy and transform it to do work and to maintain an ordered state; the ability to respond to stimuli from the environment; the reproduction, growth, and development of organisms as directed by heritable programs; and the evolution of a population as it accumulates adaptations to its specific environment.

Following the principle of holism, the higher organizational levels of life are studied to comprehend the properties that arise from the interactions and integration of the system's parts. Reductionism, in contrast, breaks down complex systems to simple components that are more manageable. The reductionist approach to biology has been a powerful strategy for understanding the most basic components of this complex phenomenon called life. The exposition of one level of organization gives insight into the emergent properties of the next level.

The Cellular Basis of Life

The cell is the simplest level of structure capable of performing all the activities of life; it is life's basic unit of structure and function. Hooke first described and named cells in 1665, when he observed a slice of cork with a simple microscope. Leeuwenhoek developed lenses that permitted him to discover the world of microscopic organisms. In 1839, Schleiden and

Schwann concluded that all living things consist of cells.

The recent advent of the electron microscope has revealed the complex structural organization of cells. Two major types of cells are recognized. The simpler prokaryotic cell, found only in bacteria, lacks both a nucleus to enclose its DNA and most membrane-bound organelles. The eukaryotic cell, with its nucleus and numerous cytoplasmic compartments, is typical of all other living organisms. The chemical processes of all cells, however, are remarkably similar.

The Correlation of Structure and Function

A study of the form of a biological structure gives information on its functioning, and a study of function provides insight into structural organization. The principle that form fits function is illustrated in each of the many structural levels of biological organization.

The Interaction of Organisms with Their Environment

Organisms affect and are affected by both the physical and biological environments with which they interact. Within an ecosystem, nutrients cycle between the abiotic and biotic components, and energy flows from sunlight to photosynthetic organisms to the life forms that feed on plants and other organisms.

The Inheritance of Biological Information

The biological instructions for the development and functioning of organisms are coded in the arrangement of the four chemical letters, or nucleotides, of DNA molecules, and transmitted from parents to offspring in the units of inheritance called genes. All forms of life use essentially the same genetic code to write and transcribe their heritable script.

Unity in Diversity

The diversity of life is estimated to be as great as 30 million species. Taxonomy, the branch of biology that names and classifies species, groups organisms into hierarchical classes that reflect their relationships. The diverse forms of life are organized into five kingdoms: Monera, Protista, Plantae, Fungi, and Animalia.

The kingdom Monera, which contains bacteria, is distinguished on the basis of the simple prokaryotic cell type. All other kingdoms have eukaryotic cells. Protists are mostly unicellular, or simple multicellular

forms. The other three kingdoms contain multicellular organisms that are characterized to a large extent by their mode of nutrition. Plants are photosynthetic; fungi are mostly decomposers that absorb their nutrients from dead organisms or organic wastes; and animals obtain their food by ingestion.

Underlying the diversity of life forms are a universal genetic code, similarities in metabolic pathways, and commonalities in cell structure.

Evolution: The Core Theme

Evolution is the one biological theme that connects all of life by common ancestry. The history of living forms extends back over 3 billion years to the ancient prokaryotes. Each species is the tip of an evolutionary branch that connects closely related species with their common ancestors.

In *The Origin of Species*, published in 1859, Charles Darwin presented his case for "descent with modification," the theory of the evolution of present forms from a succession of ancestral forms as a result of a mechanism called natural selection. Darwin synthesized the concept of natural selection from three generalized observations: (1) individuals vary in many heritable traits; (2) many more young are produced within a population than can be supported by the environment; and (3) individuals with traits best suited for the environment leave a larger proportion of offspring than do less fit individuals. This differential reproductive success results in the gradual accumulation of adaptations within a population.

According to Darwin, new species originate as unique traits and combinations of traits that arise randomly are selected for by differing environments. Descent with modification makes sense of both the unity and diversity of life; evolution is the core theme of biology.

Science as a Way of Knowing

Science emerges from our curiosity and drive to understand ourselves, the world, and the universe. Asking questions about nature and believing that those questions are answerable provide the basis of science. Few scientists rigidly follow the prescribed steps of the "scientific method," but researchers do focus on gathering evidence in experiments or by observations to test hypotheses—tentative answers to specific questions. Scientific studies usually include a control, an unaltered system or group of organisms that permits the testing of the effect of a single variable in the experimental group.

Science is characterized as progressive and self-correcting. Scientists build on the work done by others, refining or refuting their ideas, both cooperating and competing with each other.

Facts, in the form of observations and experimental results, are prerequisites of science, but new ways of organizing and relating those facts advance science. Newton, Darwin, and Einstein are outstanding scientists because they synthesized theories with great explanatory power. A theory is broader in scope and more widely accepted than a hypothesis. Good theories generate testable hypotheses, which in turn lead to new questions and hypotheses.

Biology is a demanding science—partly because living systems are so complex and partly because biology incorporates concepts from chemistry, physics and math. This book presents a wealth of information. The basic themes of biology will help you understand, appreciate, and structure your growing knowledge of biology.

STRUCTURE YOUR KNOWLEDGE

1. This chapter presents eight unifying themes of biology. Briefly describe each of these themes.
   A hierarchy of organization
   Emergent properties
   The cellular basis of life
   The correlation of structure and function
   The interaction of organisms with their environment
   The inheritance of biological information
   Unity in diversity
   Evolution: the core theme

2. Biology is the scientific study of life. How would you characterize a scientific approach to developing biological knowledge?

CHAPTER 1
INTRODUCTION: THEMES IN THE STUDY OF LIFE

Suggested Answers to Structure Your Knowledge

1. a. Living things exhibit a hierarchy of structural levels of organization, ranging all the way from the organization of particles in atoms, through molecules, macromolecules, cellular organelles, cells, tissues, organs, organ systems, whole organisms, populations, and communities, to ecosystems.

   b. Unique properties emerge at each structural level as a result of the interactions and organization of components.

   c. All living forms are composed of cells, the basic unit of structure and function that exhibits the characteristics of life.

   d. Within each level of organization is a correlation between structure and function—that is, between the physical arrangement of parts and the function they serve.

   e. Organisms influence and are influenced by the physical and biological components of their environment. Ecosystems are characterized by the cycling of nutrients and the flow of energy.

   f. The structures and processes of life are coded for by DNA, the universal language of inheritance.

   g. The diversity of life, as illustrated by the five-kingdom classification of organisms, is unified by the genetic code, similar metabolic pathways, and common cellular characteristics.

   h. Evolution, the connection of all life forms through common ancestry and the origin of new life forms through the accumulation of adaptations to specific environments by natural selection, is the core theme of biology.

2. The scientific study of life rests on the belief that natural phenomena have natural causes and that questions about life are answerable. Scientists develop hypotheses, which are tentative answers to questions, that are then tested through the accumulation of evidence from experiments and observations. Scientists share evidence and conclusions, then repeat, reinforce, and refute each other's ideas. Theories are broad explanations of various phenomena that serve to organize biological knowledge, generate new questions, and direct future studies.

THE CHEMISTRY OF LIFE

THE FAR SIDE   By GARY LARSON

"What the? ... This is lemonade! Where's my culture of amoebic dysentery?"

CHAPTER 2

ATOMS, MOLECULES, AND CHEMICAL BONDS

FRAMEWORK

This chapter considers the basic principles of chemistry that explain the behavior of atoms and molecules and that form the basis for our modern understanding of biology. The emergent properties associated with each new level of structural organization are evident even as subatomic particles are organized into atoms and atoms into molecules. The following concept map sketches out the relationships among the key concepts included in this chapter.

CHAPTER SUMMARY

An understanding of the structure and interaction of the atoms and molecules that constitute living organisms is essential to the study of the phenomenon we call life. A reductionist approach to the study of biology focuses on lower levels of organization. The principles of chemistry that govern how atoms and molecules behave lay the foundation for each successive structural level in the organization of life and for the emergent properties that accompany each level.

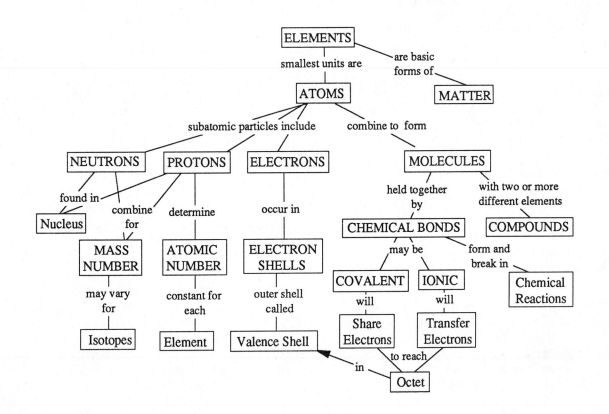

Matter: Elements and Compounds

Chemistry is the study of *matter*, anything that takes up space and has mass. Mass is a measure of the amount of matter within an object; weight is the measure of how strongly that mass is pulled by gravity. On earth, weight is a measure of mass or the quantity of matter in an object.

The basic forms of matter are *elements*, substances that cannot be broken down to other types of matter by ordinary chemical means. Chemists have identified 92 naturally occurring elements and created a dozen synthetic ones. Each element is symbolized by one or two letters.

A *compound* is made up of two or more elements combined in a fixed ratio. A compound usually has characteristics quite different from its constituent elements, an example of the emergence of novel properties in higher levels of organization.

Elements Essential to Life   About 25 elements are essential to life. Of these, carbon (C), oxygen (O), hydrogen (H), and nitrogen (N) make up 96% of living matter. The remaining 4% is composed of phosphorus (P), sulfur (S), calcium (Ca), and a few others. Some elements, like iron (Fe) and iodine (I), are required in extremely minute quantities and are called trace elements.

The Structure and Behavior of Atoms

An *atom* is the smallest unit of an element retaining the physical and chemical properties of that element. Each element has its own unique type of atom.

Subatomic Particles   Physicists have determined that atoms are composed of more than a hundred types of subatomic particles. Three stable subatomic particles are important to our understanding of atoms: Uncharged *neutrons* and positively charged *protons* are packed tightly together to form the *nucleus* of an atom. Negatively charged *electrons* orbit rapidly about the nucleus, electrically attracted to the positively charged nucleus.

Protons and neutrons have similar masses of about 1 dalton each. A *dalton* is the measurement unit for atomic mass. Electrons have negligible mass.

Atomic Number and Atomic Weight   The *atomic number* of an atom refers to the number of protons in its nucleus. Each element has a characteristic and consistent atomic number. Unless otherwise shown, the number of protons in an atom is equal to the number of electrons, and the atom has a neutral electrical charge.

A subscript to the left of the symbol for an element indicates its atomic number; a superscript indicates mass number. The *mass number* of an element is equal to the number of protons and neutrons in its nucleus and approximates the mass of an atom of that element in daltons. The term *atomic weight* is often used to refer to the mass of an atom. The number of neutrons in an atom is equal to the difference between the mass number and the atomic number. An atom of sodium, $_{11}^{23}$Na, contains 11 protons, 11 electrons, and 12 neutrons. The atomic weight of sodium is 23 daltons.

Isotopes   Although the number of protons is constant, the number of neutrons can vary within the atoms of an element, creating different *isotopes*. An element occurs in nature as a mixture of its isotopes. Some isotopes are unstable or *radioactive*; their nuclei spontaneously decay, giving off particles and energy that can be measured with electronic instruments. The characteristic *half-life* for radioactive isotopes (the length of time it takes 50% of the radioactive atoms in a sample to decay) provides a basis for dating fossils. When an organism dies, radioactive isotopes from the environment are no longer taken into its body, and the radioactive isotopes in its body decay at their fixed rate. The half-life of carbon-14 (^{14}C) is about 5600 years. The ratio of carbon-14 to carbon-12 in a fossil is compared to the environmental ratio and indicates when the organism died.

Radioactive isotopes are important tools in biological research and medicine because they can be introduced into organisms and detected in minute quantities. Molecules can be labeled with radioactive isotopes and followed through the steps of a metabolic pathway or traced throughout an organism. Techniques using scintillation counters or autoradiography can determine the quantity and location of radioactively labeled molecules within a cell. Too great an exposure to radiation from decaying isotopes poses a significant hazard to life.

Energy Levels   The nucleus of an atom is extremely small compared to the large area in which the electrons orbit. The chemical behavior of an atom—the type of interactions it has with other atoms—is determined by the number and location of its electrons.

Energy is defined as the ability to do work. *Potential energy* is energy stored in matter as a consequence of the relative position of masses or charges. Matter naturally tends to move toward a lower level of potential energy, which is a more stable state, and requires the input of energy to return to a higher potential energy. The negatively charged electrons are attracted to the positively charged nucleus; their potential energy increases as their distance from the nucleus increases.

Changes of the potential energy of electrons occur in steps of discrete amounts.

Electrons can orbit in several different potential energy states, called *energy levels* or *electron shells*, surrounding the nucleus. The closer the shell to the nucleus, the lower the potential energy. To move to a shell farther from the nucleus, an electron must absorb energy.

Electron Orbitals The three-dimensional space or volume within which an electron is most likely to be found is called an *orbital*. No more than two electrons can occupy the same orbital. The first electron shell or energy level can contain two electrons in a single spherical orbital, called the 1s orbital. The second electron shell's four orbitals, each with the potential of containing two electrons, include a 2s spherical orbital and three dumbbell-shaped p orbitals located along the x, y, and z axes. Higher electron shells contain additional orbitals, but the outermost energy shell of an atom never contains more than eight electrons, located in its s and three p orbitals.

Electron Configuration and Chemical Properties The chemical behavior of an atom is a function of its electron configuration—in particular, the number of *valence electrons* in its outermost energy shell, or *valence shell*. Electrons are distributed to orbitals in the following order: two in the s orbital, one in each of the p orbitals, then a second in each p orbital. The octet rule states that a valence shell of eight electrons is complete, resulting in an unreactive or inert atom. Atoms with incomplete valence shells are chemically reactive. The periodic table of elements is arranged in order of the sequential filling of electron orbitals. Atoms with the same number of electrons in their valence shell have similar chemical properties.

Chemical Bonds and Molecules

Atoms with incomplete valence shells can either completely transfer electrons to or from or share electrons with other atoms such that each atom is able to complete its valence shell. These interactions usually result in attractions, called *chemical bonds*, that hold the atoms together. A *molecule* consists of two or more atoms held together by chemical bonds.

Covalent Bonds When two atoms share a pair of valence electrons, a *covalent bond* is formed. Their valence shells overlap, and the electrons circulate about both nuclei. A *structural formula* represents the atoms and bonding within a molecule. Thus, H–H indicates a hydrogen molecule of two atoms of hydrogen held together by a covalent bond. O=O represents an oxygen molecule in which two pairs of valence electrons are shared between oxygen atoms, forming a *double covalent bond*. A *molecular formula* (such as O_2) only indicates the kinds and numbers of atoms in a molecule. A *compound*, such as H_2O, is a molecule formed from more than one kind of element.

The *valence* or bonding capacity of an atom is a measure of the number of covalent bonds it must form to complete its outer shell. The valences of the four most common elements of living matter are hydrogen 1, oxygen 2, nitrogen 3, and carbon 4.

The function of a molecule is often dependent on its shape or geometry. The s and p orbitals in the valence shell are rearranged in covalent bonds; they hybridize to form four teardrop-shaped orbitals extending from the nucleus in a pyramidal, three-dimensional array called a tetrahedron. A methane molecule (CH_4) has the carbon atom in the center with each of the four hydrogen atoms sharing a pair of electrons at the corners of the carbon's valence orbitals. A water molecule (H_2O) is in the shape of a V: two hydrogen atoms sharing electrons at the corners of two of oxygen's valence orbitals, with oxygen's remaining two pairs of electrons found in the other two orbitals.

Electronegativity is the attraction of an atom for shared electrons. If the atoms in a molecule have similar electronegativities, the electrons remain equally shared between the two nuclei, and the covalent bond is said to be *nonpolar*. If one atom is more electronegative, it pulls the shared electrons closer to itself, creating a *polar* covalent bond. This unequal sharing of electrons results in a partial negative charge associated with the more electronegative atom and a partial positive charge associated with the atom from which the electrons are pulled.

Ionic Bonds If two atoms are very different in their attraction for the shared electrons, the more electronegative atom may completely transfer an electron from another atom, forming an *ionic bond*. This transfer of a negatively charged electron from one atom to another results in the formation of charged atoms called *ions*. The atom that lost the electron is positively charged and called a *cation*. The atom that gained the electron is negatively charged and called an *anion*. The transfer of electrons allows atoms to achieve complete valence shells. The atoms are held together because of the attraction of their opposite charges. A salt is an ionic compound in which ions form a three-dimensional crystalline lattice arrangement held together by electrical attraction. A salt crystal is formed of atoms in specific ratios (such as one-to-one Na to Cl in table salt); the number of ions present is not fixed.

Ion also refers to whole covalent molecules that are

electrically charged. Ammonium (NH_4^-) is a positively charged ion, a cation.

The sharing of electrons between atoms in the completion of their valence shells can fall on a continuum from nonpolar covalent bonds in which electrons are equally shared, through polar covalent bonds, to ionic bonds in which electrons are shared so unequally as to be actually transferred from one atom to another. Covalent bonds are strong; they are hard to break. Whereas ionic bonds are strong in a dry salt crystal, they break apart easily when the salt dissolves. In water, ionic bonds are weak; anions and cations easily separate from each other.

Some Important Weak Bonds   Hydrogen bonds, van der Waals interactions, and hydrophobic interactions are weak chemical bonds within and between molecules that are important to cellular chemistry. They determine the spatial arrangements of molecules and thus influence the functioning, interactions, and emergent properties of these cellular molecules.

When a hydrogen atom is covalently bonded with an electronegative atom and thus has a partial positive charge, it can be attracted to another electronegative atom and form a *hydrogen bond*. Hydrogen bonds are responsible for many of the unusual properties of water.

Van der Waals interactions occur between atoms and within molecules in the form of weak attractions between transient regions of negative and positive charge that result from the unequal distribution of orbiting electrons.

Hydrophobic molecules do not dissolve in water. *Hydrophobic interactions* between molecules result in the clumping or coalescing of hydrophobic molecules due to their repulsion or exclusion from water.

Chemical Reactions

Chemical reactions involve the making or breaking of chemical bonds in the transformation of matter into different forms. Matter is conserved in chemical reactions; the same number and kind of atoms are present in both *reactants* and *products*, although the rearrangement of electrons and atoms causes the properties of these molecules to differ. Thus molecules of hydrogen and oxygen can combine to form water. The reaction is written: $2H_2 + O_2 \rightarrow 2H_2O$.

Some reactions go to completion, but most are reversible—the products of the forward reaction can become reactants in the reverse reaction. The rate of a reaction is speeded by increasing the concentrations of reactants. As products accumulate, collisions resulting in the reverse reaction become more frequent.

Eventually, *chemical equilibrium* may be reached when the forward and reverse reactions proceed at the same rate, and the concentrations of reactants and products no longer change. These relative concentrations will vary depending on the reaction; chemical equilibrium does not mean that reactants and products are equal in concentration.

STRUCTURE YOUR KNOWLEDGE

Take the time to write out or discuss your answers to the following questions. Then refer to the suggested answers at the end of this unit.

1. Describe an atom. Include the concepts of particle, mass, charge, and spatial arrangement.

2. Atoms can have various numbers associated with them. Explain the following: atomic number, mass number, atomic weight, valence. Which of these numbers is most related to the chemical behavior of an atom? Explain.

3. Explain what is meant by the statement that no distinct line divides covalent bonds and ionic bonds.

TEST YOUR KNOWLEDGE

MULTIPLE CHOICE: *Choose the one best answer.*

1. Each element has its own characteristic atom in which
   a. the atomic weight is constant.
   b. the atomic number is constant.
   c. the mass number is constant.
   d. two of the above are correct.

2. Isotopes can be used in studies of metabolic pathways because
   a. their half-life allows a researcher to time an experiment.
   b. they are more reactive.
   c. the cell does not recognize the extra protons in the nucleus, so isotopes are readily used by the cell.
   d. their location or quantity can be experimentally determined.

3. At equilibrium in a reaction
   a. the forward and reverse reactions are occurring at the same rate.
   b. the reactants and products are in equal concentration.

c. the forward reaction has gone farther than the reverse reaction.

d. both a and b are correct.

4. Oxygen has eight electrons. You would expect these to be found arranged in the orbitals in the following way:
   a. two in 1s, two in 2s, two in 2px, two in 2py, zero in 2pz
   b. two in 1s, two in 2s, two in 2px, one in 2py, one in 2pz
   c. two in 2s, two in 2px, two in 2py, two in 2pz
   d. two in 1s, two in 2px, two in 2py, two in 2pz

5. A covalent bond between two atoms is likely to be polar if
   a. one of the atoms is much more electronegative than the other.
   b. the two atoms are equally electronegative.
   c. the two atoms are of the same element.
   d. the bond is part of a tetrahedrally shaped molecule.

6. Of the three weak bonds, which is unique to compounds with polar covalent bonds?
   a. hydrogen bonds
   b. van der Waals interactions
   c. hydrophobic interactions
   d. None are found in association with polar compounds.

7. Which of these classes of substances would be least soluble in the polar compound water?
   a. ionic compounds
   b. polar compounds
   c. hydrophobic compounds
   d. salts

8. The octet rule states that
   a. a valence shell with eight electrons is stable and complete.
   b. no energy level can contain more than eight electrons.
   c. a valence shell with eight electrons produces a neutral atom.
   d. no more than eight electrons can be shared between atoms.

9. The most accurate structural formula for water is
   a. H $\overset{\diagup O \diagdown}{}$ H
   b. H–O–H
   c. H_2O
   d. $H^+ O^= H^+$

10. A triple covalent bond would
    a. be very polar.
    b. involve the bonding of three atoms.
    c. produce a triangularly shaped molecule.

d. involve the sharing of six electrons.

11. It is difficult to speak of a molecule of the salt NaCl because
    a. each sodium ion is attracted to four chloride ions.
    b. salt occurs as a crystalline lattice of many sodium and chloride ions.
    c. the ratio of sodium and chlorine atoms may vary.
    d. the bonds in a salt crystal are weak and break easily.

12. A cation
    a. has gained an electron.
    b. can easily form hydrogen bonds.
    c. is hydrophobic.
    d. has a positive charge.

The six elements most common in living organisms are:

$^{12}_{6}C$ $^{16}_{8}O$ $^{1}_{1}H$ $^{14}_{7}N$ $^{32}_{16}S$ $^{31}_{15}P$

Use this information to answer questions 13 through 20.

13. How many electrons does phosphorus have in its valence shell?
    a. 15
    b. 5
    c. 7
    d. 8

14. Which of these atoms does not undergo sp3 hybridization when it forms covalent bonds?
    a. H
    b. O
    c. C
    d. P

15. What is the atomic weight of phosphorus?
    a. 15
    b. 16
    c. 31
    d. 46

16. A radioactive isotope of carbon has the mass number 14. How many neutrons does this isotope have?
    a. 6
    b. 8
    c. 12
    d. 14

17. What is the valence of nitrogen?
    a. 1
    b. 2
    c. 3
    d. 4

18. How many covalent bonds is a phosphorus atom most likely to form?
    a. 1
    b. 2
    c. 3
    d. 4

19. Based on electron configuration, which of these elements would have chemical behavior most like that of oxygen?

    a. C
    b. N
    c. P
    d. S

20. How many of these elements are found next to each other (side by side) on the periodic chart?
    a. one group of two
    b. two groups of two
    c. one group of two and one group of three
    d. all of them

CHAPTER 3

WATER AND THE FITNESS OF THE ENVIRONMENT

FRAMEWORK

In this chapter you will be introduced to how the chemistry and emergent properties of water contribute to the biological fitness of the external and internal environment of living organisms.

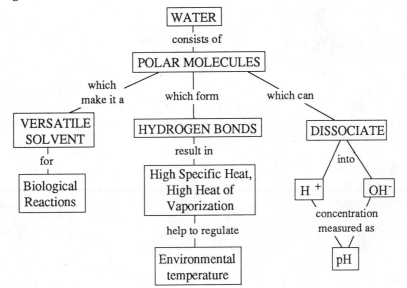

CHAPTER SUMMARY

Water is the biological medium that makes life possible. It makes up 70% to 95% of the content of the cells of living organisms and covers 75% of the earth's surface. The extraordinary properties of water significantly contribute to the fitness of the external environment for life and the fitness of an organism's internal environment for the chemical and physical processes of life. The properties of water can be traced to the structure and interactions of its molecules.

Water Molecules and Hydrogen Bonding

Water molecules consist of two hydrogen atoms covalently bonded to an oxygen atom. The molecule has a tetrahedral shape; the two hydrogen atoms and the two valence orbitals of oxygen that each contain a pair of unshared electrons form the four corners. The electronegative oxygen nucleus pulls the shared electrons away from the hydrogen nucleus, creating a polar covalent bond and a partial positive charge on the hydrogen atom. The two unshared orbitals possess a

13

partial negative charge. This asymmetric shape and the polarity of the bonds make water a *polar molecule* and create the potential for the formation of hydrogen bonds, in which a hydrogen atom covalently bonded to an oxygen atom is attracted to the negative oxygen atom of another molecule. This bonding orders water molecules into a higher level of structural organization and accounts for the emergent properties of this extraordinary substance.

Some Extraordinary Properties of Water

The Cohesiveness of Liquid Water   Liquid water is unusually cohesive due to the hydrogen bonds that tend to hold the molecules together. This *cohesion* creates a more structured organization to the liquid and contributes to the movement of water against gravity in plants. Hydrogen bonding between water molecules produces a high *surface tension* at the interface between water and air. Small insects are able to "walk on water" due to this surface tension. Water's hydrogen bonds and polarity also result in *adhesion*, in which water molecules cling to *hydrophilic* (water loving) substances. *Imbibition* is the soaking of water into a porous hydrophilic substance such as wood or the seed coat of a germinating seed.

Water's High Specific Heat   The ability of water to stabilize air temperatures is related to the high specific heat of water. A large amount of heat is absorbed or released during a slight change in water temperature.

In a body of matter, *heat* is the total quantity of *kinetic energy*, the energy created by the movement of atoms and molecules. *Temperature* measures the average kinetic energy of the molecules in a substance.

Temperature is measured by a *Celsius scale*. Water freezes at 0°C and boils at 100°C. Heat is measured by the *calorie* (cal). A calorie is the amount of heat energy it takes to raise 1 gram of water 1 degree Celsius. A *kilocalorie* (kcal, also designated as C) is 1000 calories, the amount of heat required to raise 1 kilogram of water 1 degree Celsius.

Specific heat is the amount of heat absorbed or lost when 1 gram of a substance changes temperature by 1 degree Celsius. Water's specific heat of 1 cal/gm/°C is unusually high compared with that of other common substances.

Water must absorb or lose a relatively large quantity of heat for its temperature to change. This property arises from the hydrogen bonding of water molecules. To break hydrogen bonds, heat must be absorbed. Heat energy must be used to disrupt hydrogen bonds before water molecules can move faster and temperature can

rise. When hydrogen bonds form, heat is released. As the temperature of water drops slightly, many hydrogen bonds form and release a considerable amount of heat energy. A large body of water absorbs and stores a huge amount of heat when it warms and releases this heat when it cools. The high proportion of water in the environment and within organisms, coupled with water's high specific heat, keeps temperature fluctuations within limits that permit life.

Water's High Heat of Vaporization   The transformation from a liquid to a gas is called vaporization or evaporation. Molecules with sufficient kinetic energy overcome their attraction to other molecules and enter the air as a gas. The addition of heat increases the rate of evaporation by increasing the kinetic energy of molecules. The *heat of vaporization* is the quantity of heat that must be absorbed for 1 gram of a liquid to be converted to a gaseous state. Water has a high heat of vaporization (540 cal/gm) because a large amount of heat is needed to break the hydrogen bonds holding water molecules together. This property of water helps moderate the climate on earth. Solar heat is dissipated from tropical seas during evaporation and, as tropical air circulates poleward, the heat is released as moisture condenses to form rain.

As a substance vaporizes, the liquid left behind loses the kinetic energy of the escaping molecules and cools down. *Evaporative cooling* helps to protect terrestrial organisms from overheating and contributes to the stability of temperatures in lakes and ponds.

Freezing and Expansion of Water   As water cools below 4°C, it expands. By 0°C, each water molecule becomes hydrogen bonded to four other molecules, creating a crystalline lattice and spacing the molecules farther apart. Ice is about 10% less dense than liquid water at 4°C and therefore floats. The floating ice insulates the liquid water below, allowing it to be at a temperature warmer than the air above it. The layer of ice prevents bodies of water from freezing solidly from the bottom upward in the winter.

The formation and melting of ice temper the transitions between seasons. Heat is released into the air when hydrogen bonds form as water solidifies into ice and snow. The melting of ice and snow absorbs heat as hydrogen bonds are broken. This release and absorption of heat make temperature fluctuations within the environment less abrupt.

Water as a Versatile Solvent   Water is the most versatile solvent known. A *solution* is a homogeneous mixture of two or more substances; the dissolving agent is called the *solvent*, and the substance that is dissolved is the *solute*. An *aqueous solution* is one in which water is the solvent. The positive and negative regions of water molecules are attracted to oppositely charged ions or charged regions of polar molecules. Thus, solute molecules become surrounded by water molecules and dissolve into solution. Nonpolar compounds, however, will not dissolve in water.

Aqueous Solutions

Most of the chemical reactions involved in life take place in water. Biological chemistry requires an understanding of aqueous solutions.

Solute Concentration   A *mole* is the amount of a substance that has a mass in grams numerically equivalent to its *molecular weight* (sum of the weight of all atoms in the molecule) in daltons. A mole of any substance has exactly the same number of molecules—6.02×10^{23}, called Avogadro's number. The *molarity* of a solution (abbreviated M) refers to the number of moles of a solute dissolved in 1 liter of solution.

Acids, Bases, and pH   A water molecule can *dissociate* into a *hydrogen ion*, H^+ (which is transferred to another water molecule to form a *hydronium ion*, H_3O^+), and a *hydroxide ion* (OH^-). Although reversible and statistically rare, this dissociation is very important in the chemistry of life.

In pure water, the concentrations of H^+ and OH^- ions are the same; both are equal to 10^{-7} M. When acids or bases dissolve in water, the H^+ and OH^- balance shifts. An *acid* adds H^+ to a solution, whereas a *base* reduces H^+ in a solution by accepting hydrogen ions or by adding hydroxide ions. A strong acid or strong base is a substance that dissociates completely when mixed with water. A weak acid dissociates reversibly to release or accept H^+. A weak base reversibly binds and releases H^+.

A solution with a higher concentration of H^+ than of OH^- is considered acidic. A basic solution has a higher concentration of OH^- than of H^+. In any solution, the product of the H^+ and OH^- concentrations is constant at 10^{-14} M. In a *neutral* solution, $[H^+] = 10^{-7}$ and $[OH^-] = 10^{-7}$ M; so the product is 10^{-14} M. Brackets, [ ], indicate molar concentration. If the $[H^+]$ increases, then the $[OH^-]$ decreases, due to the tendency of excess hydrogen ions to combine with the hydroxide ions in solution and form water. Likewise, an increase in $[OH^-]$ causes a decrease in $[H^+]$. The product of the concentrations remains constant at 10^{-14}; an increase in the concentration of one ion results in an equivalent decrease in the other.

The hydrogen and hydroxide ion concentrations can vary by many orders of magnitude. The *pH scale* compresses the range of concentrations through the use of logarithms. The pH of a solution is defined as the negative log (base 10) of the $[H^+]$: pH = $-\log [H^+]$. For a neutral solution, $[H^+]$ is 10^{-7} M, and the pH equals 7. As the $[H^+]$ increases in an acidic solution, the pH value decreases. A pH value below 7 indicates an acidic solution; a value above 7 denotes a basic solution. Each pH unit represents a ten-fold difference in ion concentration. A slight change in pH reflects a substantial change in $[H^+]$ and $[OH^-]$.

The chemical reactions of a cell often produce acids and bases, yet a shift in pH can be harmful. *Buffers* within the cell minimize the changes in concentrations of hydrogen and hydroxide ions and thus maintain a constant pH. A buffer is a substance that accepts H^+ ions when they are in excess and donates H^+ ions to a solution when their concentration decreases. The carbonic acid/bicarbonate buffering system is an important biological buffer. Carbonic acid acts as an acid to donate H^+ ions when the pH starts to increase. When the pH falls, the excess H^+ ions are accepted by the bicarbonate ion, which acts like a base. An acid and base in equilibrium with each other are typical of most buffering systems.

The environmental problem of *acid rain* emphasizes the sensitivity of life to pH. Acid rain, with a pH lower than the normal pH (5.6) of rain, is due to the reaction of water in the atmosphere with the sulfur oxides and nitrogen oxides released by the combustion of fossil fuels.

Acid rain has harmful effects on both terrestrial and freshwater ecosystems. Lowering the pH of the soil solution affects the solubility of minerals needed by plants. A lowered pH of lakes and ponds affects many species of fishes, amphibians, and aquatic invertebrates. The decline of European forests and the loss of fish populations from many of the lakes in the Adirondacks of New York are the results of acid rain. The reduction of acid rain depends on the development of industrial controls and antipollution devices.

STRUCTURE YOUR KNOWLEDGE

You need to be able to relate two major clusters of concepts to have a good understanding of water and its contribution to the fitness of the environment. Suggested concept maps are included at the end of this unit, but remember that your concept map should represent your own understanding. The value of this exercise is in your process of organizing these concepts for yourself.

1. The tendency of water to form hydrogen bonds between its polar molecules has a profound effect

on temperature regulation within the environment. Select the key concepts that relate to this important property of water and create a concept map showing how the breaking and formation of hydrogen bonds are related to temperature.

2. To become proficient in the use of the concepts relating to pH, develop a concept map to organize your understanding of the following terms: pH, [H⁺], [OH⁻], acidic, basic, neutral, buffer, 1–14, acid-base pair. Remember to label connecting lines and add additional concepts as you need them.

TEST YOUR KNOWLEDGE

MULTIPLE CHOICE: *Choose the one best answer.*

1. Water contributes to the fitness of the environment because
   a. plants need water to grow.
   b. life evolved in water.
   c. the making and breaking of H bonds helps regulate temperature.
   d. the surface tension of water creates a niche for water organisms.

2. The polarity of the water molecule
   a. promotes the formation of hydrogen bonds.
   b. helps water to dissolve nonpolar solutes.
   c. lowers the heat of vaporization.
   d. makes water most dense at 4°C.

3. The surface tension of water is the result of
   a. the adhesion of water molecules to hydrophilic substances.
   b. hydrogen bonds creating cohesion between water molecules.
   c. the process of imbibition.
   d. the pull of gravity on the surface of water.

4. A low specific heat would mean that
   a. little heat must be absorbed or released to effect a temperature change.
   b. breaking hydrogen bonds releases a small amount of heat.
   c. breaking hydrogen bonds absorbs a small amount of heat.
   d. boiling temperature would probably be high.

5. Temperature is a measure of
   a. specific heat.
   b. average kinetic energy of molecules.
   c. total kinetic energy of molecules.
   d. Celsius degrees.

6. Evaporative cooling is a result of
   a. a low heat of vaporization.
   b. a high heat of melting.
   c. a reduction in the average kinetic energy of the liquid remaining after molecules enter the gaseous state.
   d. sweating on a humid day.

7. Ice floats because
   a. air is trapped in the crystalline lattice.
   b. the formation of hydrogen bonds releases heat; warmer objects float.
   c. it has a larger surface area than liquid water.
   d. hydrogen bonding spaces the molecules farther apart, creating a less dense structure than water has.

8. The molarity of a solution is equal to
   a. Avogadro's number of molecules in 1 liter of solvent.
   b. number of moles of a solute in 1 liter of solution.
   c. molecular weight of a solute in 1 liter of solution.
   d. number of solute particles in 1 liter of solvent.

9. A solution with a pH of 2, compared to a solution with pH 4,
   a. is one-half as acidic.
   b. is 100 times more acidic.
   c. is 1000 times more acidic.
   d. has two times more [H⁺].

10. A buffer
    a. maintains pH at 7.
    b. absorbs excess H⁺.
    c. releases excess H⁺.
    d. is often a weak acid-base pair.

11. Which of the folowing is least soluble in water?
    a. polar compounds
    b. nonpolar compounds
    c. ionic compounds
    d. hydrophilic molecules

12. What factor accounts for the movement of water up xylem vessels in a plant?
    a. cohesion
    b. hydrogen bonding
    c. adhesion
    d. all of the above

13. What bonds must be broken for water to vaporize?
    a. polar covalent bonds
    b. nonpolar covalent bonds
    c. hydrogen bonds
    d. all of the above

14. How would you make a 0.1 M solution of acetic acid ($C_2H_4O_2$)? The mass numbers for these elements are C = 12, O = 16, H = 1.

a. Mix 60 g of acetic acid with enough water to yield 1 liter of solution.
b. Mix 2 g carbon, 4 g hydrogen and 2 g oxygen in 1 liter of water.
c. Mix 6 g of acetic acid with enough water to make 1 liter of solution.
d. Mix 0.1 mol of acetic acid with 1 mol of water.

15. How many molecules of acetic acid would be in the solution in question 14?
    a. 0.1
    b. 6
    c. 0.6×10^{23}
    d. 60

TRUE OR FALSE: *Mark T or F; then change the false statements so that they are true.*

1. _____ Heat is a measure of the total kinetic energy of the molecules within a substance or body.

2. _____ The high surface tension of water is a result of adhesion.

3. _____ The formation of ice in winter slows the transition to cold weather because the formation of hydrogen bonds absorbs cold.

4. _____ Ionic substances dissolve in water because the polar water molecules surround the charged ions.

5. _____ Kinetic energy is the energy of motion.

6. _____ A calorie is the amount of heat it takes to raise the temperature of 1 gram of a substance by 1°C.

7. _____ The high heat of vaporization of water is a result of the heat energy needed to break hydrogen bonds.

8. _____ A base accepts H^+ ions or donates OH^- ions to a solution.

9. _____ The breaking down of the crystalline structure when ice melts releases heat to the environment.

10. _____ The dissociation into H^+ and OH^- ions makes water a good solvent for biological materials.

FILL IN THE BLANKS: *Complete the following table on pH.*

$[H^+]$	$[OH^-]$	pH	acidic, basic, or neutral?
	10^{-11}	3	acidic
10^{-8}		8	
10^{-12}			
	10^{-5}		
		1	
	10^{-7}		

CHAPTER 4

CARBON AND MOLECULAR DIVERSITY

FRAMEWORK

This chapter presents the basics of *organic chemistry*, the study of carbon-containing compounds. Carbon atoms, which form four covalent bonds, can link together to make straight or branching chains and rings. These carbon skeletons serve as the basis for the complex organic molecules found in living organisms. The diversity of molecules is increased due to the existence of *isomers*—molecules with the same molecular formula but different architecture, and thus different properties. The *functional groups* that may be bound to carbon skeletons confer specific properties to these organic compounds. It is this diversity of structure and function of organic molecules that allows for the myriad of life forms on earth today.

CHAPTER SUMMARY

The ability of carbon to form large and complex molecules is central to the existence of life. The critical molecules associated with living matter—proteins, DNA, carbohydrates, and fats—all consist of carbon atoms, bound to each other and other atoms. The diverse functional properties of carbon-based molecules emerge from the structure and interactions of these molecules.

The Foundations of Organic Chemistry

Organic chemistry specializes in the study of carbon-containing, or organic, molecules. Early organic chemists could not synthesize the complex molecules found in living organisms. They attributed the existence of life and the formation of these molecules to a life force that was independent of physical and chemical laws, a belief known as *vitalism*. As chemists learned to synthesize organic compounds from inorganic substances in the 1800s, the foundation of vitalism was questioned. *Mechanism* is the belief that physical and chemical laws and explanations are sufficient to account for all natural phenomena, even the processes of life. This philosophy replaced vitalism as the basis for organic chemistry.

Various chemical techniques are used to study organic compounds. Chromatography is a commonly used procedure for separating such compounds. A molecule's relative solubility in a mobile solvent compared to its attraction or binding to a support medium (such as paper, gel on glass plate, or glass column of powder or resin) affects the movement of molecules through a chromatographic system. Polarity and molecular size are usually the basis for differential solubility and rate of movement.

The Versatility of Carbon in Molecular Architecture

Carbon has six electrons, with four in its valence shell. To complete its valence shell, carbon forms four covalent bonds with other atoms. This covalent bonding capacity is at the center of carbon's ability to form large and complex molecules with characteristic three-dimensional shapes and properties. When carbon forms four single covalent bonds, its electron orbitals hybridize into four teardrop-shaped orbitals angling out into the shape of a *tetrahedron*. The spatial arrangement of atoms can determine the function of a molecule in a cell.

Variation in Carbon Skeletons

The four covalent bonds of carbon can include single, double, or even triple bonds. Carbon atoms readily bond with each other, producing chains or rings of carbon atoms. These molecular backbones can vary in length, branching, placement of double bonds, and location of atoms of other elements. The simplest organic molecules are *hydrocarbons*, consisting of only carbon and hydrogen. Fossil fuels are composed of hydrocarbons.

Isomers   Isomers are molecules with the same molecular formula but different structural arrangements and thus, different properties. *Structural isomers* differ in the arrangement of atoms and often in the location of double bonds. Ethyl alcohol and dimethyl ether (Figure 4.1a) have the same number and kinds of atoms but a different bonding sequence and very different properties.

Geometric isomers have the same sequence of covalently bonded atoms but differ in spatial arrangement due to the inflexibility of double bonds. Atoms rotate freely about a single covalent bond. Double bonds restrict that rotation and thus fix the geometry of a molecule to a specific spatial arrangement. Maleic acid and fumaric acid (Figure 4.1b) are geometric isomers. The difference in spatial arrangements of geometric isomers can dramatically affect the biological properties of organic molecules.

Optical isomers are molecules that are mirror images of each other. An *asymmetric* carbon is one that is covalently bound to four different atoms or groups of atoms. Due to the tetrahedral shape of the asymmetric carbon, the four groups can be attached in spatial arrangements that are not superimposable on each other. Optical isomers are left- and right-handed versions of each other, and they differ greatly in their biological activity. Although *L*- and *D*-lactic acid look similar in a flat representation of their structures, they are not superimposable due to the tetrahedral arrangement of carbon bonds (Figure 4.1c). Cells can differentiate between left- and right-handed versions of a molecule; usually, only one form is biologically active.

Functional Groups

The properties of organic molecules depend not only on the arrangement of the carbon skeleton, but also on specific groups of atoms, known as *functional groups*, bonded to the carbon backbone. These functional groups are usually involved in the chemical reactions within a cell and behave consistently from one molecule to another. There are six important functional groups in the chemistry of life.

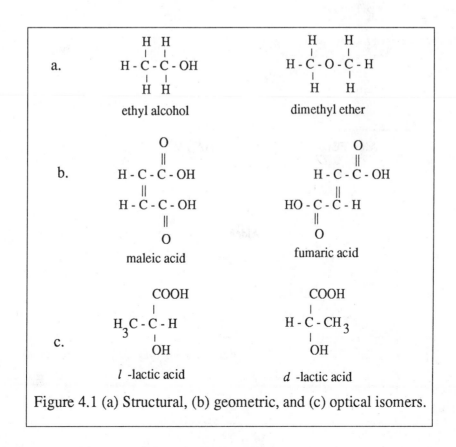

Figure 4.1 (a) Structural, (b) geometric, and (c) optical isomers.

The *hydroxyl group* consists of an oxygen and hydrogen (–OH) covalently bonded to the carbon skeleton. The polarity of the hydroxyl group, due to the electronegative oxygen, increases a compound's solubility in water. Organic molecules with hydroxyl groups are called *alcohols*, and their names often end in *-ol*.

Carbonyl groups consist of a carbon double bonded to an oxygen (–CO). If the carbonyl group is at the end of the carbon skeleton, the compound is called an *aldehyde*. If the carbonyl group is not at the end, the compound is called a *ketone*. The electronegative oxygen of this group makes it polar. Sugars have both carbonyl and hydroxyl groups.

A *carboxyl group* consists of a carbon double bonded to an oxygen and also attached to a hydroxyl group (–COOH). Compounds with a carboxyl group are called *carboxylic acids* or organic acids because they tend to dissociate to release H^+. The two electronegative oxygens of this group pull the shared electrons away from the hydrogen.

An *amino group* consists of a nitrogen atom bonded to two hydrogens (–NH$_2$). Compounds with an amino group, called *amines*, can act as bases. The nitrogen, with its pair of unshared electrons, can attract a hydrogen ion, becoming –NH$_3^+$. Amino acids, the building blocks of proteins, contain both an amino and a carboxyl group (hence the name *amino acid*). Amino acids can exist as zwitterions, molecules with cationic and anionic groups. Such amino acids can act as buffers, losing H^+ or binding H^+ as the pH value in the cell varies.

The *sulfhydryl group* consists of a sulfur atom bound to a hydrogen (–SH). Compounds with sulfhydryl groups are called *thiols*. The cross-linkings of sulfhydryl groups help to stabilize protein structure.

A *phosphate group* binds to the carbon skeleton by an oxygen attached to a phosphorus atom bound to three other oxygen atoms (–OPO$_3$H$_2$). The dissociation of hydrogen ions produces a phosphate ion. An organic compound with a phosphate group can store energy and pass it to another molecule by the transfer of that phosphate group.

The Elements of Life: A Review

Carbon, oxygen, hydrogen, nitrogen, and smaller quantities of sulfur and phosphorus, all capable of forming strong covalent bonds, are combined into the complex organic molecules of living matter. The versatility of carbon in forming four covalent bonds, linking readily with itself to produce chains and rings, and binding with other elements and functional groups, makes possible the incredible diversity of organic molecules.

STRUCTURE YOUR KNOWLEDGE

1. One characteristic of organic compounds is that they can exist in the form of isomers—molecules with the same molecular formula but different spatial arrangements of atoms and different properties. Construct a concept map that illustrates your understanding of the characteristics and significance of the three types of isomers.

2. Fill in the table below on the functional groups.

NAME OF GROUP	MOLECULAR FORMULA	CHARACTERISTICS CONFERRED TO ORGANIC COMPOUND
	-OH	
		Aldehyde or ketone, polar group
Carboxyl		
	-NH$_2$	
		Forms crosslinks in 3-D protein conformation
Phosphate Group		

TEST YOUR KNOWLEDGE

MULTIPLE CHOICE: *Choose the one best answer.*

1. A zwitterion is
   a. an optical isomer.
   b. a high-energy phosphate ion.
   c. a molecule with both carbonyl and hydroxyl groups.
   d. a molecule with both cationic and anionic groups.

2. A hydrocarbon and an alcohol molecule could be separated by chromatography because
   a. a nonpolar solvent would carry the hydrocarbon farther in the system.
   b. a nonpolar solvent would carry the alcohol farther in the system.
   c. they could be separated on the basis of color.
   d. the alcohol is a heavier molecule.

3. Which functional group forms cross-links that stabilize protein structure?
   a. amino
   b. carboxyl
   c. phosphate
   d. sulfhydryl

4. Which of the following is *not* true of an asymmetric carbon atom?
   a. It is attached to four different atoms or groups.
   b. It can result in geometric isomers.
   c. It can result in optical isomers.
   d. Its bonds are in the shape of a tetrahedron.

5. A reductionist approach to considering the structure and function of organic molecules would be based on
   a. mechanism.
   b. holism.
   c. determinism.
   d. vitalism.

6. The functional group that can cause an organic molecule to act as a base is
   a. –COOH.
   b. –OH.
   c. –SH.
   d. –NH$_2$.

7. The functional group that confers acidic properties on organic molecules is
   a. –COOH.
   b. –OH.
   c. –SH.
   d. –NH$_2$.

8. Which is *not* true about geometric isomers?
   a. They have different chemical properties.
   b. They have the same molecular formula.
   c. Their atoms and bonds are arranged in different sequences.
   d. The rigidity of the double carbon bond restricts movement.

9. The tetrahedral shape of the four covalent bonds of a carbon atom is due to
   a. the hybridization of its *sp*^{3} orbitals.
   b. the difference in electronegativity of the bonds.
   c. the asymmetric nature of the carbon atom.
   d. geometric isomerism.

10. Which of the following is *not* true of amino acids?
    a. They can act as buffers.
    b. They are the building blocks of nucleic acids.
    c. They can be zwitterions.
    d. They are named for their amine and carboxylic functional groups.

11. How many asymmetric carbons are in galactose?

    a. 2
    b. 4
    c. 5
    d. 6

    $$
    \begin{array}{c}
    H \diagdown \\
    \quad C \hspace{-2pt}=\hspace{-2pt}O \\
    | \\
    H - C - OH \\
    | \\
    HO - C - H \\
    | \\
    HO - C - H \\
    | \\
    H - C - OH \\
    | \\
    H - C - OH \\
    | \\
    H
    \end{array}
    $$

12. Which functional groups can act as acids?
    a. amine and sulfhydryl
    b. carbonyl and carboxyl
    c. carboxyl and phosphate
    d. alcohol and aldehyde

MATCHING: *Match the formulas on the following page to the terms below. Choices may be used more than once; more than one right choice may be available.*

1. ____&____ structural isomers

2. ____&____ geometric isomers

3. ____&____ optical isomers

4. _____ carboxylic acid

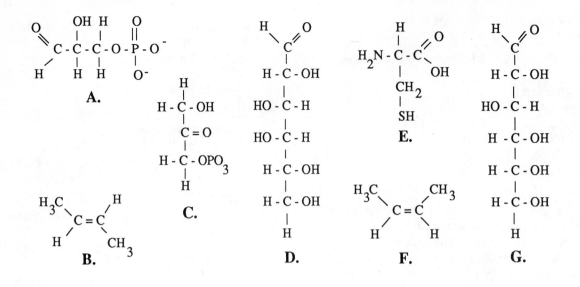

5. _____ thiol

6. _____ alcohol

7. _____ aldehyde

8. _____ amino acid

9. _____ organic phosphate

10. _____ hydrocarbon

11. _____ amine

12. _____ ketone

CHAPTER 5

STRUCTURE AND FUNCTION
OF MACROMOLECULES

FRAMEWORK

This chapter introduces the major classes of macromolecules that constitute living organisms. The chapter's central ideas are that molecular function relates to molecular structure and that the diversity of molecular structure is the basis for the diversity of life. The large variety of macromolecules is created by combining a small number of monomers or subunits into unique sequences and three-dimensional structures. The chart below briefly summarizes the major characteristics of the four classes of macromolecules.

CHAPTER SUMMARY

The four classes of large molecules found in living matter are carbohydrates, lipids, proteins, and nucleic acids. These giant molecules, called *macromolecules*, represent another level in the hierarchy of biological organization. The functions of these molecules are related to their complex and unique architecture.

Polymers

Polymers are large molecules formed from the linking together of many similar or identical small molecules, called *monomers*.

Polymers and Molecular Diversity Macromolecules are constructed from about 40 to 50 common monomers and a few rarer molecules. The limitless variety of polymers arises from the almost infinite number of possibilities in the sequencing and arrangement of these basic subunits. The molecular basis for the diversity of life rests on the ordering of common small molecules into distinctive and unique macromolecules.

Making and Breaking Polymers The same chemical process forms macromolecules of all four classes. Monomers are joined by *dehydration synthesis*, the removal of a molecule of water in a condensation reaction. One monomer loses a hydroxyl (–OH) and the other contributes a hydrogen (–H) to the water molecule, and a covalent bond between the monomers is

CLASS	MONOMERS	FUNCTIONS
Carbohydrates	Monosaccharides	Energy, raw materials, energy storage, structural compounds
Lipids	Glycerol, fatty acids, phosphates, carbon rings	Energy storage, membranes, hormones
Proteins	Amino acids	Enzymes, structural compounds, movement, transport, hormones
Nucleic Acids	Nucleotides (pentose, nitrogenous bases, phosphate group)	Information coding, heredity, ATP energy molecule

formed. Energy is required to join monomers, and the process is facilitated by enzymes.

Hydrolysis is the breaking of bonds between monomers through the addition of water molecules. A hydroxyl is joined to one monomer while a hydrogen is bonded with the other. The disassembling of polymers is necessary in the digestion of food so that small molecules can be absorbed and transported through the blood stream. Enzymes also control hydrolysis.

Carbohydrates

Carbohydrates include simple sugars, disaccharides, and polysaccharides.

Monosaccharides Monosaccharides have the general formula of $(CH_2O)_n$. The number of these units forming a sugar varies from three to seven, with hexoses $(C_6H_{12}O_6)$, trioses $(C_3H_6O_3)$, and pentoses $(C_5H_{10}O_5)$ found most commonly. Glucose, a hexose, is a molecule of key importance in the energy transformations central to life. It is formed by green plants in photosynthesis and broken down to yield energy in respiration in almost all living organisms.

Sugars are aldehydes or ketones, depending on the location of the carbonyl group (–CO). Because the other carbon atoms characteristically bear hydroxyl groups (–OH), sugars are alcohols as well. Additional diversity among sugar molecules is provided by the spatial arrangement of parts around asymmetric carbons. For each asymmetric carbon in a molecule there are two mirror-image isomeric forms (optical isomers). The hexoses glucose and galactose, for instance, differ only in the arrangement around one asymmetric carbon. These small structural differences affect the recognition and interaction of molecules within cells. In aqueous solutions, most monosaccharides are in a ring structure.

Monosaccharides serve not only as the main fuel for cellular work, but also as the raw materials for synthesis of other small organic molecules such as amino acids and fatty acids, and as monomers that are synthesized into polysaccharides that serve as storage or structural molecules.

Disaccharides A *glycosidic linkage* is the bond formed by dehydration synthesis between two monosaccharides. Maltose, a *disaccharide* formed from two glucose molecules, has a 1–4 linkage between the number one carbon of one glucose and the number four carbon of the other glucose. Sucrose, the common disaccharide we know as table sugar, is formed from the joining of a glucose and fructose molecule.

Polysaccharides Polysaccharides are macromolecules made from a few hundred to a few thousand monosaccharides joined together by dehydration synthesis. They can function as storage or structural molecules. *Starch*, a storage molecule in plants, is a polymer made of glucose molecules joined by 1–4 linkages. Starch has a helical shape resulting from the angle of these bonds. Amylose is an unbranched, simple form of starch; amylopectin, a more complex starch, is a branching polymer. Plants store sugar for later use in starch molecules. Most animals have enzymes to hydrolyze plant starch into glucose. Animals produce *glycogen*, a highly branched polymer of glucose, as their food storage form.

Cellulose, the major component of plant cell walls, is the most abundant organic compound on earth. It differs from starch by the configuration of the ring form of glucose. In cellulose, the glucose monomers are in the beta (ß) configuration; in starch, there are 1–4 linkages between the alpha (a) glucose molecules. The resulting geometry of the glycosidic bonds is responsible for the different three-dimensional shapes and properties of these plant polysaccharides. Enzymes that digest starch are unable to hydrolyze the ß linkages of cellulose. Only a few organisms (some bacteria, microorganisms, and fungi) have enzymes that can digest cellulose. Some of these bacteria and microorganisms are housed in a cow's rumen and digest the cellulose in a cow's diet. Humans do not digest cellulose, but it provides fiber to the diet.

In a plant cell wall, hydrogen bonds between hydroxyl groups hold a thousand or more parallel cellulose molecules together to form a microfibril. Several microfibrils are in turn intertwined to form a cellulose fibril, which may supercoil with other fibrils to form strong structural cables.

Chitin is a structural polysaccharide formed from modified amino sugar monomers and found in the exoskeleton of arthropods and the cell walls of many fungi.

Lipids

Fats, phospholipids, and steroids are a diverse assemblage of macromolecules, classed together as *lipids* because they are all insoluble in water.

Fats Fats are composed of fatty acids attached to the three-carbon alcohol, glycerol. A *fatty acid* consists of a long carbon skeleton (often 16 or 18 carbons in length) with a carboxyl group at the "head" end. The long hydrocarbon tail is nonpolar and thus hydrophobic, making fats insoluble in water.

Fatty acids are linked to glycerol in a dehydration synthesis by an ester linkage, a bond that forms between a hydroxyl and a carboxyl group. In a *triacylglycerol*, or fat, three fatty acids attach by dehydration synthesis to glycerol.

Fatty acids with double bonds in their carbon skeletons are called *unsaturated*. The double bonds create a kink in the shape of the molecule and prevent the fat molecules from packing close together and becoming solidified at room temperature. *Saturated* fats have fatty acids with no double bonds in their carbon skeletons; each carbon atom is "saturated" with hydrogen. Most animal fats are saturated and solid at room temperature. Plant fats are generally unsaturated and are called oils. Diets rich in animal fats have been linked to cardiovascular disease.

Fats are excellent energy storage molecules, containing twice the energy reserves as do carbohydrates such as starch. Adipose tissue, made of fat storage cells, also cushions organs and insulates the body.

Phospholipids Phospholipids consist of a glycerol linked to two fatty acids and a negatively charged phosphate group. Other small molecules may be attached to the phosphate group. The phosphate head of this molecule is hydrophilic and water soluble, whereas the two fatty acid chains have nonpolar and hydrophobic tails. This unique structure of phospholipids makes them ideal constituents of cell membranes. Arranged in a bilayer, the hydrophilic heads face toward the aqueous solutions inside and outside the cell, and the hydrophobic tails interact in the center to hold the molecules of the membrane together.

Steroids Steroids are a class of lipids distinguished by four fused carbon rings with various functional groups attached. *Cholesterol* is an important steroid that is a common component of animal cell membranes and a precursor for most other steroids, including many hormones.

Proteins

Proteins are crucial macromolecules, serving such diverse functions as structural support (fibers), storage (of amino acids and energy), transport (hemoglobin), internal coordination (hormones), movement (actin and myosin in muscle), defense (antibodies), and, perhaps most important, control of metabolism (enzymes). This wide range of functions is made possible by the highly structured and unique three-dimensional shapes of these macromolecules. Twenty amino acids are the monomers from which the huge variety of protein molecules is composed.

Amino Acids Most amino acids are composed of an asymmetric carbon (called the a carbon) bonded to a carboxyl group, an amino group, a hydrogen, and a variable *side chain* called the R group. The R group confers the unique physical and chemical properties of each amino acid. Side chains may be nonpolar and hydrophobic; polar or charged and thus hydrophilic; acidic and negatively charged; or basic and positively charged. Amino acids serve not only as protein monomers; they may function also as individual molecules.

With their asymmetric carbon, amino acids can exist in two isomeric forms (optical isomers) considered to be left- or right-handed. Usually only the L-form of an amino acid is used by cells to make proteins.

Polypeptide Chains A dehydration synthesis joins the amino group of one amino acid with the carboxyl group of another in a linkage known as a *peptide bond*. A polymer of amino acids is called a *polypeptide chain*, with a free carboxyl group at one end (the C-terminus) and a free amino group at the other (the N-terminus). The other amino and carboxyl groups are linked into a polypeptide backbone held together by peptide bonds. The side chains of each amino acid extend out from this chain and participate in interactions that help create the three-dimensional structure of proteins. Polypeptides may vary in length from a few to a thousand or more amino acids. Each specific polypeptide has its own unique sequence of amino acids.

Protein Conformation Proteins have unique three-dimensional shapes, or *conformations*, created by the twisting or folding of one or more polypeptide chains. Structural proteins tend to be fibrous, whereas enzymes are more spherical or globular. The unique conformation of a protein, which results from its unique sequence of amino acids, enables it to recognize and bind specifically to another molecule.

Levels of Protein Structure There are three superimposed structural levels of architecture in the conformation of a protein. A fourth level may be present when a protein consists of more than one polypeptide chain. *Primary structure* is the unique sequence of amino acids within a protein, coded for by genetic information. Even a slight deviation from the sequence of amino acids can severely affect a protein's function by altering the protein's conformation.

In the early 1950s, Sanger determined the primary structure of insulin through the laborious process of hydrolyzing the protein into small peptide chains, using chromatography to separate the small pieces, determining their sequences of amino acids, and then overlapping the sequences of the small fragments to reconstruct the whole polypeptide. Most of these steps are now automated, and the primary structures of hundreds of proteins have been determined.

Secondary structure involves the coiling or folding of the polypeptide backbone, stabilized by hydrogen bonds between the electronegative oxygen of one peptide bond and the weakly positive hydrogen attached to a nitrogen of another bond. An *alpha (a) helix*, a delicate coil produced by hydrogen bonding between every fourth peptide bond, was first described by Linus Pauling and Robert Corey in 1951. Some fibrous proteins have alpha helices along most of their length. Globular proteins are more likely to have regions of alpha helix alternating with non-helical regions.

A *beta (ß) pleated sheet* is also held by repeated hydrogen bonds along the protein's backbone. This secondary structure forms when the polypeptide chain folds back and forth or when regions of the chain parallel each other. Beta sheets are found in the dense core of many globular proteins and in some fibrous proteins.

Interactions between the various side chains of the constituent amino acids produce a protein's *tertiary structure*. Hydrophobic interactions between nonpolar side groups in the center of the molecule, hydrogen bonds, and ionic bonds between negatively and positively charged side chains produce a stable and unique shape to the protein. Strong covalent bonds, called *disulfide bridges*, may occur between the sulfhydryl side groups of cysteine monomers that have been brought close together by the folding of the polypeptide. Proteins may have a modular structure in which globular regions, called *domains*, are connected by more flexible regions of the polypeptide chain.

Quaternary structure occurs in proteins that are composed of more than one polypeptide chain. The individual polypeptide chains, called *subunits*, are held together in a precise structural arrangement. Collagen is a fibrous protein composed of three helical subunits supercoiled together. Hemoglobin is a globular protein consisting of four polypeptide chains.

What Determines Conformation?   The specific function of a protein is an emergent property developing from its intricate three-dimensional architecture. This conformation arises spontaneously and is dependent on the interactions among the amino acids making up the polypeptide chain. These interactions can be disrupted by changes in pH, salt concentration, temperature, or other aspects of the environment, and the protein may *denature*, losing its native conformation and thus its function. Any factor that disrupts the hydrogen bonding, hydrophobic interactions, ionic bonds, or disulfide bridges within a protein molecule will cause denaturation. For example, excess heat may cause enough molecular agitation to disrupt these

bonds. Some denaturation is reversible; if the protein does not precipitate out of solution, it may reform to its three-dimensional conformation when returned to its normal environment. The shape of a protein ultimately emerges from its sequence of amino acids, its primary structure.

The Protein-Folding Problem   The amino acid sequences of hundreds of proteins have been determined. Using the technique of X-ray crystallography, coupled with computer modeling and graphics, molecular biologists have established the three-dimensional shape of many of these molecules. But the rules of protein folding that could predict the conformation arising from a particular primary structure of a protein have been difficult to determine. Most proteins probably go through intermediate states on their way to their final form, and even that stable shape may alternate between several conformations.

Nucleic Acids

Functions of Nucleic Acids: An Overview   Nucleic *acids* are macromolecules with the unique ability to reproduce themselves and to carry the code that directs all the cell's activities. *DNA, deoxyribonucleic acid*, is the genetic material that is inherited from one generation to the next and is reproduced in each cell of an organism. *RNA, ribonucleic acid*, reads the instructions coded in DNA and directs the synthesis of proteins, the ultimate enactors of the genetic program.

Nucleotides   Nucleic acids are polymers of *nucleotides*, linked together by dehydration synthesis. Each nucleotide consists of a pentose (five-carbon) sugar covalently bonded to a phosphate group and to one of five nitrogenous bases. There are two families of nitrogenous bases. *Pyrimidines*, including cytosine (C), thymine (T), and uracil (U), are characterized by six-membered rings of carbon and nitrogen atoms. Thymine is found only in DNA, uracil is found only in RNA. *Purines*, adenine (A) and guanine (G), add a five-membered ring to the pyrimidine ring. These compounds are called nitrogenous *bases* because the nitrogen tends to take up H^+. The pentose sugar is either *ribose*, in RNA, or *deoxyribose* (missing a hydroxyl group), in DNA. *Nucleosides* consist of a nitrogenous base joined to a sugar. Nucleotides have a phosphate group attached to the number five carbon of the sugar and a nitrogenous base.

Nucleotides can function as monomer subunits in nucleic acids, or as individual molecules. Adenosine triphosphate, ATP, is a nucleotide involved in energy transfer in cellular processes.

Polynucleotides   Nucleotides are linked together into *polynucleotides* by *phosphodiester linkages*, which join the phosphate of one nucleotide with the sugar of the next by a covalent bond. The nitrogenous bases extend from this repeating sugar–phosphate backbone.

The Double Helix: An Introduction   DNA molecules consist of two such polynucleotide chains spiraling around an imaginary axis in a *double helix*. In 1953, Watson and Crick first proposed the double helix arrangement, which consists of two sugar–phosphate backbones on the outside of the helix with their nitrogenous bases pairing and hydrogen bonding together in the inside. Adenine pairs with only thymine; guanine always pairs with cytosine. Thus, the sequences of nitrogenous bases on the two strands of DNA are complementary, predictable counterparts of each other. Because of this specific base-pairing property, DNA can replicate itself and precisely copy the genes of inheritance.

STRUCTURE YOUR KNOWLEDGE

1. For each of the four families of macromolecules—carbohydrates, lipids, proteins, and nucleic acids—create a concept map (or a brief table, chart, or summary, if you prefer) that emphasizes the structure and functions of that group. Include the monomers involved, the type of linkage, and the unique characteristics of the structure of the polymers. A few examples in each group would probably be helpful. Organize this information so that you can relate structure to function.

TEST YOUR KNOWLEDGE

MATCHING OF FORMULAS: *Match the chemical formulas at the bottom of the page with their description. Answers may be used more than once.*

1. _____ molecules that would combine to form a fat

2. _____ molecule that would be attached to other monomers by a peptide bond

3. _____ molecules or groups that would combine to form a DNA nucleotide

4. _____ molecules that are carbohydrates

5. _____ molecule that is a purine

6. _____ monomer of a protein

7. _____ groups that would be joined by phosphodiester bonds

MATCHING: *Match the molecule with its class of macromolecules.*

1. _____ glycogen	**A.** carbohydrate	
2. _____ cholesterol	**B.** lipid	
3. _____ ATP	**C.** protein	
4. _____ collagen	**D.** nucleic acid	
5. _____ hemoglobin		
6. _____ a gene		
7. _____ triaclyglycerol		

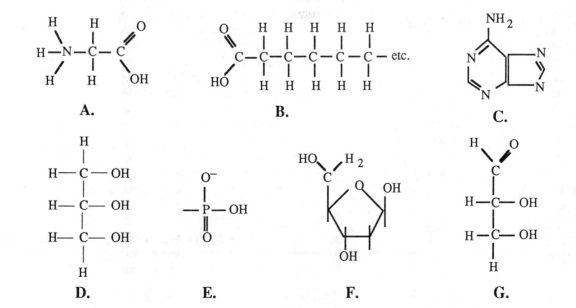

8._____ enzyme

9._____ cellulose

10._____ chitin

MULTIPLE CHOICE: *Choose the one best answer.*

1. Dehydration synthesis is a process that
   a. creates bonds between amino acids in the formation of a peptide chain.
   b. involves the removal of a water molecule.
   c. links the phosphate of one nucleotide with the sugar of the next.
   d. involves all of the above.

2. Which of the following is *not* true of pentoses?
   a. They are found in nucleic acids.
   b. They can occur in a ring structure.
   c. They have the formula $C_5H_{12}O_5$.
   d. They have hydroxyl and carbonyl groups.

3. Disaccharides can differ from each other in all of the following ways *except*
   a. in the number of their monosaccharides.
   b. in the existence of optical isomers.
   c. in the types of monomer involved.
   d. in the location of their glycosidic linkage.

4. Which of the following is *not* true of cellulose?
   a. It is the most abundant organic compound on Earth.
   b. It differs from starch because of the configuration of glucose and the geometry of the glycosidic linkage.
   c. It is a highly branched, strong structural component of cell walls.
   d. Few organisms have enzymes that hydrolyze its glycosidic linkages.

5. Plants store most of their energy as
   a. glucose.
   b. glycogen.
   c. starch.
   d. sucrose.

6. When a protein denatures, it
   a. loses its primary structure.
   b. loses its secondary and tertiary structure.
   c. becomes insoluble and precipitates.
   d. hydrolyzes into component amino acids.

7. The alpha helix of proteins is
   a. part of the tertiary structure and is stabilized by disulfide bridges.
   b. a double helix.
   c. stabilized by hydrogen bonds and commonly found in fibrous proteins.

   d. found in some regions of globular proteins and stabilized by hydrophobic interactions.

8. A fatty acid that has the formula $C_{16}H_{32}O_2$ is
   a. saturated.
   b. unsaturated.
   c. branched.
   d. hydrophilic.

9. Three molecules of the fatty acid in question 8 are joined to a molecule of glycerol ($C_3H_8O_3$). The resulting molecule has the formula
   a. $C_{48}H_{96}O_6$.
   b. $C_{51}H_{104}O_9$.
   c. $C_{51}H_{102}O_8$.
   d. $C_{51}H_{98}O_6$.

10. The molecule formed in question 9 is
    a. a triaclyglyceride.
    b. a lipid.
    c. a fat.
    d. all of the above.

11. Which of the following is hydrophobic?
    a. steroid
    b. chitin
    c. head end of phospholipid
    d. polynucleotide

12. Beta sheets are characterized by
    a. disulfide bridges between cysteine amino acids.
    b. back-and-forth folds of the polypeptide chain held together by hydrophobic interactions.
    c. folds stabilized by hydrogen bonds between segments of polypeptide chains.
    d. membrane sheets composed of phospholipids.

FILL IN THE BLANKS

1. The nitrogenous base absent in RNA is _____.

2. Cytosine always pairs with _____.

3. Adenine and guanine are _____.

4. A nitrogenous base joined to a pentose sugar is called a _____.

5. Adding a phosphate group to the molecule in question 4 yields a _____.

6. Proteins with more than one peptide chain have _____structure.

7. The conformation of a protein is determined by its _____.

8. The energy storage molecule of animals is _____.

9. Membranes are composed of a bilayer of _____.

10. The linkages between amino acids in a protein are called _____.

11. Technique used to determine the three-dimensional structure of macromolecules is _____.

12. The man who determined the amino acid sequence of insulin was _____.

CHAPTER 6

INTRODUCTION TO METABOLISM

FRAMEWORK

This chapter considers metabolism, the totality of the chemical reactions that take place in living organisms. Two key topics are emphasized: the energy transformations that underlie all chemical reactions and the role of enzymes in the "cold chemistry" of the cell.

The reactions in a cell either consume or release energy. The following illustration summarizes some of the components of the energy changes within a cell.

Enzymes are biological catalysts that lower the activation energy of a reaction and thus greatly speed up metabolic processes. An enzyme is a three-dimensional protein molecule with an active site specific for its substrate. Intricate control and feedback mechanisms produce the metabolic integration necessary for life.

CHAPTER SUMMARY

The Metabolic Map

Metabolism is the totality of an organism's chemical processes, involving the thousands of precisely coordinated, complex, efficient, and integrated chemical reactions in a cell. These reactions are ordered into metabolic pathways—sequenced and intricately branched routes controlled by enzymes. Through these pathways the cell creates and transforms the organic molecules that provide the material and energy for life. Metabolism is an emergent property arising from the organization and orderly interactions of molecules in cells.

Catabolic pathways release the energy stored in complex molecules through the breaking down or

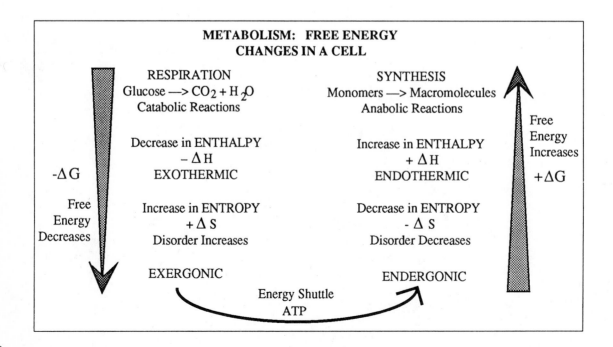

degradation of these molecules into simpler compounds. *Anabolic pathways* require energy to combine simpler molecules into more complicated ones. The coupling of catabolic and anabolic pathways in a cell often supplies this energy. All metabolic processes involve energy. Thus, an understanding of energy transformations is essential to the study of metabolism.

Energy: Some Basic Principles

Energy has been defined as the capacity to do work, to move matter against an opposing force.

Forms of Energy   *Kinetic energy* is the energy of motion, of matter that is moving. This matter does its work by transferring its motion to other matter. Heat and light are forms of kinetic energy due to the motion of molecules and photons. *Potential energy* is the capacity of matter to do work as a consequence of its location or arrangement. Chemical energy is a form of potential energy stored in the bonds holding atoms together in molecules.

Energy Transformations   Energy can be converted from one form to another: kinetic energy into potential energy and vice versa. Plants change light energy to the chemical energy of sugar, and cells release the potential energy in sugar to drive cellular processes.

Two Laws of Thermodynamics   *Thermodynamics* is the study of energy transformations. The *first law of thermodynamics* states that energy can be neither created nor destroyed. Energy can be transferred between matter and transformed from one kind to another, but the total energy of the universe is constant. According to this principle of the conservation of energy, chemical reactions that either require or produce energy are merely transforming a set amount of energy into a different form.

The *second law of thermodynamics* states that every energy transformation or transfer results in an increasing disorder within the universe. *Entropy* is the term used as a quantitative measure of disorder or randomness. Thus, every process results in an increase in the entropy in the universe.

A system, such as a cell, may become more ordered, but it does so with an attendant increase in the entropy of its surroundings. A cell can use highly ordered organic molecules as a source of the energy needed to create its own highly ordered structure, but it returns heat and the simple molecules of carbon dioxide and water to the environment. In any energy transformation or transfer, some of the energy is converted to heat, a less-ordered kinetic energy. No energy transfer is 100% efficient. The quantity of energy in the universe may be constant, but its quality is not. Every process results in an increasing disorder in energy, often into the form of the random molecular motion of heat.

Chemical Energy: A Closer Look

Chemical reactions rearrange atoms by breaking and forming chemical bonds. Energy is released when bonds form and is absorbed to break bonds. *Bond energy*, usually expressed in kilocalories per mole of bonds formed or broken, is the quantity of energy involved in the forming or breaking of a particular bond.

Heat of Reaction   The net energy released or consumed by the conversion of reactants into products is called the heat of reaction, signified by ΔH. ΔH represents the difference between the energy consumed when bonds break and that released when bonds form. *Enthalpy* is the heat content of a molecule, determined from the total potential energy stored in its bonds. The heat of reaction, H, can be thought of as the difference between the enthalpy of the reactants and of the products, between the sum of the bond energies of the reactants and the sum of the bond energies of the products. A reaction that has a negative ΔH and thus releases heat does so because the products have less enthalpy than do the reactants. Such heat-releasing reactions are called exothermic. When the products of a reaction have more enthalpy than the reactants, the reaction is called endothermic, has a positive ΔH, and absorbs heat from the surroundings.

Spontaneous Reactions   Spontaneous reactions occur without the addition of external energy. Energy-rich systems are intrinsically unstable and tend to change so as to decrease their energy. Chemical reactions tend to be spontaneous if the total energy stored in the molecules decreases as a result of the reaction. Exothermic reactions are usually spontaneous; endothermic reactions, requiring energy and increasing the chemical energy of the molecules, are usually not spontaneous.

Entropy, the trend toward disorder and randomization, also contributes to driving spontaneous reactions. The change in entropy (ΔS) of a reaction is represented by $\Delta S = S_{final\,state} - S_{initial\,state}$. Thus, a positive ΔS represents an increase in entropy and will contribute to a reaction being spontaneous. The increase in entropy may be enough to drive a reaction even though the reaction may be endothermic, as in the spontaneous evaporation of water. It is also possible for a reaction to represent a decrease in entropy but still proceed spontaneously due to the decrease in enthalpy, as in the formation of highly ordered snowflakes.

Free Energy   The balance between enthalpy and entropy can be represented by the concept of *free energy*. A change in free energy (ΔG) is directly related to the change in enthalpy but inversely related to the change in entropy. The equation, $\Delta G = \Delta H - T\Delta S$, represents these relationships. Temperature (T= °Kelvin = °C + 273) must be factored into the calculation of entropy because a higher temperature increases random molecular motion, disrupting order.

For a reaction to be spontaneous, the free energy of the system must decrease; ΔG must be negative. A reduction in free energy is dependent on the magnitude of the change in enthalpy ($-\Delta H$ is exothermic) and entropy ($+\Delta S$ increases entropy) and the temperature. At times, temperature may be the deciding factor in whether or not a reaction will proceed. For example, the denaturation of proteins is favored by an increase in entropy but retarded by the increase in enthalpy necessary to break the hydrogen bonds and other interactions creating the shape of the molecule. An increase in temperature (usually above 60°C) will weight the entropy factor in favor of a net decrease in free energy so that denaturation will proceed.

Free energy is also a measure of how much work a spontaneous reaction can do. When temperature is constant, as it is in a cell, free energy is an indication of the total energy released from the system and available to do work.

Exergonic and Endergonic Reactions   Using the reference of free energy, reactions can be classified as *exergonic* or *endergonic*. An exergonic ($-\Delta G$) reaction proceeds with a net release of free energy and is spontaneous. The magnitude of ΔG indicates the maximum amount of work the reaction can do. Endergonic reactions ($+\Delta G$) are nonspontaneous; they must absorb free energy from the surroundings. The magnitude of G indicates the minimum amount of work needed to drive the reaction. In metabolism, exergonic reactions are often coupled closely with endergonic reactions; the free energy released from the former is used to power the latter.

Free Energy and Equilibrium   At equilibrium in a chemical reaction, the forward and backward reactions are proceeding at the same rate. At equilibrium, $\Delta G=0$ because there is no net free energy change. As equilibrium is approached, the ΔG of a reaction is negative; when moving away from equilibrium, ΔG is positive. To drive a reaction away from its normal equilibrium, a cell must add free energy, usually by coupling the reaction with an exergonic reaction.

ATP and Cellular Work

A cell must perform several kinds of work: mechanical work involved in movement of the cell or parts of the cell, transport work in pumping molecules across membranes, and chemical work in driving endergonic reactions to synthesize cellular molecules. In most cases, the immediate source of the energy to perform this work comes from *adenosine triphosphate*, or *ATP*.

Structure and Hydrolysis of ATP   ATP is a nucleoside triphosphate, the purine base adenine bonded to the sugar ribose, which is connected to a chain of three phosphate groups. The bonds between the phosphate groups are unstable and can be broken by hydrolysis. Thus, ATP can be hydrolyzed to ADP (adenosine diphosphate) and an inorganic phosphate molecule, releasing 7.3 kilocalories of energy per mole of ATP. This quantity of free energy has been experimentally measured; the ΔG of the reaction in the cell is estimated to be closer to –10 to –12 kcal/mol.

Although the phosphate bonds in ATP are called high-energy bonds, they are actually weak bonds, easily hydrolyzed to yield stronger, more stable bonds.

How ATP Performs Work   The transformation to a more stable state releases energy. In a cell, this energy can be used to transfer the phosphate group from ATP to another molecule, producing a *phosphorylated intermediate* that is more reactive. The phosphorylation of other molecules by ATP forms the basis for almost all cellular work.

Regeneration of ATP   A cell regenerates ATP at a phenomenal rate; ten million molecules of ATP may be consumed and regenerated per second. The formation of ATP from ADP and inorganic phosphate is endergonic, with a ΔG of +7.3 kcal/mol. Cellular respiration (the catabolic processing of glucose and other organic molecules) provides the energy for the regeneration of ATP. Plants can also produce ATP using light energy.

Metabolic Disequilibrium   Respiration and other cellular chemical reactions are reversible and could reach equilibrium if the cell did not keep a steady supply of reactants and siphon off the products (as reactants for new processes or as waste products to be expelled). Chemical systems at equilibrium have a ΔG of 0 and can do no work. Respiration continues to produce ATP as long as the cell can provide glucose and expel CO_2.

Enzymes

Thermodynamics can indicate what reactions are spontaneous but not how fast those reactions occur. Some spontaneous reactions may take centuries to happen. *Enzymes* are used by the cell to speed and regulate metabolic reactions so that life is possible. Enzymes are biological catalysts—agents that change the speed of a reaction but are unchanged by the reaction.

Enzymes and Activation Energy   The *free energy of activation*, $\Delta G^{\ddagger}$, is the energy that must be absorbed by reactants to reach the unstable *transition state*, in which bonds are more fragile and likely to break, and from which the reaction can proceed. This state can be reached by the addition of thermal energy from the surroundings, causing the reactant molecules to collide more often and more forcefully. Even in an exergonic reaction, in which ΔG is negative, energy must first be absorbed to reach the transition state.

The activation energy barrier is essential to life because it prevents the energy-rich macromolecules of the cell from decomposing spontaneously. For metabolism to proceed in a cell, however, $\Delta G^{\ddagger}$ must be reached for selected reactions. Heat, a normal source of activation energy in reactions, would be harmful to the cell and would also speed metabolic reactions indiscriminately. Enzymes are able to lower $\Delta G^{\ddagger}$ for specific reactions so that metabolism can proceed at cellular temperatures. Enzymes do not change ΔG for a reaction; they only speed up reactions that would otherwise occur very slowly.

Specificity of Enzymes   Enzymes are proteins, macromolecules with characteristic three-dimensional shapes. The specificity of an enzyme for the particular *substrate* on which it works is determined by its unique shape. The substrate is temporarily bound to its enzyme at the *active site*, a pocket or groove, found on the surface of the enzyme molecule, that has a shape and charge arrangement complementary to the substrate molecule. When a substrate molecule enters the active site, the enzyme changes shape slightly, creating what is called an *induced fit* between substrate and active site, which enhances the ability of the enzyme to catalyze the chemical reaction.

The Catalytic Cycle of Enzymes   The substrate is held in the active site by hydrogen or ionic bonds, creating an enzyme–substrate complex. The side chains (R groups) of some of the surrounding amino acids in the active site facilitate the conversion of substrate to product. The product then leaves the active site and the enzyme can bind with another substrate molecule. The

conversion is extremely fast; an enzyme can catalyze 1000 reactions or more per second.

Enzymes can catalyze reactions involving the joining of two reactants by providing active sites in which the substrates are bound closely together and properly oriented. An induced fit can stretch or bend critical bonds in the substrate molecule and make them easier to break. An active site may provide a microenvironment that is necessary for a particular reaction. For example, acidic R groups would lower the pH of the active site. Enzymes may also actually participate in a reaction by forming brief covalent bonds with the substrate.

The rate at which an enzyme molecule works partly depends on the concentration of its substrate. The speed of a reaction will increase with increasing substrate concentration up to the point at which all enzyme molecules are saturated with substrate molecules and working at full speed.

Factors Affecting Enzyme Activity   The activity of an enzyme depends on its three-dimensional shape, and this shape depends on environmental factors that affect the weak chemical bonds maintaining protein structure. The velocity of an enzyme-catalyzed reaction may increase with rising temperature up to the point at which increased thermal agitation begins to disrupt the hydrogen and ionic bonds that stabilize protein conformation. A change in pH may denature an enzyme by disrupting the hydrogen bonding of the molecule. Each enzyme has a temperature and pH optimum at which it is most active. Enzymes are sensitive to salt concentration because inorganic ions may interfere with ionic bonds within the enzyme molecule. When the conformation of an enzyme changes through denaturation, its activity decreases.

Cofactors are small molecules that bind with enzymes and are necessary for enzyme catalytic function. They may be inorganic, such as various metal atoms, or organic molecules called *coenzymes*. Most vitamins are coenzymes or precursors of coenzymes.

Enzyme inhibitors selectively disrupt the action of enzymes either reversibly by binding with the enzyme with weak bonds or irreversibly by attaching with covalent bonds. *Competitive inhibitors* compete with the substrate for the active site of the enzyme. Increasing the concentration of substrate molecules may overcome this type of inhibition as long as the inhibitor does not bind too strongly to the active site. *Noncompetitive inhibitors* bind to a part of the enzyme separate from the active site and change the conformation of the enzyme, thus impeding enzyme action. Many pesticides are noncompetitive inhibitors of key enzymes and act as metabolic poisons. The cell itself uses selective inhibitors to control enzyme action and thus regulate metabolism.

The molecules that inhibit or activate enzyme activity may bind to an *allosteric site*, a receptor site not associated with the active site on the enzyme. Enzymes with allosteric sites are often complex molecules made of two or more polypeptide chains or subunits, each with its own active site. Allosteric sites are usually located where subunits join. The entire unit may oscillate between two conformational states, and the binding of an activator (or inhibitor) to the allosteric site stabilizes the catalytically active (or inactive) conformation. The subunits interact such that a conformational change in one is transmitted to the other(s). Through a phenomenon called *cooperativity*, the induced-fit binding of a substrate molecule to one subunit can change the conformation such that the active sites of all subunits are more active. Allosteric enzymes may be critical regulators of metabolic pathways.

The Control of Metabolism

The intricate ordering and control of metabolic pathways are made possible through the chemical regulation and structural organization of the enzymes that direct each step of a pathway.

Feedback Inhibition   *Feedback inhibition* commonly regulates metabolic pathways. The product of a pathway can act as an inhibitor of an enzyme early in the pathway and shut down the process when the cell has produced sufficient end product.

Structural Order and Metabolism   The complex internal structure of the cell serves to order metabolic pathways in space and time. Enzymes can be grouped into *multienzyme complexes* in which the enzymes regulating successive steps of a metabolic pathway are serially arranged. Specialized cellular compartments may contain high concentrations of the enzymes and substrates needed for a particular pathway. Or enzymes may be incorporated into the membranes of cellular compartments. Thus the structural organization of cells facilitates and controls metabolism.

Emergent Properties: A Reprise

This unit has illustrated how life is organized into a hierarchy of structural levels with emergent properties associated with each new level of order. The structure and interactions of atoms, molecules, monomers, and macromolecules have been linked to metabolism, the orderly chemistry characteristic of life.

STRUCTURE YOUR KNOWLEDGE

This chapter introduces many complex ideas concerning the thermodynamics of metabolism. Take the time to organize your understanding of small "chunks" of this information and then try to integrate these pieces into your picture of the energy transformations taking place within the cells of living organisms.

1. Create a simple concept map concerning energy: its definition, types, and the first two laws of thermodynamics.

2. Develop two separate concept maps, one for enthalpy and ΔH, and one for entropy and ΔS. Then try to combine these two maps under the broad umbrella of free energy and ΔG. The value in this exercise is for you to wrestle with and organize these concepts for yourself. Do not turn to the suggested concept map until you have worked on your own understanding. Remember that the concept map in the answer section is only one way of structuring these ideas—have confidence in your own organization.

3. Now take a break from concept mapping and answer these questions about enzymes. (A map on enzymes would, of course, be helpful.)
   a. Why are enzymes so critical to the existence of life?
   b. What characteristics of proteins make them good catalysts?
   c. Briefly describe the process by which an enzyme catalyzes a reaction.
   d. How does a cell control enzyme action and regulate metabolism?

TEST YOUR KNOWLEDGE

MULTIPLE CHOICE: *Choose the one best answer.*

1. When glucose is converted to CO_2 and H_2O, changes in enthalpy, entropy and free energy are as follows:
   a. $-\Delta H, -\Delta S, -\Delta G$
   b. $-\Delta H, +\Delta S, -\Delta G$
   c. $-\Delta H, +\Delta S, +\Delta G$
   d. $+\Delta H, +\Delta S, +\Delta G$

2. When water evaporates spontaneously, the following changes apply:
   a. $+\Delta H, +\Delta S, +\Delta G$
   b. $-\Delta H, +\Delta S, -\Delta G$

c. $+ \Delta H, + \Delta S, - \Delta G$
d. $+ \Delta H, - \Delta S, - \Delta G$

3. When a protein forms from amino acids, the following changes apply:
   a. $+ \Delta H, - \Delta S, + \Delta G$
   b. $+ \Delta H, + \Delta S, - \Delta G$
   c. $- \Delta H, - \Delta S, + \Delta G$
   d. $- \Delta H, + \Delta S, + \Delta G$

4. Exothermic and exergonic reactions are similar in that
   a. they both release energy.
   b. they are always spontaneous.
   c. they both result in a decrease in entropy.
   d. all of the above are true.

5. A negative ΔG means that
   a. the quantity G of energy is available to do work.
   b. the reaction is spontaneous.
   c. the reactants have more free energy than the products.
   d. all of the above are true.

6. According to the first law of thermodynamics,
   a. for every action there is an equal and opposite reaction.
   b. every energy transfer results in an increase in disorder or entropy.
   c. the total amount of energy in the universe is conserved or constant.
   d. energy can be transferred or transformed, but disorder always increases.

7. Catabolic pathways
   a. combine molecules into more complex and energy-rich molecules.
   b. are usually coupled with anabolic pathways to which they supply energy in the form of ATP.
   c. involve endergonic reactions that break complex molecules into simpler ones.
   d. are spontaneous and do not need enzyme catalysis.

8. Chemical energy
   a. is a form of potential energy.
   b. is stored in the bonds between atoms in molecules.
   c. can be transferred from one molecule to another, but some of it will be transformed to heat energy.
   d. all of the above are true.

9. A spontaneous reaction
   a. proceeds rapidly.
   b. occurs without the addition of external energy (aside from $\Delta G^{\ddagger}$).
   c. does not need to be catalyzed by enzymes.
   d. all of the above are true.

10. The formation of ATP from ADP and inorganic phosphate
    a. is an exergonic process.
    b. transfers the phosphate to another intermediate that becomes more reactive.
    c. produces an unstable, high-energy bond that can drive cellular work.
    d. has a ΔG of -7.3 kcal/mol.

11. At equilibrium,
    a. no enzymes are functioning.
    b. $\Delta G = O$.
    c. the forward and backward reactions have stopped.
    d. all of the above are true.

12. An enzyme is capable of
    a. lowering the free energy of activation of a reaction.
    b. changing the equilibrium of a reaction.
    c. increasing the free energy change for a reaction.
    d. all of the above.

13. In cooperativity,
    a. a multienzyme complex contains all the enzymes of a metabolic pathway.
    b. a product of a pathway serves as a competitive inhibitor of an early enzyme in the pathway.
    c. a molecule bound to the active site of one subunit of an enzyme affects the active site of other subunits.
    d. the allosteric site is filled with an activator molecule.

14. Substrates are held in the active site of an enzyme by
    a. hydrogen and ionic bonds.
    b. the action of coenzymes and cofactors.
    c. the matching shape of the allosteric site.
    d. the lowering of the activation energy.

15. According to the induced-fit hypothesis,
    a. the binding of the substrate depends on the shape of the active site.
    b. a competitive inhibitor can outcompete the substrate for the active site.
    c. the binding of the substrate changes the shape of the enzyme slightly and can stress or bend substrate bonds.
    d. the active site creates a microenvironment ideal for the reaction.

FILL IN THE BLANKS

1. _____ is the totality of an organism's chemical processes.

2. _____ pathways require energy to combine molecules.

3. _____ energy is the energy of motion.

4. _____ refers to the heat content or potential energy in the bonds of a molecule.

5. _____ is the term for the measure of disorder or randomness.

6. _____ is the energy that must be absorbed by molecules to reach the transition state.

7. _____ inhibitors change the enzyme's conformation by binding to an allosteric site.

8. _____ are organic molecules that bind to enzymes and are necessary for their functioning.

9. _____ is a regulatory device in which the product of a pathway binds to an enzyme early in the pathway.

10. _____ enzymes change between two conformations depending on whether an activator or inhibitor is bound to them.

CHAPTER 2
ATOMS, MOLECULES AND CHEMICAL BONDS

Suggested Answers to Structure Your Knowledge

1. An atom is the smallest unit of an element that maintains the physical and chemical properties of that element. Atoms are composed of subatomic particles, the largest and most stable of which are protons, neutrons, and electrons. Protons and neutrons are packed into the nucleus of an atom, whereas electrons travel at great speeds around the nucleus in orbitals within energy shells. Protons and neutrons both have a mass of 1 dalton; electrons have negligible mass. Neutrons do not carry a charge. Protons have a charge of +1; electrons have a charge of –1. Atoms that are not involved in chemical bonds are neutral because the number of electrons equals the number of protons. Ions are atoms that have gained or lost electrons, and their charge reflects the difference between the number of protons and electrons.

2. The atoms of each element have a characteristic number of protons in their nuclei, referred to as the *atomic number*. In a neutral atom, the atomic number also indicates the number of electrons.

   The *mass number* is an indication of the approximate mass of an atom and is equal to the number of protons and neutrons in the nucleus.

   The *atomic weight* refers to the atomic mass of an atom. It is equal to the mass number and is measured in the atomic mass unit of daltons. Protons and neutrons have a mass of approximately 1 dalton.

   The *valence* is an indication of the bonding capacity of an atom. It is the number of covalent bonds that must be formed for an atom to satisfy the

octet rule and complete its valence shell with eight electrons. The valence of an atom is most related to the chemical behavior of an atom because it is an indication of the number of bonds the atom will make, or the number of electrons the atom must share in order to reach a filled valence shell.

3. Ionic and nonpolar covalent bonds represent the two extremes along a continuum of electron sharing between atoms in a molecule. In ionic bonds the electrons are completely pulled away from one atom to the other, creating negatively and positively charged ions. In nonpolar covalent bonds the electrons are equally shared between two atoms. The middle ground of these two extremes is filled with polar covalent bonds in which a more electronegative atom pulls the shared electrons closer to it, producing a partial negative charge associated with that portion of the molecule and a partial positive charge associated with the atom from which the electrons are pulled. Because electrons are rapidly orbiting the nuclei and constantly changing positions, a very polar bond may vacillate between being covalent and ionic.

Answers to Test Your Knowledge

Multiple Choice:

1. b	5. a	9. a	13. b	17. c
2. d	6. a	10. d	14. a	18. c
3. a	7. c	11. b	15. c	19. d
4. b	8. a	12. d	16. b	20. c

CHAPTER 3
WATER AND THE FITNESS OF THE ENVIRONMENT

Suggested Answers to Structure Your Knowledge

1.

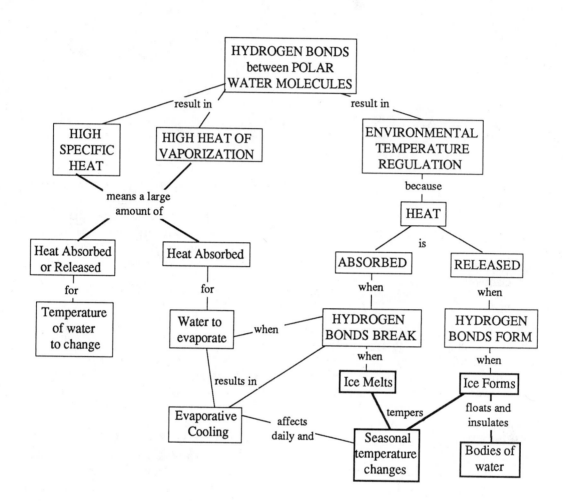

2.

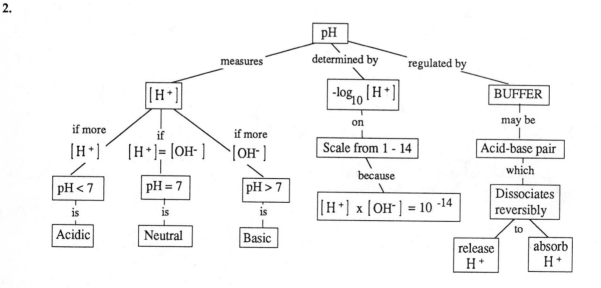

Answers to Test Your Knowledge

Multiple Choice:

1. c	4. a	7. d	10. d	13. c
2. a	5. b	8. b	11. b	14. c
3. b	6. c	9. b	12. d	15. c

True or False:

1. True
2. False, change *adhesion* to *hydrogen bonding*
3. False, change *absorbs cold* to *releases heat*
4. True
5. True
6. False, change *calorie* to *specific heat*, or change *substance* to *water*
7. True
8. True
9. False, change *releases* to *absorbs*, and *to* to *from*
10. False, change *dissociation into . . . ions* to *hydrogen bonding*

Fill in the Blanks:

$[H^+]$	$[OH^-]$	pH	acidic, basic, or neutral?
10^{-3}	10^{-11}	3	acidic
10^{-8}	10^{-6}	8	basic
10^{-12}	10^{-2}	12	basic
10^{-9}	10^{-5}	9	basic
10^{-1}	10^{-13}	1	acidic
10^{-7}	10^{-7}	7	neutral

CHAPTER 4
CARBON AND MOLECULAR DIVERSITY

Suggested Answers to Structure Your Knowledge

1.

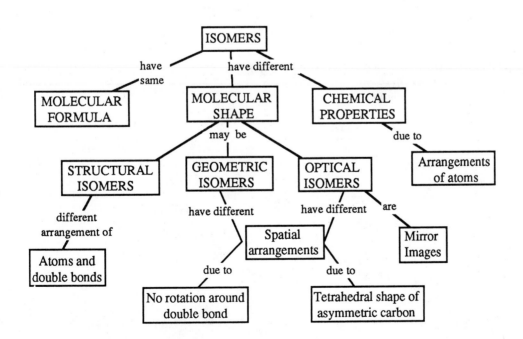

2.

FUNCTIONAL GROUPS TABLE

NAME OF GROUP	MOLECULAR FORMULA	CHARACTERISTICS CONFERRED TO ORGANIC COMPOUND
Hydroxyl	-OH	Polarity, solubility in water
Carbonyl	-C=O	Aldehyde or ketone, polar group
Carboxyl	-COOH	Acidic, can dissociate and release H^+
Amine	$-NH_2$	Basic, accepts H^+, becoming $-NH_3^+$
Sulfhydryl	-SH	Forms crosslinks in 3-D protein conformation
Phosphate Group	$-OPO_3H_2$	Acidic, use in energy transfer

Answers to Test Your Knowledge

Multiple Choice:

1.	d	4.	b	7.	a	10.	b
2.	a	5.	a	8.	c	11.	b
3.	d	6.	d	9.	a	12.	c

Matching:

1.	A, C	4.	E	7.	A, D, G	10.	B, F
2.	B, F	5.	E	8.	E	11.	E
3.	D, G	6.	A, C, D, G	9.	A, C	12.	C

CHAPTER 5
STRUCTURE AND FUNCTION OF MACROMOLECULES

Suggested Answers to Structure Your Knowledge

1.

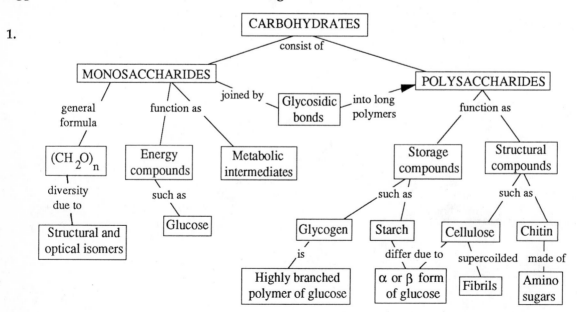

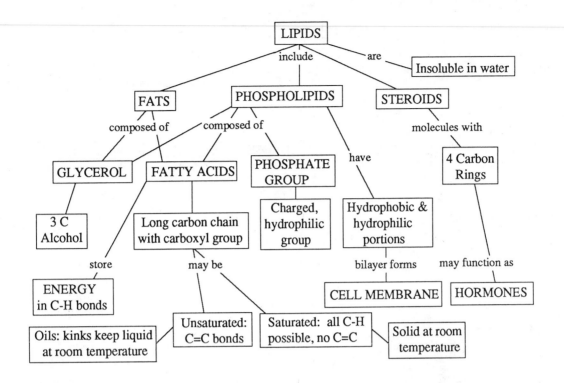

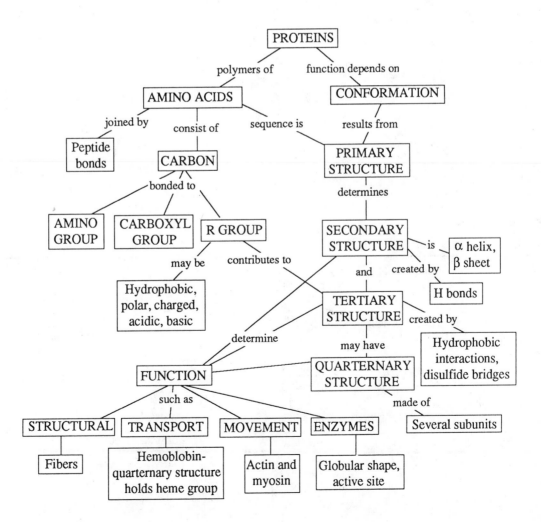

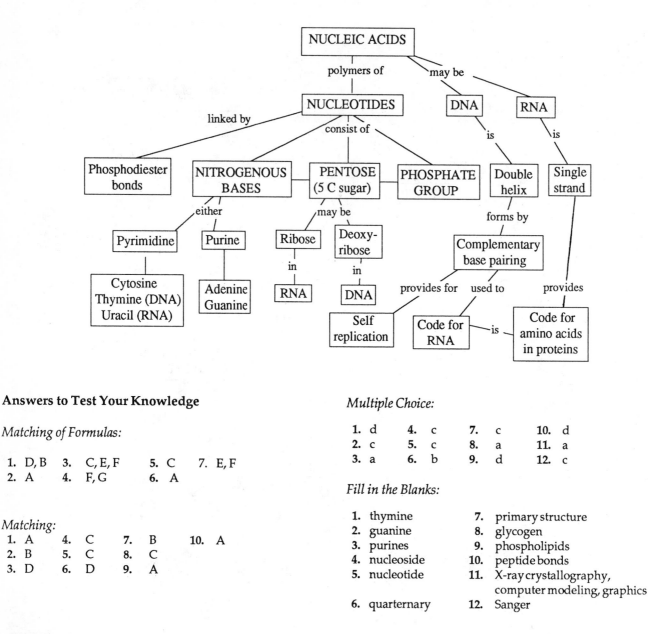

Answers to Test Your Knowledge

Matching of Formulas:

1. D, B 3. C, E, F 5. C 7. E, F
2. A 4. F, G 6. A

Matching:

1. A 4. C 7. B 10. A
2. B 5. C 8. C
3. D 6. D 9. A

Multiple Choice:

1. d	4. c	7. c	10. d
2. c	5. c	8. a	11. a
3. a	6. b	9. d	12. c

Fill in the Blanks:

1. thymine
2. guanine
3. purines
4. nucleoside
5. nucleotide
6. quarternary
7. primary structure
8. glycogen
9. phospholipids
10. peptide bonds
11. X-ray crystallography, computer modeling, graphics
12. Sanger

CHAPTER 6
INTRODUCTION TO METABOLISM

Suggested Answers to Structure Your Knowledge

1.

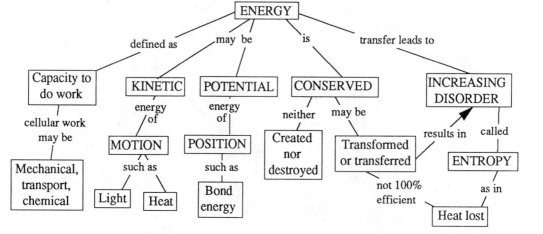

2.

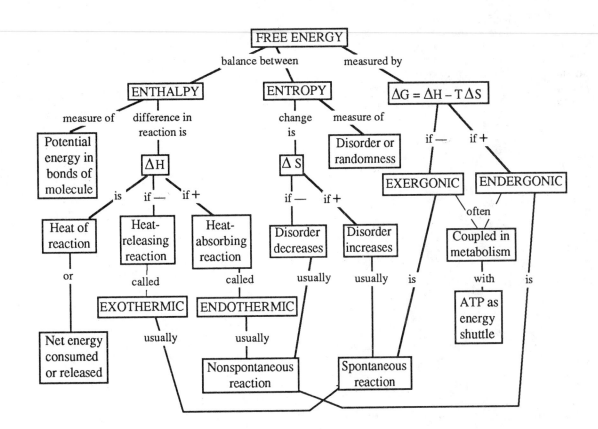

3. a. Enzymes are essential for the "cold chemistry" of life because they lower the free energy of activation of the reactions they catalyze and allow those reactions to occur extremely rapidly at a temperature conducive to life. Enzymes are also very specific, so that by regulating the enzymes it produces, the cell can regulate which of the myriad of possible chemical reactions take place at any given time.

b. The three-dimensional conformation of a protein permits it to have a characteristically shaped active site with specific chemical attributes. The shape and chemistry of the active site result in the specificity of an enzyme for a particular substrate.

c. A particular substrate binds to the active site on its enzyme with hydrogen and ionic bonds. The enzyme may change shape slightly, closing around the substrate. This induced fit strains the substrate's bonds or brings reactive groups in close proximity so that the transition state of the molecule(s) is reached and the reaction may proceed more easily. Once the reaction takes place, the substrate leaves the active site and another one enters. The speed of the reaction will increase with increasing substrate concentration until all enzyme molecules are saturated.

d. One way a cell can control enzyme action is through feedback inhibition, in which the product of a metabolic pathway then acts as an inhibitor of an enzyme operating earlier in that pathway. The cell can regulate which enzymes are produced and in what quantity through control of gene expression. The compartmentalization of the cell can order metabolic pathways and their substrates in space. Multienzyme complexes, where enzymes are sequentially arranged, may be located within cellular membranes.

Answers to Test Your Knowledge

Multiple Choice:

1. b	**5.** d	**9.** b	**13.** c
2. c	**6.** c	**10.** c	**14.** a
3. a	**7.** b	**11.** b	**15.** c
4. a	**8.** d	**12.** a	

Fill in the Blanks:

1. metabolism
2. anabolic
3. kinetic
4. enthalpy
5. entropy
6. free energy of activation
7. noncompetitive
8. coenzymes
9. feedback inhibition
10. allosteric

UNIT II

THE CELL

"He told you *that*? Well, he's pulling your flagellum, Nancy,"

CHAPTER 7

A TOUR OF THE CELL

FRAMEWORK

This chapter deals with the fundamental unit of life— the cell. The complexities in the processes of life are reflected in the complexities of the structure of the cell. It is easy to become overwhelmed by the number of new vocabulary terms and definitions for this array of cell organelles and membranes. The following concept map provides an organizational framework for the wealth of detail found in a "tour of the cell."

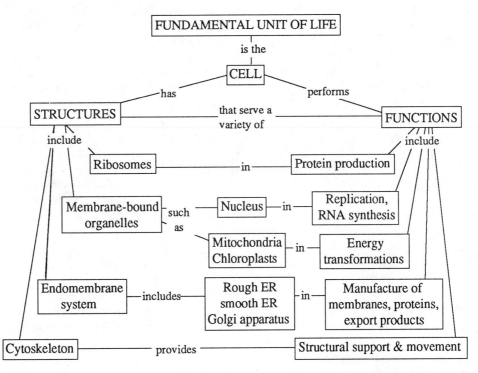

CHAPTER SUMMARY

The cell is the basic structural and functional unit of all living organisms. In the hierarchy of biological organization, the capacity for life emerges from the structural order of the cell. In this intricately integrated collection of subunits, structure is clearly correlated with function. The cell is able to sense and respond to its environment. Through the long history of evolution, environmental pressures have helped to shape the finely tuned structures and functions of this fundamental unit of life.

How Cells Are Studied

The growth of scientific knowledge and the development of new instruments and methods usually go hand-in-hand. The invention of the microscope in the 17th century led to the initial discovery and study of cells. Today's electron microscope and techniques such as cell fractionation have greatly expanded our understanding of cell structure and function.

Microscopy   The glass lenses of *light microscopes* refract (bend) the visible light passing through a specimen such that the projected image is magnified. *Resolving power* is a measure of the clarity of the image—determined by the minimum distance two points can be separated and still distinguished. The resolving power of the light microscope is limited by the wavelength of visible light. Details finer than 0.2 m cannot be resolved. Microscopes can be built to magnify specimens 1500 times, but beyond that the image just becomes more blurry—no new detail can be seen. Increasing the contrast improves the visibility of the structures large enough to be resolved.

Most subcellular structures, called *organelles*, are too small to be resolved by the light microscope. Although cells were discovered by Hooke in 1665, their details were largely unknown until the development of the *electron microscope* in the 1950s. The electron microscope (EM) focuses a beam of electrons through the specimen. The short wavelength of electrons allows a resolution of about 0.2 nanometer (nm), a thousand times greater than that of the light microscope, and exposes the *ultrastructure* of the cell.

In a *transmission electron microscope* (*TEM*) a beam of electrons is passed through a thin section of a specimen, and electromagnets, acting as lenses, focus and magnify the image. The density of regions of the specimen determines how readily electrons pass through. Contrast is increased by staining preserved cells with large atoms of metals that inhibit the passage of electrons. TEM is used to study internal ultrastructure of cells.

In a *scanning electron microscope* (*SEM*) the electron beam scans the surface of a specimen, which is usually coated with a thin gold film, exciting electrons from the specimen that are collected and focused onto a screen. The resulting image appears three-dimensional and shows the topography of the surface of a specimen.

Cytology is the study of cell structure. Modern cell biology integrates cytology with biochemistry to understand the relationships between cellular structure and function.

Cell Fractionation   *Cell fractionation* is a technique that separates major organelles of a cell so that their structures and functions can be studied. Cells are homogenized or broken open by ultrasound or grinding. The resulting cellular soup is separated into component fractions by spinning in a *centrifuge,* a machine that spins test tubes at various speeds. *Ultracentrifuges* can spin at more than 100,000 rpm. First, the homogenate is spun slowly, and nuclei and large particles settle to form a pellet. The remaining supernatant is then centrifuged at increasing speeds, each time isolating smaller and smaller cellular components in the pellet. Each cellular fraction contains a large quantity of the same cellular components and permits the isolated study of their composition and metabolic functions.

The Geography of the Cell: A Panoramic Overview

Prokaryotic and Eukaryotic Cells   The kingdom Monera, which includes bacteria and cyanobacteria (blue-green algae), is characterized by having *prokaryotic cells*, cells with no nuclear membrane or membrane-bound organelles. The DNA of prokaryotic cells is concentrated in a region called the nucleoid. *Eukaryotic cells* are found in the other four kingdoms of life: protists, plants, fungi, and animals. A eukaryotic cell is much more structurally complex. It has a true nucleus enclosed in a nuclear envelope or membrane and numerous organelles suspended in a semifluid medium called *cytosol.* The *cytoplasm* is the region between the nucleus and the membrane enclosing the cell.

Cell Size   Almost all cells are microscopic; their sizes are measured in micrometers (μm; 1 mm = 1000μm). The smallest known cells are certain bacteria with diameters of 0.1 μm—just large enough to pack in the DNA, ribosomes, and enzymes necessary to reproduce and sustain life. Most bacterial cells range from 1 to 10μm in diameter, whereas eukaryotic cells are ten times larger, ranging from 10 to 100μm.

The small size of cells is dictated by geometry and the requirements of metabolism. Area is proportional to the square of the linear dimension and volume is proportional to its cube. Eukaryotic cells, with 10 times the diameter of bacterial cells, have 100 times as much surface area but 1000 times greater volume. The *plasma membrane* must regulate sufficient exchange of oxygen, nutrients, and wastes to provide for the metabolism of the cell, and the nucleus must control the metabolic processes within the cell. The microscopic size of cells provides more surface area for exchange relative to the volume of the cell and a smaller volume of cytoplasm to be regulated by the nucleus.

The Importance of Compartmental Organization

The difference in the volume-to-surface-area ratio in eukaryotic compared to prokaryotic cells is partly compensated for by an extensive internal membrane system. These membranes compartmentalize the eukaryotic cell, providing local environments for specific metabolic functions, and participate in many metabolic processes through membrane-bound enzymes.

Membranes are composed of phospholipid molecules, in a two-layer arrangement called a bilayer, and various embedded proteins. The specific molecular composition of a membrane varies according to its functions. These structures play a key role in the organization and functioning of a cell's metabolism.

The Nucleus

The nucleus is surrounded by the *nuclear envelope,* a double membrane perforated by pores that may regulate the movement of large macromolecules between the nucleus and the cytoplasm. The inner membrane is associated with a layer of protein that helps to maintain the shape of the nucleus and may contribute to the organization of the genetic material.

Most of the cell's DNA is located in the nucleus, where, along with associated proteins, it is organized into *chromosomes.* Each eukaryotic species has a characteristic chromosomal number. Chromosomes are visible only when coiled and condensed in a dividing cell; otherwise they appear as a mass of stained material called *chromatin.*

The *nucleolus,* a round structure visible in the nondividing nucleus, consists of specialized regions of chromosomes, called *nucleolar organizers,* that have multiple copies of genes for ribosome synthesis. Ribosomal subunits are constructed in the nucleolus from RNA, produced in the nucleolus and elsewhere in the nucleus, and from proteins transported in from the cytoplasm. The subunits are transported through nuclear pores into the cytoplasm, where protein synthesis occurs.

The genetic instructions for specific proteins are transcribed from DNA into *messenger RNA (mRNA)* in the nucleus. The messenger RNA passes into the cytoplasm, where it complexes with ribosomes that translate the message into the primary sequence of proteins. The prominence of nucleoli and the number of ribosomes are related to the rate of protein synthesis within a cell.

Ribosomes

Ribosomes are composed of a large and a small subunit that join together when they attach to a messenger RNA and begin protein synthesis. Subunits consist of protein and RNA and are not enclosed in a membrane. The difference in molecular composition of the smaller ribosomes of prokaryotes has medical consequences; some antibiotics specifically inhibit the prokaryotic ribosomes of bacteria without harming the activity of the patient's ribosomes.

Ribosomes usually occur in clusters, called polyribosomes, or *polysomes,* attached to a single messenger RNA. Most of the proteins produced by *free ribosomes* are used within the cytosol. *Bound ribosomes,* attached to the endoplasmic reticulum, usually make proteins that will be included within membranes or exported from the cell.

The Endomembrane System

The *endomembrane system* of a cell consists of the nuclear envelope, endoplasmic reticulum, Golgi apparatus, lysosomes, microbodies, vacuoles, and the plasma membrane. These membranes are all related either through direct contact or by the transfer of membrane segments by membrane-bound sacs called vesicles. Although related, the membranes differ in molecular composition and structure, depending on their functions.

Endoplasmic Reticulum The *endoplasmic reticulum (ER)* is the most extensive portion of the endomembrane system. It is continuous with the outer nuclear membrane and encloses a network of interconnected compartments or tubules called *cisternae.* Ribosomes are attached to the cytoplasmic surface of *rough ER,* which manufactures membrane and secretory proteins. *Smooth ER* lacks ribosomes.

The smooth ER of various cells can serve diverse functions, such as fat, phospholipid, steroid and sex hormone synthesis; carbohydrate metabolism (as in the removal of phosphate from glucose in glycogen hydrolysis by liver cells); storage and release of calcium ions necessary for contraction of muscle cells; and detoxification of drugs and poisons. Barbituates, alcohol, and other drugs increase a liver cell's production of smooth ER and detoxification enzymes, thus leading to an increased tolerance (and thus reduced effectiveness) for these and other drugs. Another function of smooth ER is to process and transfer products formed in rough ER.

Proteins intended for secretion are manufactured by membrane-bound ribosomes and then threaded into the cisternal space of the rough ER, where they fold into their conformation. Most secretory proteins are *glycoproteins.* A carbohydrate, usually an *oligosaccharide* (a molecule made of a "few" saccharides), is covalently bonded to the protein by enzymes built into the ER membrane. The secretory proteins are transported from the rough ER in tiny membrane-bound *transport vesicles* budded off from adjoining smooth ER.

Rough ER manufactures membranes by inserting proteins formed by the attached ribosomes into the membrane. Both rough and smooth ER assemble phospholipids with the aid of enzymes built into their membranes and precursors obtained from the cytosol.

Free ribosomes and ER-bound ribosomes are identical. The polypeptide chain of a secretory protein begins with a *signal sequence* of amino acids that enables the ribosome to attach to a receptor site on the ER membrane. The growing polypeptide moves into the ER cisternal space, where the signal sequence is removed by an enzyme. Signal sequences, coded for by mRNA and attached to polypeptide chains, are thought to direct various proteins to specific sites within the cell.

The Golgi Apparatus   Transport vesicles, pinched off from the ER, travel first to the *Golgi apparatus*, where products of the ER are modified, stored, and directed to other locations. Individual Golgi apparatuses in plant cells are called *dictyosomes*.

The Golgi apparatus consists of a stack of flattened membrane sacs. One end of the complex is called the *cis face* and is closely associated with smooth ER. Vesicles that bud from the smooth ER join to a Golgi membrane at the *cis* face, adding to it their contents and membrane. Products that travel through the Golgi apparatus are usually modified or refined. Some macromolecules are manufactured within the Golgi. The other end of the Golgic apparatus is called the *trans face*, from which vesicles pinch off. Golgi products may be tagged with different oligosaccharides or phosphate groups that help direct them to various parts of the cell.

Lysosomes   Lysosomes are membrane-enclosed sacs of hydrolytic enzymes used by the cell to digest macromolecules. The membrane's inward pumping of hydrogen ions from the cytosol maintains an acidic environment within the lysosome. The cell can protect itself from unwanted digestion and also provide an optimal pH value by containing hydrolytic enzymes within lysosomes.

Lysosomes digest food particles ingested by *phagocytosis* by cells such as protozoa and macrophages. Lysosomal hydrolytic enzymes also recycle the cell's own macromolecules by engulfing organelles or small bits of cytosol, a process known as autophagy. During development or metamorphosis, lysosomes may be programmed to destroy their cells to create a specific body form. Storage diseases are inherited defects in which a lysosomal enzyme is missing, and lysosomes become packed with indigestible substances.

Microbodies   Microbodies are single-membrane enclosed compartments filled with particular enzymes for a specific metabolic pathway. *Peroxisomes* are microbodies that produce hydrogen peroxide, H_2O_2, while breaking down fats to smaller molecules or detoxifying alcohol or other poisons. Hydrogen peroxide is itself toxic, but peroxisomes contain an enzyme to convert it to water. Compartmentalizing these essential enzymes together is important to cellular function.

Glyoxysomes, commonly found in the tissues of germinating seeds, contain enzymes that convert the oils stored in the seed to sugar for the developing seedling.

Vacuoles   Vacuoles are also membrane-enclosed sacs within the cell, distinguished by being larger than vesicles. *Food vacuoles* are formed as a result of phagocytosis. *Contractile vacuoles* pump excess water out of freshwater protozoa.

A large *central vacuole* is found in mature plant cells, surrounded by a membrane called the *tonoplast*. This vacuole stores organic compounds and inorganic ions for the cell. Hydrolytic enzymes are compartmentalized in the central vacuole and digest stored macromolecules and recycle organic compounds from used organelles. Poisonous or unpalatable compounds that may protect the plant from predators, and dangerous metabolic byproducts may also be contained in the vacuole. Plant cells can increase in size with a minimal addition of new cytoplasm as their vacuoles absorb water and expand. The cytoplasm forms a thin band between the plasma membrane and the tonoplast, maintaining a good membrane surface-area-to-volume ratio even in large plant cells.

Relationships of Endomembranes: A Summary   Through the fusion of transport vesicles, or through direct continuity, the components of the endomembrane system are related. The nuclear envelope extends from the rough ER, which also connects to the smooth ER. Membranes pinched off from the ER move to the Golgi and on to lysosomes, microbodies, vacuoles, and even the plasma membrane. Though related, each type of membrane has structural modifications designed for its specific function. The structural compartmentalization of the endomembrane system organizes and facilitates the complexity of cellular function.

Energy Transducers: Mitochondria and Chloroplasts

Cellular respiration, the catabolic processing of fuels to produce ATP with the help of oxygen, occurs within the *mitochondria* of most eukaryotic cells. *Chloroplasts*, found only in plants and eukaryotic algae, produce organic compounds from carbon dioxide and water by absorbing solar energy. Mitochondria and chloroplasts are not considered part of the endomembrane system because their membrane proteins are made, not by the ER, but by ribosomes either free in the cytosol or contained within these organelles. They also contain a small amount of DNA that directs the synthesis of some of their proteins.

Mitochondria   The number of mitochondria found in a cell can range from hundreds to thousands, depending on the metabolic needs of the cell. Two membranes, each a phospholipid bilayer with unique embedded proteins, enclose a mitochondrion. The smooth outer membrane allows the passage of small molecules but blocks the passage of macromolecules from the cytosol into the *intermembrane space* between the membranes. The inner membrane is thrown into folds called *cristae*, creating a large membrane surface area and enclosing the *mitochondrial matrix*. Within this matrix are enzymes that control many of the metabolic steps of cellular respiration. Other respiration enzymes are built into the extensive inner membrane.

Chloroplasts   *Plastids* are plant organelles that include amyloplasts (or leucoplasts), which store starch; chromoplasts, which contain pigments; and chloroplasts, which contain the green pigment chlorophyll and function in photosynthesis. All three types of plastids can develop in unspecialized cells from proplastids, the fates of which depend both on the location and the environment of the cell. Thus, chloroplasts develop only when proplastids are exposed to light or when already existing chloroplasts divide.

Chloroplasts are bounded by two membranes, which enclose the viscous fluid called the *stroma* and the *thylakoids*, a membranous system of flattened sacs. Thylakoids may be stacked together to form structures called *grana*.

The Cytoskeleton

The *cytoskeleton* is a network of fibers that function to give mechanical support; maintain or change cell shape; anchor or direct the movement of organelles and cytoplasm; and control movement of cilia, flagella, pseudopods, and even contraction of muscle cells. At least three types of fibers are involved in the cytoskeleton: microtubules, microfilaments, and intermediate filaments.

Microtubules   *Microtubules* are hollow rods constructed of two kinds of globular proteins called a and ß tubulins, arranged in 13 columns and rolled into a

tube. In addition to providing the major supporting framework of the cell, microtubules serve as tracks to guide the movement of organelles, such as vesicles moving from the Golgi apparatus to the plasma membrane. All eukaryotic cells have microtubules. In many cells they radiate out from an organizing mass called the cell center. Other bundles of microtubules located near the plasma membrane contribute to cell shape.

The separation of chromosomes in a dividing cell is associated with microtubules assembled from the cell center. In animal cells, *centrioles*, composed of nine sets of triplet microtubules arranged in a ring, may help to organize microtubule assembly for cell division.

Cilia and *flagella* are extensions of eukaryotic cells, composed of and moved by microtubules. Cilia are numerous and short; flagella occur one or two to a cell and are longer. A flagellum moves with an undulating motion whereas a cilium generates force with a power stroke alternating with a recovery stroke (like an oar). Many protists use cilia or flagella to swim (or actually crawl) through aqueous media. Sperm of animals and lower plants are propelled by flagella. Cilia or flagella attached to stationary cells of a tissue sweep fluid past the cell.

Cilia and flagella both are composed of a core of nine doublets of microtubules arranged in a ring surrounding two single microtubules (a nearly universal "9 + 2" arrangement), and the core is enclosed in an extension of the plasma membrane. The sliding of the microtubule doublets past each other occurs as sidearms, composed of the protein *dynein*, alternately attach to adjacent doublets, pull down (as the conformation of dynein changes), release, and reattach further along the doublet. In conjunction with the radial spokes that anchor the doublets near the central microtubules, this action, driven by ATP, causes the bending of the flagella or cilia. A *basal body*, structurally identical to a centriole, anchors the nine doublets of the cilium or flagellum to the cell.

Microfilaments and Movement   *Microfilaments* are solid rods consisting of a helix of two chains of molecules of the globular protein *actin*. In muscle cells, thousands of microfilaments of actin interdigitate with thicker filaments made of the protein *myosin*. The sliding of actin and myosin filaments past each other, driven by ATP-powered arms extending between the filaments, causes the shortening of the cell and thus the contraction of muscles.

Microfilaments seem to be present in nearly all eukaryotic cells. They function in support, such as in the core of microvilli, which project from the surface of cells specialized for absorption of material across the plasma membrane. Miniature versions of the actin and myosin muscle arrangement are found in many cells in areas responsible for localized contractions.

Actin–myosin aggregates are responsible for the pinching apart of animal cells when they divide and the extension and retraction of *pseudopodia* during *amoeboid movement*. Cyclosis, or *cytoplasmic streaming*, in plant cells probably involves microfilaments.

Intermediate Filaments   Intermediate filaments are intermediate in size between microtubules and microfilaments. Unlike the other two fibers that are consistent in composition across all eukaryotic cells, intermediate fibers vary in protein makeup from one cell type to another. Intermediate fibers appear to be important in maintaining the shape of the cell and anchoring certain organelles. The nucleus is often securely held in a web of intermediate fibers. These tension-bearing filaments may serve as the superstructure of the entire cytoskeleton.

Current research is focusing on how the components of the cytoskeleton are integrated. High-voltage electron microscopy seems to reveal that short protein fibers cross-link all the elements of the cytoskeleton into one great network known as a microtrabecular lattice. Some researchers maintain that the cross-links are an artifact of sample preparation. The structure and integration of the microtubules, microfilaments, and intermediate filaments of the cytoskeleton are central to the organization and functions of the cell.

The Cell Surface

Most cells synthesize and secrete some sort of covering over their plasma membrane.

Cell Walls   A *cell wall* is a diagnostic feature of plant cells and is also found in prokaryotes, fungi, and some protists. Plant cell walls are composed of fibers of cellulose embedded in a matrix of other polysaccharides and some protein. The exact composition of the wall varies between species and even between cell types in the same plant.

The *primary cell wall*, secreted by a young plant cell, is relatively thin and flexible and can stretch as the cell grows. Adjacent cells are connected by the *middle lamella*, a thin layer of polysaccharides called pectins that glue the cells together. When they stop growing, some cells secrete a thicker and stronger *secondary cell wall* between the plasma membrane and primary cell wall.

Plasmodesmata are channels in plant cell walls through which strands of cytoplasm connect bordering cells, and water and small solutes can move. The plasma membranes of adjacent cells are continuous through the channel, linking most cells of a plant into a living continuum.

The Glycocalyx of Animal Cells   Animal cells often secrete a *glycocalyx* made of sticky oligosaccharides that serves to strengthen the cell surface and glue cells together. The glycocalyx probably functions in cell–cell recognition.

Intercellular Junctions   Intercellular junctions between membranes of neighboring cells improve communication and interaction between cells.

The three main types of intercellular junctions between animal cells are desmosomes, tight junctions, and gap junctions. *Desmosomes* are clumps of intercellular filaments that connect two cells and terminate inside each cell as a disk of dense material reinforced by intermediate filaments. Desmosomes connect cells into strong epithelial sheets but still permit materials to pass through the intercellular space between cells. *Tight junctions* are connections between cells that create an impermeable layer within the intercellular space. Specialized proteins built into one plasma membrane bond to similar proteins in the other membrane. Tight junctions are found in epithelial layers that separate two solutions and must prevent intercellular transport across the layer of cells. *Gap junctions* allow for the exchange of sugars, amino acids, and small molecules between cells. They consist of special donut-shaped proteins embedded in bordering plasma membranes that connect across the intercellular space.

STRUCTURE YOUR KNOWLEDGE

1. **A.** Table A lists the general functions an animal cell performs. List the structures associated with each of these functions.

    B. Table B lists structures that are unique to plant cells. Fill in the functions of these structures.

2. Create a modified concept map or flow chart to trace the development of a secretory product (such as a digestive enzyme) from the DNA code to its export from the cell.

TEST YOUR KNOWLEDGE

MULTIPLE CHOICE: *Choose the one best answer.*

1. Which of the following is/are *not* found in a prokaryotic cell?
    a. ribosomes
    b. plasma membrane

FUNCTIONS OF CELL	ASSOCIATED ORGANELLES AND STRUCTURES
Cell division	
Information storage and transferal	
Energy conversions	
Manufacturing: membranes and products	
Digestion, recycling	
Specific pathways	
Structural integrity	
Movement	
Exchange with environment	
Cell-cell interaction	

Table A

STRUCTURES UNIQUE TO PLANT CELLS	FUNCTIONS
Cell wall	
Central vacuole	
Plastids: chloroplasts amyloplasts	
Dictyosomes	
Glyoxysomes	
Plasmodesmata	

Table B

c. mitochondria
d. a and c

2. Eukaryotic cells can be larger than prokaryotic cells because
   a. their plasma membrane is more permeable.
   b. their internal membrane system allows compartmentalization of functions and extra surface area for exchange and enzyme location.
   c. their DNA is localized in the nucleus whereas protein synthesis occurs in the cytoplasm.
   d. they have a better surface-area-to-volume ratio.

3. Which of the following is *not* a similarity among the nucleus, chloroplasts, and mitochondria?
   a. They all contain DNA.
   b. They all are bounded by a double phospholipid bilayer membrane.
   c. They can all divide to reproduce themselves.
   d. They all are derived from the endoplasmic reticulum system.

4. The pores in the nuclear envelope provide for the movement of
   a. proteins into the nucleus.
   b. ribosome subunits out of the nucleus.

c. mRNA out of the nucleus.
d. all of the above.

5. The nucleolus functions in
   a. synthesis and storage of components of ribosomes.
   b. formation of spindle fibers.
   c. condensation of chromosomes.
   d. direction of DNA synthesis.

6. The largest number of ribosomes most likely would be found in a cell
   a. with a high metabolic rate.
   b. that produces secretory products.
   c. with many cilia.
   d. that is actively dividing.

7. Which structure is *not* considered to be part of the endomembrane system?
   a. mitochondrion
   b. smooth ER
   c. nuclear envelope
   d. lysosome

8. A plant cell grows larger primarily by
   a. increasing the number of vacuoles.
   b. synthesizing more cytoplasm.
   c. taking up water into its central vacuole.
   d. synthesizing more cellulose.

9. The innermost portion of a mature plant cell wall is the
   a. primary cell wall.
   b. secondary cell wall.
   c. middle lamella.
   d. glycocalyx.

10. The contractile elements of muscle cells are
    a. smooth ER.
    b. centrioles.
    c. microtubules.
    d. microfilaments.

11. Microtubules are components of all of the following *except*
    a. centrioles.
    b. the spindle apparatus for separating chromosomes in cell division.
    c. the pinching apart of the cytoplasm in animal cell division.
    d. flagella and cilia.

12. Of the following, which is probably the most common route for membrane flow in the endomembrane system?
    a. rough ER→Golgi→lysosomes→vesicles →plasma membrane
    b. rough ER→smooth ER→Golgi→vesicles→plasma membrane

c. nuclear envelope→rough ER→Golgi→smooth ER→lysosomes
d. rough ER→vesicles→Golgi→smooth ER→plasma membrane

13. Proteins to be secreted by the cell are generally synthesized
    a. with signal sequences that result in the ribosomes binding to ER.
    b. by free polysomes.
    c. by the nucleolus.
    d. within the Golgi apparatus.

14. Many of the proteins made by rough ER become part of
    a. the ribosomes.
    b. the cytosol.
    c. membranes.
    d. the cytoskeleton.

15. Plasmodesmata in plant cells are similar in function to
    a. desmosomes.
    b. tight junctions.
    c. gap junctions.
    d. glycocalyx.

16. The ultrastructure of a chloroplast could be best seen using
    a. transmission electron microscopy.
    b. scanning electron microscopy.
    c. phase contrast light microscopy.
    d. cell fractionation.

17. Resolving power of a microscope is a measure of
    a. the distance between two separate points.
    b. the sharpness or clarity of an image.
    c. the degree of magnification of an image.
    d. the depth of focus on a specimen's surface.

18. Dyneine is involved in
    a. the movement of microtubules.
    b. the contraction of microfilaments.
    c. the attachment of the microtrabecular lattice.
    d. the tensile strength of intermediate filaments.

FILL IN THE BLANKS *with the appropriate cellular organelle or structure.*

1. _____ transport membranes and products to various locations

2. _____ infolding of mitochondrial membrane with attached enzymes

3. _____ sticky, supportive coat on animal cells

4. _____ small sacs with enzymes for specific metabolic pathway

5. _____ system of flattened sacs inside chloroplasts

6. _____ anchoring structure for cilia and flagella

7. _____ semifluid medium between nucleus and plasma membrane

8. _____ system of fibers that maintains cell shape, anchors organelles

9. _____ connection between cells that creates impermeable layer

10. _____ membrane surrounding central vacuole of plant cells

MATCHING: *Match the function and structure with the proper organelle or cellular component.*

FUNCTION

A. produce rRNA, assemble ribosomes

B. manufacture proteins

C. contain chromosomes, control center

D. site of cellular respiration

E. site of photosynthesis

F. store food, pump water, found in center of plant cells

G. maintain cell shape

H. house enzymes that synthesize lipids, detoxify drugs, store and release ions

I. digest macromolecules

J. manufacture membranes and secretory products

K. process products of rough ER

STRUCTURE

a. network of membranes with attached ribosomes

b. double membrane sac containing stacks of membrane sacs

c. large and small subunits, attach to mRNA

d. sac with hydrolytic enzymes, acid pH

e. round structure with multiple copies of genes for rRNA

f. single membrane bound sac

g. surrounded by double membrane, inner layer of protein, with pores

h. large sac formed from small vesicles

i. double membrane sac with cristae

j. stack of flattened sacs, *cis, trans* faces

k. membrane network without attached ribosomes

FUNCTION	STRUCTURE	ORGANELLE
1. _____	_____	chloroplast
2. _____	_____	Golgi apparatus
3. _____	_____	lysosome
4. _____	_____	mitochondria
5. _____	_____	nucleolus
6. _____	_____	nucleus
7. _____	_____	ribosome
8. _____	_____	rough ER
9. _____	_____	smooth ER
10. _____	_____	vacuole

CHAPTER 8

MEMBRANE STRUCTURE AND FUNCTION

FRAMEWORK

This chapter presents the fluid mosaic model of membrane structure and relates the molecular arrangement of biological membranes to their function of regulating the passage of substances into the cell, out of the cell, and between intracellular compartments.

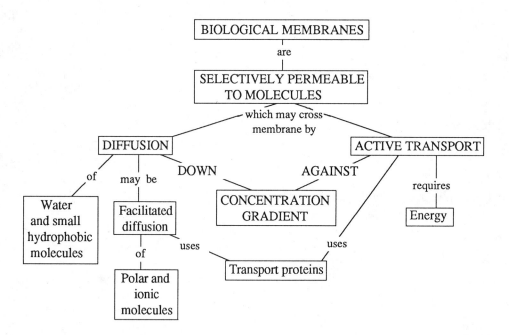

CHAPTER SUMMARY

The plasma membrane is the boundary of life, providing the cell with the potential to maintain a unique internal environment and to control the movement of materials into and out of the cell. Biological membranes are *selectively permeable*; their ability to discriminate in the chemical exchanges they allow is based on their unique structure.

Models of Membrane Structure

Two Generations of Membrane Models Over the years, scientists have created a series of models to explain membrane structure and function. The observation that lipid-soluble molecules rapidly entered cells led to the conclusion that membranes were composed of lipids. In the early 1900s, membranes isolated from red blood cells were found to consist of proteins as well as lipids.

The *phospholipid* components of membranes are *amphipathic*, having both a hydrophobic region consisting of two long, nonpolar hydrocarbon tails, and a hydrophilic head created by the phosphate group. A bilayer of phospholipids, with the hydrophobic tails in the center and the hydrophilic heads facing the aqueous solution on both sides and covered with a coat of globular proteins, was proposed as the molecular model of membranes by Davson and Danielli in 1935. This sandwich model was consistent with the observed thickness of plasma membranes (adjusting the globular proteins to layers of protein in the pleated-sheet configuration) and the apparent triple-layer staining results seen with the electron microscope in the 1950s. The Davson–Danielli model was widely accepted by the 1960s for all cellular membranes.

The molecular variability among cellular membranes and the amphipathic nature of membrane proteins created problems in the Davson–Danielli model. For example, where would the hydrophobic regions of such proteins fit between the aqueous environment and the hydrophilic heads of the phospholipids? In 1972, Singer and Nicolson proposed their *fluid mosaic model* in which the membrane proteins are embedded in the phospholipid bilayer with their hydrophilic regions extending out into the aqueous environment. The phospholipid bilayer is envisioned as fluid with a mosaic of individually inserted protein molecules shifting laterally within it. This model is consistent with the known properties of membranes and is supported by evidence from *freeze–fracture* electron microscopy, which shows the interior of the bilayer with protein bumps dispersed in a smooth matrix. Freeze–fracture is a technique of freezing a specimen, fracturing it with a cold knife, etching the fractured surface by removing water by sublimation to enhance contrast, and then coating the fractured surface with platinum and carbon. The resulting replica of the surface is examined in the electron microscope.

The fluid mosaic model is currently the most accepted and useful model for organizing the existing knowledge of and extending further research on membrane structure.

The Fluid Mosaic Model: A Closer Look
Membranes are held together primarily by hydrophobic attractions that allow the lipids and some of the proteins to drift about laterally, keeping their hydrophobic regions in contact. The intermingling of membrane proteins of a hybrid mouse–human cell illustrates the fluid nature of membranes. Some proteins do not drift due to their attachment to the cytoskeleton.

An abundance of phospholipids with unsaturated hydrocarbon tails maintains membrane fluidity at lower temperatures. The steroid cholesterol, common in animal cell membranes, also prevents the close packing of lipids and thus enhances fluidity. Some cells may change the lipid composition of their membranes in response to changing temperature.

Each membrane has its own unique complement of membrane proteins, which determine most of the specific functions of that membrane. *Integral proteins* extend into the hydrocarbon regions of the lipids. They may be unilateral and reach partway through the membrane or transmembrane, having two hydrophilic ends and a hydrophobic midsection. *Peripheral proteins* are attached to the surface of the membrane. Membranes are asymmetric; they have distinct inner and outer faces related to the directional orientation of their proteins, the composition of the lipid bilayers, and, in the plasma membrane, the attachment of carbohydrates to the exterior surface. The manufacture of the plasma membrane by ER (with the fusion of vesicles expanding the membrane) means that the exterior of the plasma membrane corresponds to the interior of the ER.

Cell–cell recognition, the ability of a cell to determine if a cell is similar to or different from itself, is most likely based on the recognition of membrane carbohydrates. These branched oligosaccharides (fewer than 15 sugar units) are covalently bonded either to lipids or, most often, to proteins. The identity of these glycolipids and glycoproteins varies from species to species, individual to individual, and even among cell types.

The critical ability of membranes to regulate the passage of substances into and out of the cell and between cellular compartments is determined by their supramolecular structure, the unique architecture determined by the arrangement of many molecules into a higher level of organization.

Traffic of Small Molecules

The plasma membrane permits a regular exchange of nutrients, waste products, oxygen, and inorganic ions. This steady traffic of small molecules is regulated in a variety of ways.

Selective Permeability
Biological membranes are selectively permeable; there is a difference in the ease and rate with which small molecules pass through them. The hydrophobic center of the lipid bilayer impedes passage of ions and polar molecules. The rates at which hydrophobic molecules cross an artificial membrane are related to their lipid solubilities and size. Small polar, but uncharged, molecules, such as H_2O and CO_2, can cross the synthetic membrane rap-

idly; however, large polar molecules, such as glucose, and all ions cannot permeate easily.

Transport proteins span the plasma membrane and provide the mechanisms for movement of ions and moderately sized polar molecules (such as sugars). These highly specific proteins provide either a channel or physical binding and transport for the particular molecules they move.

Diffusion and Passive Transport

Diffusion is the movement of a substance down its *concentration gradient* due to random molecular motion. This spontaneous process decreases free energy and increases entropy (disorder, randomness) in a system. Much of the exchange across membranes occurs by diffusion. The cell does not expend energy when substances diffuse across membranes down their concentration gradient; the process is called *passive transport*.

Osmosis: A Special Case of Passive Transport

Osmosis is the diffusion of water across a selectively permeable membrane. Water will move from a *hypoosmotic* solution across a membrane into a *hyperosmotic* solution—in other words, from a region with a lesser concentration of solutes into a region with a greater concentration of solutes. *Isosmotic* solutions have equal solute concentrations, and there is no net movement of water across a membrane separating them. Water diffuses down its own concentration gradient, which is affected by the binding of water molecules to solute particles. The association between water and solute molecules lowers the proportion of unbound water that is free to cross the membrane. The direction of osmosis is determined by the difference in total solute concentration, regardless of the kinds of solute molecules involved.

Osmotic pressure, which can be determined by an instrument called an osmometer, is a measure of the tendency for a solution to take up water when separated from pure water by a selectively permeable membrane. This pressure is proportional to *osmotic concentration*; it increases as the concentration of solute particles increases. Water moves from a solution with a lesser osmotic pressure (fewer solute particles—hypoosmotic) to a solution with a greater osmotic pressure (hyperosmotic).

An animal cell placed in a hyperosmotic environment will lose water and shrivel. If placed in a hypoosmotic environment, the cell will gain water, swell, and possibly lyse (burst). Cells without rigid walls must either live in an isosmotic environment (salt water or isosmotic body fluids) or have adaptations for *osmoregulation*, the control of water balance. Protozoans in fresh water environments may have membranes that are less permeable to water and contractile vacuoles that expel excess water.

The cell walls of plants, fungi, prokaryotes, and some protists play a role in water balance within hypoosmotic environments. The inward movement of water is offset by building pressure within the cell against its cell wall, resulting in a dynamic equilibrium in the movement of water and creating a *turgid* cell. Turgid cells provide mechanical support for non-woody plants. Plant cells in an isosmotic surrounding are *flaccid*. In a hyperosmotic medium, plant cells undergo *plasmolysis*, the pulling away of the plasma membrane from the cell wall as the cell shrivels, and usually die.

Facilitated Diffusion

Facilitated diffusion involves the diffusion of polar molecules and ions across a membrane with the aid of transport proteins. Transport proteins are similar to enzymes in several ways: they are specialized for the solute they transport, presumably with a specific binding site for that solute; they speed up the transport of solute; they can reach a maximum rate of transport when all transport proteins are saturated by a high concentration of the solute; and they can be inhibited by molecules that resemble their normal solute. A proposed model for the mechanism of transport-protein function involves a change in conformation of the protein, caused by the binding of the solute, that serves to translocate the binding site (and attached solute) from one side of the membrane to the other. Facilitated diffusion is passive transport; it speeds diffusion but cannot change the direction of movement of a solute down its concentration gradient.

Active Transport

Active transport, requiring the expenditure of energy by the cell, can move solutes against their concentration gradients. Active transport is essential for a cell to be able to maintain internal concentrations of small molecules that are different from the environment. Transport proteins that engage in active transport use cellular energy, usually in the form of ATP, to pump molecules against their concentration gradient. The terminal phosphate group of ATP may be transferred to the protein, inducing it to change its conformation and translocate the bound solute across the membrane. The *sodium–potassium pump*, also called sodium–potassium ATPase because it acts as an enzyme to hydrolyze ATP, allows the cell to exchange Na^+ and K^+ across the membrane in animal cells, creating a greater concentration of potassium ions and a lesser concentration of sodium ions within the cell.

The Special Case of Ion Transport

Cells have a *membrane potential*, a voltage across their plasma membrane created by the electrical potential energy resulting from the separation of opposite charges. The cytoplasm of a cell is negatively charged compared to the extracellular fluid. The membrane potential favors the diffusion of cations (positively charged) into the cell and anions out of the cell. Diffusion of an ion is affected by both the membrane potential and its own concentration gradient; thus an ion diffuses down its *electrochemical gradient*.

Why is cytoplasm usually negatively charged? At cellular pH, most proteins and other macromolecules are negatively charged. The sodium–potassium pump actually pumps three Na^+ ions out of the cell for every two K^+ ions it pumps in. Also, ions diffuse through the selectively permeable plasma membrane at different rates, and K^+ leaks out of the cell faster than Na^+ leaks back in. An *electrogenic pump* is a transport protein that generates voltage across a membrane by active transport of ions. Whereas the sodium–potassium pump is the major pump in animals, a proton pump that transports H^+ out of the cell helps to generate membrane potential in plants, fungi, and bacteria. Proton pumps are also important in the membranes of mitochondria and chloroplasts.

Cotransport Cotransport is a mechanism through which the active transport of one solute is indirectly driven by an ATP-powered pump that transports another solute against its gradient. Energy is stored by concentrating a substance on one side of the membrane. As the solute diffuses back across the membrane through specific transport proteins, other solutes may be cotransported against their own concentration gradients. The proton pump of plant cells drives the active transport of amino acids and sugars by coupling them with hydrogen ions that diffuse through specific symport proteins. A symport translocates two different molecules simultaneously in the same direction.

Traffic of Large Molecules: Endocytosis and Exocytosis

Large molecules, such as proteins and polysaccharides, are transported across the plasma membrane through the processes of exocytosis and endocytosis. In *exocytosis*, the cell secretes macromolecules by the fusion of vesicles containing these molecules with the plasma membrane. In *endocytosis*, a region of the plasma membrane sinks inward and pinches off to form a vesicle containing material that had been outside the cell. *Phagocytosis* is a form of endocytosis in which pseudopodia wrap around a food particle and create a vacuole. The vacuole then fuses with a lysosome containing hydrolytic enzymes. In *pinocytosis*, droplets of extracellular fluid are taken into the cell

in small vesicles. *Receptor-mediated endocytosis* allows a cell to acquire specific substances that bind with receptor sites clustered in *coated pits* on the cell surface. A layer of the fibrous protein *clathrin* on the cytoplasmic side of the membrane reinforces the coated pits. *Ligands*, molecules that bind specifically to receptor sites, are carried into the cell when the coated pit buds off to form a *coated vesicle*.

Animal cells use receptor-mediated endocytosis to take in cholesterol for the synthesis of membranes and other steroids. Familial hypercholesteremia is an inherited disease in which receptor sites are missing, and cholesterol accumulates in the blood, leading to early atherosclerosis.

Membranes and ATP Synthesis

The existence of concentration gradients across a membrane can serve as an energy source for the generation of ATP. The diffusion of H^+ down its concentration gradient through a transport protein can cause the protein to act as an ATPase and add a phosphate group to ADP. The membranes of mitochondria and chloroplasts make ATP by using energy from food or light, respectively, to generate concentration gradients of H^+.

STRUCTURE YOUR KNOWLEDGE

1. Relate the structure of membranes as depicted by the fluid mosaic model to their permeability to nonpolar hydrophobic molecules and polar or ionic molecules.

2. Compare facilitated diffusion and active transport in relation to the type of molecules carried, the membrane components involved, and the energy required.

3. Create a concept map to illustrate your understanding of the important process of osmosis.

TEST YOUR KNOWLEDGE

MULTIPLE CHOICE: *Choose the one best answer.*

1. Glycoproteins and glycolipids are important for
   a. facilitated diffusion.
   b. active transport.
   c. cell–cell recognition.
   d. cotransport.

2. Amphipathic molecules
   a. have hydrophilic and hydrophobic regions.
   b. include phospholipids and integral membrane proteins.
   c. may be held together by hydrophobic interactions.
   d. include all of the above.

3. Which of the following is *not* true about osmosis?
   a. It increases free energy in a system.
   b. Water moves from a hypoosmotic to a hyperosmotic solution.
   c. Osmotic pressure increases with increasing osmotic concentration.
   d. It increases the entropy in a system.

4. One of the problems with the Davson–Danielli sandwich model for membranes was that
   a. it did not explain the easy permeability by hydrophobic molecules.
   b. it did not account for the actual thickness of the plasma membrane.
   c. it could not explain where the hydrophobic regions of the membrane proteins would align.
   d. it could not account for the membrane potential.

5. Evidence for the fluid mosaic model of membrane structure came from
   a. the freeze–fracture technique of electron microscopy.
   b. the movement of proteins in hybrid cells.
   c. the amphipathic nature of membrane proteins.
   d. all of the above.

6. Membrane proteins that function in active transport are likely to be
   a. peripheral proteins.
   b. integral proteins.
   c. ATPases.
   d. both b and c.

7. Ions diffuse across membranes down their
   a. electrochemical gradients.
   b. chemical gradients.
   c. electrical gradients.
   d. concentration gradients.

8. The fluidity of membranes in cold weather may be maintained by
   a. increasing the number of phospholipids with saturated hydrocarbon tails.
   b. activating an H^+ pump.
   c. increasing the concentration of cholesterol in the membrane.
   d. increasing the proportion of integral proteins.

9. Water, a polar molecule, can easily pass through membranes because it

   a. is small enough to pass between the lipids of the membrane.
   b. always has a strong concentration gradient.
   c. creates enough pressure to force its way through the membrane.
   d. has specialized carrier molecules that facilitate its diffusion.

10. A plant cell placed in a hypoosmotic environment will
    a. plasmolyze.
    b. lyse.
    c. become turgid.
    d. become flaccid.

11. Which of the following is *not* true of the carrier molecules involved in facilitated diffusion?
    a. They increase the speed of transport across a membrane.
    b. They can concentrate solute molecules on one side of the membrane.
    c. They have specific binding sites for the molecules they transport.
    d. They may undergo a conformational change upon binding of solute.

12. The membrane potential of a cell may be created by
    a. the proton pump.
    b. the selective permeability of membranes to various ions.
    c. the sodium–potassium pump.
    d. all of the above.

13. Cotransport may involve
    a. passive transport of two solutes.
    b. transport of one solute in tandem with another that is diffusing down its concentration gradient.
    c. ion diffusion against the electrochemical gradient.
    d. receptor-mediated endocytosis.

14. Exocytosis involves all of the following *except*
    a. clathrin coated pits.
    b. the fusion of a vesicle, usually from the ER or Golgi apparatus, with the plasma membrane.
    c. a mechanism to transport carbohydrates to the outside of plant cells during the formation of cell walls.
    d. a mechanism to rejuvenate the plasma membrane.

15. The proton pump in plant cells is the functional equivalent of an animal cell's
    a. cotransport mechanism.
    b. sodium–potassium pump.
    c. ligand.
    d. ATP synthase.

16. Cell walls provide an adaptive advantage for certain organisms by
    a. providing structural support.
    b. maintaining cell turgidity.
    c. resisting excessive water influx.
    d. all of the above.

17. Pinocytosis may involve
    a. the fusion of a newly formed food vacuole with a lysosome.
    b. receptor-mediated endocytosis and the formation of coated vesicles.
    c. the pinching in of the plasma membrane around droplets of external fluid.
    d. both b and c.

Use the U-tube setup below to answer questions 18 through 20.

18. Initially, the solution in side A, with respect to that in side B, is
    a. hypoosmotic.
    b. hyperosmotic.

    c. isosmotic.
    d. saturated.

19. After the system reaches equilibrium, what changes are observed?
    a. The water level is higher in side A than in side B.
    b. The water level is higher in side B than in side A.
    c. The molarity of glucose is higher in side A than in side B.
    d. The molarity of sucrose has increased in side A.

20. During the period before equilibrium is reached, which molecule(s) will show net movement through the membrane?
    a. water
    b. glucose
    c. water and glucose
    d. water and sucrose

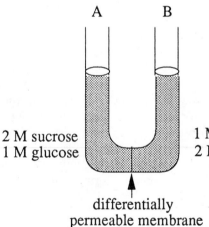

A   B

2 M sucrose
1 M glucose

1 M sucrose
2 M glucose

differentially
permeable membrane

The solutions in the two arms of this U-tube are separated by a membrane that is permeable to water and glucose but not to sucrose. Side A is filled with a solution of 2 M sucrose and 1 M glucose. Side B is filled with 1 M sucrose and 2 M glucose.

CHAPTER 9

RESPIRATION: HOW CELLS HARVEST CHEMICAL ENERGY

FRAMEWORK

This chapter details the intricate processes through which cells harvest energy from organic molecules. The catabolic pathways of glycolysis and respiration capture the chemical energy in glucose and other fuels and store it in ATP. Glycolysis, occurring in the cytosol, produces a small amount of ATP, pyruvic acid, and NADH, which then enter the mitochondrion for respiration. The mitochondrion consists of a matrix in which the enzymes of the Krebs cycle are localized, a highly folded inner membrane in which enzymes and the molecules of the electron transport chain are embedded, and an intermembrane space between the two membranes to temporarily house H+ pumped across the inner membrane during the redox reactions of the electron transport chain. This proton motive force drives oxidative phosphorylation as protons move back through ATP synthases located in the membrane. The following diagram illustrates the key components of these integrated processes.

CHAPTER SUMMARY

Cells require energy to perform the work associated with living: synthesis of macromolecules, pumping of substances across membranes, movement, reproduction, and maintenance of their complex, ordered structure. For most cells, this energy ultimately comes from the sun through the organic molecules produced by photosynthetic organisms. Cells obtain the chemical energy stored in their food by systematically degrading, with the help of enzymes, the energy-rich molecules into simpler, energy-poor waste products.

Fermentation, which occurs without oxygen, is the partial degradation of organic molecules to release energy. *Cellular respiration*, the most common catabolic pathway, breaks down glucose (or other energy-rich organic compounds) with the use of oxygen to obtain energy in the usable form of ATP. This exergonic process has a free energy change of -686 kcal/mol glucose. The waste products of respiration, CO_2 and H_2O, are the raw materials chloroplasts use for photo-

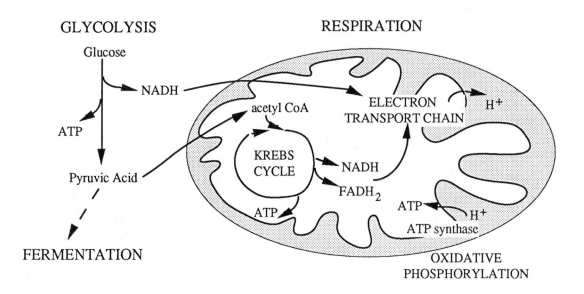

synthesis to produce glucose and return oxygen to the air. The chemical elements essential to life are recycled. Energy, however, is not. It is converted from the ordered form of light, stored temporarily in chemical bonds, to the disordered form of heat.

How Cells Make ATP: An Introduction

ATP is very reactive because of the instability of the bonds between its three phosphate groups. Hydrolysis of the terminal high-energy phosphate bond of one mole of ATP releases 7.3 kcal ($\Delta G = -7.3$ kcal/mol). Generally, the phosphate group with its unstable bond is shifted by enzymes to another molecule. This phosphorylated molecule is then energized and can perform work. The cell must regenerate its supply of ATP by phosphorylating ADP—an endergonic reaction requiring 7.3 kcal/mol. Respiration supplies the energy for ATP synthesis.

An Overview of Cellular Respiration Respiration occurs in three stages: *glycolysis*, the *Krebs cycle*, and the *electron transport chain and oxidative phosphorylation*. Glycolysis and the Krebs cycle generate some ATP, but they primarily provide energized electrons for oxidative phosphorylation. For each molecule of glucose respired to carbon dioxide and water, the cell makes as many as 36 to 38 molecules of ATP.

Substrate-level Phosphorylation *Substrate-level phosphorylation* involves the direct transfer of phosphate to ADP from compounds with high-energy phosphate bonds. Both ADP and a high-energy intermediate compound, formed in fermentation or respiration, are bound to an enzyme, and the phosphate group is shifted to ADP. This reaction is energetically possible because the $-\Delta G$ for the hydrolysis of the intermediate compound is greater than the $+\Delta G$ for the phosphorylation of ADP. A small percentage of ATP is formed this way in most cells.

Chemiosmotic Coupling: The Basic Principle Oxidative phosphorylation produces most cellular ATP through a process called *chemiosmosis*. A proton gradient stores energy across a membrane, and as the H^+ diffuse through *ATP synthase*, a protein complex inserted in the membrane, ADP is phosphorylated. Mitchell received the Nobel Prize for his formulation of the chemiosmotic theory of ATP synthesis. The energy for the generation of the proton gradient that powers the ATP synthases comes either from light in photosynthesis or from catabolism of food in respiration.

The molecules of the electron transport chain, located in the inner membrane of the mitochondrion,

pump H^+ from the mitochondrial matrix into the intermembrane space. The proton gradient thus formed stores energy that can be used by the ATP synthases to make ATP. The cristae of the inner membrane greatly increase the surface area for the insertion of the electron transport chain and ATP synthases, an excellent example of the correlation of structure and function.

Redox Reactions in Metabolism

Oxidation-reduction or *redox reactions* involve the partial or complete transfer of one or more electrons from one reactant to another. The substance that loses electrons is *oxidized* and acts as a *reducing agent* to the substance that gains electrons and is thus *reduced*. This reduced substance acts as an *oxidizing agent* by accepting electrons from the substance that is oxidized. These relationships can be illustrated as follows:

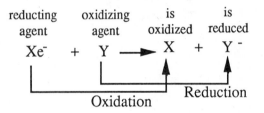

Oxygen, which is highly electronegative, attracts electrons strongly and is one of the most powerful oxidizing agents. As electrons move to electronegative atoms, they give up potential energy. Spontaneous redox reactions release energy most often as heat. In respiration, the energy from the oxidation of glucose and the reduction of oxygen is captured in ATP.

Respiration as a Redox Process Organic molecules with an abundance of hydrogen are rich in high-energy electrons. Respiration is a redox process that transfers these electrons of hydrogen to oxygen, releasing energy along the way in the gradual steps of glycolysis and the Krebs cycle.

NAD+ and the Oxidation of Glucose The electrons that are removed in the catabolic breakdown of glucose are usually transferred to a special oxidizing agent, a coenzyme called *NAD+* (nicotinamide adenine dinucleotide). *FAD* (flavin adenine dinucleotide) is another coenzyme used in the redox reactions of respiration. Dehydrogenases, the enzymes that catalyze these redox reactions, remove two hydrogen nuclei with two electrons from the substrate. The two electrons and one of the hydrogen nuclei are passed to

NAD[+], forming the reduced compound *NADH*, and the other H[+] is released to the surroundings. NADH captures most of the energy of the high-energy electrons and is used in oxidative phosphorylation, the redox reactions of the electron transport chain that lead to the chemiosmotic production of ATP.

Glycolysis

Glycolysis, the breakdown of glucose to pyruvic acid, consists of two major segments. In the first series of steps, two molecules of ATP are consumed as glucose is split into two phosphorylated three-carbon molecules. These smaller sugars are then oxidized in their conversion to two molecules of pyruvic acid; two NADH molecules are produced; and four ATPs are formed by substrate-level phosphorylation. The net gain of ATP from glycolysis is two, since two ATPs are used to prime the process. Most of the energy of this exergonic pathway ($\Delta G = -140$ kcal/mol glucose) is conserved in the high-energy bonds of NADH and ATP.

The Krebs Cycle

The two molecules of pyruvic acid still hold more than three-fourths of the chemical energy stored in glucose. If oxygen is present, pyruvic acid enters the mitochondria to be transformed into *acetyl CoA*, which is further oxidized in the Krebs cycle.

Formation of Acetyl CoA: Linking Glycolysis to the Krebs Cycle   Within a multienzyme complex, pyruvic acid is transformed into acetyl coenzyme A, or acetyl CoA, by the removal of the carboxyl group and release of CO_2; the oxidation of the remaining two-carbon group to acetic acid, accompanied by the reduction of NAD[+] to NADH; and the attachment of a coenzyme to the acetic acid by a high-energy bond. This molecule then shunts its acetyl group into the Krebs cycle.

How the Krebs Cycle Works   The Krebs cycle, also known as the tricarboxylic acid cycle (TCA) or citric acid cycle, has nine steps, each catalyzed by a specific enzyme in the mitochondrial matrix. For each turn of the Krebs cycle, two carbons enter in the reduced state in an acetyl group, two carbons exit completely oxidized as CO_2, three molecules of NAD[+] are reduced to NADH, one molecule of reduced $FADH_2$ is formed, and one ATP molecule is made by substrate-level phosphorylation. The NADH and $FADH_2$ proceed to the electron transport chain where most of the ATP formed from respiration is made by chemiosmosis.

The Electron Transport Chain and Oxidative Phosphorylation

The electron transport chain is a series of electron carrier molecules located in the inner membrane of the mitochondrion. Through a sequence of redox reactions, the electrons from NADH are passed from molecule to molecule until they finally reach oxygen. The chain uses the electron flow to pump protons across the inner membrane, creating the proton gradient that then powers ATP synthases to phosphorylate ADP. Oxidative phosphorylation is the production of ATP powered by the oxidative transfer of electrons in the electron transport chain.

Electron Transport   Most components of the chain are proteins with tightly bound cofactors (prosthetic groups) that shift between reduced and oxidized states as they accept and donate electrons. NADH passes its high-energy electrons to a flavoprotein (FMN), which becomes oxidized as it passes the electrons to the next electron carrier, an iron–sulfur protein. This molecule passes electrons to ubiquinone (Q), which passes the electrons on to a series of molecules called *cytochromes*, proteins with an iron-containing *heme group*. The last cytochrome of the chain passes its electrons to oxygen, which picks up a pair of H[+] from the aqueous medium and forms water.

Electron transfer from NADH to oxygen is very exergonic ($\Delta G = -53$ kcal per mol). The electron transport chain releases this energy in a series of shorter steps as electrons are passed on to the next, more electronegative acceptor. $FADH_2$, the other reduced electron acceptor of the Krebs cycle, adds its electrons to the chain at a lower energy level, and thus one-third less energy for ATP synthesis is derived from $FADH_2$ than from NADH.

Generation of the Proton Gradient   Some electron carriers in the chain accept and release H[+] along with the electrons they are transporting and thus generate a proton gradient. These carriers accept H[+] (protons) from the matrix side of the inner membrane and release them to the intermembrane space. The electron carrier molecules are organized into three complexes, each of which spans the membrane. Carriers that are mobile in the membrane transfer electrons between the complexes. Protons are pumped across the membrane at three points along the electron transport chain.

The Proton-Motive Force and ATP Synthesis
The proton gradient stores potential energy, referred to as the *proton-motive force*. This force consists of an electrochemical gradient created by both the concentration gradient and the voltage across the membrane

due to the uneven H^+ concentrations. The proton-motive force drives H^+ across the membrane through ATP synthase complexes that span the membrane. This chemiosmotic mechanism couples the exergonic passage of protons with the endergonic phosphorylation of ATP and accounts for most of the ATP formed from cellular respiration.

Respiratory poisons provide evidence for the chemiosmotic model. Some poisons, such as cyanide, block the passage of electrons along the electron transport chain. No protons are pumped, and no ATP is made. A class of poisons called uncouplers make the membrane leaky to H^+, destroying the proton gradient so that no ATP is made. Another group of poisons inhibits the ATP synthase, resulting in a higher proton gradient than usual because protons are not moving through the ATP synthase and in no production of ATP.

The ATP Ledger for Respiration   The tally for ATP formed from the oxidation of a molecule of glucose to six molecules of carbon dioxide includes the substrate-level phosphorylation of ATP in glycolysis (4 ATP) and the Krebs cycle (2 ATP), and the ATP generated by chemiosmotic mechanisms in the electron transport chain (34 ATP—approximately 3 for every 2 NADH and 2 for every $FADH_2$), minus the 2 ATP required initially to phosphorylate glucose and a loss of 2 ATP for the passage of the 2 NADH formed during glycolysis across the mitochondrial membrane. The total is 36 ATP. In prokaryotes, the yield is 38 because NADH from glycolysis does not have to pass its electrons across a mitochondrial membrane. The yield of ATP per molecule of glucose oxidized is only an estimate because electron transport and ATP synthesis are loosely linked by the proton gradient. The use of the gradient to transport solutes across the mitochondrial membrane and differences in the permeability of the membrane to H^+ may affect the yield of ATP from respiration.

Fermentation: The Anaerobic Alternative

When oxygen is present, glycolysis serves as the first step of respiration, with the Krebs cycle and electron transport continuing in the mitochondria. Glycolysis produces 2 ATP per sugar molecule, whether under *aerobic* or *anaerobic* conditions. Without oxygen, glycolysis is part of fermentation—the anaerobic conversion of organic nutrients to a waste product that regenerates NAD^+, the oxidizing agent for glycolysis, and captures a small percentage of the energy stored in glucose.

In *alcohol fermentation*, pyruvic acid is converted into a two-carbon acetaldehyde and CO_2 is released. Acetaldehyde is then reduced by NADH to form etha-nol (ethyl alcohol). The alcohol fermentation of yeast, a fungus, is used in brewing. In *lactic acid fermentation*, pyruvic acid is reduced by NADH to form lactic acid and recycle NAD^+; no CO_2 is released. The dairy industry uses the lactic acid fermentation of certain fungi and bacteria to make cheese and yogurt. Muscle cells make ATP by lactic acid fermentation in vigorous exercise when energy demand is high and oxygen supply is low.

Comparison of Aerobic and Anaerobic Catabolism

In the three major catabolic schemes—aerobic respiration, fermentation, anaerobic respiration—glucose or other fuels are oxidized and their high-energy electrons are passed to NAD^+. The ultimate acceptor of these electrons differs for the three schemes. In *aerobic respiration*, the final electron acceptor is oxygen. Organisms that rely on aerobic respiration for energy are called *strict aerobes*; they must have oxygen to survive. *Strict anaerobes*, found in a few groups of bacteria, are poisoned by oxygen. The final electron acceptor in *anaerobic respiration* is a substance such as sulfate or nitrate. Both anaerobic and aerobic respiration are considered to be cellular respiration; they involve catabolic pathways that use electron transport chains to make ATP.

Fermentation produces ATP by substrate-level phosphorylation without the use of an electron transport chain. The final electron acceptor is pyruvic acid or one of its derivatives. Some strict anaerobes rely entirely on fermentation; they do not have electron transport chains. *Facultative anaerobes*, such as yeasts and some bacteria, can make ATP by fermentation or respiration, depending on whether oxygen is available. When oxygen is available, pyruvic acid is converted to acetyl CoA, which is oxidized in the Krebs cycle. Without oxygen, pyruvic acid serves as the final electron acceptor.

Evolutionary Significance of Glycolysis   Glycolysis is common to fermentation and respiration. This metabolic process, occurring in all organisms, probably evolved as the method of ATP formation by ancient prokaryotes before oxygen was available in the atmosphere. The cytoplasmic location of glycolysis, not using any of the membrane-bound organelles of eukaryotic cells, also implies great antiquity.

Catabolism of Other Molecules

Fats, proteins, and carbohydrates can all be used by cellular respiration to make ATP. Starch and glycogen are hydrolyzed to glucose. Digestion of disaccharides provides glucose and other monosaccharides that can be enzymatically converted to glucose. Proteins can be used for fuel after they are digested into their constituent amino acids, which are then deaminated by the removal of their amino groups. Depending on its structure, an amino acid can enter into respiration at several sites (pyruvic acid, acetyl CoA, or an intermediate of the Krebs cycle). The digestion of fats yields glycerol, which is converted to an intermediate of glycolysis, and fatty acids, which are broken down to two-carbon fragments that enter the Krebs cycle as acetyl CoA. Fats are important fuels because they are rich in hydrogen, which has high-energy electrons. A gram of fat produces more than twice as much ATP as a gram of carbohydrate.

Biosynthesis

Not all the organic molecules of food are oxidized as fuel to make ATP; many provide the carbon skeletons the cell needs for biosynthesis. Some monomers from digestion, such as amino acids, can be directly incorporated into the cell's macromolecules. Many of the intermediates of glycolysis and the Krebs cycle serve as precursors for the cell's anabolic pathways. The molecules of carbohydrates, fats and proteins can all be interconverted through the intermediates of metabolism to provide for the cell's needs. Metabolism is a very intricate and adaptable network.

Control of Respiration

Metabolic pathways are strictly controlled under the basic principles of supply and demand. Through feedback inhibition, the end product of a pathway inhibits the enzyme that initiates the pathway. Respiration is controlled by the supply of ATP in the cell. The allosteric enzyme that catalyzes an early step of glycolysis, phosphofructokinase, is inhibited by ATP and activated by ADP or AMP. Phosphofructokinase is also inhibited by citric acid transported from the mitochondria into the cytosol, thus synchronizing the rates of glycolysis and the Krebs cycle. Other allosteric enzymes located at key intersections help to maintain metabolic balance. The processes through which cells harvest energy and raw materials from food molecules are intricately connected and controlled.

STRUCTURE YOUR KNOWLEDGE

1. The following table lists the major divisions of the energy-producing metabolic pathways of the cell. Fill in the location of the pathways and the major molecules that enter and exit for each molecule of glucose processed.

2. There are two ways in which ATP is produced by the cell during respiration. Create a concept map to organize your understanding of substrate-level phosphorylation and oxidative phosphorylation.

TEST YOUR KNOWLEDGE

MULTIPLE CHOICE: *Choose the one best answer.*

1. Which of the following is *not* true of a substance that is phosphorylated?

PROCESS	LOCATION	MOLECULES IN	MOLECULES OUT
Glycolysis			
Fermentation			
Krebs Cycle			
Electron Transport Chain Oxidative Phosphorylation			

a. It has an unstable phosphate bond.
b. Its reactivity has been increased; it is primed to do work.
c. It has entered the oxidative phosphorylation process.
d. None of the above are true.

2. Chemiosmosis involves
   a. the diffusion of water down a concentration gradient.
   b. a proton-motive force that then drives ATP formation.
   c. a proton gradient that drives the redox reactions in the cristae.
   d. an ATP synthase that pumps H^+ across the mitochondrial membrane.

3. In a redox reaction,
   a. the substance that is reduced gains energy.
   b. the substance that is oxidized gains energy.
   c. the oxidizing agent donates electrons.
   d. the reducing agent is the most electronegative.

4. Which of the following is *not* true of oxidative phosphorylation?
   a. It uses oxygen as the ultimate electron donor.
   b. It involves the redox reactions of the electron transport chain.
   c. It involves an ATP synthase located in the inner mitochondrial membrane.
   d. It produces three ATP for every two NADH that are oxidized.

5. Substrate-level phosphorylation
   a. involves the shifting of a phosphate group from ATP to a substrate.
   b. can use NADH or $FADH_2$.
   c. accounts for all the ATP formed by anaerobic respiration.
   d. takes place both in the cytosol and in the mitochondrial matrix.

6. Facultative anaerobes
   a. are poisoned by oxygen.
   b. make ATP by fermentation or aerobic respiration, depending on availability of oxygen.
   c. must use some molecule other than oxygen as the final electron acceptor.
   d. include plants that can produce their own oxygen but undergo anaerobic respiration at night.

7. Glycolysis is not as energy-productive as respiration because
   a. NAD^+ is regenerated by alcohol or lactic acid production, without the high-energy electrons passing through the electron transport chain.
   b. it is the pathway common to fermentation and respiration.

c. it does not take place in a specialized membrane-bound organelle.
d. pyruvic acid is more reduced than CO_2; it still contains much of the energy from glucose.

8. The electron carrier molecules Q and cytochrome c
   a. are reduced as they pass electrons on to the next molecule.
   b. contain heme prosthetic groups.
   c. are mobile carriers that transfer electrons between the electron carrier complexes.
   d. transport H^+ into the mitochondrial matrix, establishing the proton-motive force.

9. When pyruvic acid is converted to acetyl CoA
   a. CO_2 and ATP are released.
   b. a multienzyme complex removes the carboxyl group and attaches a coenzyme.
   c. one turn of the Krebs cycle is completed.
   d. NAD^+ is regenerated so that glycolysis can continue.

10. How many molecules of CO_2 are generated for each turn of the Krebs cycle?
    a. 2
    b. 3
    c. 4
    d. 6

11. Support for the chemiosmotic theory comes from
    a. the production of only two ATP from the oxidation of FADH.
    b. the difference in the membrane potential between the inner mitochondrial membrane and the plasma membrane.
    c. the study of poisons called uncouplers that prevent the formation of the proton gradient and result in no synthesis of ATP.
    d. the presence of the electron transport molecules on the inside of the cristae where they can bind to H^+.

12. Which of the following reactions is incorrectly paired with its location?
    a. ATP synthesis—inner membrane of the mitochondrion
    b. fermentation—cell cytoplasm
    c. glycolysis—cell cytoplasm
    d. Krebs cycle—cristae of mitochondrion

13. In the reaction $C_6H_{12}O_6 + 6 O_2 \rightarrow 6 CO_2 + 6 H_2O$,
    a. oxygen becomes reduced.
    b. glucose becomes reduced.
    c. oxygen becomes oxidized.
    d. water is a reducing agent.

14. What do muscle cells in oxygen deprivation gain from fermentation of pyruvic acid?
    a. ATP and lactic acid

b. NAD⁺ and lactic acid
c. CO_2 and lactic acid
d. ATP and alcohol

15. Glucose, made from six radioactively labeled carbon atoms, is fed to yeast cells in the absence of oxygen. How many molecules of radioactive alcohol (C_2H_5OH) are formed from each molecule of glucose?
a. 0
b. 1
c. 2
d. 3

16. Which of the following produces the most ATP per gram?
a. glucose, because it is the starting place for glycolysis
b. glycogen or starch, because they are polymers of glucose
c. fats, because they are richer in hydrogen with high-energy electrons
d. proteins, because of the energy stored in their tertiary structure

17. Fats and proteins can be used as fuel in the cell because
a. they are converted to glucose by enzymes.
b. they are converted to intermediates of glycolysis or the Krebs cycle.

c. they can pass through the mitochondrial membrane to enter the Krebs cycle.
d. they contain high-energy phosphate bonds.

18. Which is *not* true of the enzyme phosphofructokinase? It is
a. an allosteric enzyme.
b. inhibited by citric acid.
c. the pacemaker of glycolysis and respiration.
d. inhibited by ADP or AMP.

19. We can produce and store fat in our bodies, even if out diet is fat-free because
a. carbohydrates such as breads and potatoes are fattening.
b. a genetic predisposition may interfere with feedback inhibition.
c. excess carbohydrates and proteins can be converted to fat through intermediates of glycolysis and the Krebs cycle.
d. fats contain more energy than proteins or carbohydrates.

20. What is the functional importance of NAD⁺ in respiration?
a. It acts as a coenzyme.
b. It captures energy from electrons for use in ATP synthesis.
c. It is an oxidizing agent.
d. all of the above

CHAPTER 10

PHOTOSYNTHESIS

FRAMEWORK

This chapter describes how the process of photosynthesis captures light energy from the sun and stores it in high-energy chemical bonds of organic molecules. In the *light reactions*, the absorption of photons of light by *chlorophyll* and other pigments, assembled into *photosystems* in the *thylakoid membranes* of *chloroplasts*, transfers high-energy electrons to *NADP+* and passes other excited electrons down a redox electron transfer chain that results in *photophosphorylation* by the chemiosmotic synthesis of *ATP*. *Water* is split to replace the electrons of chlorophyll trapped in NADPH, and *oxygen* is evolved. In the *Calvin cycle* taking place in the *stroma*, CO_2 is fixed to the five-carbon RuBP, which splits and is reduced, with the aid of ATP and NADPH from the light reactions, to form *glyceraldehyde phosphate*, an energy-rich sugar. C_4 and CAM plants have adaptations that supply a high concentration of CO_2 to *rubiscocarboxylase* in order to avoid *photorespiration*.

CHAPTER SUMMARY

In photosynthesis, the chloroplasts of plants capture the light energy of the sun and convert it into chemical energy stored in the bonds of organic molecules made from carbon dioxide and water. Organisms obtain the organic molecules they require for energy and carbon skeletons by *autotrophic* or *heterotrophic nutrition*. Autotrophs "feed themselves" in the sense that they make their own organic molecules from inorganic raw materials. Plants, algae, and certain protists and prokaryotes are *photoautotrophs*, which use light as the source of energy for the manufacture of organic molecules. Some bacteria are *chemoautotrophs*, organisms that produce their organic compounds using energy obtained from oxidizing inorganic substances.

Heterotrophs use the organic compounds of other organisms to obtain energy and organic molecules. They may eat plants or animals or decompose organic litter, but almost all are ultimately dependent on plants for food and oxygen.

Chloroplasts: Sites of Photosynthesis

Chloroplasts, found mainly in the *mesophyll* tissue of the leaf, contain *chlorophyll*, the green pigment that absorbs the light energy that drives photosynthesis. A chloroplast consists of a double membrane surrounding the dense fluid called the stroma and an elaborate thylakoid membrane system enclosing the thylakoid space. Thylakoid sacs may be layered to form grana. Chlorophyll is contained in the thylakoid membrane; the conversion of light energy into chemical energy occurs in the thylakoids. The production of sugar from this chemical energy and CO_2 occurs in the stroma. CO_2 enters and O_2 leaves the leaf through *stomata*. Water for photosynthesis is supplied through veins or vascular bundles extending from the root.

Photosynthetic prokaryotes lack chloroplasts, but chlorophyll is built into their plasma membranes or other membranes, an exception to the generalization that prokaryotes do not have cellular membranes other than the plasma membrane.

How Plants Make Food: An Overview

The process of photosynthesis is represented as follows: $6 CO_2 + 12 H_2O + \text{light energy} \rightarrow C_6H_{12}O_6 + 6 O_2 + 6 H_2O$. This equation shows that water is both consumed and created during photosynthesis. If only the net consumption is considered, the equation for photosynthesis is the reverse of respiration. Although these two reactions are distinct, they have in common many of the same intermediate compounds, an electron transport chain, and a chemiosmotic mechanism of ATP formation. Both processes occur in plant cells.

The Splitting of Water   Using evidence from bacteria that use H_2S rather than water for photosynthesis and create sulfur instead of oxygen as a byproduct, van Niel hypothesized that all photosynthetic organisms need a hydrogen source and that plants split water as their hydrogen source, releasing oxygen.   Twenty years later scientists confirmed this hypothesis by using a heavy isotope of oxygen (^{18}O) to label the CO_2 and water used in photosynthesis. Labeled O_2 was produced only when the water was the radioactively labeled molecule.

The most important result of photosynthesis is the movement of hydrogen from a low potential energy state in water to sugar, where its high-energy electrons store energy.

Photosynthesis as a Redox Process   Photosynthesis is a redox process like respiration but differs in the direction of electron flow. The electrons increase their potential energy when they travel from water to sugar. This endergonic process (G = +686 kcal/mol) uses light as the energy source.

The Two Stages of Photosynthesis   The *light reactions* are the steps of photosynthesis that convert solar energy into chemical energy. The light energy absorbed by chlorophyll drives the transfer of electrons from water (which is split and releases oxygen) to the electron acceptor *NADP+* (the coenzyme nicotinamide adenine dinucleotide phosphate) to reduce it to form NADPH, a source of energized electrons. ATP is also formed during the light reactions in a process called *photophosphorylation.*

In the *Calvin cycle*, a process known as *carbon fixation* incorporates carbon into organic compounds that are then reduced to form carbohydrate. NADPH and ATP, from the light reactions that occur in the thylakoids, supply the reducing power and chemical energy needed for the Calvin cycle. These "dark reactions" reduce CO_2 to sugar in the stroma.

How the Light Reactions Capture Solar Energy

The Nature of Sunlight   In the giant thermonuclear reactor of the sun, fusion reactions of hydrogen atoms combining to form helium convert some mass to energy. This *electromagnetic energy*, also called *radiation*, travels as rhythmic wave disturbances of electrical and magnetic fields. The distance between the crests of the electromagnetic waves, called *wavelength*, ranges from less than a nanometer to more than a kilometer. Within this *electromagnetic spectrum*, the narrow band of radiation from 400 to 700 nm is called *visible light*, because humans can see it as various colors.

Light behaves as if it consists of discrete particles called quanta or *photons*, which have a fixed quantity of energy. The amount of energy in a photon of light is inversely related to its wavelength. Thus a photon of red light, with a long wavelength, has less energy than a photon of violet light.

Photosynthetic Pigments   Light may be either reflected, transmitted, or absorbed when it strikes matter. Substances that absorb visible light are called *pigments*. A *spectrophotometer* measures the ability of a pigment to absorb various wavelengths of light. The *absorption spectrum* of *chlorophyll a* shows that it absorbs blue and red light best. Chlorophyll is green because it reflects or transmits green wavelengths of light.

An *action spectrum* shows the relative rates of photosynthesis under different wavelengths of light. The action spectrum for photosynthesis does not exactly match the absorption spectrum of chlorophyll *a*. Although only chlorophyll *a* can directly channel energy into the light reactions, other pigments in the chloroplasts can absorb light of different wavelengths and transfer the energy to chlorophyll *a*. These accessory pigments include chlorophyll *b*, a green pigment structurally similar to *a*, and the *carotenoids*, yellow and orange pigments that extend the spectrum of light that can drive photosynthesis.

The Photooxidation of Chlorophyll   When a pigment molecule absorbs the energy from a photon, one of the molecule's electrons is elevated to an orbital where it has more potential energy, and the pigment is said to be in an *excited state*. Only photons that have an energy equal to the difference in energy between the excited state and *ground state*, when the electron is in its normal orbital, are absorbed by a molecule.

The excited state is unstable. Generally, the potential energy of the electron, created from the light energy of the photon, is released as heat as the electron drops back to its ground-state orbital. Some pigments also emit a lower energy photon of light, called *fluorescence*, as their electrons return to ground state.

Chlorophyll removed from chloroplasts will fluoresce red light upon illumination. However, when chlorophyll bound in a thylakoid membrane absorbs light, a nearby molecule called a *primary electron acceptor* traps the high-energy electron. In a redox reaction, chlorophyll is photooxidized, transferring an electron to the primary electron acceptor, which becomes reduced.

The Two Photosystems Chlorophyll *a* and the accessory pigments are clustered together in the thylakoid membrane in assemblies of several hundred pigment molecules. The one chlorophyll *a* molecule that can donate its excited electron to the primary electron acceptor is located in the *reaction center* of the pigment assembly. Other pigment molecules in the assembly act as light-gathering antennae to absorb photons and pass the energy from molecule to molecule until it reaches the reaction center. This light-harvesting unit of the thylakoid membrane, consisting of the antenna complex, reaction-center chlorophyll *a*, and the primary electron acceptor is called a *photosystem*.

There are two types of photosystems in the thylakoid membrane. The chlorophyll molecule at the reaction center of *photosystem I* is called *P700*, after the wavelength of light (700 nm) it absorbs best. At the reaction center of *photosystem II* is a chlorophyll *a* molecule called *P680*, absorbing light best at 680 nm. The differences between these two molecules, and between these molecules and the rest of the chlorophyll *a* molecules in the photosystem, are the specific proteins to which they are bound in the thylakoid membrane and the close proximity to their respective primary electron acceptors. Structure, as always, relates to function.

Cyclic Electron Flow Of the two possible routes of electron flow in the light reactions, *cyclic electron flow* is the simplest, involving only photosystem I and generating only ATP, no NADPH or O_2. Electrons, often traveling in pairs, are excited by the absorption of two photons of light and are passed by a series of redox reactions along an electron transport chain until they return to P700. Certain electron carriers in the chain transport hydrogen ions along with the electrons, moving them across the thylakoid membrane and storing energy in the form of a proton-motive force of an H^+ gradient. This energy is tapped by ATP synthase enzymes built into the thylakoid membrane through a chemiosmotic mechanism similar to that found in mitochondria. The synthesis of ATP in chloroplasts is called photophosphorylation because it is driven by light energy. *Cyclic photophosphorylation* refers to ATP formation during cyclic electron flow.

Noncyclic Electron Flow In *noncyclic electron flow*, electrons pass continuously from water to NADPH. Both photosystems are involved in this process. Two electrons excited from P700 of photosystem I are not returned to P700 but are trapped in their high-energy state in NADPH. The oxidized P700 now becomes a potent oxidizing agent, attracting electrons supplied from photosystem II. A pair of electrons, elevated when P680 absorbs two photons, is trapped by the primary electron acceptor of photosystem II and passed through the electron transport chain used in cyclic electron flow to P700. As the electrons lose their potential energy, the transport chain pumps H^+ across the thylakoid membrane, creating the proton-motive force to drive ATP synthesis in the process called *noncyclic photophosphorylation*.

Oxidized P680 has such a great affinity for electrons that it can remove them from water, splitting it into two hydrogen ions and an oxygen atom. The oxygen atom immediately combines with another oxygen to form O_2, the byproduct of photosynthesis. The net effect of noncyclic electron flow is to move electrons from their low potential-energy state in water to their high potential-energy state in NADPH. For each pair of electrons transferred, four photons of light must be absorbed—two by each photosystem—an NADPH molecule is formed, and at least one molecule of ATP is created.

It is not certain whether cyclic photophosphorylation, observed in certain experimental conditions, is important in plants in nature. The ATP produced by the light reactions is usually a result of noncyclic electron flow.

Comparison of Chemiosmosis in Chloroplasts and Mitochondria Chemiosmosis in mitochondria and chloroplasts is very similar. The key difference is that in respiration chemical energy from food is being converted into ATP, whereas in chloroplasts light energy is the source of the chemical energy stored in ATP.

In the thylakoids, the components of the electron transport chain pump protons from the stroma into the thylakoid compartment. As they diffuse back through the ATP synthase, ATP is formed on the stroma side, where it is available for the Calvin cycle. In the light, the proton gradient across the thylakoid membrane is as great as 3 pH units. The gradient is eliminated when the lights are turned off. Additional evidence for the chemiosmotic theory came from the work of Jagendorf in which he induced thylakoids to make ATP in the dark by artificially creating a pH gradient across the membrane.

How the Calvin Cycle Makes Sugar

In the Calvin cycle, the ATP and NADPH formed by the light reactions drive the reduction of CO_2 to sugar. *Glyceraldehyde 3-phosphate* is the three-carbon sugar formed by three turns of the cycle. First, CO_2 is fixed by being added to ribulose bisphosphate (RuBP), a five-carbon sugar with phosphate groups on both ends. This reaction is catalyzed by the enzyme *RuBP carboxylase (rubisco)*, perhaps the most abundant protein on earth. The unstable six-carbon intermediate formed splits immediately into two molecules of 3-phosphoglyceric acid, which are then phosphorylated by ATP to form two molecules of 1,3-diphosphoglyceric acid. Hydrolysis of this phosphate bond helps to transfer a pair of high-energy electrons from NADPH to create glyceraldehyde phosphate—the reduced, energy-rich sugar formed by the Calvin cycle (and also created when glucose is split in glycolysis).

The cycle must turn three times, reducing three CO_2 molecules, to create one molecule of glyceraldehyde phosphate and regenerate the three molecules of the five-carbon RuBP used in the cycle. The rearrangement of five molecules of glyceraldehyde phosphate into the three molecules of RuBP requires three more ATPs. A total of nine ATPs and six molecules of NADPH are consumed in the formation of one glyceraldehyde phosphate. Glucose, usually thought of as the product of photosynthesis, is formed from two molecules of glyceraldehyde phosphate—the Calvin cycle turns six times, and 18 ATPs and 12 molecules of NADPH, formed in the light reactions, are used. The process of photosynthesis involves the integration of the light reactions and Calvin cycle within the thylakoid membrane and stroma of intact chloroplasts.

Photorespiration   Rubisco, the enzyme that initiates the Calvin cycle, accepts O_2 in place of CO_2 when the air spaces in the leaf have a high concentration of O_2. (Oxygen is a competitive inhibitor of rubisco.) When oxygen is added to RuBP, the five-carbon product splits into a three-carbon molecule that remains in the cycle, and a two-carbon compound, glycolic acid, which leaves the chloroplasts, enters a peroxisome and is broken down to release CO_2. This seemingly wasteful process that removes fixed carbon from the Calvin cycle and produces no ATP is called *photorespiration*. It is most common on hot, dry, bright days when the leaves close their stomata to conserve water, and oxygen formed from the light reactions collects in the leaf.

One hypothesis holds that photorespiration is an evolutionary relic from the time when there was little O_2 in the atmosphere and the ability of rubisco to distinguish between O_2 and CO_2 was not so critical. Photorespiration can result in the loss of as much of 50% of the carbon fixed in the Calvin cycle, and as such, represents a considerable agricultural liability.

C_4 *Plants* C_3 *plants* incorporate CO_2 directly into the Calvin cycle and are susceptible to photorespiration. C_4 *plants* have evolved a different mode of carbon fixation that serves to minimize photorespiration. CO_2 is first added to phosphoenol pyruvic acid (PEP) with the aid of an enzyme (*PEP carboxylase*) that has no affinity for O_2 and a greater affinity for CO_2 than rubisco. The four-carbon oxaloacetic acid formed by this carbon fixation

in the *mesophyll cells* of the leaf is usually converted to malic acid and transported through plasmodesmata into the *bundle-shealth cells* tightly packed around the veins of the leaf. This four-carbon compound is broken down to release CO_2, which is then in high enough concentrations that rubisco will accept CO_2 and initiate the Calvin cycle. C_4 plants illustrate the important connection between structure and function in their ability to photosynthesize efficiently in hot, arid regions.

CAM Plants   *CAM* plants illustrate another adaptation to very dry climates. These plants, mostly cacti, close their stomata during the day to prevent water loss but open them at night to take up CO_2 and incorporate it into a variety of organic acids. These carbon-containing compounds are stored in the vacuoles of the mesophyll cells during the night and then broken down to release CO_2 for photosynthesis during the daylight. Unlike the C_4 pathway, the CAM pathway (crassulacean acid metabolism named after the plant family in which it was first discovered) does not structurally separate carbon fixation from the Calvin cycle; instead, the two processes are separated in time.

The Fate of Photosynthetic Products

Carbohydrate is transported through the plant in the form of sucrose, a disaccharide. About 50% of the organic material produced by photosynthesis is used as fuel for cellular respiration in the mitochondria of plant cells; the rest is used for carbon skeletons for synthesis of organic molecules (mainly cellulose), stored as starch, or lost through photorespiration. The incredible productivity of photosynthesis, producing about 160 billion metric tons of carbohydrate per year, provides food, shelter, fuel, and oxygen for heterotrophs.

STRUCTURE YOUR KNOWLEDGE

1. In a diagrammatic form, outline the key events of photosynthesis. Trace the flow of electrons through photosystem II and I, the transfer of ATP and NADPH formed by the light reactions into the Calvin cycle, and the major steps in the production of glyceraldehyde phosphate.

2. Describe the difference between an absorption and an action spectrum. What is the significance of the fact that the absorption spectrum of chlorophyll *a* does not completely correspond to the action spectrum of photosynthesis?

3. Create a concept map to help you develop your understanding of the chemiosmotic synthesis of ATP in photophosphorylation.

TEST YOUR KNOWLEDGE

MULTIPLE CHOICE: *Choose the one best answer.*

1. Chemoautotrophs
   a. use hydrogen sulfide as their hydrogen source for the photosynthesis of their organic compounds.
   b. "feed themselves" by obtaining energy in the chemical bonds of organic molecules.
   c. oxidize inorganic compounds to obtain energy to drive the synthesis of their organic compounds.
   d. live as decomposers of inorganic chemicals in organic litter.

2. Which of the following processes and locations is mismatched?
   a. light reactions—grana
   b. electron transport chain—thylakoid membrane
   c. Calvin cycle—stroma
   d. ATP synthase—double membrane surrounding chloroplast

3. Photosynthesis is a redox process in which
   a. CO_2 is reduced and water is oxidized.
   b. $NADP^+$ is reduced and RuBP is oxidized.
   c. CO_2, $NADP^+$, and water are reduced.
   d. O_2 acts as an oxidizing agent, and water acts as a reducing agent.

4. Which of the following is *not* true of a photon?
   a. The energy it contains is inversely related to its wavelength.
   b. It has a wavelength of either 680 or 700 nm.
   c. Its energy may be converted to heat or light of a longer wavelength when the electron it excites drops back to its ground-state orbital.
   d. It is thought of as a discrete package or unit of radiant energy.

5. A spectrophotometer measures
   a. the absorption spectrum of a substance.
   b. the action spectrum of a substance.
   c. the amount of energy in a photon.
   d. fluorescence.

6. Accessory pigments within chloroplasts are responsible for
   a. driving the splitting of water molecules.
   b. absorbing photons of different wavelengths of light and passing that energy to reaction center molecules.
   c. reducing the primary electron acceptor.
   d. extending the absorption spectrum of chlorophyll *a*.

7. Which of the following is *not* true of the reaction center molecule?
   a. It is always a chlorophyll *a* molecule.
   b. It is named for the wavelength of light it absorbs best.
   c. It is bound to specific proteins in the thylakoid membrane and located close to its primary electron acceptor.
   d. It accepts energized electrons from the antenna complex.

8. Noncyclic electron flow in the chloroplast results in the production of
   a. ATP.
   b. ATP and NADPH.
   c. ATP, NADPH, and O_2.
   d. Glyceraldehyde 3-phosphate.

9. The chlorophyll known as P680 is reduced by electrons from
   a. photosystem I.
   b. photosystem II.
   c. water.
   d. NADPH.

10. CAM plants avoid photorespiration by
    a. fixing CO_2 into organic acids during the night.
    b. fixing CO_2 into four-carbon compounds in the bundle-sheath cells.
    c. fixing CO_2 into four-carbon compounds in the mesophyll.
    d. using an enzyme, PEP carboxylase, that outcompetes rubisco for CO_2.

11. Electrons that flow through the two photosystems have their lowest potential energy in
    a. P700.
    b. P680.
    c. the primary electron acceptor of photosystem II.
    d. water.

12. Chloroplasts can make carbohydrate in the dark if provided with
    a. ATP and NADPH and CO_2.
    b. an artificially induced proton gradient.
    c. organic acids or four-carbon compounds such as malic acid.
    d. a source of hydrogen.

13. In the chemiosmotic synthesis of ATP, H^+ diffuses through the ATP synthase

a. from the stroma into the thylakoid compartment
b. from the thylakoid compartment into the stroma
c. from the cytoplasm into the matrix
d. from the cytoplasm into the stroma

14. How many "turns" of the Calvin cycle are required to produce one molecule of glyceraldehyde phosphate?
    a. 1
    b. 2
    c. 3
    d. 6

15. In green plants, most of the ATP for synthesis of proteins, cytoplasmic streaming, and other cellular activities comes directly from
    a. Photosystem I and II.
    b. the Calvin cycle.
    c. photophosphorylation.
    d. oxidative phosphorylation.

16. The six molecules of glyceraldehyde phosphate formed from three turns of the Calvin cycle are converted into
    a. three molecules of glucose.
    b. one glyceraldehyde phosphate and three molecules of RuBP.
    c. one molecule of glucose and two molecules of RuBP.
    d. one glyceraldehyde phosphate and three four-carbon intermediates.

17. NADPH is used in the Calvin cycle to reduce
    a. an unstable six-carbon intermediate.
    b. RuBP.
    c. 1,3-diphosphoglyceric acid.
    d. glyceraldehyde phosphate.

18–25. Indicate if the following events occur during
    a. respiration
    b. photosynthesis
    c. both respiration and photosynthesis
    d. neither respiration nor photosynthesis

18. ____   Chemiosmotic synthesis of ATP

19. ____   Reduction of oxygen

20. ____   Reduction of CO_2

21. ____   Reduction of NAD^+

22. ____   Oxidation of water

23. ____   Oxidation of $NADP^+$

24. ____   Oxidative phosphorylation

25. ____   Electron flow along a cytochrome chain

CHAPTER 11

REPRODUCTION OF CELLS

FRAMEWORK

This chapter details the process through which cells create new cells, with identical genetic material. The following concept map summarizes the crucial process of cell division.

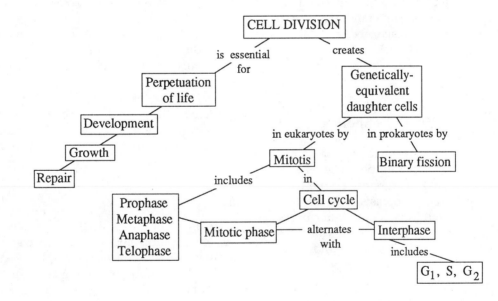

CHAPTER SUMMARY

Cell division, the process through which a cell reproduces itself, is the basis for the perpetuation of all life. Cell division creates duplicate offspring in unicellular organisms and provides for growth, development, and repair in multicellular organisms. The process of recreating a structure as intricate as a cell necessitates the exact duplication and equal division of the DNA and genes that contain the cell's genetic program.

Bacterial Reproduction

Prokaryotes, such as bacteria, reproduce themselves by a process known as *binary fission*. First, two copies of the single circular DNA molecule are produced and attached to the plasma membrane. Growth of the membrane separates the duplicate chromosomes. After the cell has reached about twice its size, the membrane pinches in between the chromosomes and a cell wall develops, creating two daughter cells.

Eukaryotic Chromosomes and Their Duplication

Copying and dividing the tens of thousands of genes of a eukaryotic cell involves a precise process. Through DNA replication, the cell duplicates the genes on its chromosomes prior to cell division. Grouping genes on chromosomes simplifies their management and distribution in cell division. The duplicated chromosomes separate during *mitosis* (the division of the nucleus) and then the cytoplasm divides (*cytokinesis*), producing two separate, genetically equivalent daughter cells.

Each chromosome is a very long DNA molecule with associated proteins that help to structure the chromosome and control the activity of the genes. This DNA-protein complex, called *chromatin*, is organized into a coiled and folded fiber. Somatic cells (cells that are not sperm or egg) all have the same number of chromosomes in their nuclei. Sperm and egg cells have one-half the chromosome complement of somatic cells. Each species has a characteristic number of chromosomes.

Replicated chromosomes consist of two identical *sister chromatids*, joined together at a specialized region called a *centromere*. Toward the end of mitosis, the two sister chromatids separate from each other, and daughter cells that have one copy of each chromosome are formed.

Reproduction of Cellular Organelles

In addition to replicating its genetic material, a cell usually increases its mass and supply of organelles (especially ribosomes) and membranes before division. Mitochondria and chloroplasts both contain small circular DNA molecules that replicate and divide in a manner similar to bacterial DNA. Lipids for these organelles are supplied by the cell's ER, and some proteins essential for their functioning are coded for by the cell's chromosomal DNA. Centrioles, associated with the microtubule apparatus involved in mitosis, begin to separate and then replicate before cell division. The resulting centriole pairs move to opposite ends of the cell so that each daughter cell receives one pair of centrioles.

The Cell Cycle

The *cell cycle* includes the *M phase* (mitotic phase), in which the chromosomes and cytoplasm are divided, and *interphase*, during which most of the cell's growth, metabolic activities and chromosome replication oc-

cur. The length of the cell cycle varies greatly depending on the cell type and its physiological state.

Interphase, usually lasting 90% of the time of the cell cycle, includes the G_1 *phase* (the "first gap" period), the *S phase* (synthesis phase), and the G_2 *phase* ("second gap" period). Most organelles and cell components are produced continuously throughout the G_1, S, and G_2 phase; DNA synthesis occurs only during the S phase.

Phases of Mitosis   Although mitosis is a dynamic process of chromosome separation followed by cytoplasm division, it is conventionally described in four stages: *prophase, metaphase, anaphase,* and *telophase*.

During prophase, the nucleoli disappear and the chromatin fibers coil and fold into visible chromosomes consisting of sister chromatids joined at the centromere. The centriole pairs (present in animal cells) move to opposite poles of the cell and the microtubules of the spindle (and aster in animal cells) become arranged. The nuclear membrane disintegrates, and microtubules attach to the chromosomes.

In metaphase the centromeres of all the chromosomes align on the *metaphase plate*, a plane equidistant between the spindle's two poles.

Anaphase involves the physical separation of the sister chromatids; these split at the centromeres and the individual chromosomes move along the spindle apparatus toward opposite poles. The poles of the cell move farther apart.

By telophase, equivalent sets of chromosomes have gathered at the two poles of the cell. Nuclear envelopes form from fragments of the parent nuclear envelope and contributions from the endomembrane system. Nucleoli reappear, the chromatin fibers unravel, and cytokinesis occurs.

The Mitotic Spindle   The *mitotic spindle* consists of fibers made of microtubules and proteins. In animal cells, the spindle begins as two *asters*—radial arrays around each centriole pair. Both plant and animal cells develop *polar fibers* extending from the poles to the equator of the cell. The organizing center of the spindle is called the *microtubule organizing center* (MTOC). This amorphous cloud of material, visible only with electron microscopy, is present in both plants and animal cells and is associated with the centrioles in animal cells. Little is known of the details of its action.

The chromosomes interact with the spindle by special links at the centromere called *kinetochores*, one for each sister chromatid. Special microtubules called *kinetochore fibers* attach to each kinetochore. Interaction between the kinetochore fibers and the polar fibers sets the chromosomes into agitated motion until they line up on the metaphase plate.

The extension of the spindle poles away from each

other as the cell elongates may be due to the addition of subunits to the polar fibers and the sliding of fibers past each other where they overlap at the equator. This sliding movement is thought to involve the hydrolysis of ATP by an enzyme similar to dynein, the protein responsible for movement of cilia and flagella.

The mechanism for the movement of newly separated chromosomes along the spindle is not yet known. It appears to be different from the force generated for pole separation, because blocking dynein's action does not prevent chromosome movement. Perhaps a low-energy process such as the dissociation of the tubulin subunits of kinetochore microtubules may be involved. Recent experiments suggest that the kinetochore and its chromosome may move poleward just ahead of the depolymerizing end of the kinetochore fiber.

Cytokinesis During late anaphase or early telophase, cytokinesis begins. *Cleavage* is the process that separates the two daughter cells in animals. A shallow groove in the cell surface along the metaphase plate, known as the *cleavage furrow,* forms as the *contractile ring* of microfilaments on the cytoplasmic side of the membrane begins to contract. The cleavage furrow deepens until the dividing cell is pinched in two.

In plant cells, a *cell plate* forms at the site of the old metaphase plate from the fusion of membrane vesicles derived from the Golgi apparatus. The two sides of the membrane thus formed join with the plasma membrane, creating two daughter cells with their own continuous plasma membrane. A new cell wall develops between these membranes.

Mitosis may not always be followed by cytokinesis. Multinucleated cells can thus be formed, as in the *plasmodia* of slime molds.

Control of Cell Division

Control of the timing and rate of cell division is critical to normal growth, development, and maintenance. Control mechanisms have been studied through the use of *tissue culture,* or *cell culture.* In this technique, cells are grown in the laboratory in glass or plastic containers supplied with a nutrient solution called a *growth medium.*

Researchers using cell culture have identified factors that stimulate or inhibit cell division. Certain hormones and other chemicals are essential for normal cell division. Inhibiting protein synthesis, depleting essential nutrients, or overcrowding serves to block division. Most normal vertebrate cells will stop dividing when they come in contact with other cells, a phenomenon known as *contact inhibition.*

An important turning point seems to occur during the G_1 phase of the cell cycle. Some mechanism determines whether the cell will switch into a nondividing state (called G_0) or continue past a so-called *restriction point* into the S, G_2, and M phases. Normal cells that stop dividing usually do so before the restriction point in the G_1 phase.

Abnormal Cell Division: Cancer Cells

Cancer cells escape from the body's normal control mechanisms. If they do stop dividing, they seem to be at random points in the cell cycle. Cancer cells do not exhibit contact inhibition when grown in cell culture. Moreover, most cancer cells in cell culture appear to be "immortal;" they continue to divide indefinitely instead of stopping after the typical 20 to 50 divisions of normal mammalian cells. Much remains to be learned about the cell division processes of both normal and cancerous cells.

STRUCTURE YOUR KNOWLEDGE

1. Describe the life of one chromosome as it proceeds through an entire cell cycle.

2. Draw a sketch of a mitotic spindle that includes the components of the spindle and lists the functions of each part.

TEST YOUR KNOWLEDGE

MATCHING: *Match the event described with its place in the cell cycle.*

1. ___ most cells that have finished dividing stop here **A.** Anaphase

2. ___ sister chromatids separate and chromosomes move apart **B.** G_1 phase

3. ___ mitotic spindle forms **C.** G_2 phase

4. ___ cell plate forms or cleavage furrow pinches cells apart **D.** Metaphase

5. ___ chromosomes duplicate **E.** Prophase

6. ___   chromosomes line up F. S Phase
           at equatorial plane

7. ___   phase after DNA   G. Telophase
           replication

8. ___   chromosomes become visible

9. ___   nuclear envelope reforms

10. ___   poles of cell move farther apart

MULTIPLE CHOICE: *Choose the one best answer.*

1. One of the major differences in the cell division of prokaryotic cells compared to eukaryotic cells is that
   a. cytokinesis does not occur.
   b. genes are not duplicated on chromosomes.
   c. no cell plate forms.
   d. the chromosomes do not separate along a mitotic spindle.

2. Chloroplasts and mitochondria
   a. reproduce during prophase.
   b. move along the polar fibers during mitosis.
   c. replicate by a process similar to binary fission.
   d. contain all the DNA needed for their growth and functioning.

3. The longest part of the cell cycle is
   a. interphase.
   b. G_1 phase.
   c. G_2 phase.
   d. mitosis.

4. Cytokinesis may involve
   a. the separation of sister chromatids.
   b. the contraction of the contractile ring of microfilaments.
   c. interactions between kinetochore and polar fibers.
   d. the formation of the microtubule organizing center.

5. Humans have 46 chromosomes. That number of chromosomes will be found in
   a. cells in anaphase.
   b. the egg and sperm cells.
   c. the somatic cells.
   d. all the cells of the body.

6. Polar fibers
   a. are found only in animal cells.
   b. attach to the kinetochore region of the centromere.
   c. seem to be involved in moving the chromosomes toward the poles.
   d. seem to be involved in pushing the poles of the cell apart.

7. Contact inhibition is seen in
   a. cancer cells.
   b. normal cells growing in cell culture.
   c. cells in tissue culture that contact the surface of the culture vessel.
   d. cells that stop growing in the G_2 phase.

8. A cell that passes the restriction point will most likely
   a. continue to divide.
   b. stop dividing.
   c. continue to divide only if it is a cancer cell.
   d. die.

9. Which of the following is *not* true of a cell plate?
   a. It forms at the site of the metaphase plate.
   b. It results from the fusion of microtubules.
   c. It fuses with the plasma membrane.
   d. A cell wall is laid down between its membranes.

10. Sister chromatids
    a. have one-half the amount of genetic material as does the original chromosome.
    b. move together toward one pole of the cell.
    c. each have their own kinetochore.
    d. are formed during prophase.

CHAPTER 7
TOUR OF THE CELL

Suggested Answers to Structure Your Knowledge

1. A. Functions and structures of animal cells.

FUNCTIONS OF CELL	ASSOCIATED ORGANELLES AND STRUCTURES
Cell division	Chromosomes, centrioles, microtubules (spindle), microfilaments (actin-myosin aggregates pinch apart cell)
Information storage and transferal	Nucleus, chromosomes, DNA —> RNA —> enzymes and other proteins
Energy conversions	Mitochondria
Manufacturing: membranes and products	Ribosomes, rough and smooth ER, Golgi apparatus, vesicles
Digestion, recycling	Lysosomes, food vacuoles
Specific pathways	Microbodies, perioxosomes
Structural integrity	Cytoskeleton: microtubules, microfilaments, intermediate filaments
Movement	Cilia and flagella (microtubules), microfilaments (actin and myosin)
Exchange with environment	Plasma membrane
Cell-cell interaction	Desmosomes, tight and gap junctions, glycocalyx

1. B. Structures and functions unique to plant cells

STRUCTURES UNIQUE TO PLANT CELLS	FUNCTIONS
Cell wall	Structural support, middle lamella glues cells together
Central vacuole	Storage, digestion, growth of cell by water absorption
Plastids: chloroplasts amyloplasts	Photosynthesis, produce carbohydrates Storage of starch
Dictyosomes	Golgi apparatus
Glyoxysomes	Contain enzymes, supply energy for embryo
Plasmodesmata	Cytoplasmic connection between cells

2.

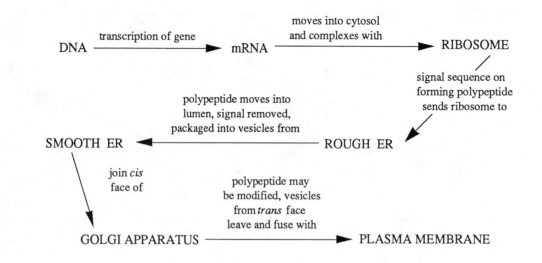

Answers to Test Your Knowledge

Multiple Choice:

1. c	4. d	7. a	10. d	13. a	16. a
2. b	5. a	8. c	11. c	14. c	17. b
3. d	6. b	9. b	12. b	15. c	18. a

Matching:

1. E	b	5. A	e	9. H	k
2. K	j	6. C	g	10. F	h
3. I	d	7. B	c		
4. D	i	8. J	a		

Fill in the Blanks:

1. vesicles	5. grana	9. tight junction
2. cristae	6. basal body	
3. glycocalyx	7. cytosol	10. tonoplast
4. microbodies	8. cytoskeleton	

CHAPTER 8
MEMBRANE STRUCTURE AND FUNCTION

Suggested Answers to Structure Your Knowledge

1. The fluid mosaic model depicts a membrane composed of a phospholipid bilayer with embedded and surface proteins. Small hydrophobic molecules easily dissolve in the hydrophobic interior of the membrane and pass through by diffusion. The rate of permeability is determined by the size and relative lipid solubility of the molecule. Small polar or ionic compounds cannot pass easily through the hydrophobic center of the membrane. They are transported by specific transport proteins either through channels in the protein or by a conformational change in the protein caused by the binding of the solute, which shifts the solute from one side of the membrane to the other.

2. Facilitated diffusion and active transport both involve transport proteins in the membrane that are specific for solute molecules. In facilitated diffusion, polar or ionic molecules that could not easily permeate the membrane are moved down their concentration gradient. This passive transport does not require energy from the cell. Active transport is used to move any type of small molecule against its concentration gradient. The transport protein often acts as an enzyme in hydrolyzing ATP to provide the energy for this transport.

3.

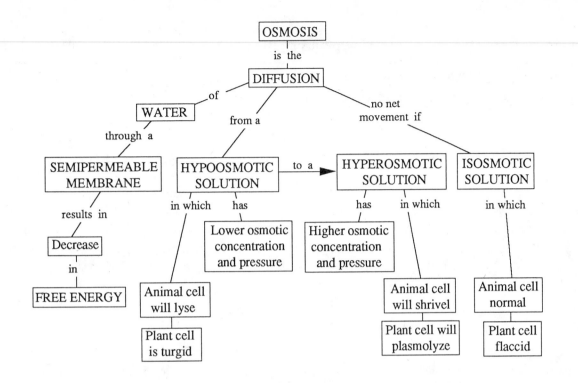

Answers to Test Your Knowledge

Multiple Choice:

1. c	**5.** d	**9.** a	**13.** b	**17.** d
2. d	**6.** d	**10.** c	**14.** a	**18.** c
3. a	**7.** a	**11.** b	**15.** b	**19.** a
4. c	**8.** c	**12.** d	**16.** d	**20.** c

CHAPTER 9
RESPIRATION: HOW CELLS HARVEST ENERGY

Suggested Answers to Structure Your Knowledge

1.

PROCESS	LOCATION	MOLECULES IN	MOLECULES OUT
Glycolysis	Cytosol	Glucose, 2 ATP [NAD^+, ADP, Ⓟ]	2 pyruvic acid, 2 NADH, 4 ATP (2 net ATP)
Fermentation	Cytosol	Glucose, 2 ATP [NAD^+, ADP, Ⓟ]	2 alcohol, 2 CO_2, or 2 lactic acid, 4 ATP (2 net ATP)
Krebs cycle	Matrix of mitochondria	2 acetyl CoA (from pyruvic adic), [NAD^+, FAD, ADP]	4 CO_2, 6 NADH, 2 $FADH^+$, 2 ATP
Electron transport chain Oxidative phosphorylation	Mitochondrial inner membrane	10 NADH (2 from glycolysis use 2 ATP to enter mitochondria), 2 $FADH^+$, 6 O_2	12 H_2O, 34 ATP (32 net ATP)

2.

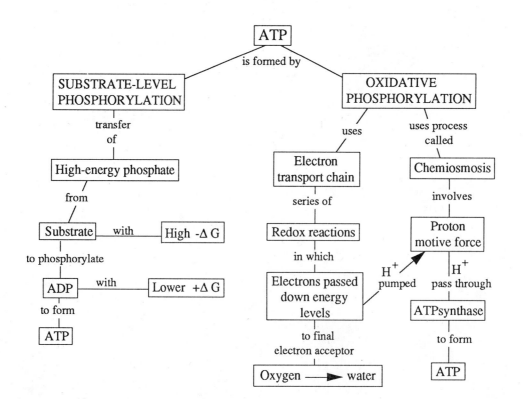

Answers to Test Your Knowledge

Multiple Choice:

1. c	**5.** d	**9.** b	**13.** a	**17.** b
2. b	**6.** b	**10.** a	**14.** b	**18.** d
3. a	**7.** d	**11.** c	**15.** c	**19.** c
4. a	**8.** c	**12.** d	**16.** c	**20.** d

CHAPTER 10
PHOTOSYNTHESIS

Suggested Answers to Structure Your Knowledge

1.

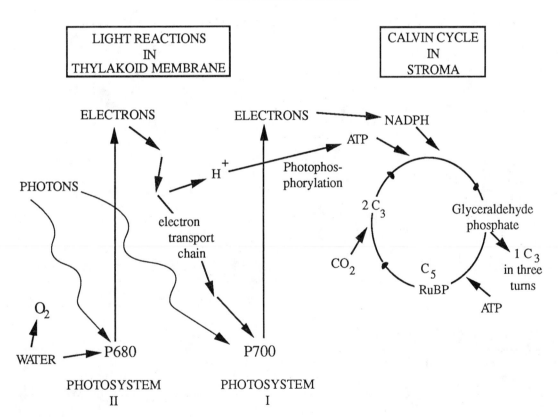

2. An absorption spectrum shows the wavelengths of light that are best absorbed by a particular pigment. An action spectrum presents the effectiveness or rate of a reaction at various wavelengths. The absorption spectrum of chlorophyll *a*, the key photosynthetic pigment, and the action spectrum for photosynthesis are not identical. Some wavelengths of light, particularly in the yellow and orange range, result in a higher rate of photosynthesis than would be indicated by the absorption of those wavelengths by chlorophyll *a*. These differences are accounted for by the existence of accessory pigments that absorb light energy from different wavelengths and pass that energy on to the reaction center molecules.

3.

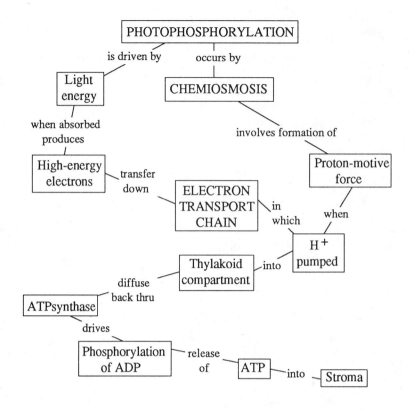

Answers to Test Your Knowledge

Multiple Choice:

1. c	**6.** b	**11.** d	**16.** b	**21.** a
2. d	**7.** d	**12.** a	**17.** c	**22.** b
3. a	**8.** c	**13.** b	**18.** c	**23.** d
4. b	**9.** c	**14.** c	**19.** a	**24.** a
5. a	**10.** a	**15.** d	**20.** b	**25.** c

CHAPTER 11
REPRODUCTION OF CELLS

Suggested Answers to Structure Your Knowledge

1. Journey of one chromosome through cell cycle:

   Interphase: 90% of cell cycle; growth, metabolism, DNA replication.
   - G₁ Phase—first gap phase: the chromosome, consisting of chromatin fiber made of DNA and associated proteins, is diffuse throughout the nucleus. Proteins, stabilizing the coiling and folding of the fiber, may contribute to the control of gene activity. RNA molecules are being transcribed from genes that are switched on.

   - S Phase—synthesis of DNA: the chromosome is replicated; two exact copies, called sister chromatids, are produced and held together at a region called the centromere.
   - G₂ Phase—second gap phase: growth and metabolic activities of cell continue; RNA transcription continues.

   Mitosis: cell division
   - Prophase—the chromosome, consisting of two sister chromatids, becomes tightly coiled and folded. Kinetochore fibers of the mitotic spindle

attach to kinetochore on each chromatid. When kinetochore fibers interact with polar fibers, the chromosome begins to move.

- Metaphase—the centromere of the chromosome is aligned at the metaphase plate along with the centromeres of the other chromosomes.
- Anaphase—the sister chromatids separate (now considered to be individual chromosomes) and move to opposite poles.
- Telophase—chromatin fiber of chromosome uncoils and is surrounded by reforming nuclear membrane.

2.

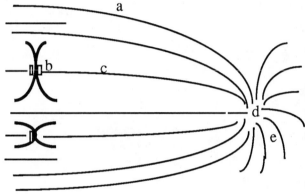

Answers to Test Your Knowledge

Matching:

1.	B	4.	G	7.	C	10.	A
2.	A	5.	F	8.	E		
3.	E	6.	D	9.	G		

Multiple Choice:

1.	d	4.	b	7.	b	10.	c
2.	c	5.	c	8.	a		
3.	a	6.	d	9.	b		

a. POLAR FIBER—microtubules from pole to center; push poles apart by sliding at region of overlap at equator and addition of tubulin molecules.

b. KINETOCHORE—region of centromere where kinetochore fiber attaches.

c. KINETOCHORE FIBERS—interact with polar fibers to move chromosomes to metaphase plate, pull chromosomes apart.

d. MICROTUBULE ORGANIZING CENTER —region of organization of mitotic spindle (associated with centrioles in animal cell)

e. ASTER—first part of spindle formed (only in animal cells).

UNIT III

THE GENE

Embarrassing moments at gene parties

CHAPTER 12

MEIOSIS AND SEXUAL LIFE CYCLES

FRAMEWORK

This chapter introduces the process of meiosis and the concept of genetic variation that results from crossing over, independent assortment, and random fertilization. The sexual life cycles of animals, plants, algae, fungi, and protists all include meiosis and fertilization but vary in the timing of these events.

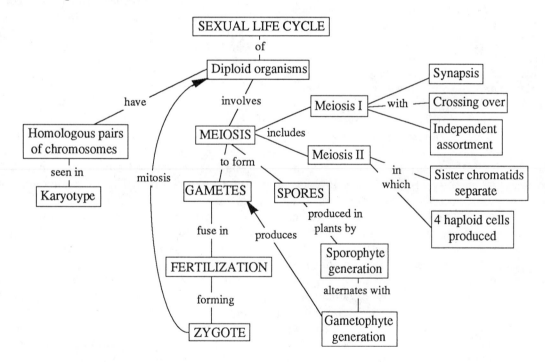

CHAPTER SUMMARY

The Scope of Genetics

Genetics is the study of heritable information. The inheritance of traits from parents to offspring involves the transmission of discrete units of information, known as *genes*, which are coded in the genetic material DNA. The possession of a genetic program, which is transmitted from one generation to the next, expressed by translation of genes into specific traits, and subject to modification through mutation and natural selection, is a characteristic unique to life.

Sexual Versus Asexual Reproduction

In *asexual reproduction*, a single parent passes on its genes to its offspring, which are usually exact genetic copies of the parent. In *sexual reproduction*, an individual receives a unique combination of genes inherited from two parents. The genetic variation created by sexual reproduction is a result of the management of chromosomes during the sexual life cycle.

Introduction to Sexual Life Cycles: The Human Example

In *somatic cells*, chromosomes are present in pairs known as *homologous chromosomes* or homologues. Homologues have the same length, centromere position, and staining pattern. A gene controlling a particular trait is found at the same *locus*, or location, on each chromosome of a homologous pair.

A *karyotype* is an ordered display of an individual's chromosomes. It is made by cutting the individual chromosomes out of a photograph taken of white blood cells that were stimulated to undergo mitosis, arrested in metaphase, and stained. The chromosomes are then arranged into homologous pairs by size and shape.

The exception to the rule of homologous chromosomes is found in the *sex chromosomes*, the *X* and *Y* chromosomes that determine the sex of a person. Females have two homologous *X* chromosomes; males have one *X* and one *Y* chromosome. The remaining homologous pairs of chromosomes are called *autosomes*.

Diploid cells contain a set of chromosomes from each parent, brought together by the fusion of *gametes*—egg and sperm—in the process of *fertilization*. The *haploid* number (N) of chromosomes for humans is 23. Somatic cells contain the diploid number (2N) of 46 chromosomes. The combined chromosome sets present in the fertilized egg or *zygote* are passed on to all somatic cells of the body with precision by mitosis. The sex organs then produce new cells of the *germ line*, the gametes, to pass the genetic program on to the next generation.

Meiosis is a special type of cell division that halves the chromosome number in gametes to compensate for the doubling that occurs at fertilization. An alternation between diploid and haploid conditions, involving the processes of meiosis and fertilization, is characteristic of the life cycles of all sexually reproducing organisms.

Meiotic Cell Division

Meiosis, like mitosis, is preceeded by chromosome replication. This replication is then followed by two consecutive cell divisions, *meiosis I* and *meiosis II*, resulting in four haploid daughter cells.

In interphase I, each chromosome replicates, producing two genetically identical chromatids attached at their centromeres. Prophase I lasts more than 90% of the time required for meiosis. Homologous chromosomes *synapse*, producing a *bivalent* (or tetrad), and corresponding segments of nonsister chromatids may *cross over*, breaking and reattaching to the other chromatid. The homologous chromosomes are held together by these *chiasmata* as they move toward the metaphase plate.

In metaphase I, the chromosome bivalents line up on the metaphase plate, with their centromeres pointing toward opposite poles. The homologous chromosomes separate in anaphase I and move toward opposite poles. By telophase I, each pole has a haploid set of chromosomes; each chromosome is composed of two chromatids. Cytokinesis may be followed by a brief period of interkinesis, when nuclei reform, or the two cells may proceed directly into meiosis II. There is no replication of genetic material prior to this second division.

Meiosis II looks like a regular mitotic division, in which chromosomes line up individually on the metaphase plate, and sister chromatids separate and move apart in anaphase II. At the end of telophase II, there are four haploid daughter cells.

Comparison of Mitosis and Meiosis

Mitosis produces daughter cells that are genetically identical to the parent cell. Meiosis produces haploid cells that differ genetically, both from their parent cell and from each other. The unique events that produce this result occur during meiosis I.

In prophase I, when homologous chromosomes synapse and form bivalents, genetic material is exchanged between nonsister chromatids by crossing over, which is visible during this stage by the appearance of x-shaped regions called *chiasmata*. During anaphase I, the homologous pairs, which had lined up together on the metaphase plate, separate, and one homologue goes to each pole. The centromeres do not divide, and sister chromatids remain together.

Meiosis II is identical to mitosis in that the sister chromatids separate. It is not preceded by a replication and results in the formation of four haploid cells.

Sexual Sources of Genetic Variation

The sorting out and recombining of chromosomes during the processes of meiosis and fertilization are responsible for most of the genetic variation found from one generation to the next. This variation is the raw material on which natural selection acts.

Independent Assortment of Chromosomes The first meiotic division results in the independent assortment of chromosomes. Each homologous pair lines up independently at the metaphase plate, and the orientation of the maternal and paternal chromosomes is random. The gametes produced by meiosis can have various combinations of maternal and paternal chromosomes. The number of combinations possible is equal to 2^N, where N is the haploid number. For humans, this number of possible combinations of chromosomes is 2^{23}—8 million distinct gametes that each individual can produce.

Crossing Over Crossing over also adds to the variability created by meiosis. The exchange of segments of nonsister chromatids results in new genetic combinations of maternal and paternal genes on the same chromosome. This *genetic recombination* results in gene combinations different from those inherited from the previous generation.

The mechanisms of synapsis and crossing over involve a *synaptonemal complex*, a protein structure that appears in electron micrographs to look somewhat like a zipper. Protein nodules scattered along the zipper are correlated with chiasmata formation.

Random Fertilization The random nature of fertilization adds to the genetic variability established in meiosis. A given pair of parents will produce a zygote with any of 64 trillion (8 million x 8 million) diploid combinations.

Independent assortment of chromosomes, genetic recombination by crossing over, and random fertilization all contribute to genetic variability in a sexually reproducing population. *Mutations*, rare changes in the DNA of genes, are the ultimate source of this genetic diversity.

The Variety of Sexual Life Cycles

In most animals, meiosis occurs in the formation of gametes, which are the only haploid cells in the life cycle. In many fungi and some protists, the only diploid stage is the zygote. Meiosis occurs right after the gametes fuse, and mitosis creates a multicellular haploid organism. Gametes will be produced by mitosis in these organisms.

Plants and some species of algae have a third type of life cycle, called an *alternation of generations*, that includes both diploid and haploid multicellular stages. The multicellular diploid *sporophyte* stage produces haploid *spores* by meiosis. These spores develop into a multicellular haploid plant, the *gametophyte*, which produces gametes by mitosis. Gametes fuse to form a diploid zygote that develops into the next sporophyte generation.

STRUCTURE YOUR KNOWLEDGE

1. Create a diagram of an animal sexual life cycle, including the major events and processes.

2. Fill in the following table, listing the key events that occur in the stages of meiosis.

STAGE	KEY EVENTS OF MEIOSIS
Interphase I	
Prophase I	
Metaphase I	
Anaphase I	
Telophase I	
Interkinesis	
Prophase II	
Metaphase II	
Anaphase II	
Telophase II	

TEST YOUR KNOWLEDGE

MULTIPLE CHOICE: *Choose the one best answer.*

1. The restoration of the chromosome number after halving in meiosis is due to
   a. synapsis.
   b. fertilization.
   c. mitosis.
   d. DNA replication.

2. A karyotype is a
   a. genotype of an individual.
   b. unique combination of chromosomes found in a gamete.
   c. blood type determination of an individual.
   d. picture of homologous chromosomes of an individual.

3. Autosomes are
   a. sex chromosomes.
   b. chromosomes that occur singly.
   c. chromosomes that move independently during meiotic cell divisions.
   d. none of the above.

4. A synaptonemal complex would be found during
   a. prophase I of meiosis.
   b. fertilization or fusion of gametes.
   c. metaphase II of meiosis.
   d. metaphase of mitosis.

5. During the first meiotic division (meiosis I)
   a. homologous chromosomes separate.
   b. the chromosome number becomes haploid.
   c. crossing over between nonsister chromatids occurs.
   d. all of the above occur.

6. A cell with a diploid number of 6 could produce gametes with how many different combinations of chromosomes?
   a. 6
   b. 8
   c. 12
   d. 64

7. The DNA content of a cell is measured in the G_2 phase. After meiosis I, the DNA content of such a cell would be
   a. equal to that of the G_2 cell.
   b. twice that of the G_2 cell.
   c. one-half that of the G_2 cell.
   d. one-fourth that of the G_2 cell.

8. In many fungi and some protists,
   a. the zygote is the only haploid stage.
   b. gametes are formed by meiosis followed by mitosis.
   c. the adult organism is haploid.
   d. the gametophyte generation produces gametes by mitosis.

9. In the alternation of generations found in plants,
   a. the sporophyte generation produces spores by meiosis.
   b. the gametophyte generation produces gametes by mitosis.
   c. the zygote will develop into a sporophyte generation by mitosis.
   d. all of the above are correct.

10. Which of the following is not a source of genetic variation in sexually reproducing organisms?
    a. chiasmata
    b. replication of DNA during S phase.
    c. independent assortment of chromosomes
    d. random fertilization of gametes

11. Meiosis II is similar to mitosis because
    a. sister chromatids separate.
    b. homologous chromosomes separate.
    c. DNA replication preceeds the division.
    d. they both take the same amount of time.

12. Homologous chromosomes
    a. have identical genes.
    b. have genes for the same traits at the same loci.
    c. pair up in prophase II.
    d. are found in haploid cells.

CHAPTER 13

MENDEL AND THE GENE IDEA

FRAMEWORK

Through his work with garden peas, Mendel developed the fundamental principles of inheritance and the laws of segregation and independent assortment. This chapter describes the basic monohybrid and dihybrid crosses that Mendel performed to establish that inheritance involves particulate genetic factors (genes) that segregate independently in the formation of gametes and recombine to form offspring. The laws of probability can be applied to predict the outcomes of genetic crosses.

Various complications in the expression of genotypes as phenotypes include intermediate inheritance, multiple alleles, pleiotropy, penetrance and expressivity, epistasis, and polygenic traits. Genetic screening and counseling for recessively inherited genetic disorders use Mendelian principles in the analysis of human pedigrees.

CHAPTER SUMMARY

From ancient Greece through the nineteenth century, pangenesis was the prevailing theory of inheritance. According to this theory, particles, called pangenes, from all parts of the body come together to form eggs or semen, and changes to various parts of the body during an organism's life could be passed on to the next generation. Two additional theories, developed in the seventeenth century with the early microscopic observations of sperm and ovarian follicles, were that the sperm or the egg contained a miniature human being.

By the early nineteenth century, biologists believed that both parents contribute to the characteristics of offspring with the blending or irreversible mixing of hereditary materials. In the 1860s, Gregor Mendel developed the fundamental principles of inheritance that eventually refuted this "blending" theory.

Mendel's Model of Inheritance

Mendel's work demonstrated that discrete, particulate heritable factors (genes) are passed from parents to offspring and retain their individuality from generation to generation.

Mendel's Methods   Mendel worked with garden peas, a good choice of study organism because it was easy to grow them, to control their fertilization, and to quantify their offspring.

The seven characteristics Mendel chose occurred in two alternative forms that could be found in *true-breeding* varieties, in which self-fertilizing parents always produced offspring with the parental form of the characteristic. To follow the transmission of these well-defined traits, Mendel cross-pollinated plants that were true breeding for alternate forms of the same characteristic, and then he self-pollinated the next generation. The true-breeding parental plants are known as the *P generation* (for parental); the results of the first *cross* are called the F_1 *generation* (for first filial); and the next generation, from the self-cross of the F_1, is known as the F_2 *generation*.

Mendel's Law of Segregation   Mendel found that the F_1 offspring did not show a blending of the parental characteristics. Instead, only one of the parental forms of the trait was found in the hybrid offspring. In the F_2 generation, however, the missing parental form reappeared in the ratio of 3:1—three offspring with the parental trait shown by the F_1 to one offspring with the reappearing trait.

Mendel's explanation for this phenomenon

contains four parts: (1) there are alternate forms, now called *alleles*, for the heritable factors called genes, (2) an organism has two alleles for each inherited trait, one received from each parent, (3) gene pairs separate (segregate) during the formation of gametes, so an egg or sperm only carries one allele for each inherited trait, and (4) when two different alleles occur together, one allele, called the *dominant allele*, may be completely expressed while the other, *recessive allele,* is masked.

This hypothesis explains the 3:1 ratio observed in the F_2 plants. During the separation of genes in the formation of gametes in the F_1 generation, one-half of the gametes would receive one allele while the other half would receive the alternate allele. Random fertilization of gametes would result in one-fourth of the plants having two dominant alleles, one-half having one dominant and one recessive allele, and one-fourth receiving two recessive alleles. Since the dominant allele would be the one expressed in the half of the offspring that are hybrid, the ratio of plants showing the dominant or recessive trait would be 3:1. A *Punnett square* can be used to predict the results of simple genetic crosses.

The mechanism of inheritance that accounts for these results is stated by Mendel's *law of segregation*: allele pairs segregate during gamete formation, and the paired condition is reestablished by the random fusion of gametes at fertilization.

Dominant alleles are often symbolized by a capital letter, recessive alleles by a small letter. An organisms that has a pair of identical alleles for a trait is said to be *homozygous.* If the organism has two different alleles, it is said to be *heterozygous* for that trait. Homozygotes are true breeding; heterozygotes are not, since they produce gametes with one or the other allele that can recombine to produce homozygous-dominant, heterozygous, and homozygous-recessive offspring, in the genotypic ratio of 1:2:1. The *phenotype* is an organism's expressed traits; its *genotype* is its genetic makeup.

The phenotypic ratio is often different from the genotypic ratio, since homozygous-dominants and heterozygotes may have the same phenotype. A *testcross*, in which an organism of unknown genotype is bred with a recessive homozygote, can determine the genotype of organisms expressing the dominant phenotype. The recessive homozygote can only contribute a recessive allele; if there is a 1:1 phenotypic ratio of offspring, the unknown parent must have been heterozygote.

Inheritance as a Game of Chance   General rules of probability apply to the segregation of alleles and the reconstitution of pairs at fertilization. The probability of an event occurring is the ratio of the number of possible events that fulfill that criterion over all the possible events; for example, the probability of getting a head in a coin toss is 1/2.

The outcome of independent events is not affected by previous events. For the probability of independent events occurring simultaneously, however, the *rule of multiplication* states that the probability of a compound event is equal to the product of the separate probabilities of the independent events. The probability of a particular genotype being formed by fertilization is equal to the product of the probability of the formation of each type of gamete needed to produce that genotype. The probability of getting a homozygous-recessive offspring from a cross of heterozygotes is equal to the probability of getting one gamete with the recessive allele (1/2) times the probability of getting the other gamete with the recessive allele or $1/2 \times 1/2 = 1/4$.

If the genotype can be formed in more than one way, then the *rule of addition* states that the probability of an event is equal to the sum of the separate probabilities of the different ways in which the event can occur. Thus, a heterozygote offspring can occur if the egg contains the dominant allele and the sperm the recessive (1/4 probability) or vice versa (1/4). A heterozygote offspring would be predicted to result from a hybrid cross 1/4 + 1/4, or 1/2 (50%) of the time.

These rules of probability account for the 1:2:1 ratio of genotypes seen in the offspring of a hybrid cross. Large samples will usually show these expected distributions more precisely because the influence of individual events is averaged out.

Mendel's Law of Independent Assortment   Mendel deduced the law of segregation from working with *monohybrid crosses*, crosses that involved parents that differed in only a single trait. He used *dihybrid crosses*, involving parents that differed in two traits, to determine whether the two traits were transmitted together (in the same combination in which they were inherited from the parental plants) or independently of each other.

If the two pairs of genes segregate independently, then gametes from an F_1 hybrid generation (AaBb) should contain four combinations of genes in equal quantities (AB, Ab, aB, ab). The random fertilization of these four classes of gametes should result in 16 (4 x 4) gamete combinations that produce four phenotypic categories in a ratio of 9:3:3:1 (nine offspring showing both dominant traits, three showing dominant of one trait and recessive of the other, three showing recessive of the one and dominant of the other, and one showing both recessive traits). Mendel obtained these ratios when he scored the F_2 progeny of dihybrid crosses, and thus obtained evidence for his *law of independent assortment*. This principle states that genes assort independently from each other when they segregate in the formation of gametes.

Fairly complex genetics problems can be solved by applying the rules of probability to segregation and independent assortment. The probability of a particular genotype arising from a cross can be determined by considering each gene involved as a separate monohybrid cross and then multiplying the probabilities of all the independent events involved in the final genotype.

Not all genes follow the case of simple dominance that Mendel discovered; nor do genotypes always predict phenotypes in a rigid manner. However, more complicated inheritance patterns do not alter the particulate theory of inheritance, with genes being transmitted according to Mendel's two laws and the rules of chance.

From Genotype to Phenotype: Some Complications

Intermediate Inheritance Some characteristics show intermediate inheritance, in that the F_1 hybrids have a phenotype intermediate between that of the parents. In *incomplete dominance*, neither allele masks the expression of the other, and the phenotype of heterozygotes is distinguishable from the two homozygous conditions. Heterozygote snapdragons are pink because they have only half the red pigment as do red homozygotes.

The human inherited disease familial hypercholesterolemia (FH) is an example of intermediate inheritance. Heterozygotes, who have one defective allele for the production of LDL receptors, accumulate higher blood cholesterol levels, potentially leading to atherosclerosis and increased risks of heart attack or stroke. The cells of heterozygotes have half the normal number of LDL surface receptors, and homozygotes for the FH allele have no receptors.

Multiple Alleles More that two alleles may exist for some genes. Some of the alleles may be *codominant*, meaning that both alleles are expressed. The gene that determines human blood groups has three alleles; the alleles I^A and I^B are dominant over i but codominant with each other. Individuals with the ii genotype have type O blood; the $I^A I^B$ genotype results in type AB blood. Blood type is critical in transfusions because the antigens found on the red blood cells of unmatched donated blood may interact with the recipient's antibodies and cause agglutination, or clumping, of blood cells within the blood vessels.

Pleiotropy Pleiotropy is the characteristic of a single gene having multiple phenotypic effects in an individual. Many hereditary diseases, with complex sets of symptoms, are caused by a single gene.

Penetrance and Expressivity The phenotype of an individual is the result of complex interactions between its genotype and the environment. *Penetrance* is the proportion of individuals that express the phenotype for a particular genotype. A gene with incomplete penetrance is not expressed in all individuals who carry the gene. The *expressivity*, or degree to which a gene is expressed, can also vary. Occasionally, a strong environmental influence will produce a phenotype, called a *phenocopy*, that simulates the effect of a gene that is not present in the genotype.

Epistasis In *epistasis*, one gene may interfere with the expression of another gene. An F_2 phenotypic ratio of 9:7 can be explained as the result of one gene being epistatic to another. Unless the first gene is present in its dominant form, the other gene cannot be expressed. Often both genes code for enzymes in the same biosynthetic pathway, and the expression of the second gene may depend on the successful enzymatic completion of an earlier step of the pathway controlled by the first gene.

Polygenic Traits Quantitative traits, such as height or skin color, may vary in a population in a continuous way. In *polygenic inheritance*, two or more genes have an additive effect on a single phenotypic trait. Each dominant allele contributes one "unit" to the expressed trait. A polygenic trait, combined with the influence of environmental factors, may result in a normal distribution, forming a bell-shaped curve, of the trait within a population.

Mendelian Inheritance in Humans

Human Pedigrees A family *pedigree* is a family tree with the history of a particular trait shown across the generations. By convention, circles represent females, squares are used for males, and solid symbols indicate individuals that express the phenotype in question. Parents are joined by a horizontal line and offspring are listed below parents from left to right in order of birth. The genotypes of individuals in the pedigree can often be deduced by following the patterns of inheritance.

For genetic disorders inherited as simple Mendelian traits, geneticists, physicians, and genetic counselors use pedigree analysis to make statistical predictions concerning the probability that a trait may be inherited.

Recessively Inherited Disorders For the 1,000 or so genetic disorders that are inherited as simple recessive traits, only homozygous-recessive individuals express the phenotype. *Carriers* of the disease are heterozygotes who are phenotypically normal but may transmit the recessive allele to their offspring. A mating between two heterozygotes has a 1/4 chance of producing an offspring with the homozygous-recessive disorder. A normal child produced by this mating has a 2/3 chance of being a heterozygote and thus, being a carrier.

Genetic disorders are usually unevenly distributed among racial or cultural groups due to the different genetic histories of such groups. Cystic fibrosis is the most common lethal genetic disease in the United States; it is found more frequently in Caucasians than in other races. This recessive allele causes excessive secretions of mucus in various organs, leading to blockage of the digestive tract, cirrhosis of the liver, pneumonia, or other infections. Most afflicted children die before adolescence.

Tay-Sachs disease is a lethal disorder in which the brain cells are unable to metabolize a type of lipid that then accumulates in and damages the brain, resulting in an early death. There is a disproportionately high incidence of this disease among Jewish people whose ancestors lived in Central Europe.

Sickle cell anemia is the most common inherited disease among blacks. Due to a single amino acid substitution in the hemoglobin protein, the molecules tend to crystallize, causing the red blood cells to deform into a sickle shape. The sickled cells block blood vessels and cause severe pain. Heterozygous individuals are said to have sickle cell trait but are usually healthy. The resistance to malaria that accompanies the sickle cell trait may explain why this lethal recessive allele remains in relatively high frequency in areas where malaria is common.

The likelihood of two mating individuals carrying the same rare deleterious allele increases when the individuals have common genetic ancestors. *Consanguineous* matings, between siblings or close relatives, are more likely to produce offspring homozygous for various traits, including a harmful lethal trait. Matings between "blood" relatives are indicated on pedigrees with double lines.

There is some debate over the extent to which human consanguinity increases the risk of inherited diseases. Inbreeding among zoo animals and domesticated animals has resulted in a higher incidence of harmful recessive traits but has enabled some endangered species to avoid extinction.

Dominantly Inherited Disorders   A few human disorders are due to dominant genes. In achondroplasia, dwarfism is due to a single copy of a mutant allele. Homozygosity for this dominant allele is lethal. This rare allele illustrates that the dominance or recessiveness of an allele is not determined by its frequency in the population.

Dominant lethal alleles are more rare than recessive lethals because the harmful allele cannot be masked in the heterozygote. Most dominant lethal alleles are the result of mutations and kill the developing organism before it can reproduce and pass on the new form of the gene. A late-acting lethal dominant allele, however, can be passed on if the symptoms do not develop until after an organism is old enough to reproduce. Huntington's disease is a degenerative disease of the nervous system that does not develop until the individual is 35 to 45 years old. Children of a person who develops Huntington's have a 50% chance of having inherited the dominant lethal allele and developing the disease later in life. Medical researchers have recently developed a method to detect the lethal gene without waiting for the symptoms to begin. One type of Alzheimer's disease and some cases of manic-depressive disease have been linked with dominant alleles.

Genetic Screening and Counseling   The risk of a genetic disorder being transmitted to offspring can sometimes be determined through genetic counseling and testing before a child is conceived or in the early stages of pregnancy.

The probability of a child having a genetic defect can be determined by considering the family history of the disease. If two prospective parents both have siblings who had the disorder, both sets of prospective grandparents must have been carriers. The parents each have a 2/3 chance that they are heterozygote carriers. The probability that both parents are carriers is 2/3 x 2/3; the chance that two heterozygotes will have a recessive homozygous child is 1/4. The overall chance that a child will inherit the disease (following the rule of multiplication) is 1/9. Should this couple have a baby that has the disease, this establishes that they are both carriers, and the chance that a subsequent child would have the disease is 1/4.

Determining whether parents are carriers is important for determining the risk of a genetic disorder. Biochemical tests that permit carrier recognition have been developed for several heritable disorders. The use of new methods for analyzing DNA should help in the detection of harmful genes.

Amniocentesis, a procedure done between the fourteenth and sixteenth week of pregnancy, allows detection of several genetic disorders. Amniotic fluid removed from the sac surrounding the baby in the uterus is analyzed biochemically, and fetal cells present in the fluid are cultured for karyotyping to check for certain chromosomal defects.

Chorionic villi sampling is a technique in which a small amount of the fetal tissue making up the placenta is suctioned off. These rapidly growing cells can be

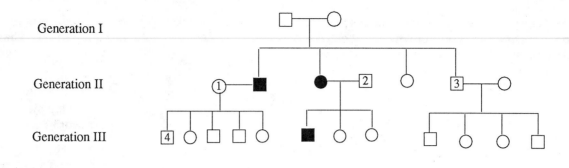

Generation I

Generation II

Generation III

karyotyped immediately, and the procedure can be performed at only 8 to 10 weeks of pregnancy.

Ultrasound is a simple noninvasive procedure that can reveal major abnormalities. It uses sound waves to produce an image of the fetus. Fetoscopy, the insertion of a needle-thin viewing scope and light into the uterus, also allows the fetus to be checked for anatomical problems.

Some genetic disorders can be detected at birth. The recessively inherited disorder phenylketonuria (PKU) is caused by the lack of an enzyme needed to break down the amino acid phenylalanine. An accumulation of phenylalanine and its by-product causes severe mental retardation. If the disease is detected in the newborn by a routine screening test, the diet of the child can be adjusted and normal development is possible.

STRUCTURE YOUR KNOWLEDGE

1. Relate Mendel's two laws of inheritance to the behavior of chromosomes in meiosis that you studied in Chapter 12.

2. Tall red-flowered plants are crossed with short white-flowered plants. The resulting F_1 generation consists of all tall pink-flowered plants. Assuming that height and flower color are each determined by a single gene locus, predict the results of an F_1 cross of TtRr plants. Fill in the gametes and genotypes in the following Punnett square and the phenotypes and numbers of the F_2 generation.

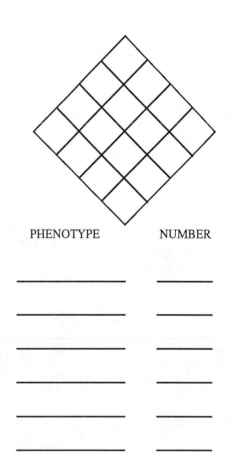

PHENOTYPE   NUMBER

_____   _____

_____   _____

_____   _____

_____   _____

_____   _____

_____   _____

3. Albinism (lack of skin pigmentation) is caused by a recessive allele. Consider the human pedigree shown at the top of the page for this trait (solid symbols represent individuals who are albinos). From your knowledge of Mendelian inheritance, determine the probable genotypes of the parents in generation I, the mates in generation II, and son 4

in generation III. Can you determine the genotype of son 3 in generation II? Why or why not? (Let *AA* and *Aa* represent normal pigmentation and *aa* be the albino genotype.)

Generation I

Genotype of father _____

of mother _____

Generation II

Genotype of mate 1 _____

of mate 2 _____

Generation III

Genotype of son 4 _____

TEST YOUR KNOWLEDGE

MATCHING: *Match the definition with the correct term.*

1. ____ codominance
2. ____ homozygous
3. ____ heterozygous
4. ____ expressivity
5. ____ polygenic trait
6. ____ pleiotropy
7. ____ phenocopy
8. ____ testcross
9. ____ dihybrid cross
10. ____ penetrance
11. ____ epistasis
12. ____ incomplete dominance

A. true breeding variety

B. cross between two hybrids

C. cross that involves two different gene pairs

D. the degree to which a gene is expressed

E. the expressed characteristics of an individual

F. containing two different alleles for same locus

G. containing multiple alleles

H. one gene masks the expression of another gene

I. both alleles are expressed in heterozygote

J. single gene with multiple phenotypic effects

K. proportion of individuals that express genotype

L. phenotype induced by environment, not coded for in genotype

M. heterozygote is intermediate between homozygous phenotypes

N. two or more genes with additive effect on phenotype

O. cross with homozygous recessive to determine genotype of unknown

MULTIPLE CHOICE: *Choose the one best answer.*

1. The genotypic ratio resulting from a cross
   a. will be equal to the phenotypic ratio.
   b. can be predicted with a Punnett square.
   c. will change if the gene involved has multiple alleles.
   d. will indicate which allele is dominant and which is recessive.

2. The phenotypic ratio of a cross
   a. will be equal to the genotypic ratio.
   b. will vary from what is expected if the gene involved is pleiotropic.
   c. will vary from what is expected if the gene involved has multiple alleles.
   d. will include all the different phenotypic combinations of gametes.

3. Recessive alleles
   a. are not expressed.
   b. are expressed when present as both alleles of a gene pair.
   c. are always the least frequently found alleles in a population.
   d. code for mutations.

4. According to the theory of pangenesis
   a. the characteristics of the offspring are a blending of parental traits.
   b. the sperm contains a miniature human being called the homunculus.

c. particles from various parts of the body come together to form gametes, and thus changes to an organism's body can be passed to the next generation.

d. particulate heritable factors separate in the formation of gametes.

5. The F_2 generation
   a. has a phenotypic ratio of 3:1.
   b. is the result of the self-fertilization or crossing of F_1 individuals.
   c. can be used to determine the genotype of individuals with the dominant phenotype.
   d. will show up in a dihybrid cross.

6. According to Mendel's law of segregation
   a. there is a 50% probability that a gamete will get a dominant allele.
   b. gene pairs segregate independently of other genes in gamete formation.
   c. allele pairs separate in gamete formation.
   d. the laws of probability determine gamete formation.

7. A 1:1 phenotypic ratio in a testcross indicates that
   a. the alleles are dominant.
   b. one parent must have been homozygous recessive.
   c. the dominant phenotype parent was a heterozygote.
   d. the alleles segregated independently.

8. According to Mendel's law of independent assortment,
   a. an individual heterozygous for three genes should produce six different gametic combinations of alleles.
   b. allele pairs segregate in gamete formation.
   c. the F_2 phenotypic ratio of a dihybrid cross of two true breeding varieties should be 9:3:3:1.
   d. all of the above are correct.

9. Quantitative traits
   a. are found in large numbers of offspring.
   b. may exist in a normal distribution within a population.
   c. may be the result of pleiotropic genes.
   d. may be the result of varying penetrance of a gene.

10. The probability that a particular genotype may result from a cross
    a. can be determined as the product of the probabilities of the formation of the gametes needed to produce the genotype.
    b. can be determined from the genotypic ratio for the cross.
    c. will depend on the genotypes of the parents.
    d. is related to all of the above.

11. Carriers of a disease
    a. are indicated by solid symbols on a family pedigree.
    b. are heterozygotes for the gene that can cause the disease.
    c. will produce children with the disease.
    d. usually are involved in consanguineous matings.

12. A lethal recessive allele is more likely to be maintained in a population
    a. if it is somehow beneficial in the heterozygous condition.
    b. if it is not expressed until late in life.
    c. if it can be identified by genetic screening.
    d. both a and b are correct.

13. Dominant lethal alleles
    a. are expressed late in life.
    b. are lethal in both the homozygous and heterozygous condition.
    c. are more likely to occur in consanguineous matings.
    d. are responsible for cystic fibrosis, Tay-Sachs disease and sickle cell anemia.

14. With amniocentesis,
    a. a couple is tested to see if they are carriers of a genetic disorder.
    b. a fetus is observed for anatomical problems.
    c. fetal cells of the placenta can be karyotyped.
    d. fetal cells can be cultured and amniotic fluid is biochemically tested.

15. If both parents are carriers of a lethal recessive gene, the probability that their child will inherit and express the disorder is
    a. $2/3 \times 2/3 \times 1/4$, or $1/9$.
    b. $1/2$.
    c. $1/4$.
    d. $1/2 \times 1/2 \times 1/4$, or $1/16$.

16. A green budgie (Fig. 13.11) is crossed to another green budgie. Which of the following results is not possible?
    a. all green offspring
    b. all blue offspring
    c. all white offspring
    d. all of the above are possible, but with different probabilities

17. After obtaining two heads from two tosses of a coin, the probability of tossing the coin and obtaining a head is
    a. $1/2$.
    b. $1/4$.
    c. $1/8$.
    d. $1/16$.

18. The probability of tossing three coins and obtaining three heads is
    a. 1/2.
    b. 1/4.
    c. 1/8.
    d. 1/16.

GENETIC PROBLEMS:

1. Blood typing is often used as evidence in paternity cases, when the blood type of the mother and child may indicate that a man alleged to be the father could not possibly have fathered the child. For the following mother and child combinations, indicate which blood groups of potential fathers would be exonerated.

Blood group of mother	Blood group of child	Man is exonerated if he belongs to blood group(s)
AB	A	_____
O	B	_____
A	AB	_____
O	O	_____
B	A	_____

2. Polydactyly (extra fingers and toes) is due to a dominant gene. A father is polydactyl, the mother has the normal phenotype, and they have had one normal child. What is the genotype of the father? Of the mother? What is the probability that a second child will have the normal number of digits?

3. For the following genotypes, indicate what proportion of the gametes will have the indicated genes.

Parental genotype	Type of Gamete	Proportion expected
AABb	AB	_____
AaBb	ab	_____
AABbcc	ABc	_____
AaBbCc	ABc	_____

4. For the following crosses, indicate the probability of obtaining the indicated genotype in an offspring. Remember it is easiest to treat each gene separately as a monohybrid cross and then combine the probabilities.

Cross	Offspring	Probability
AAbb x AaBb	AAbb	_____
AaBB x AaBb	aaBB	_____
AABbcc x aabbCC	AaBbCc	_____
AaBbCc x AaBbcc	aabbcc	_____

5. In dogs, black (B) is dominant to chestnut (b), and solid color (S) is dominant to spotted (s). What are the genotypes of the parents that would produce a cross with 3/8 black solid, 3/8 black spotted, 1/8 chestnut solid, and 1/8 chestnut spotted puppies? (Hint: first determine what genotypes the offspring must have before you deal with the fractions.)

6. The height of spike weed is a result of polygenic inheritance involving three genes, each of which can contribute 5 cm to the plant. The base height of the weed is 10 cm, and the tallest plant can reach 40 cm. If a tall plant (AABBCC) is crossed with a base-height plant (aabbcc), what is the height of the F_1 plants? How many phenotypic classes will there be in the F_2?

7. In guinea pigs, the gene for production of melanin is epistatic to the gene for the deposition of melanin. The dominant allele M causes melanin to be produced; mm individuals cannot produce the pigment. The dominant allele B causes the deposition of a lot of pigment and produces a black guinea pig, whereas only a small amount of pigment is laid down in bb animals, producing a light-brown color. Without an M allele, no pigment is produced so the allele B has no effect, and the guinea pig is white. A homozygous black guinea pig is crossed with a homozygous recessive white: MMBB x mmbb. Give the phenotypes of the F_1 and F_2 generations.

8. When hairless hamsters are mated with normal-haired hamsters, about one-half the offspring are hairless and one-half are normal. When hairless hamsters are crossed with each other, the ratio of normal-haired to hairless is 1:2. How do you account for the results of the first cross? How would you explain the unusual ratio obtained in the second cross?

9. If two medium-tailed pigs were mated and the liter produced included three stub-tailed piglets, six medium-tailed, and four long-tailed piglets, what would be the simplest explanation of these results?

10. The ability to taste phenylthiocarbamide (PTC) is controlled in humans by a single dominant allele (T). A woman nontaster married a man taster and

they had three children—two boy tasters and a girl nontaster. All the grandparents were tasters. Create a pedigree for this family for this trait. (Solid symbols should signify nontasters.) Where possible, indicate whether tasters are TT or Tt.

11. Fur color in rabbits is determined by a single gene locus for which there are four alleles. Four phenotypes are possible: black, Chinchilla (gray color caused by white hairs with black tips), Himalayan (white with black patches on extremities), and white. The black allele (C) is dominant over all other alleles, the Chinchilla allele (C^{ch}) is dominant over Himalayan (C^h), and the white allele (c) is recessive to all others. A black rabbit is crossed with a Himalayan, and the F_1 consisted of a ratio of 2 black: 2 Chinchilla. Can you determine the genotypes of the parents?

A second cross was done between a black rabbit and a Chinchilla. The F_1 contained a ratio of 2 black: 1 Chinchilla: 1 Himalayan. Can you determine the genotypes of the parents of this cross?

12. A dominant allele B produces bristles in fruit flies when it is present in the heterozygote. When it is homozygous (BB), it is lethal. Homozygous recessive (bb) flies are nonbristled. Another gene S acts to suppress the action of B, but it is also lethal when homozygous (SS). The ss genotype has no effect on bristles. Two nonbristled flies in which the B allele is being suppressed are crossed. What is the phenotypic ratio of the F_1? If the bristled flies from the F_1 are backcrossed to parental flies, what phenotypic ratio would be predicted for the offspring?

CHAPTER 14

THE CHROMOSOMAL BASIS OF INHERITANCE

FRAMEWORK

```
                              ┌──────────────┐
                              │ CHROMOSOMES  │
                              └──────────────┘
              may be        are location of      may have
        ┌──────────────┐    ┌─────────┐    ┌──────────────────┐
        │Sex chromosomes│    │ GENES  │    │   MUTATIONS &    │
        └──────────────┘    └─────────┘    │   ALTERATIONS    │
                                           └──────────────────┘
          can carry          may be              may include
    ┌──────────┐   ┌──────────┐ ┌────────┐ ┌────────────┐ ┌──────────┐ ┌──────────┐
    │Sex-linked│   │ Unlinked │ │ Linked │ │ Structural │ │Aneuploidy│ │Polyploidy│
    │  traits  │   └──────────┘ └────────┘ │alterations │ └──────────┘ └──────────┘
    └──────────┘   do not        may       └────────────┘   such as      seen in
           show    show                       such as                  ┌────────┐
      ┌──────────────┐  ┌────────┐     ┌────────────┐ ┌──────────┐    │ Plants │
      │ INDEPENDENT  │  │ Cross  │     │ Deletions  │ │ Trisomy  │    └────────┘
      │ ASSORTMENT   │  │  over  │     │Duplications│ │monosomy  │
      └──────────────┘  └────────┘     │Translocations│└──────────┘
        results in   gives   provide   │ Inversions │   can cause
    ┌──────────┐ ┌────────────┐ ┌──────────┐ └────────────┘  ┌────────┐
    │ Genetic  │ │Recombination│ │Cytological│              │ Human  │
    │recombination│ frequency  │ │ evidence │               │disease │
    └──────────┘ └────────────┘ └──────────┘                └────────┘
                              can use to
                           ┌────────────┐
                           │Map location│
                           │ of genes   │
                           └────────────┘
```

CHAPTER SUMMARY

The Chromosome Theory of Inheritance

By 1900, three botanists had independently arrived at the same genetic principles that Mendel had discovered 35 years previously. These principles, combined with the cytological evidence of the processes of mitosis and meiosis gathered in the late 1800s, led to the development of the *chromosome theory of inheritance*. According to this theory, Mendelian factors, or genes, are located on chromosomes that undergo segregation and independent assortment in the process of gamete formation.

Morgan and the Drosophila School   T. H. Morgan, working with the fruit fly, *Drosophila melangaster*, first associated a specific gene with a specific chromosome. Fruit flies are excellent organisms for genetic studies because they are prolific breeders and have only four pairs of chromosomes, which are easily distinguishable with a microscope. The sex chromosomes occur as *XX* in females and *XY* in male flies.

Traits that are the normal phenotypes found most commonly in nature are called *wild type*, whereas alternative traits, assumed to have arisen as *mutations* in a wild-type gene, are called *mutant phenotypes*. The genetic notation commonly used by *Drosophila* geneticists employs small letters to signify the mutant allele and the small letter with a superscript plus sign for the wild allele.

Morgan discovered a mutant white-eyed male fly that he mated with a wild-type red-eyed female. The F_1 were all red-eyed and the F_2 showed the 3:1 phenotypic ratio typical of a simple dominant trait. In the F_2, however, all female flies were red-eyed, whereas half of the males were red-eyed and half were white-eyed.

Morgan deduced that the gene for eye color was *sex-linked*, occurring only on the *X* sex chromosome. Males only have one *X* and their phenotype is determined by the eye-color allele they inherit from their mother. The association of a specific gene with a chromosome provided evidence for the chromosome theory of inheritance.

Linked Genes: Exceptions to Independent Assortment

Linked genes are genes that are located on the same chromosome. Because chromosomes are inherited as a unit, the law of independent assortment does not usually apply to linked genes.

In 1908, Bateson and Punnett observed an unusual pattern of inheritance of two traits in the sweet pea. The results of a dihybrid F_2 cross did not give the expected 9:3:3:1 ratio. A large number of offspring showed either both dominant or both recessive traits. The explanation for this phenomenon emerged later from the work of Morgan: the genes for the two traits were linked on the same chromosome and inherited together, unless a crossover event occurred in the formation of eggs or pollen.

The Chromosomal Basis of Recombination

The production of offspring with new combinations of traits inherited from two parents is called *genetic recombination*.

Recombination of Unlinked Genes: Independent Assortment

In a dihybrid cross between a heterozygote (AaBb) and a homozygous recessive (aabb), one-half of the offspring, called *parental types*, will have phenotypes like one or the other parent (both dominant traits or both recessive traits), and one-half of the offspring, called *recombinants*, will have combinations of the two traits that are unlike the parents (Aabb or aaBb). This 50% frequency of recombination is observed when two genes are located on different chromosomes. This recombination is the result of the random alignment of homologous chromosomes at metaphase I and the resulting independent assortment of alleles.

Recombination of Linked Genes: Crossing Over

Linked genes do not assort independently, and one would not expect to see recombination of parental alleles in the offspring. In a testcross (cross using homozygous recessive individuals) with flies heterozygous for two genes, Morgan found that most of the offspring resembled the parental phenotypes, indicating that the genes were linked. Seventeen percent of the offspring, however, showed a recombination of parental traits. The reciprocal trade between nonsister chromatids during prophase I, called crossing over, accounts for the recombination of linked genes.

Mapping Chromosomes

Morgan's group was the first to work out methods to *map* genes in sequence on particular chromosomes.

Maps Based on Crossover Data

Sturtevant suggested that recombination frequencies reflect the relative distances between genes. If genes are located farther apart on the chromosome, there is a greater probability that a crossover event will occur between them. Sturtevant used recombination data to locate genes on a chromosomal map and defined one *map unit* as equal to a 1% recombination frequency.

The sequence of genes on a chromosome can be determined by finding the recombination frequency between different pairs of genes. Thus, if a and b are 12 map units apart, b and c are 3 map units apart, and c and a are 9 units apart, the sequence of genes must be a-c-b.

It is impossible to determine linkage if genes are 50 or more map units apart because they would show the 50% frequency of recombination typical of unlinked genes. Distant genes on the same chromosome may be mapped by adding the recombination frequencies determined with intermediate genes.

Sturtevant and his co-workers clustered the genes for the various known mutations of *Drosophila* into four groups of linked genes. Combined with the fact that there are four sets of chromosomes in *Drosophila*, these results provided additional evidence for the location of genes on chromosomes.

Cytological Maps

Crossover data provide only relative distances for the location of genes on chromosomes. *Cytological mapping* is a technique that associates a mutant phenotype with a visible chromosomal defect or other feature and can thus pinpoint the location of a particular gene.

X-rays can cause *point mutations*, in which a single part of a gene is altered, or *chromosomal mutations*, in which breaks in the chromosome may be visible in the microscope. Muller discovered in the 1920s that the use of X-rays greatly increased the number of mutants of *Drosophila*, providing more raw material for genetic studies, but also indicating the dangers of X-rays.

The *polytene chromosomes*, found in the salivary glands of *Drosophila* larvae, consist of hundreds of aligned chromatids on which differences in staining density produce characteristic patterns of dark and light bands. Homologous chromosomes closely pair in these cells, a rare event in nonmeiotic cells. When one homologue has a structural alteration, the pairing is incomplete and the normal homologue loops out at the unpaired segment.

The chromosomal location of a gene can be determined by matching a specific mutant phenotype with a localized deformity in polytene chromosome pairing. By inducing mutations with X-rays and other mutagens and associating mutant phenotypes with chromosomal defects, geneticists have developed extensive cytological maps of the genes of *Drosophila* and other organisms. The spacings between genes on these maps differ from those determined from crossover data, because the frequency of crossing over varies for different regions of a chromosome.

The Chromosomal Basis of Sex

Sex is a phenotypic characteristic usually determined by the presence or absence of special chromosomes.

Systems of Sex Determination Humans, other mammals, and fruit flies all have an *X–Y* system of sex determination. Males produce two kinds of gametes— sperm with either an *X* or a *Y* chromosome—and are called the *heterogametic sex*. Females are the *homogametic sex*; all eggs contain an *X* chromosome. Maleness is determined in mammals by the presence of the *Y* chromosome; whereas the ratio of *X* chromosomes to autosomes, regardless of the presence or absence of a *Y* chromosome, determines maleness in *Drosophila*.

An *X–O* system is found in grasshoppers, crickets, roaches, and some other insects. The female has two X chromosomes. There is no *Y* chromosome—males are *XO*, having only one sex chromosome. In the *Z–W* system seen in birds, some fishes, and butterflies and moths, the female is the heterogametic sex, and males are designated as ZZ.

Bees and ants have no sex chromosomes. Males develop from unfertilized eggs (called *parthenogenesis*)

and are haploid. Females develop from fertilized eggs and are diploid.

The chromosomal basis of sex determination of plants that produce male and female flowers on two separate plants (dioecious) is usually an *X–Y* system. Most plant species and some animals (such as earthworms and snails) are monoecious; a single individual produces both sperm and eggs. In these cases, all individuals of the species have the same complement of chromosomes.

Sex-Linked Inheritance

Sex chromosomes may carry genes for traits that are not related to maleness or femaleness. In humans, the *X* chromosome is much larger than the *Y* one, and most *X*-linked genes (coding for what are usually called sex-linked traits) do not have corresponding loci on the *Y* chromosome. Males inherit their sex-linked alleles from their mothers; fathers can pass sex-linked alleles only to their daughters.

Recessive sex-linked traits are seen more often in males, who are hemizygous for sex-linked genes. Females need to inherit recessive alleles on both their X chromosomes to express a recessive trait. *Hemophilia* is a sex-linked recessive trait that is characterized by excessive bleeding due to the malfunction of a blood-clotting factor. Color-blindness is also a sex-linked trait.

Most genes on the *Y* chromosome code for traits found only in males and have no counterparts on the *X*. In 1987, a gene that determines maleness in humans and other mammals was located on the *Y* chromosome. This *TDF* gene codes for testis-determining factor, a protein that induces testes-development in embryos.

Gene Dosage Compensation Normal diploid cells have two copies of autosomal genes and thus a double dose of each gene's product. According to the *Lyon hypothesis*, only one of the *X* chromosomes is fully active in most mammalian female somatic cells. The other *X* chromosome is inactive and contracted into a *Barr body*. With only one functional *X* chromosome, the dosages of sex-linked genes are equal in both females and males. Which of the two *X* chromosomes is inactivated is a random event occurring in embryonic cells. A female who is heterozygous for a sex-linked trait will express one allele in approximately one-half her cells and the alternate in the other cells.

Sex-Limited and Sex-Influenced Traits Sex-limited traits, such as beard growth in humans, are expressed exclusively in one sex, although the genes for them may be carried on autosomes, and thus be present in and transmitted by both sexes.

With *sex-influenced traits*, the penetrance or expressivity of autosomal genes is sex dependent, probably because of the different hormonal conditions of males and females. Baldness may be expressed in a man who is heterozygous for the "baldness" allele, whereas a woman will express baldness only if she is homozygous for the allele.

Chromosomal Alterations

Chromosomal mutations, caused by errors in meiosis or mutagens, can alter either the number of chromosomes in the cell or the structure of individual chromosomes.

Alterations of Chromosome Number In *nondisjunction*, a pair of homologous chromosomes does not separate properly in meiosis I, or sister chromatids do not separate in meiosis II, and a gamete may receive either two or none of that chromosome. A zygote formed with one of these aberrant gametes has a chromosomal alteration known as *aneuploidy*, a nontypical number of chromosomes. If one chromosome is present in triplicate, the organism is said to be *trisomic* for that chromosome, and the chromosome number is $2N + 1$. A $2N - 1$ chromosome number indicates that an individual is *monosomic* for a chromosome. Aneuploid organisms usually have a set of symptoms caused by the abnormal dosage of genes on the extra or missing chromosome.

Polyploidy is a chromosomal alteration in which an organism has more than two complete chromosome sets, as in *triploid* ($3N$) or *tetraploid* ($4N$) organisms. Polyploidy is common in the plant kingdom and has played an important role in the evolution of plants. In animals, complete polyploids are rare. More common are mosaic polyploids, in which patches of tetraploid tissue grow from a cell in which sister chromatids did not separate in mitosis.

Alterations of Chromosome Structure Breakage of chromosomes can lead to *deletion*, in which a fragment of the chromosome is lost; *duplication*, in which a broken fragment may join to the homologous chromosome; *translocation*, in which the fragment may join a nonhomologous chromosome; or *inversion*, in which the fragment may rejoin the original chromosome in the reverse orientation. Errors in crossing over can result in deletions and duplications caused by nonequal exchange of chromatids.

A homozygous deletion is usually lethal; most genes are necessary for an organism's existence. Duplications and translocations are typically harmful. Even though all the genes are present in proper quantities in inversions and translocations, the phenotype may be altered due to the *position effects* of neighboring genes on the expression of the relocated genes.

Chromosomal Alterations in Human Disease The frequency of aneuploid zygotes may be fairly high in humans, but development is usually so disrupted that the embryos spontaneously abort long before birth. Some genetic diseases, expressed as "syndromes" of characteristic traits, are the result of aneuploidy.

Down syndrome, affecting one out of every 700 children born, is the most common serious birth defect in the United States. Trisomy of chromosome 21 results in characteristic facial features, short stature, heart defects, and mental retardation. Meiotic nondisjunction seems to become more common in older women, and the incidence of trisomy 21 and other major chromosomal defects increases for older mothers.

Patau syndrome, caused by trisomy of chromosome 13, and Edwards syndrome, caused by a trisomy of chromosome 18, produce serious defects, and victims of both of these syndromes rarely live more than a year.

Most sex-chromosome aneuploidies upset the genetic balance less than do autosomal aneuploidies, perhaps because so few genes are located on the Y chromosome, and extra X chromosomes are inactivated as Barr bodies. XXY males will exhibit *Klinefelter syndrome*, a condition in which the individual has abnormally small testes, is sterile, may have feminine body contours, and is usually of normal intelligence. Additional X or Y chromosomes (XXXY, XXYY, etc.) usually result in individuals with Klinefelter syndrome who may be mentally retarded.

Males with a single extra Y chromosome may be somewhat taller than average males, but they do not exhibit any well-defined syndrome. Trisomy X results in *metafemales* with limited fertility and intelligence. Monosomy X individuals exhibit *Turner syndrome*, phenotypically female, sterile individuals with short stature, and usually normal intelligence.

In humans, the presence of a Y chromosome determines "maleness" and the absence of a Y chromosome determines "femaleness," regardless of the number of X chromosomes.

Structural alterations of chromosomes, such as deletions or translocations, are associated with specific human disorders and often with severe physical and mental defects. The cri du chat syndrome is caused by a deletion in chromosome 5; chronic myelogenous leukemia is a cancer that is associated with a chromosomal translocation; and some individuals with Down syndrome have an extra part of chromosome 21 attached to another chromosome.

SYSTEMS OF SEX DETERMINATION					
	X - Y	X - O	Z - W	HAPLO-DIPLOIDY	NO SEX CHROMOSOMES
FEMALE					
MALE					
EXAMPLES					

Mapping Human Chromosomes

Sex-linked traits and syndromes associated with observable chromosomal alterations have led to the mapping of some genes to particular chromosomes. New DNA technology is being used to map human genes on chromosomes.

Extranuclear Inheritance

Exceptions to Mendelian inheritance are found in the case of extranuclear genes located in cytoplasmic organelles, such as mitochondria and plant plastids, that are usually transmitted to offspring in the cytoplasm of the egg cell. In 1909, Correns observed the inheritance of variegated leaf coloration from the maternal plant. Maternal inheritance of mitochondrial genes in mammals is due to the large cytoplasmic contribution of the egg cell.

STRUCTURE YOUR KNOWLEDGE

1. Fill in the above table of the various chromosomal systems of sex determination.

2. Two of the genes Mendel used to support his law of independent assortment were actually located on the same chromosome. Explain why genes located more than 50 map units apart behave as though they are not linked. How can one determine whether these genes are linked and what the relative distance between them is?

3. You have found a new mutant phenotype in fruit flies that you suspect is recessive and sex-linked. What is the best cross you could make to confirm your predictions?

4. Fill in the following table concerning human diseases or syndromes related to chromosomal abnormalities. What explanation can you give for the

NAME OF DISEASE OR SYNDROME	CHROMOSOMAL ALTERATION INVOLVED	SYMPTOMS OR ASSOCIATED TRAITS
	Trisomy 21	
		Sterility, small testes, enlarged breasts or feminine body contours, normal intelligence or mental retardation
Turner Syndrome		
Metafemale	Trisomy X	
	Deletion in chromosome 5	Mental retardation, small head, unusual facial features, unusual cry

adverse phenotypic effects associated with these chromosomal alterations?

TEST YOUR KNOWLEDGE

MULTIPLE CHOICE: *Choose the one best answer.*

1. The chromosomal theory of inheritance states that
   a. genes are located on chromosomes.
   b. chromosomes and their associated genes undergo segregation during meiosis.
   c. chromosomes and their associated genes undergo independent assortment in gamete formation.
   d. all of the above are correct.

2. A wild type is
   a. the phenotype found most commonly in nature.
   b. the dominant allele.
   c. designated by a small letter.
   d. your basic party animal.

3. Sex-linked traits
   a. are carried on an autosome but expressed only in males.
   b. are coded for by genes located on a sex chromosome.
   c. are found in only one or the other sex, depending on the sex-determination system of the species.
   d. are always inherited from the mother in mammals and fruit flies.

4. Genetic and cytological maps for the same chromosome
   a. are both based on mutant phenotypes and recombination data.
   b. may have different sequences of genes.
   c. have both the same sequence of genes and intergenic distances.
   d. have the same sequence of genes but different intergenic distances.

5. Genetic recombination
   a. results in recombinant offspring.
   b. occurs in the fertilization process.
   c. occurs by independent assortment and crossing over in the formation of gametes.
   d. involves all of the above.

6. A 1:1:1:1 ratio of offspring from a dihybrid testcross indicates that
   a. the genes are linked.
   b. the genes are not linked.
   c. crossing over has occurred.
   d. the genes are 25 map units apart.

7. In grasshoppers, the heterogametic sex is
   a. females that produce gametes with either a Z or W chromosome.
   b. males that produce gametes with either an X or no sex chromosome.
   c. males that produce gametes with either an X or Y chromosome.
   d. neither sex, since sex depends on haploidy or diploidy.

8. Dioecious plants
   a. are usually polyploid.
   b. produce male and female flowers on separate plants.
   c. do not have a chromosomal basis for sex determination.
   d. all of the above are true.

9. A son inherits color-blindness from his
   a. mother.
   b. father.
   c. mother only if she is color-blind.
   d. father only if he is color-blind.

10. Baldness is a sex-influenced trait in that
    a. its expressivity varies between men and women.
    b. it is expressed exclusively in men even though its gene is carried on an autosome.
    c. it is inherited only from fathers.
    d. it is carried on the X chromosome and thus shows up more often in men.

11. Nondisjunction
    a. occurs when homologous chromosomes do not separate properly in meiosis I.
    b. occurs when sister chromatids do not separate in meiosis II.
    c. can result in both trisomic and monosomic offspring.
    d. may involve any of the above.

12. In translocation,
    a. a fragment of a chromosome joins to its homologue.
    b. a fragment of a chromosome joins to a nonhomologous chromosome.
    c. the breakage and refusion of a chromosome does not affect phenotype.
    d. the offspring shows aneuploidy.

13. A triploid individual
    a. is the result of a duplication.
    b. has a $2N + 1$ chromosome number.
    c. has a $3N$ chromosome number.
    d. results from cells in which chromosomes did not separate during mitosis.

14. Extra dosages of genes
  a. usually have deleterious effects.
  b. can be caused by duplications.
  c. are found in Down, Patau, and Edwards syndromes.
  d. all of the above

15. Sex-chromosome aneuploidies upset genetic balance less than do autosomal aneuploidies because
  a. extra X chromosomes are inactivated as Barr bodies.
  b. the Y chromosome only determines sex; it carries no genes.
  c. many more individuals with sex-chromosome abnormalities lead normal lives.
  d. all of the above

16. Which of the following is *not* true of maternal inheritance?
  a. It is the result of the larger contribution of egg cytoplasm.
  b. It gave evidence for the existence of cytoplasmic genes.
  c. Mitochondrial, ribosomal, and plant plastid genes come from the mother.
  d. It does not follow Mendelian principles of inheritance.

17. Which of the following chromosomal mutations does not alter genic balance but may alter phenotype because of position effects?
  a. deletion
  b. inversion
  c. duplication
  d. aneuploidy

18. According to the Lyon hypothesis,
  a. females are a genetic mosaic due to random nonseparation of chromatids during mitosis.
  b. males inherit their sex-linked traits from their mothers.
  c. equal dosages of X-linked genes for males and females occur because of the formation of Barr bodies.
  d. maternal inheritance is a result of the large amount of egg cytoplasm.

19. A sex-linked lethal recessive gene occurs in pigeons. What would be the sex ratio in the offspring of a cross between a male heterozygous for the lethal gene and a normal female?
  a. 2:1 male to female
  b. 1:2 male to female
  c. 1:1 male to female
  d. 4:3 male to female

20. The sex determination system of *Drosophila* resembles that of humans in that
  a. an *XXX* is a viable female in both species.
  b. an *XO* is a viable female in both species.
  c. sex is determined by the number of *X* chromosomes present.
  d. male sex is determined by the presence of the *Y* chromosome.

GENETICS PROBLEMS

1. Two normal color-sighted individuals produce the following children and grandchildren. Fill in the probable genotype of the indicated individuals in the pedigree at the bottom of the page. Solid symbols represent color-blindness. Choose an appropriate notation for the genotypes.

2. The following recombination frequencies were found. Determine the order of these genes on the chromosome.

a—c	10%	b—c	4%	c—d	20%
a—d	30%	b—d	16%		
a—e	6%	b—e	20%		

3. In guinea pigs, black (B) is dominant to brown (b), and solid color (S) is dominant to spotted (s). A heterozygous black, solid-colored pig is mated with a brown, spotted pig. The total offspring for several litters are

black solid	16
black spotted	5
brown solid	5
brown spotted	14

GENOTYPES

1 _____   5 _____

2 _____   6 _____

3 _____   7 _____

4 _____

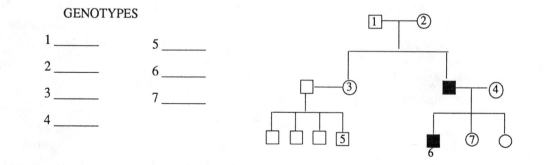

Are these genes linked or nonlinked? If they are linked, how many map units are they apart?

4. A woman is a carrier for a sex-linked lethal gene that causes spontaneous abortions. She has nine children. How many of these children do you expect to be boys?

5. A recessive sex-linked gene in fruit flies produces vermillion eye color (vv females and vY males have vermillion eyes). An autosomal recessive produces brown eye color (bwbw). Flies that are homozygous recessive for both vermillion and brown genes have white eyes. Determine the outcome of a cross between a wild type female (v⁺v⁺ bw⁺bw⁺) and a white-eyed male (vYbwbw). What would be the F₁ of a reciprocal cross between a white-eyed female and a red-eyed male? Determine the phenotypic ratio of the F₂ for the first of these crosses.

6. A dominant sex-linked gene B produces white bars on black chickens, as seen in the Barred Plymouth Rock breed. A clutch of chicks has equal numbers of black and barred chicks. If only the females are found to be black, what were the genotypes of the parents? If males and females are evenly represented in the black and barred chicks, what were the genotypes of the parents? (Remember how sex is determined in birds.)

CHAPTER 15

THE MOLECULAR BASIS OF INHERITANCE

FRAMEWORK

This chapter outlines the key evidence that was gathered to establish DNA as the molecular basis of inheritance. Watson and Crick's double helix, with its rungs of specifically paired nitrogenous bases and twisting side ropes of phosphate and sugar groups, provided the three-dimensional model of DNA that explained DNA's ability to encode a great variety of information and produce exact copies of itself through semiconservative replication. The replication of DNA is an extremely fast and accurate process involving many enzymes and proteins, including: proteins that bind to origins of replication; helicases, topoisomerases, and single-strand binding proteins that unwind and unkink DNA and keep it apart; primases that form the RNA primer needed to begin replication; DNA polymerases and DNA ligase that direct the addition of nucleotides and link fragments of DNA together; and various enzymes that proofread and repair the DNA molecule.

CHAPTER SUMMARY

Deoxyribonucleic acid, DNA, is the genetic material, the substance of genes, the basis of heredity. Nucleic acids' unique ability to direct their own replication allows for the precise copying and transmission of DNA to all the cells in the body and from one generation to the next. DNA encodes the blueprints for the proteins and enzymes that direct and control the developmental, biochemical, anatomical, physiological, and behavioral traits of organisms.

The Search for the Genetic Material

The role of DNA in heredity was first established through work with microorganisms—bacteria and viruses. By the 1940s, chromosomes were known to carry hereditary information and to consist of proteins and DNA. Most scientists believed that the proteins carried the genetic program because of the known specificity and heterogeneity of proteins. The simple, repetitious nature of nucleic acids did not seem capable of coding the wealth of information needed for heredity.

Evidence That DNA Can Transform Bacteria The work of Griffith in 1928 provided the first evidence that the genetic material was some sort of heat-stable chemical. Griffith worked with two strains of *Streptococcus pneumoniae*—a smooth strain (S) that synthesized a polysaccharide coat or capsule and a rough strain (R) that did not form a capsule. Only live S strain cells injected in mice caused pneumonia; mice injected with heat-killed S cells or R cells did not develop pneumonia. However, when Griffith injected a mixture of heat-killed S cells and live R cells, the mice died. Griffith was able to isolate live S cells from the blood of these mice, even though only dead S cells had been injected. R cells had somehow acquired the ability to make polysaccharide coats from dead S cells. This transfer of external genetic material is called *transformation*.

Avery worked for a decade to identify this transforming agent by purifying chemicals from heat-killed S cells. In 1944, he and his colleagues McCarty and MacLeod announced that DNA was the molecule that transferred the genetic information. Their claim, however, was met with scepticism.

Evidence That Viral DNA Can Program Cells Viruses consist of little more than DNA, or sometimes RNA, contained in a protein coat. They can reproduce only by infecting another cell and commandeering that cell's metabolic machinery. *Bacteriophages*, or phages for short, are viruses that infect bacteria. In 1952, Hershey and Chase showed that DNA was the genetic material of a phage known as T2 that infects the bacterium *Escherichia coli* (*E. coli*).

Hershey and Chase devised an experiment using radioactive isotopes to tag the phage's DNA and proteins to determine which was transferred to the bacteria. One batch of T2 was grown with radioactive sulfur that became incorporated into protein; another was grown with radioactive phosphorus that labelled the DNA.

The labelled T2 cells were allowed to infect separate samples of *E. coli*. The cultures were blended to shake off parts of the phages that remained outside the bacteria cells and then centrifuged to separate the heavier bacterial cells from the lighter viral particles. In the samples with the labeled proteins, the radioactivity was found in the supernatant, indicating that the phage protein did not enter the bacterial cells. In the samples with the labeled DNA in the T2 phage, most of the radioactivity was contained in the bacterial cell fraction. When these *E. coli* cells were returned to culture, they lysed and released phages.

Additional Evidence That DNA Is the Genetic Material of Cells Circumstantial evidence that DNA was the genetic material came from the observations that, prior to mitosis, a eukaryotic cell doubles its DNA content and that diploid sets of chromosomes have twice as much DNA as haploid sets.

Chargaff, in 1947, reported that DNA composition is species-specific. Using paper chromatography to separate bases, he found that each species he studied had a different ratio of nitrogenous bases. Chargaff also determined that the number of adenines and thymines were equal and the number of guanines and cytosines were equal in the DNA from all the organisms he studied. The A=T and G=C property of DNA, known as *Chargaff's Rules*, was not explained until the double helix was discovered.

Discovery of the Double Helix

By the early 1950s, the arrangement of covalent bonds in a nucleic acid polymer was established, and the race was on to determine the three-dimensional structure of DNA. Linus Pauling in California and Maurice Wilkins and Rosalind Franklin in London were working on the problem, but James D. Watson and Francis Crick were the first to solve the DNA puzzle.

Crick was studying protein structure using a technique called *X-ray crystallography*. An X-ray beam passed through a crystal can expose photographic film to produce a pattern of spots that a crystallographer can interpret into information about the three-dimensional atomic structure of the crystal. Watson saw an X-ray photo produced by Franklin that clearly showed the basic shape of DNA to be a helix. He and Crick deduced that the helix had a uniform width of 2 nanometers with its purine and pyrimidine bases stacked 0.34 nanometers apart. This width suggested that the helix consisted of two strands, thus the term *double helix*.

Watson and Crick developed models of wire to build a double helix that would conform to the X-ray measurements and the known chemistry of DNA. They finally arrived at a model that paired the nitrogenous bases on the inside of the helix with the sugar–phosphate chains on the outside. The helix makes one full turn every 3.4 nanometers; thus ten layers of nucleotide pairs, stacked 0.34 nanometers apart, are present in each turn of the helix. The sugar–phosphate chains are oriented in opposite directions, said to be *antiparallel*.

The idea that there was specific base pairing meant that one strand of the double helix would complement the information contained on the other strand. At first Watson assumed that bases paired with themselves. To produce the molecule's 2 nanometer width, however, a purine must pair with a pyrimidine. The molecular arrangement of the side groups of the bases permit two hydrogen bonds to form between adenine and thymine, and three hydrogen bonds between guanine and cytosine. This complementary pairing explains Chargaff's rules (A=T and G=C). The cumulative effect of a multitude of these weak interactions, coupled with the van der Waals forces between the stacked bases, contributes to the stability of the double helix.

The sequence of nucleotides along the length of a DNA strand is not controlled by any pairing rules. The infinite variety of sequences possible would support the encoding of unique and detailed genetic information.

In April 1953, Watson and Crick published a paper in *Nature* reporting the double helix as the molecular model for DNA.

DNA Replication

The Template Concept A template theory had already been advanced for the explanation of how genetic material could be precisely copied and passed on to the next generation. The specific base pairing of the Watson and Crick model immediately suggested a template mechanism of DNA replication.

The Semiconservative Nature of DNA Replication In a second paper, Watson and Crick suggested that DNA replication was accomplished by the separation of the two DNA strands and the creation of two new complementary strands as nucleotides paired up along the two exposed strands. This *semiconservative model* predicted that the two daughter DNA molecules would each have one old strand from the parent DNA and one newly formed strand. A conservative model would mean that the parent strand would remain intact and the duplicate molecule would be totally new.

Meselson and Stahl provided the evidence for the semiconservative model. They grew *E. coli* in a medium with ^{15}N, a heavy isotope that the bacteria incorporated into their nitrogenous bases. When ultracentrifuged in a cesium chloride density gradient, this DNA could be separated by density from nonlabeled DNA. Cells with labeled DNA were transferred to a medium with the lighter isotope, ^{14}N. After one generation of bacterial growth, the DNA extracted from the culture was all of intermediate density; it was a hybrid of the parental heavier DNA strand and the newly formed lighter DNA strand.

Close-up on Replication   DNA replication is extremely rapid and accurate. More than a dozen enzymes and other proteins are involved in this intricate process.

Replication seems to begin at special sites, called *origins of replication*, where specific proteins that initiate replication bind. Replication spreads in both directions from these origins. Bacterial and viral DNA molcules have only one replication origin, but the large eukaryotic DNA molecules may have hundreds to thousands of origin sites. The points along the DNA molecule where replication is occurring are called *replication forks* because of their helix shape.

Enzymes called *helicases* unwind the helix and *single-strand binding proteins* help keep the separated strands apart. The kinks formed from separating the strands of the double helix are untangled by enzymes called *topoisomerases* that temporarily break or nick one strand so that the molecule can rotate freely to "unkink."

A *primer* of RNA is needed to initiate DNA replication. An enzyme called *primase* pairs about five RNA nucleotides to a short portion of the DNA strand that serves as the primer template. Nucleotides then line up along the exposed DNA strands, held by hydrogen bonds between complementary base pairs. Enzymes called *DNA polymerases* catalyze the synthesis of a new DNA strand by connecting the nucleotides. The energy for creating the covalent bonds between nucleotides is provided by the release of a pyrophosphate group from each incoming nucleotide.

DNA is replicated in a $5' \rightarrow 3'$ direction. The 5' end of the strand is the one with a phosphate group attached to the number five carbon of the terminal sugar. At the 3' end, a free hydroxyl group is attached to the number

three carbon of the end sugar. New nucleotides are added to the 3' end.

Because the DNA strands run in an antiparallel direction, the simultaneous synthesis of both strands presents a problem. The *leading strand* is the new $5' \rightarrow 3'$ strand being polymerized as a single polymer. The *lagging strand* is created as a series of short segments, called *Okazaki fragments*, that are formed in the $5' \rightarrow 3'$ direction as the DNA molecule unzips. Each fragment requires an RNA primer. A continuous strand of DNA is produced after a DNA polymerase removes the RNA primer and replaces it with DNA, and a linking enzyme called *DNA ligase* joins the 3' end of each new fragment to the 5' end of the growing chain.

Initial pairing errors in nucleotide placement may occur as often as 1 per 10,000 bases. The amazing accuracy of DNA replication is achieved by proofreading and correction of pairing errors. In bacteria, DNA polymerase checks each nucleotide against its template after insertion and backs up and replaces incorrect nucleotides. It is not known whether DNA polymerase or other enzymes perform this proofreading function in eukaryotic cells.

DNA Repair

DNA molecules may be accidentally altered by the action of reactive chemicals, radioactive emissions, X-rays and ultraviolet (UV) light. These changes, or mutations, may be corrected through the action of many types of DNA repair enzymes. In *excision repair*, the damaged strand is cut out by a repair enzyme and the gap is correctly filled through the action of a DNA polymerase and DNA ligase. This type of repair is particularly important in skin cells to repair the DNA damage (such as thymine dimers) induced by ultraviolet rays of sunlight.

Alternative Forms of DNA

The elegantly simple architecture of the double helix reaffirms the fundamental structure-fits-function theme of biology.

Recent evidence indicates that not all DNA corresponds to the double helix model. DNA molecules may be linear or circular, supercoiled, or occasionally single-stranded. Even linear DNA may show variations in its helical structure, with different three-dimensional forms known as A-DNA, C-DNA, and Z-DNA. Z-DNA twists in the left-hand direction, opposite that of the most common B-DNA represented by the Watson-Crick double helix. The biological functions of these non-B forms of DNA are being investigated.

STRUCTURE YOUR KNOWLEDGE

1. Fill in the chart below about these key investigators and the evidence they provided as to the structure and function of DNA.

2. Create a concept map to illustrate the key participants and events involved in DNA replication.

TEST YOUR KNOWLEDGE

MULTIPLE CHOICE: *Choose the one best answer.*

1. One of the reasons most scientists believed proteins were the carriers of genetic information was that
   a. proteins were more heat stable than nucleic acids.
   b. DNA was not always double stranded.
   c. proteins were much more complex molecules than nucleic acids.
   d. early experimental evidence pointed to proteins as the hereditary material.

2. Transformation involves
   a. the transfer of genetic material, often from one bacterial strain to another.
   b. the creation of a strand of RNA from a DNA molecule.
   c. the infection of bacterial cells by bacteriophages.
   d. the type of replication shown by DNA.

3. According to Chargaff's Rules,
   a. heavier labelled fractions will be found in the pellet of a centrifuge tube.
   b. DNA replication must be semiconservative.
   c. A=T and G=C.
   d. each species has a different ratio of nitrogenous bases.

4. For his work with DNA, Chargaff used
   a. X-ray crystallography.
   b. paper chromatography.
   c. heavy isotopes of sulfur and phosphorus.
   d. ultracentrifugation and ^{15}N.

INVESTIGATOR	ORGANISMS USED	TECHNIQUES, EXPERIMENTS	CONCLUSIONS
Griffith			
Avery			
Hershey & Chase			
Chargaff			
Franklin			
Watson & Crick			
Meselson & Stahl			

5. In his work with pneumonia-causing bacteria and mice, Griffith found that
   a. DNA was the transforming agent.
   b. the R and S strains mated.
   c. heat-killed S cells could cause pneumonia when mixed with heat-killed R cells.
   d. that some heat-stable chemical was transferred to R cells to transform them into S cells.

6. When T2 phages are grown with radioactive sulfur,
   a. their DNA is tagged.
   b. their proteins are tagged.
   c. their DNA is found to be of medium density in a centrifuge tube.
   d. they transfer their radioactivity to *E. coli* when they infect them.

7. Meselson and Stahl
   a. provided evidence for the semiconservative replication of DNA.
   b. were able to separate phage protein coats from *E. coli* by using a blender.
   c. found that DNA labelled with ^{15}N was of intermediate density.
   d. grew *E. coli* on labelled phosphorus and sulfur.

8. Watson and Crick concluded that each base could not pair with itself because
   a. there would not be room for the helix to make a full turn every 3.4 nanometers.
   b. the width of 2 nm would not permit two purines to pair together.
   c. the bases could not be stacked 0.34 nm apart.
   d. identical bases could not hydrogen bond together.

Use these centrifuge tubes showing density bands of DNA to answer questions 9 and 10.

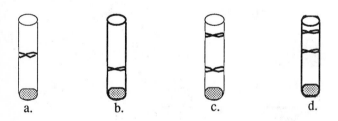

9. *E. coli* grown on ^{15}N medium are transferred to ^{14}N medium for one generation of growth. DNA extracted from these cells is mixed with cesium chloride and placed in an ultracentrifuge. What density distribution of DNA would you predict for this experiment from the possible DNA density bands illustrated above?

   a.
   b.
   c.
   d.

10. In an experiment similar to that in question 9, the cells are allowed to grow on the ^{14}N medium for another generation. What density distribution would you predict for this experiment from those illustrated above?
    a.
    b.
    c.
    d.

11. The joining of nucleotides in the polymerization of DNA requires
    a. the hydrolysis of GTP.
    b. the hydrolysis of ATP.
    c. the loss of a pyrophosphate from incoming nucleoside triphosphates.
    d. hydrogen bonding.

12. The continuous elongation of DNA along one strand of DNA
    a. occurs on the leading strand.
    b. occurs because DNA ligase can only elongate in the 5' → 3' direction.
    c. does not require an RNA primer.
    d. all of the above

13. Topoisomerases
    a. unwind DNA strands prior to replication.
    b. unkink single-stranded DNA by nicking and resealing the polymer.
    c. provide excision repair of damaged DNA.
    d. create the RNA primer to initiate replication.

14. DNA ligase is needed to
    a. join Okazaki fragments on the leading strand.
    b. replace the RNA primer.
    c. join the 3' end of DNA repaired segments to the 5' end of the gap.
    d. do all of the above.

15. Which of the following statements about DNA polymerase is incorrect?
    a. It is found only in eukaryotes.
    b. It is able to proofread and correct for errors in its base pairing.
    c. It is unable to join free nucleotides unless an RNA primer is present.
    d. It only works in the 5' → 3' direction.

16. Thymine dimers, covalent links between adjacent thymine bases in DNA, may be induced by UV light. When they occur, they are repaired by
    a. excision enzymes that cut out the damaged region.

b. DNA polymerase.
c. igase.
d. all of the above.

17. Okazaki fragments are produced by
a. RNA primase.
b. DNA polymerase working in 5'→ 3' direction.
c. excision enzymes.
d. both a and b.

18. Not all DNA conforms to the double helix model because
a. some DNA may be single-stranded.
b. some DNA may be supercoiled.
c. Z-DNA twists in the left-hand direction.
d. all of the above

Use this diagram of replicating DNA to answer questions 19 through 20. Each letter is used only once.

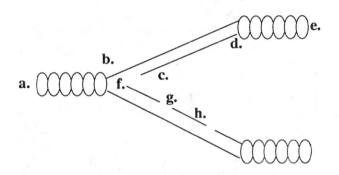

19. A helicase would be found at letter _____.

20. An RNA primase would be found at letter _____.

21. A Okazaki fragment is indicated by letter _____.

22. A ligase would be found at letter _____.

23. On the leading strand, DNA polymerase would be moving from letter _____ to _____.

24. The parental strand of DNA is indicated by letter

_____.

25. One of the daughter strands is indicated by letter

_____.

Use the following diagram to answer questions 26 through 28.

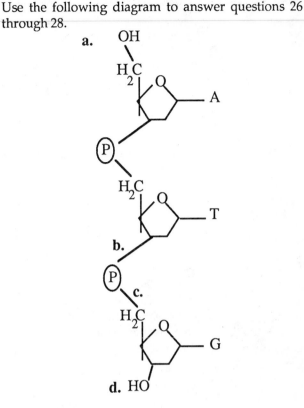

26. Which letter indicates the 5' end of this single DNA strand?
a.
b.
c.
d.

27. Which letter indicates a phosphodiester bond formed by DNA polymerase?
a.
b.
c.
d.

28. The base sequence of the DNA strand made from this template would be (from top to bottom)
a. A T C
b. C G A
c. T A C
d. U A C

CHAPTER 16

FROM GENE TO PROTEIN

FRAMEWORK

This chapter deals with the central dogma of molecular biology, the sequence of DNA to RNA to proteins. The outline at the bottom of the page lists some important concepts involved in each step of the pathway from gene to protein.

CHAPTER SUMMARY

The sequence of DNA → RNA→ protein has been called the central dogma of molecular biology. The DNA inherited by an organism directs the activities of each cell by controlling the synthesis of enzymes and other proteins through instructions coded into RNA.

How Genes Control Metabolism

In 1909, British physician Garrod first suggested that genes determine phenotype through the action of enzymes that control chemical processes in the cell. Garrod reasoned that inherited diseases were attributable to an inability to make certain enzymes.

Evidence That Genes Specify Enzymes In the 1930s, Beadle and Ephrussi speculated that the mutations causing the various eye colors in *Drosophila* were a result of a nonfunctioning enzyme at some point in the metabolic pathway leading to pigment formation. A short time later, Beadle and Tatum, working with mutants of the orange bread mold, *Neurospora crassa*, demonstrated the relationship between genes and enzymes.

Neurospora is a useful organism for genetic studies. In its haploid form, its phenotype is a direct expression of its genotype—the effects of a gene that is present in a mutant form can be readily observed. The diploid zygote formed from sexual mating undergoes meiosis to produce haploid spores. These ascospores are contained within a thin-walled sac called an ascus from which they can be removed and individually grown on artificial growth medium.

Beadle and Tatum studied several nutritional mutants, called *auxotrophs*, that could not grow on the *minimal medium* of agar mixed with inorganic salts, sucrose, and biotin used by wild-type *Neurospora*. By growing a spore from irradiated fungi on *complete growth medium* and then transferring bits of this fungus to various combinations of minimal medium and added nutrients, Beadle and Tatum were able to identify the specific metabolic pathway for which the auxotroph was defective. Three classes of *Neurospora* mutants that were unable to synthesize arginine were tested by providing them with the different precursors of the pathway. Beadle and Tatum reasoned that each class was blocked at a different enzymatic step. Assuming that each mutant was defective in a single gene and

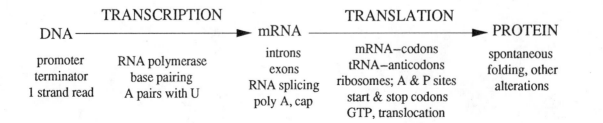

that its metabolic pathway was blocked by lack of a specific enzyme, Beadle and Tatum formulated the *one gene—one enzyme hypothesis*: the function of a gene is to control the production of a specific enzyme.

One Gene–One Polypeptide Molecular biologists revised this hypothesis to one gene–one protein, because, with a very few exceptions, all enzymes are proteins, but not all proteins are enzymes. Many proteins consist of more than one polypeptide chain, so the axiom has been further revised to the one gene–one polypeptide hypothesis.

The Languages of Macromolecules

Nucleic acids and proteins are similar in that they are constructed of many monomers linked together in specific linear sequences. DNA and RNA are long sequences of only four different monomers: four nucleotides differing in their nitrogenous bases. The sequence of the 20 different amino acid monomers that may be linked together to form a polypeptide chain is coded for by the base sequence of DNA.

Transcription is the transfer of information from DNA to RNA—the same "language" of nucleic acids is used. *Translation* is the term for the transfer of information from RNA to a polypeptide—the language changes from nucleotides to amino acids.

The translation of nucleotides into amino acids involves a sequence of three nucleotides, called a *codon*, to specify each amino acid. Since there are only four possible bases in a nucleotide sequence, a two-base code would provide only 16 (4^2) unique arrangements. A triplet code provides 64 (4^3) possible codons, more than enough to specify the 20 amino acids.

Transcription, the Synthesis of RNA

The genetic information in DNA is transcribed by the synthesis of *messenger RNA (mRNA)* along a DNA template. As in DNA replication, the double helix unwinds and separates, and the RNA molecule grows along one of the DNA strands in the 5' → 3' direction. The nucleotides of RNA take their places along the DNA template by forming hydrogen bonds with the nucleotides there, following the same base-pairing rules except that uracil, rather than thymine, pairs with adenine. *RNA polymerase* links the RNA nucleotides together.

RNA polymerase attaches to a *promoter* sequence and starts transcribing; a *terminator* sequence signals the polymerase to stop synthesis and release the RNA molecule. As the RNA is made, it peels away from the DNA template, and the hydrogen bonds between DNA strands reform. For any gene, only one strand of DNA is signaled to be read by a promoter region. Different genes, however, may be read from different strands of a DNA molecule.

Several RNA molecules may be transcribed simultaneously from a single gene through the action of multiple molecules of RNA polymerase. The mRNA of eukaryotes must be modified before it crosses the nuclear membrane to serve for protein synthesis in the cytoplasm.

Translation, the Synthesis of Protein

Transfer RNA Transfer RNA (tRNA) pairs the appropriate amino acid with its codon on mRNA. These single-stranded, short RNA molecules are arranged into a clover-leaf shape by four regions of hydrogen bonding between complementary base sequences and then folded into a three-dimensional, roughly L-shaped structure. The loop protruding from one end of the L holds the specialized base triplet called the *anticodon*, which is complementary to and base pairs with a particular codon on mRNA. The 3' end on the tRNA molecule on the other end of the L is the site of attachment of its amino acid.

Amino Acid Activating Enzymes *Amino acid activating enzymes*, or *aminoacyl-tRNA synthetases*, have specific active sites that bind one type of amino acid with its appropriate tRNA molecule. Each enzyme first binds an amino acid to its active site along with an ATP molecule that loses a pyrophosphate and joins to the amino acid as AMP. The appropriate tRNA then displaces the AMP and covalently bonds to the amino acid. This activated amino acid–tRNA complex can furnish its amino acid to the growing polypeptide chain according to the sequence of codons on mRNA.

Ribosomes Ribosomes, consisting of two subunits composed of proteins and a specialized form of RNA, called *ribosomal RNA (rRNA)*, have a binding site for mRNA, a *P site* that holds the tRNA carrying the growing polypeptide chain, and an *A site* that binds to the tRNA carrying the next amino acid. The transfer of this amino acid from its tRNA to the carboxyl end of the growing polypeptide chain is catalyzed by the ribosome.

The Process of Protein Synthesis The three stages of protein synthesis—chain initiation, chain elongation, and chain termination—all require enzymes, and the first two stages need phosphate-bond energy provided by GTP.

Initiation requires several proteins, called *initiation factors*—GTP, the mRNA to be read, the two subunits of a ribosome, and the first amino acid attached to its tRNA. Translation must begin with the correct nucleotide so that the grouping of bases into codons, called the *reading frame*, produces the proper sequence of amino acids.

The mRNA and an initiator tRNA, usually carrying methionine, are bound to the small subunit of the ribosome. The tRNA is bound to the start codon, usually AUG, on the mRNA. It has been shown in prokaryotes that the nucleotides to the 5' side of the start codon on mRNA constitute a recognition signal and base-pairing region for rRNA in the ribosome. Next, the large subunit of the ribosome attaches to the small one. The initiator tRNA fits into the P site of the now functional ribosome.

The addition of amino acids, involving the use of several proteins called *elongation factors*, occurs in a three-step cycle: (1) with the hydrolysis of a high-energy phosphate bond of GTP, the mRNA codon in the A site forms hydrogen bonds with the appropriate anticodon of a tRNA carrying its amino acid; (2) an enzyme called *peptidyl transferase* catalyzes the formation of a peptide bond between the polypeptide held in the P site and the amino acid in the A site; the polypeptide is released from the tRNA that was holding it and is now held by the tRNA of the amino acid in the A site; and (3) the tRNA, carrying the growing polypeptide, is *translocated* to the P site. This process requires the hydrolysis of a GTP molecule and makes the next mRNA codon available in the A site. Translation and translocation occur in the 5' → 3' direction.

Termination occurs when a *termination codon*—UAA, UAG, or UGA—reaches the A site of the ribosome and binds with a protein called *release factor*. Peptidyl transferase attaches a water molecule to the polypeptide chain, freeing the completed polypeptide from the tRNA in the P site. The ribosome then separates into its small and large subunits.

Creation of a Functional Protein　During and following translation, a polypeptide undergoes several changes in preparation for its function in the cell. One or more amino acids at the beginning, amino end of the chain may be enzymatically removed; whole segments of the polypeptide may be excised; or the chain may be cleaved into several pieces. Disulfide bridges between cysteine molecules may form as the chain spontaneously folds and coils. Several polypeptides may associate into a quaternary structure.

Sites of Protein Synthesis　In prokaryotic cells, which lack a nucleus, ribosomes may begin translation even before transcription is complete. In eukaryotic cells, proteins destined to become part of membranes or to be exported from the cell are produced by ribosomes bound to the ER. Proteins that are to function in the cytoplasm are usually produced by free ribosomes. Usually several ribosomes are reading a mRNA at one time, creating a cluster known as a polyribosome.

The Genetic Code

In the early 1960s the genetic code was "cracked" by a series of experiments that determined the amino acid translations of each of the codons of nucleic acids. Nirenberg synthesized artificial mRNA by linking together uracil RNA nucleotides. Adding this "poly-U" to an in-vitro system of a test tube containing all the biochemical ingredients necessary for protein synthesis, Nirenberg obtained a polypeptide containing the single amino acid phenylalanine. The codons AAA, GGG, and CCC were deciphered the same way.

Using more elaborate techniques to decode the triplets with mixed bases, scientists had deciphered all 64 triplets by the mid-1960s. Sixty-one of the triplets code for amino acids. The three remaining codons function as stop signals. AUG both codes for methionine and functions as the start signal for translation.

The code is often redundant; more than one codon may specify a single amino acid. Often this redundancy, called "degeneracy," occurs as differences in the third base of the triplet. The code is never ambiguous; no codon specifies two different amino acids.

In general, a particular nucleotide sequence on DNA is read in only one reading frame, starting at a start triplet and reading each triplet sequentially.

The Wobble Phenomenon　Sixty-four codons can be read from mRNA, but there are only about 40 different tRNA molecules. In a phenomenon known as *wobble*, the third nucleotide (5' end) of some tRNA molecules can form hydrogen bonds with more than one kind of base in the codon. In several tRNA, the unusual base inosine (I) is found in the third position and can pair with U, C, or A. Thus one tRNA can recognize three mRNA codons, all of which code for the same amino acid carried by that tRNA.

Universality of the Genetic Code　The genetic code of codons and their corresponding amino acids is universal for almost all organisms. A bacterial cell can translate the genetic messages of human cells and vice versa. This universality lends compelling evidence to the evolutionary connection of all living organisms.

Recent exceptions to the constancy of the genetic code have been found in several single-celled ciliates. The genetic code found in mitochondrial DNA varies with different organisms.

Split Genes and RNA Processing in Eukaryotes

A surprising and recent development in molecular biology involves split genes, in which much of the DNA does not code for amino acid sequences of proteins. Long segments of noncoding sequences of bases, known as *introns* or intervening sequences, have been found within the boundaries of eukaryotic genes. The remaining coding regions are called *exons*, since they are the regions that are expressed. An entire mRNA transcript is made of the gene, and then the introns are removed and the exons joined before the RNA leaves the nucleus. This process, known as *RNA splicing* or *RNA processing*, also occurs in the production of tRNA and mRNA.

Signals for RNA splicing are sets of a few nucleotides at either end of each intron. *Small nuclear ribonucleoproteins*, or *snRNPs*, composed of proteins and *small nuclear RNA (snRNA)*, are nuclear particles that are involved in RNA splicing. Several kinds of snRNPs are components of a molecular complex called a *spliceosome*, in which the ends of an intron are brought close together and the intron is snipped out of the RNA transcript, and the now adjacent exons are joined. Several other schemes of RNA processing have been identified, and some splicing occurs without proteins or extra RNA—the intron RNA catalyzes the process itself.

Before leaving the nucleus, mRNA is modified by the addition of a *cap* of modified guanosine triphosphate at the 5' end and a string of adenine nucleotides, called *poly A*, at the 3' end. These additions attach to untranslated leader and trailer sequences, respectively, at the ends of the mRNA molecule. The cap enhances translation, and both cap and poly A apparently protect the ends of the mRNA.

Several hypotheses concern the functions of introns. One is that they are somehow involved in regulation of gene activity or in the flow of mRNA into the cytoplasm. Another hypothesis is that they facilitate recombination between exons to create a diversity of proteins. Exons may generally code for polypeptide *domains*, functional segments of a protein, such as binding sites. The crossing over between homologous chromosomes of domains within a gene, coupled with mutational changes, can give rise to new genes.

Almost all mammalian genes are split, and introns are often longer than exons. Simpler organisms have genes containing fewer and shorter introns. The evolutionary significance of these differences remains an interesting question.

RNA: A Review

RNA differs from DNA in that it is single-stranded rather than double-stranded, has the sugar ribose rather than deoxyribose, and has the nitrogenous base uracil in place of thymine. RNA comes in three types in both prokaryotic and eukaryotic cells: mRNA, tRNA, and rRNA. Eukaryotic cells also contain snRNA and probably other types of small RNA molecules. Three-dimensional shapes, created by base-pairing between regions of the RNA chain, are important to the functions of tRNA, rRNA, and snRNA. Indeed, the functions of the various types of RNA are more than just protein synthesis; RNA has been found to act as an enzyme (both with and without the help of proteins) and even as a template for DNA synthesis in reverse transcription.

Mutations and Their Effects on Proteins

Mutations, changes in the nucleotide sequence of DNA, may involve large portions of a chromosome or just a single nucleotide pair, as in a *point mutation*. These changes may be reflected in the proteins translated from a gene.

Types of Mutations Mutations within a gene are of two general types: base-pair substitutions and base-pair insertions or deletions. *Base-pair substitutions* involve the replacement of one nucleotide and its complementary partner with another pair of nucleotides.

Due to the redundancy of the genetic code, some base-pair substitutions have no effect on the translation of the gene. An exchange in the third nucleotide of a codon may still result in the insertion of the same amino acid.

A substitution may result in the insertion of a different amino acid without altering the character of the protein, if the new amino acid is similar in properties or is located in an area of the protein not crucial to that protein's function.

A base-pair substitution that results in the insertion of a different amino acid in a critical portion of a protein, such as the active site of an enzyme, may significantly affect protein function. Occasionally such a mutation results in an improved protein, but much more frequently the mutation is harmful to the cell.

An incorrectly coded amino acid is called a *missense mutation*. When the point mutation changes a codon for an amino acid into a stop codon, the translation of the polypeptide chain is prematurely halted. Such *non-*

sense mutations almost always lead to nonfunctional proteins.

Base-pair *insertions* or *deletions* that are not in multiples of three nucleotides alter the reading frame, and all nucleotides downstream from the mutation will be improperly grouped into codons, creating extensive missense and ending usually in nonsense—premature termination. These *frameshift mutations* almost always produce nonfunctional proteins.

Conditional Mutations A *conditional mutation* is harmful or fatal under some conditions but not others. *Temperature-sensitive mutations* involve an amino acid substitution that is not harmful at a permissive temperature but changes the stability of the protein at a different, usually higher, temperature. Geneticists make use of temperature-sensitive mutations to grow mutants at permissive temperatures and then study the effects of their altered enzymes at nonpermissive temperatures.

Mutagenesis *Mutagenesis*, the generation of mutations, may occur in a number of ways. *Spontaneous mutations* include base-pair substitutions, insertions, or deletions that may occur during DNA replication or repair. Physical and chemical agents called *mutagens* can cause mutations in DNA. X-rays can cause breaks in DNA and UV radiation creates pyrimidine dimers. Chemical mutagens include *base analogues* that substitute for normal bases in DNA synthesis and result in mispairing and base-pair substitutions. Reactive and intercalating chemicals are other types of mutagens. Regions of DNA that tend to move from one section of DNA to another are called *transposons* and may act as mutagens by disrupting DNA sequences.

Redefining the Gene

The definition of a gene has evolved from Mendel's heritable factors, to Morgan's loci along chromosomes, to Beadle and Tatum's one gene–one polypeptide theory. Research continually refines our understanding of the structural and functional aspects of genes, which now include introns and some overlapping genes. The best working definition of a gene is a sequence of nucleotides with a specific functional product.

STRUCTURE YOUR KNOWLEDGE

1. Describe transcription and translation, including the molecules involved, enzymes needed, and

products made. Make a chart of these processes to help you organize your knowledge. How are these two processes similar? How are they different?

2. What is the genetic code? Explain redundancy and the wobble hypothesis. What is meant by saying that the genetic code is almost universal?

3. Prepare a concept map dealing with the topic of mutation.

TEST YOUR KNOWLEDGE

MULTIPLE CHOICE: *Choose the one best answer.*

1. Beatle and Tatum's study of *Neurospora* showed that
   a. auxotrophs could not grow on minimal medium.
   b. mutants had defective enzymes in their metabolic pathways.
   c. the occurrence of different mutants with defective metabolic pathways supported a one gene–one enzyme hypothesis.
   d. all of the above.

2. Transcription involves
   a. the transfer of information from DNA to RNA.
   b. the transfer of information from RNA to DNA.
   c. the transfer of information from mRNA to an amino acid sequence.
   d. the transfer of information from DNA to an amino acid sequence.

3. Which of the following is *not* true of an anticodon?
   a. It consists of three nucleotides.
   b. It is the basic unit of the genetic code.
   c. It extends from one end of a tRNA molecule.
   d. It may pair with more than one codon, especially if it has the base inosine in its third position.

4. RNA polymerase
   a. is the protein responsible for the production of ribonucleotides.
   b. is the enzyme that creates hydrogen bonds between nucleotides on the DNA strand and their complementary RNA nucleotides.
   c. only transcribes exons.
   d. begins transcription at a promoter sequence and moves along one of the DNA strands in a 5" → 3" direction.

5. Transfer RNA

a. is the nucleic acid that forms the small subunit of the ribosome.

b. binds to its specific amino acid by replacing AMP in the active site of an aminoacyl–tRNA synthetase.

c. uses GTP as the energy source to bind its amino acid.

d. forms hydrogen bonds with the anticodon in the A site of a ribosome.

6. Translocation involves
   a. the hydrolysis of a GTP molecule.
   b. the movement of the tRNA in the A site to the P site.
   c. the movement of mRNA over one triplet in the A site.
   d. all of the above.

7. Changes in a polypeptide following translation may involve
   a. the formation of disulfide bridges between cysteine molecules.
   b. the action of enzymes to add amino acids at the beginning of the chain.
   c. the removal of poly A from the end of the chain.
   d. all of the above.

8. A single mRNA may produce several proteins at a time by
   a. the action of several ribosomes in a cluster called a polyribosome.
   b. several RNA polymerase molecules working sequentially.
   c. association with rough ER.
   d. containing several promoter regions.

9. Transposons
   a. remove introns to produce functional polypeptides.
   b. are mutagens but cause little damage due to the redundancy of the genetic code.
   c. act in the recognition of the codon by the anticodon.
   d. are mobile segments of DNA that can produce nonsense mutations.

10. In RNA processing,
    a. exons are excised before the mRNA is translated.
    b. assemblies of protein and snRNPs, called spliceosomes, may catalyze splicing.
    c. the RNA transcript that leaves the nucleus may be much longer than the original transcript.
    d. large quantities of rRNA are assembled into nucleosomes.

11. Base-pair substitutions may have little effect on the resulting protein for all of the following reasons except which one?

a. The redundancy of the code may result in no change in translation.

b. As long as the substitution is three nucleotides the reading frame is not altered.

c. The missense mutation may not occur in a critical part of the protein.

d. The new amino acid may have similar properties to the replaced one.

12. A conditional mutation
    a. may only occur under high temperatures.
    b. may harm the organism at permissive temperatures.
    c. may revert back under permissive conditions.
    d. may involve temperature-sensitive mutations.

13. Base analogues
    a. may be created by UV radiation.
    b. are regions of DNA that can move from one section of DNA to another.
    c. may substitute for normal bases and result in base-pair substitutions.
    d. are chemical mutagens that spontaneously occur.

14. A base deletion early in the coding sequence of a gene may result in a protein that is
    a. functionally disrupted by the mutation.
    b. prematurely terminated.
    c. largely missense coding due to a frameshift mutation.
    d. all of the above.

15. The bonds between the anticodon of a tRNA molecule and the complementary codon of mRNA are
    a. formed by the input of energy from GTP.
    b. catalyzed by peptidyl transferase.
    c. hydrogen bonds.
    d. formed by the input of energy from ATP.

16. A prokaryotic gene 600 nucleotides long can code for a polypeptide chain of about how many amino acids?
    a. 200
    b. 300
    c. 600
    d. 1800

17. Which of the following mutations would be expected to have the least harmful effect on the resulting protein?
    a. a base deletion near the start of the coding sequence
    b. a base deletion in the middle of an exon.
    c. a three-base addition within an intron
    d. a base-pair substitution in the first position of a codon

18. Peptidyl transferase

a. translocates the tRNA holding the polypeptide chain from the A to the P site.
b. hydrolyzes a GTP to pair the codon and anti-codon in the A site.
c. catalyzes the formation of a peptide bond between a polypeptide and the amino acid in the A site.
d. binds the initiator tRNA to the start codon in the P site.

19. How many amino acids could be coded for if there were six instead of four different nucleotide bases and codons consisted of doublets (two nucleotides) instead of triplets?
   a. 6
   b. 2
   c. 216
   d. 432

20. Frameshift mutations can be caused by
   a. base-substitution.
   b. transposons.
   c. base analogues.
   d. a and b.

FILL IN THE BLANKS: *Complete the table of DNA sequences, codons, anticodons, and corresponding amino acids. Use the following portion of the genetic code to help you answer these questions.*

Codon	Amino Acid	Codon	Amino Acid
AUG	Methionine	GGG	Glysine
AAG	Lysine	GCA	Alanine
CCA	Proline	UGU	Cysteine
GUC	Valine	UAG	Stop

DNA			T T C	
mRNA codon				U A G
Anticodon		C A G		
Amino acid	methionine			

CHAPTER 17

THE GENETICS OF VIRUSES AND BACTERIA

FRAMEWORK

The concept maps below outline the major genetic characteristics of viruses and bacteria.

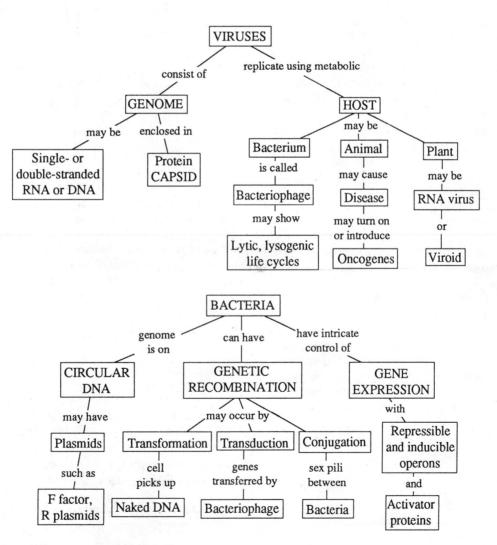

CHAPTER SUMMARY

The study of viruses and bacteria has provided information about the molecular genetics of all organisms, an appreciation for the special genetic features of microbes, an understanding of how viruses and bacteria cause disease, and new powerful techniques of manipulating genes that have had an immense impact on basic research and biotechnology.

The Discovery of Viruses

The search for the cause of tobacco mosaic disease led to the discovery of viruses. Mayer, in 1883, found that he could transmit the disease by spraying sap from an infected plant onto a healthy plant. He concluded that this contagious disease was caused by an unusually small bacteria, since he could not see the infectious agent with a microscope. Ivanowsky filtered sap from infected plants through a filter designed to remove bacteria and found the sap still transmitted the disease.

Beijerinck determined that a filterable bacterial toxin could not be the cause of tobacco mosaic disease because the infectious agent in filtered sap could not only cause the disease in one plant, but also reproduce in that plant and be passed through filtered sap to other plants. Unlike bacteria, the infectious agent could not be cultivated on nutrient media, nor was it killed by alcohol. In 1935, Stanley crystallized the infectious particle, now known as tobacco mosaic virus (TMV). Since that time, many viruses have actually been seen with the electron microscope.

Viral Structure

A *virion*, or viral particle, may consist simply of nucleic acid enclosed in a protein shell.

Viral Genomes   Viral *genomes* may be single- or double-stranded DNA, or single- or double-stranded RNA. As few as four or as many as several hundred genes may be contained on a linear or circular single molecule of nucleic acid.

Capsids and Envelopes   The *capsid*, or protein shell built from a large number of often identical protein subunits, may be rod-shaped (helical), polyhedral, or complex in structure. Membranous *envelopes*, derived from membranes of the host cell and including viral proteins and glycoproteins, may cloak the capsids of viruses found in animals.

The most complex capsids are found among viruses that infect bacteria. The seven *bacteriophages* that infect the bacterium *Escherichia coli* were the first phages to be discovered. The three "T-even" phages—T2, T4, and T6—have a similar capsid structure consisting of an icosahedral (20-sided) head enclosing the genetic material and a protein tailpiece with tail fibers for attaching to a bacterium.

The Replication of Viruses

Viruses are obligate intracellular parasites that lack the metabolic equipment needed to express their genes and reproduce. They produce hundreds or thousands of progeny in each generation by using the host cell's enzymes, ribosomes, and other resources to make copies of their genome and capsid proteins.

Genome Replication   The replication of the viral genome depends on the form of the viral nucleic acid. The replication of double-stranded DNA is similar to that of cellular genes. Most RNA viruses have a gene for an *RNA replicase*, an enzyme that uses viral RNA as a template for making complementary RNA strands. Some RNA viruses have a gene for an enzyme called *reverse transcriptase*, which uses viral RNA as a template for DNA synthesis. This DNA is then transcribed into viral RNA. The three patterns of viral genome replication are DNA→DNA, RNA→RNA, and RNA→DNA→RNA.

Self-Assembly of Virus Particles   Viral nucleic acid and capsid proteins spontaneously assemble to form new virus particles within the host cell.

Host Specificity   Each virus type has a *host range*, a limited group of host cells that it can infect. This specificity is the result of proteins on the outside of the virion that recognize *receptor sites* on the surface of the host's cell.

Bacterial Viruses

Bacteriophages are the best understood of viruses. The study of the T phages helped demonstrate that DNA is the genetic material. The study of lambda (l), another phage of *E. coli* that was discovered in 1951 by Lederberg, revealed the lytic and lysogenic cycles of double-stranded DNA viruses.

The Lytic Cycle   *Virulent* bacteriophages cause their host cells to lyse during their replication cycle, which is known as the *lytic cycle*. The T4 phage uses its tail fibers to stick to a receptor site on the surface of an *E. coli* cell. The sheath of the tail contracts through the expenditure of ATP stored in the tailpiece, and thrusts the viral DNA into the cell, leaving the empty capsid attached to the outside of the cell. The *E. coli* cell begins to transcribe and translate the hundred or so genes of the phage. One of the first genes translated codes for an enzyme that chops up the host cell's DNA. Nucleotides from the host cell's degraded DNA are used to create viral DNA. Three sets of capsid proteins are made and assembled into phage tails, tail fibers, and polyhedral heads. The viral components spontaneously assemble into 100 to 200 phage particles that are released after a lysozyme is manufactured that digests the bacterial cell wall.

The lytic cycle takes only 20 to 30 minutes at 37°C, during which time the T4 population increases more than a 100-fold. The concentration of phage particles in a sample can be determined by mixing a dilute sample with bacteria, spreading the mixture on solid medium in a Petri dish, and counting the number of *plaques*, or clear holes that develop in the growing bacterial lawn. Each plaque indicates the lysis of bacterial cells by successive generations of a single phage particle.

Bacteria may avoid viral infection when mutations change the bacterial receptor sites used by a phage. Bacteria may also defend against infection when bacterial *restriction enzymes* chop up viral DNA after it enters the cell. Bacterial hosts and their viral parasites are continually co-evolving in response to each other's new defenses and offenses.

The Lysogenic Cycle When the phage lambda binds to the surface of a bacterium and injects its DNA, it can enter either a lytic cycle or a *lysogenic cycle*. A *temperate virus* can reproduce without killing its host when its DNA inserts into the bacterial chromosome and is replicated along with the cellular DNA, thus following a lysogenic cycle.

Most of the genes of the inserted phage genome, known as a *prophage*, are inactive. One prophage gene coding for a *repressor protein* remains active and keeps the other genes switched off. Reproduction of the host cell replicates the prophage along with the bacterial DNA. The prophage may be excised from the bacterial chromosome spontaneously or through the action of environmental stimuli, and start the phage's lytic cycle.

A host cell carrying a prophage is called a *lysogenic cell*. Expression of a few of the prophage's genes may cause a change in the phenotype of the bacterium, called *lysogenic conversion*. Several disease-causing bacteria would be harmless except for the expression of prophage genes coding for toxins.

Plant Viruses and Viroids

Most plant viruses are RNA viruses. Plant viral diseases may spread through *horizontal transmission*, in which the virus invades from an external source. Injuries to the plant increase susceptibility to viral infections, and insects can act as carriers or *vectors* of viruses, passing disease from plant to plant. The inheritance of a viral infection from a parent plant is called *vertical transmission*.

Viral particles spread easily throughout a plant through the plasmodesmata, the cytoplasmic connections between cells. Reducing the spread of disease and breeding resistant varieties are the best preventions for plant viral infections; cures for most viral diseases have not been found.

Viroids are very small molecules of naked RNA that can disrupt the metabolism of a plant cell and severely stunt plant growth. Viroid RNA has been found to be very similar to sequences of self-excising introns in some eukaryotic rRNA genes. This observation has led to the theory that viroids originated as "escaped introns."

Animal Viruses

Replication Cycles of Animal Viruses A membranous envelope surrounding the capsid is present in several groups of animal viruses. Paramyxoviruses, including the viruses that cause measles and mumps, have single-stranded RNA genomes. Glycoproteins extending from the membranous envelope of the virus attach to receptor sites on the host cell's plasma membrane, and the envelope fuses with the plasma membrane, transporting the capsid into the cell. Viral enzymes replicate the RNA genome and make mRNA, which the host cell's ribosomes use for protein synthesis. New capsids form around viral genomes and leave the cell by budding off within an envelope from the host's plasma membrane. This process is called a *productive cycle* because the viruses exit without destroying the host cell.

The envelopes of herpes viruses come from the host's nuclear membrane. Clinical evidence indicates that the herpes virus' double-stranded DNA can integrate into the cell's genome as a *provirus*, similar to a bacterial prophage.

Viral Diseases in Animals The symptoms of a viral infection may be caused by toxins produced by infected cells, toxic components of the virions themselves, cells killed or damaged by the virus, or the body's defense mechanisms fighting the infection. The ability of the infected tissue to regenerate or repair itself may determine the severity of the damage of a viral infection.

Vaccines are harmless derivatives of pathogens that induce the immune system to produce antibodies against the actual disease agent. Vaccinations have greatly reduced the incidence of many viral diseases.

Since viruses use the host's cellular machinery to replicate, few drugs have been found to treat or cure viral infections. Some antiviral drugs that are effective against a number of human viruses are analogues of purine nucleosides and thus interfere with viral nucleic acid synthesis.   A few other drugs have been shown to be effective in inhibiting the influenza virus.

Viruses and Cancer   *Tumor viruses* have been found that can cause cancer in animals. When tumor viruses infect cells growing in tissue culture, the cells are transformed—they assume rounded shapes and lose the contact inhibition regulation of growth. There is strong evidence that links viruses, such as the Epstein-Barr virus and the *retroviruses*, HTLV-I and HIV, to certain types of human cancer.

Tumor viruses transform cells through integration of viral nucleic acid into the host cell genome. Retroviruses must use reverse transcription to transcribe DNA from their RNA genetic material. DNA viruses can be directly inserted into a chromosome.

Oncogenes are genes responsible for triggering cancerous transformation in cells. Surprisingly, these genes have been found not only in tumor viruses but also within the genomes of normal cells of many species. These oncogenes generally code for cellular growth factors or receptor proteins for growth factors. Some tumor viruses lack oncogenes but transform cells simply by turning on the cells' oncogenes. Carcinogens may also act by turning on cellular oncogenes.

The Origin of Viruses

Viruses inhabit a gray area between life and non-life—they are inert molecules containing a genetic program they can express only within other living cells.

Different families of viruses are genetically more similar to their host cells than to each other. It is most likely that viruses evolved from fragments of cellular nucleic acid that acquired special packaging. Viral genomes also show similarities to plasmids—self-replicating circles of DNA that can move between cells—and transposons—segments of DNA that can change location on a chromosome.

The Bacterium and Its Genome

The single, circular *bacterial chromosome* is simpler in structure than eukaryotic chromosomes, which have more associated proteins and carry a thousand times more DNA. Transcription and translation can occur simultaneously because the bacterial chromosome is not separated from the rest of the cell. Smaller rings of DNA, called *plasmids*, carry accessory genes and are found in many bacteria.

Replication proceeds bidirectionally from a single origin during cellular reproduction. Binary fission can occur as rapidly as once every 20 minutes.

Transfer and Recombination of Bacterial Genes

Rapid growth, easily observable phenotypes, and mechanisms for the transfer of genes make bacteria ideal subjects for genetic studies.

Some bacteria can take up segments of naked DNA in a process called *transformation*. If the foreign DNA is integrated into the bacterial chromosome by recombination, a new combination of genes is produced and passed on to progeny.

Bacteriophages can transfer genes from one bacterium to another by *general transduction*, when a random piece of host DNA is accidentally packaged within a phage capsid, or by *restricted transduction*, when a prophage includes some bacterial genes as it excises from the bacterial chromosome. In restricted transduction, also called specialized transduction, most of the phage genes are also included in the capsid, and the bacterial genes transduced are those adjacent to the prophage insertion site on the chromosome.

Lederberg and Tatum discovered the third type of genetic transfer between bacteria. In *conjugation*, two cells temporarily join by appendages called sex pili, and DNA transfers from one cell to the other. A plasmid called the fertility factor, or *F factor*, carries the genes both for pili production and for functions involved with DNA transfer. Cells containing the F factor are called F^+. The F factor replicates along with the bacterial chromosome, and F^+ cells pass on the trait to daughter cells. The plasmid also replicates before conjugation and is transferred to the recipient cell, changing it from an F^- cell to an F^+ cell.

Cells in which the F factor inserts into the bacterial chromosome are called *Hfr* for "high frequency of recombination." When these cells undergo conjugation, the F factor is transferred along with the attached bacterial chromosome. Movement may disrupt the mating, resulting in a partial transfer of genes. Recombination between the new DNA and the recipient cell's chromosome produces new genetic combinations in this cell and its offspring.

Mapping the Bacterial Chromosome Hfr bacteria of a given strain will always transfer chromosomal genes in the same order, beginning with the F factor. The loci of genes can be determined by experiments using interrupted mating. Hfr and F⁻ strains with different alleles are mixed together, and samples are removed and agitated to disrupt conjugation at different time intervals. Using various nutritional selective media and an antibiotic-resistance marker, the order of and relative distance between transferred genes can be determined.

Interrupted mating experiments also provided the first evidence that the *E. coli* chromosome was circular. Experiments using Hfr strains with the F factor integrated at different places on the chromosome showed that the sequence of genes remained the same, but each transfer started with a different gene and the genes could run in either forward or backward order. More detailed mapping of close genes has been done using the transducing phage P1, which transfers only short chromosomal fragments at one time.

Plasmids Plasmids are small, circular DNA molecules that replicate separately from the bacterial chromosome. Plasmids, such as F factor, that can integrate into the bacterial chromosome are called *episomes*.

R plasmids carry genes that code for antibiotic-destroying enzymes. R plasmids can be transferred to non-resistant cells, creating the medical problem of antibiotic-resistant pathogens.

Transposons Transposons—mobile segments of DNA—include two types of nucleotide sequences: a sequence for transposase, the enzyme responsible for the cutting and ligating of DNA—and a pair of "inverted repeat" sequences at the ends of the transposon that serve as recognition sites for transposase. *Insertion sequences (IS)* are the simplest transposons, consisting of only a transposase gene and inverted repeats. Their insertion into a gene as they move about the chromosome often serves to inactivate the gene.

Complex transposons contain additional DNA and may serve to move genes from one chromosome to another, and occasionally from one species to another. Transposon insertion is not dependent on extensive DNA sequence homology and thus can occur almost anywhere.

The F factor and R plasmids both contain transposons. Also, the DNA version of the retrovirus genome is a transposon. There is evidence that reverse transcription is involved in the movement of some eukaryotic transposons. In the human genome there are nonfunctional copies of genes lacking introns, which may have arisen from reverse transcription of mRNA.

The Control of Gene Expression in Prokaryotes

A cell is able to adjust its metabolism in response to environmental conditions by controlling either enzyme activity, through feedback inhibition, or enzyme synthesis, through regulation of gene expression.

Constitutive Genes and Their Control Constitutive genes are continually involved in RNA synthesis. These unregulated genes generally produce proteins that are always needed by the cell. The rate at which they are transcribed, however, is determined by the efficiency of their promoter region for binding RNA polymerase, which is related to the relative need for their encoded proteins. Regulated genes can be switched on and off as metabolic needs change. The mechanism for gene regulation was first described by Jacob and Monod in 1961 for the lactose genes in *E. coli*.

Operons and Their Control: A Repressible Operon Operons are clusters of genes on a bacterial chromosome or phage genome that have related functions. The *structural genes* within an operon may code for the different enzymes of a single metabolic pathway. The tryptophan (*trp*) operon of *E. coli* contains five structural genes for the enzymes of the tryptophan pathway. Because a single promoter region serves all the genes of an operon, an RNA polymerase will tend to transcribe all the genes at one time, producing a *polycistronic* mRNA molecule with the transcript of all the enzymes for the pathway.

An *operator* is a segment of DNA that overlaps the promoter region and controls the movement of RNA polymerase onto the promoter and along the structural genes of the operon. A *repressor* is a protein with an "active site" that binds to a specific operator and blocks attachment of RNA polymerase. An *activator* is a protein that binds near or within the promoter and promotes transcription.

Regulatory genes code for repressor and activator proteins. Most regulatory genes are constitutive, constantly producing repressor proteins at a slow rate. The activity of the repressor protein is determined by the presence or absence of a metabolite (a precursor, intermediate, or end product of a metabolic pathway). In the *trp* operon, tryptophan is the metabolite that binds to an allosteric site of the repressor protein, changing its conformation into its active state, which has a high affinity for the operator and switches off the *trp* operon.

In this regulatory system, tryptophan is a *co-repressor*. Both the repressor protein and tryptophan are needed to turn off the operon. Should the tryptophan concentration in the cell fall too low, repressor proteins are no longer bound with tryptophan, the *trp* operon is no longer repressed, RNA polymerase attaches to the promoter, and mRNA for the enzymes for tryptophan synthesis is produced. The synthesis of *repressible enzymes* is inhibited or repressed by a metabolite.

An Inducible Operon   The synthesis of *inducible enzymes* is stimulated by a metabolite, called an *inducer*. The *lac* operon, controlling lactose metabolism in *E. coli*, is an inducible operon that contains three structural genes. One of these genes codes for ß-galactosidase, an enzyme that cleaves lactose, and another gene codes for a transport protein that moves lactose into the cell. Unlike the *trp* repressor, which is inactive until bound with its co-repressor, the *lac* repressor is innately active, binding to the *lac* operator and switching off the operon. The presence of an early metabolite of lactose metabolism inactivates the repressor protein, and the operon can be transcribed.

Operon Regulation and the E. coli Economy   Repressible enzymes are usually found in anabolic pathways, where the pathway's end product serves to switch off enzyme synthesis and prevent overproduction of metabolic products. Genes for repressible enzymes are switched on until the repressor is activated. Inducible enzymes are found in catabolic pathways, where the presence of nutrient molecules stimulates the production of enzymes necessary for their breakdown. Genes for inducible enzymes are switched off until they are derepressed when a metabolite inactivates the repressor. Both types of regulation are examples of *negative control*: they involve a repressor protein that, when active, binds with the operator and blocks transcription.

Catabolite Activator Protein: An Example of Positive Control   A regulatory system that uses *positive control* is one in which an activator molecule interacts directly with the genome to turn on transcription. Even when lactose inactivates the repressor of the *lac* operon, little mRNA is produced because the *lac* promoter has a low affinity for RNA polymerase. The

binding of a *catabolite activating protein (CAP)* to the promoter region stimulates gene expression by increasing the promoter's ability to associate with RNA polymerase. *Cyclic AMP (cAMP)*, derived from ATP, accumulates in the cell when glucose is missing and binds with the activating protein. The cAMP-CAP complex attaches to the *lac* promoter region and stimulates transcription. A low level of cAMP in the cell signifies that sufficient glucose is present for an energy source; the cAMP-CAP complex disassociates; the activating protein releases from the promoter; and transcription of the *lac* operon slows down.

The regulation of the *lac* operon includes negative control by the repressor protein that is inactivated by the presence of lactose and positive control by CAP that functions when complexed with cAMP. *E. coli* is able to control its gene expression and conserve its RNA and protein synthesis depending on the presence of glucose, its primary energy and carbon source, and the availability of secondary energy sources (such as lactose) when glucose supply is limited. Prokaryotic cells have complex genetic and metabolic regulatory mechanisms.

STRUCTURE YOUR KNOWLEDGE

1. Create a concept map or diagram that describes the lytic and lysogenic cycles of bacteriophage.

2. In the diagram below of an operon located on a stretch of a bacterial chromosome, identify components A through I.

3. Develop a concept map to illustrate your understanding of the fascinating mechanisms by which bacteria are able to regulate their gene expression in response to varying needs. Try to distinguish repressible and inducible operons, which are both examples of negative control, and activator proteins, which illustrate positive control of gene expression.

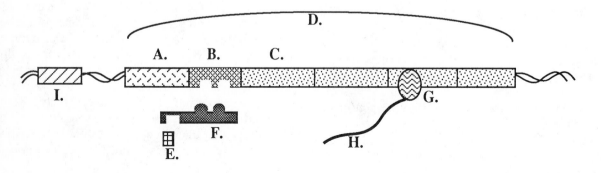

TEST YOUR KNOWLEDGE

MATCHING: *Match the term with its description.*

1. ____ viral enzyme that uses RNA as a templatefor DNA synthesis
2. ____ viral enzyme that uses RNA as a template for RNA synthesis
3. ____ bacterial enzyme that chops up foreign DNA
4. ____ change in phenotype of bacterium as result of expression of prophage genes
5. ____ transfer of genes through bridge joining two bacteria
6. ____ enzyme that cuts and ligates DNA at inverted repeats
7. ____ uptake of naked DNA by bacterial cell
8. ____ involves budding of viruses from host cell, surrounded by membraneous envelope
9. ____ injection of host cell DNA incorporated into phage capsid into a new cell
10. ____ a mRNA molecule that is a transcript of several genes

A. conjugation

B. lysogenic conversion

C. reverse transcriptase

D. lysozyme

E. lysogenic cycle

F. plaque

G. polycistronic

H. restriction enzyme

I. RNA replicase

J. productive cycle

K. transduction

L. transformation

M. transposase

MULTIPLE CHOICE: *Choose the one best answer.*

1. The study of the genetics of viruses and bacteria has done all the following except

   a. provide information on the molecular biology of all organisms.
   b. illuminate the sexual reproductive cycles of viruses.
   c. develop new techniques for manipulating genes.
   d. develop an understanding of the causes of cancer.

2. Beijerinck concluded that the cause of tobacco mosaic disease was not a filterable toxin because
   a. the infectious agent could not be cultivated on nutrient media.
   b. a plant sprayed with filtered sap would develop the disease.
   c. the infectious agent could be crystallized.
   d. the infectious agent reproduced in and could be transmitted in filtered sap from a plant infected with filtered sap.

3. Viral genomes may be any of the following except
   a. several molecules of single-stranded DNA.
   b. double-stranded RNA.
   c. a circular DNA molecule.
   d. a linear single-stranded RNA molecule.

4. Retroviruses have a gene for reverse transcriptase that
   a. uses viral RNA as a template for making complementary RNA strands.
   b. uses viral RNA as a template for DNA synthesis.
   c. destroys the host cell's RNA.
   d. translates RNA into proteins.

5. Virus particles assemble from capsid proteins and nucleic acid molecules
   a. spontaneously by the formation of weak bonds.
   b. at the direction of viral enzymes.
   c. by using host cell enzymes.
   d. through the addition of ATP stored in the tail-piece.

6. The study of lambda, a phage of *E. coli*,
   a. led to the discovery of the *lac* operon.
   b. showed that DNA was the genetic material.
   c. led to the discovery of the lytic and lysogenic replication cycles.
   d. revealed the structure of an icosahedral head and tail fibers attached to the tailpiece.

7. Vertical transmission of a plant viral disease may involve
   a. the movement of viral particles through the plasmodesmata.
   b. the inheritance of an infection from a parent plant.

c. the spread of an infection from the roots up through the plant.

d. insects as vectors carrying viral particles between plants.

8. Bacteria defend against viral infection through the action of
   a. antibiotics that they produce.
   b. restriction enzymes that chop up foreign DNA.
   c. their R plasmids.
   d. reverse transcriptase.

9. Drugs that are effective in treating viral infections
   a. induce the body to produce antibodies.
   b. inhibit the action of viral ribosomes.
   c. interfere with the synthesis of viral nucleic acid.
   d. change the cell-recognition sites on the host cell.

10. Which of the following is *not* true of tumor viruses?
    a. They can integrate viral nucleic acid into the host cell genome.
    b. They transform cells growing in tissue culture into rounded cells that lose their contact inhibition.
    c. They may not always contain oncogenes but may turn on the host cell's oncogenes.
    d. They are transferred by bacteriophage.

11. Transcription and translation can occur simultaneously in bacteria because
    a. the mRNA does not have to leave the nucleus to reach the ribosomes.
    b. the same enzyme serves both functions.
    c. replication proceeds bidirectionally around the circular chromosome.
    d. one promoter region serves all the structural genes in an operon.

12. The F factor of bacteria
    a. carries genes for sex pili production.
    b. may be transferred and convert a cell from F⁻ to F⁺.
    c. is an episome.
    d. all of the above

13. Tiny molecules of naked RNA are
    a. retroviruses.
    b. transposons.
    c. viroids.
    d. episomes.

14. Regulatory genes are genes that
    a. code for repressor and activator proteins.
    b. are usually constitutive.
    c. are not contained in the operon they control.
    d. all of the above

15. Inducible enzymes
    a. are usually involved in anabolic pathways.
    b. are produced when a metabolite inactivates the repressor protein.
    c. are produced when an activator molecule enhances the attachment of RNA polymerase with the promoter.
    d. all of the above

16. In *E. coli*, tryptophan switches off the *trp* operon by
    a. binding to the promoter.
    b. binding to the operator.
    c. binding to the repressor and increasing the latter's affinity for the operator.
    d. inactivating the repressor protein.

17. A mutation that renders the regulatory gene of a repressible operon inactive would result in
    a. continuous transcription of the structural genes.
    b. inhibition of transcription of the structural genes.
    c. irreversible binding of the repressor to the promoter.
    d. no difference in transcription rate when an activator protein was present.

18. The role of the specific metabolite in an inducible operon is to
    a. bind to the promoter region and increase the affinity of RNA polymerase for the promoter.
    b. bind to the operator region and block the attachment of RNA polymerase to the promoter.
    c. bind to the repressor protein and inactivate it.
    d. increase the production of inactive repressor proteins.

MATCHING: *Match these components of the lac operon with their functions:*

1. _____ beta-galactosidase   A. is inactivated when attached to lactose

2. _____ cAMP-CAP complex   B. codes for synthesis of repressor

3. _____ lactose   C. hydrolyzes lactose

4. _____ operator   D. stimulates gene expression

5. _____ promoter   E. repressor attaches here

6. _____ regulator gene   F. RNA polymerase attaches here

7. _____ repressor

8. _____ structural gene

G. acts as inducer that inactivates repressor

H. codes for an enzyme

CHAPTER 18

CONTROL OF GENE EXPRESSION AND DEVELOPMENT IN EUKARYOTES

FRAMEWORK

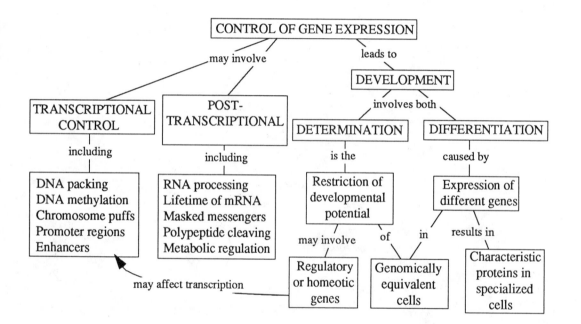

CHAPTER SUMMARY

Eukaryotic cells in multicellular organisms must control their gene expression, not only in response to their external and internal environments, but also to direct the *cellular differentiation* necessary to create specialized cells.

The mechanisms for the control of gene expression began to be known in the mid-1970s, aided by the techniques of gene cloning and nucleic acid sequencing. The understanding of the molecular basis of development is just beginning.

In eukaryotes, the greater complexity of chromo-some structure, gene organization, and cellular structure provides more avenues for control of gene activity than are present in prokaryotes.

Packing of DNA in Eukaryotic Chromosomes

Each chromosome consists of a single, extremely long DNA molecule (about 2 meters of DNA are present in the 46 chromosomes of each human somatic cell). Multilevel folding compacts the DNA in the chromosomes and helps to determine the activity of its genes.

"Beads on a String"   *Histones* are small, positively charged proteins; they bind tightly to DNA to make up chromatin. Partially unfolded chromatin appears as a string of beads, each bead a *nucleosome* consisting of DNA wound around a protein core of eight histone molecules. Nucleosomes may limit the access of transcription proteins to DNA.

Higher Levels of DNA Packing   The *30 nanometer chromatin fiber* is the next order of DNA packing. It apparently consists of a repeating array of six nucleosomes organized around histone molecules, called H1.

The *looped domain* is a fold in the 30 nm chromatin fiber that may contain genes that are expressed in a coordinated fashion. Looped domains may also coil and fold, further compacting the chromatin. Chromatin visible with the light microscope during interphase is called *heterochromatin*, a highly compacted DNA that is not actively transcribed. The more open form of chromatin, called *euchromatin*, can be transcribed.

The Control of Gene Expression

The Role of DNA Packing and Methylation   One of the two *X* chromosomes in female mammalian somatic cells is present as a highly compacted Barr body, an example of heterochromatin that is not transcribed. Its inactivity may be related to the *DNA methylation* of cytosine residues. Genes are more heavily methylated in cells in which they are not expressed; drugs that inhibit methylation can induce gene reactivation. DNA methylation may be a long-term control mechanism for gene expression.

DNA methylation may also cause the double helix to assume the left-hand Z form, which some molecular biologists speculate functions in gene regulation.

Transcriptional Control   As in prokaryotes, some control of gene expression occurs at the transcription stage. Chromosome puffs appear along the polytene chromosomes seen in the salivary glands of *Drosophila* larvae. These DNA loops may make the DNA more accessible to RNA polymerase. Autoradiography has shown that the puffs correspond to regions of active RNA synthesis. The shifting locations of puffs, especially during critical times in development, indicate that genes are turned on and off. Ecdysone, the insect hormone that initiates molting, can induce these changes in puff patterns, showing that gene regulation is responsive to chemical signals.

Because so little DNA is expressed in most eukaryotic cells, scientists speculate that genes are usually inactive and that most gene-regulatory proteins inter-

act with the nucleotide sequences called regulatory sites to activate transcription of genes.

Two types of promoter regions have been identified that are specifically bound by RNA polymerase. There are three different RNA polymerases, which transcribe genes for rRNA, mRNA, and tRNA or other small RNAs.

Enhancers are DNA sequences that greatly increase the activity of nearby genes. Enhancer sequences, which can move about within regions of DNA, are thought to be recognition sites for proteins that increase the DNA's accessibility to RNA polymerase.

Posttranscriptional Control   Gene expression, measured by the number and types of proteins (or tRNA or rRNA) that are made, can be blocked or stimulated at any posttranscriptional step. Translation is separated from transcription by the nuclear envelope.

RNA processing involves the removal of the coded intron DNA segments and the splicing together of exons. Different splicing patterns of the same RNA transcript can create different proteins, providing extra genetic flexibility. The addition of nucleotides to both ends of mRNA during processing influences the exit of mRNA from the nucleus and its destiny in the cytoplasm.

Prokaryotic mRNA molecules are degraded after only a few minutes of activity, and thus prokaryotic organisms can vary their protein synthesis rapidly in response to environmental changes. Eukaryotic mRNA can last hours or even weeks; the length of time before they are degraded by cellular enzymes is related to the quantity of protein synthesis they can direct.

The translation of certain mRNA can be delayed until a control signal initiates it. A great deal of mRNA is synthesized and stored in egg cells as "masked messenger," waiting to be translated during the active protein synthesis period following fertilization.

Initiation factors may be necessary for translation within eukaryotic cells and serve as control factors for gene expression. Following translation, polypeptides are often extensively cleaved to produce an active protein. The selective degradation of proteins and metabolic regulation of enzymes also serve as control mechanisms in the cell.

Eukaryotic Gene Organization and its Evolution

Unlike prokaryotic genes, which are often organized into operons controlled by the same regulatory sites and transcribed together, eukaryotic genes coding for enzymes in the same metabolic pathway are often scattered throughout the chromosomes and separately transcribed. The integrated control of these scattered genes may involve specific nucleotide sequences associated with each gene that serve as recognition signals for regulatory molecules, which then transcribe or repress all the genes in synchrony.

Multigene Families   *Multigene families* are collections (often clustered but occasionally scattered in the genome) of similar or identical genes, that usually code for RNA products. The identical genes coding for the major rRNA molecules are arranged one after another, hundreds to thousands of times, forming huge *tandem arrays* that enable cells that are actively synthesizing proteins to produce the millions of ribosomes they need.

Examples of multigene families of nonidentical genes are the two families of genes that code for the a and ß polypeptide chains of hemoglobin. Different versions of each subunit are clustered together and are expressed at the appropriate time during development.

Families of identical genes most likely arose by repeated gene duplication, a phenomenon that seems to occur due to mistakes in DNA replication and recombination. Nonidentical gene families probably arose from mutations in duplicated genes. *Pseudogenes*, sequences of DNA similar to real genes but lacking signals for gene expression, may be present within gene families. The globin pseudogenes, which lack introns and have poly-A tails, may have been transcribed from RNA by reverse transcriptase and may have moved about the genome by transposition.

Highly Repetitive Sequences   Highly repetitive short sequences may make up 10% to 25% of the total DNA in complex eukaryotes. Called *satellite DNA* because their base compositions may be sufficiently different from the rest of the cell's DNA to isolate them by ultracentrifugation, this DNA is primarily located at the centromeres and may serve a structural rather than a genetic role in the cell.

The Program of Development

A major challenge of biology is to understand the developmental processes that produce a complex organism from a single fertilized egg. The long-term regulation of gene expression leads to the differentiation or specialization of cells that express a certain subset of the organism's genes to produce their particular set of proteins.

Determination of Embryonic Cells   The genetic program for development is contained in the DNA. The zygote is said to be *totipotent* because it can give rise to a complete adult organism. For many species, embryonic cells remain totipotent through the first few cell divisions. As the embryo develops, the cells lose the ability to develop into all the tissues of the adult. *Determination*, the restriction of developmental potential, occurs progressively as the cells develop the particular molecular and structural characteristics that determine their functions.

Differentiation   Differentiation of cells is signaled by changes in cellular structure, the appearance of characteristic proteins, or initially, by the accumulation of specific mRNA molecules. Environmental cues or chemicals turn on the genes that allow cells to become specialists at making the particular proteins associated with their functions.

Genomic Equivalence   Even though they do not express most of their genes, all cells contain the full complement of genes; they exhibit *genomic equivalence*. Experiments involving the transplantation of nuclei from differentiated cells into zygotes support genomic equivalence and the reversibility of differentiation. In the 1950s, Briggs and King transplanted nuclei from embryonic and tadpole cells into enucleated eggs. The ability of the transplanted nucleus to direct normal development was inversely related to the age of the cell from which it was taken. Determination affects the ability of a nucleus to express all its genome, but this change may be reversible, indicating that all the genes are still present in the genome.

Although nuclei from differentiated cells can produce whole organisms when placed into the cytoplasmic environment of an egg cell, no one has been able to induce a differentiated cell from a vertebrate to form an embryo. This feat is done routinely, however, with differentiated plant cells. Carrots were the first plants to be produced from mature somatic cells. This technique is used to clone plants and is an important technique in genetic engineering work.

Regeneration is the replacement of lost parts in an organism. Many invertebrates have extensive abilities to regenerate complete organisms from fragments of their bodies. Salamanders are able to regenerate lost limbs through the formation of a mass of undifferentiated cells, called a blastema, that can develop into the tissues of a new limb. The blastema cells, derived from muscle and bone cells, undifferentiate and then become redetermined and follow new developmental pathways.

As supported by evidence from nuclear transplantation, somatic cell development of plants, and regeneration, the specialized cells of an organism contain all the genes present in the zygote. The mechanism for differentiation of cells must then rely on the selective and differential expression of identical genomes.

Genes that Control Development

Cancer: Cells Out of Control   When normal cells differentiate, control mechanisms, such as contact inhibition, limit their growth and division. A cell transformed to a cancer cell escapes from these controls and may form a mass called a *tumor* within the body. In culture, tumor cells continue to divide indefinitely. The HeLa cell line has been reproducing in culture for 30 years.

Benign tumors remain at their original site in the body and can be completely removed by surgery. *Malignant tumors* are formed by cells that not only multiply excessively, but also have abnormal cell surfaces that cause them to lose the recognition and sticking qualities of normal cells. These cells may travel by way of blood and lymph vessels to form tumors in new locations in the body. This spread of cancer cells is called *metastasis*.

Cancer may be caused by physical or chemical agents, called *carcinogens*, or induced by certain viruses. The onset of cancer involves the activation of oncogenes, either native to the cell or introduced by viruses. The cellular genes corresponding to oncogenes (called *proto-oncogenes* when the cell is non-cancerous) are thought to be key genes that control cell growth and differentiation. The study of cancer provides insight into the control of normal development and vice versa.

A gene that has an essential function in a normal cell may become an oncogene if it is present in more copies than normal; undergoes transposition or chromosomal translocation (both of which may separate it from its normal control regions); or has a mutated nucleotide sequence that creates a more active or resilient protein.

Homeotic Genes and Homeoboxes *Homeotic genes*, first identified in fruit flies, appear to play a major role in the control of developmental patterns in all organisms. A sequence of 180 nucleotides, called a *homeobox*, was found in each of the fruit-fly homeotic genes. Similar sequences have now been identified in every organism examined.

The homeobox nucleotide sequences are translated into amino acid sequences, called homeodomains, which are able to bind to DNA. The proteins containing homeodomains have turned out to be regulatory proteins that control the transcription of other genes by binding to DNA. Thus, homeotic genes code for pro-

teins that may coordinate the transcription of developmental genes.

Eukaryotic homeodomain amino acid sequences are quite similar, and show some relationship to DNA-binding domains of prokaryotic regulatory proteins. These sequences must have arisen early and been conserved through evolution as important regulators of gene expression and development.

Genome Modification

Gene Amplification   Multiple copies of rRNA genes are present in the genomes of most eukaryotic cells, where they form the core of the nucleolus. In the oocyte of amphibians and some insects, a million or more extra copies of the rRNA genes are synthesized and contained as extrachromosomal circles of DNA, enabling the developing egg to make the huge numbers of ribosomes needed for protein synthesis after fertilization. The selective synthesis of extra DNA is called *gene amplification*.

Selective Gene Loss   *Chromosome diminution* is the elimination of whole or parts of chromosomes from some cells early in development. At the 16-cell stage of a developing gall midge, 14 of the cells lose 32 of their initial 40 chromosomes. The remaining two cells are the germ cells that give rise to gametes.

Rearrangements in the Genome   Rearrangements of the DNA on chromosomes can activate or inactivate genes. The two mating types found in yeasts are determined by the presence of the *alpha* (a) or *a* gene at the *mating-type locus* on a chromosome. Yeast cells can change their mating type by switching the gene with a copy of the other gene. Unexpressed, "silent" copies of both genes are kept within the genome so that mating type can be changed by this *cassette mechanism*.

Immunoglobulin genes develop permanent rearrangements of DNA segments in their differentiation. B-lymphocytes are white blood cells that produce proteins, known as antibodies or *immunoglobulins*, that recognize and bind to viruses, bacteria, and other invading molecules. Each differentiated B-lymphocyte and its descendants produce one specific antibody. During differentiation, the antibody gene is pieced together from a number of different, widely separated, DNA sequences. Much of the variation in antibodies arises from different combinations of variable and constant regions of the four polypeptides that form the complete antibody molecule. The rearrangement and deletion of DNA in the formation of antibody genes is a unique example of genomic nonequivalence.

Aging as a Stage in Development

Following embryonic development, organisms may continue to change by such processes as growth and maturation, metamorphosis, wound healing, and regeneration of body parts. Aging and death may also be considered to be forms of developmental change.

Genes program the aging and death of selected parts of organisms throughout development. The death of cells in the webbing between the fingers of a human embryo, the production of mature xylem vessels by the digestion of the living cells, and the loss of tadpole tails are examples of the programmed death of cells.

Human embryonic fibroblasts in cell culture have been found to stop reproducing, degenerate and die after about 50 mitotic divisions. Other cell types also stop dividing after a characteristic number of divisions, providing evidence that cells may be subject to programmed obsolescence.

One explanation of the predictability of aging is that the cumulative effects of mutations and other insults cause the functional decline of cells within an average number of cell divisions. Another hypothesis suggests that aging and death are programmed either by the expression of specific genes or by scheduled changes, such as a decline in the ability of DNA to replicate or repair itself, that affect all genes. All somatic cells die. Gametes, however, may be thought of as immortal in that their DNA may continue to replicate through the generations.

Epigenesis Versus Preformation

The idea of *preformation* came to include the notion that the egg or sperm, believed to contain a preformed, miniature individual that only needed to grow during its development, also contained future preformed, miniature individuals. *Epigenesis* is the view that form emerges gradually through progressive development. With the aid of the microscope, epigenesis replaced preformation as the theory of embryology.

The gradual emergence of the embryo's form is programmed by the genome of the zygote. An excellent model for studying development has been found in the small, transparent nematode, *Caenorhabditis elegans*. The developmental lineage of all of the organism's 959 somatic cells has been established. Its small number of genes, rapid and prolific reproduction, and ability to both self- and cross-fertilize have made this nematode an excellent genetic subject.

The concept of preformation, while certainly not literally true, is valid in that the total plan for development is present in the combination of the genome of the zygote and the cytoplasm of the egg. Nuclear–cytoplasmic interactions are fundamental to differentiation. The selected control of gene expression involves the chemical dialogue between nucleus and cytoplasm; chemical messages influence expression of genes, and the products of genes influence the chemical environment. Somehow these complex and intricate control mechanisms direct the development of multicellular eukaryotes.

STRUCTURE YOUR KNOWLEDGE

1. You have learned that the eukaryotic chromosome is far from a simple linear sequence of genes. Create a concept map or table that helps you to organize your knowledge of the physical structure of chromosomes, regulatory and structural nucleotide sequences, and the organization and functions of genes, all of which contribute to the complex organization of eukaryotic DNA.

2. Discuss development, determination, and differentiation, and relate them to genomic equivalency and the control of gene expression.

3. Develop a concept map covering the major transcriptional and posttranscriptional mechanisms that control the expression of eukaryotic genes.

4. Create a concept map that organizes your understanding of homeotic genes, homeoboxes, and homeodomains, and their relationship to development.

TEST YOUR KNOWLEDGE

MULTIPLE CHOICE: *Choose the one best answer.*

1. The control of gene expression is a more complicated process in eukaryotic cells because
   a. chromosomes are contained in a nucleus.
   b. gene expression is involved in the differentiation of specialized cells as well as in response to the environment.
   c. the chromosomes are linear and more numerous.
   d. operons are controlled by more than one promoter region.

2. Histones are
   a. small, positively charged proteins that bind tightly to DNA.

b. small bodies in the nucleus involved in rRNA synthesis.

c. basic units of DNA packing consisting of DNA wound around a protein core.

d. repeating arrays of six nucleosomes organized around an H1 molecule.

3. Heterochromatin
    a. has a higher degree of packing than does euchromatin.
    b. is visible with the light microscope during interphase.
    c. is not actively involved in transcription.
    d. all of the above.

4. DNA methylation of cytosine residues
    a. can be induced by drugs that reactivate genes.
    b. may contribute to the formation of the inactive Barr bodies found in female somatic cells.
    c. produces the promoter regions that specifically bind RNA polymerase.
    d. all of the above.

5. Chromosome puffs along the polytene chromosomes of *Drosophila*
    a. are loops of the hundreds of parallel chromatids that may make the DNA more accessible to RNA polymerase.
    b. are looped domains, or folds in the chromatin fiber containing genes that are expressed in a coordinated fashion.
    c. are regions of DNA methylation causing highly compacted and inactive areas on the chromosome.
    d. contain collections of multigene families.

6. Which of the following is *not* true of enhancers?
    a. They can move about within regions of DNA.
    b. They greatly increase the activity of nearby genes.
    c. They are promoter regions that specifically bind one of the three different RNA polymerases.
    d. They are thought to be recognition sites for proteins that make DNA more accessible to RNA polymerase.

7. Which of the following is *not* an example of posttranscriptional control of gene expression?
    a. masked messenger—mRNA that needs a control signal to initiate translation.
    b. the length of time mRNA lasts before it is degraded.
    c. RNA processing with different splicing patterns that can result in mRNA for different proteins.

d. the blocking of mRNA from translation by regulatory proteins.

8. Pseudogenes are
    a. tandem arrays of rRNA genes that enable actively synthesizing cells to create enough ribosomes.
    b. highly repetitive short sequences found around the centromeres.
    c. genes that can become oncogenes when induced by carcinogens.
    d. sequences of DNA that are similar to real genes but lack signals for gene expression.

9. The differentiation of cells is the result of
    a. the genomic equivalency of developing cells.
    b. the expression of multigene families.
    c. the long-term regulation of gene expression.
    d. homeotic genes in segmented animals.

10. A tumor that metastasizes
    a. is not malignant.
    b. contains proto-oncogenes.
    c. spreads cancer cells to new locations in the body.
    d. has gone into remission.

11. In order to maintain genomic equivalency with other cells, a cell must
    a. have not yet undergone differentiation.
    b. have not yet undergone determination.
    c. contain the full complement of its genes.
    d. be totipotent.

12. Which of the following is *not* true? Proto-oncogenes
    a. are thought to be genes that control cell growth and differentiation.
    b. may be activated by carcinogens.
    c. are introduced into the cell by viruses.
    d. are noncancerous.

13. A gene can develop into an oncogene when it
    a. is present in more copies than normal.
    b. undergoes a transpostition or translocation that removes it from its control region.
    c. developes a mutation that creates a more active or resistant protein.
    d. does any of the above.

14. Gene amplification involves
    a. the production of loud genes.
    b. the enlargement of chromosomal areas undergoing transcription.
    c. the synthesis of extra DNA to meet particular metabolic needs.
    d. the transformation of genes into oncogenes.

15. The specific number of mitotic divisions observed in fibroblasts in cell culture

a. provides evidence of the programmed obsolescence of cells.
b. indicates that the cells have been transformed by cancer viruses.
c. provides evidence for the determination of the genetic program.
d. is equal to the number of mitotic divisions found in most cells in culture.

16. Epigenesis
    a. is the theory that egg or sperm contains a miniature individual.
    b. is the view that form emerges gradually in development.
    c. is the result of programmed obsolescence.
    d. is the expression of extrachromosomal DNA.

17. Homeodomains
    a. are coded for by homeobox DNA sequences.
    b. are part of regulatory proteins that bind to DNA and activate or repress transcription.
    c. have been found in regulatory proteins in every eukaryotic organism studied and are similar to DNA-binding domains of prokaryotic regulatory proteins.
    d. all of the above

18. Gene expression
    a. controls the chemical environment of the cytoplasm.
    b. is controlled by the chemical environment of the cell and cytoplasm.
    c. involves both a and b.
    d. is a result of preformation.

19. The rearrangements of immunoglobulin genes that result in the wide variety of mammalian antibodies
    a. is an example of genomic nonequivalence.
    b. is a reversible rearrangement of the genome.
    c. is an example of posttranscriptional control of gene expression.
    d. occurs in response to exposure to different antigens.

20. *Caenorhabditis elegans* is a valuable experimental organism because it
    a. is grown easily, has a short generation time, and can reproduce by either self-fertilization or mating.
    b. is transparent so that individual cells can be identified and traced during development.
    c. has extensive cell-to-cell interactions and thus helps to answer questions concerning the development of complex organisms.
    d. both a and b are correct.

CHAPTER 19

RECOMBINANT DNA TECHNOLOGY

FRAMEWORK

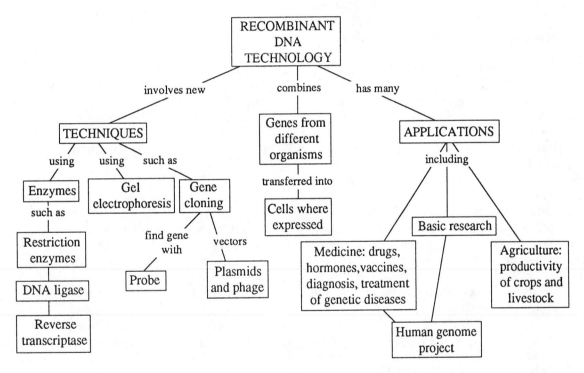

CHAPTER SUMMARY

Recombinant DNA technology involves a set of newly developed techniques for recombining genes from different organisms and then introducing this recombinant DNA into living cells in which the genes may be expressed. Using techniques developed in basic research on the biochemistry and molecular biology of bacteria, scientists can now isolate and produce large quantities of particular genes and their products. This methodology has begun an industrial revolution in *biotechnology*—the use of living systems to manufacture practical products. The human genome project to determine the complete nucleotide sequence of human DNA using the tools of recombinant DNA may be completed in the next 10 to 20 years.

Basic Strategies of Gene Manipulation

Before 1975, the technology for altering genes in organisms involved finding natural mutants or creating new mutations with mutagenic radiation or chemicals. The *selection* of the desired mutants involved methods that would permit only the desired organism to survive and grow. Bacteria and their phages can transfer genes from one bacterial strain to another by the processes of transformation, conjugation, and transduction by phage. Geneticists learned to use these processes for the detailed molecular study of prokaryotic and phage genes. The study of eukaryotic genes has been made possible with recombinant DNA techniques.

Restriction Enzymes   *Restriction enzymes* protect bacteria from foreign DNA by cutting up nonbacterial DNA in a process called *restriction*. Most restriction enzymes recognize short nucleotide sequences and cut at specific points within them. The cell protects its own DNA from restriction by *modification*, the methylation of nucleotide bases within recognition sequences. Recognition sequences are symmetrical sequences of four to eight nucleotides running in opposite directions on the two strands, such as GAATTC on one strand and CTTAAG on the other. The restriction enzyme usually cuts phosphodiester bonds in a staggered way, such as between the G and A on both strands. "Sticky ends" of short single-stranded sequences occur on both sides of the cut.

DNA from different sources can be recombined in the laboratory when the strands are cut by the same restriction enzyme and the complementary bases on the resulting sticky ends of the restriction fragments pair by hydrogen bonding and are then joined together by *DNA ligase*.

Restriction fragments can be separated by *gel electrophoresis*, a technique that separates macromolecules on the basis of their size and electrical charge. Negatively charged nucleic acids move through the electric field produced in a thin slab of gel at a rate inversely proportional to their size. The band patterns of restriction fragments produced in the gel can be used to identify specific viral, plasmid, and chromosomal DNA. Individual fragments can be isolated from the gel and still retain their biological activity.

Gene-Cloning Vectors   Small DNA fragments can be inserted into plasmid DNA, which can then serve as *vectors* for introducing recombinant DNA into cells. The plasmid method of gene cloning involves isolating plasmids that carry identification genes, such as antibiotic-resistance genes, treating the plasmids with a restriction enzyme that will cut the DNA ring at a single site, and mixing them with foreign DNA that has been treated with the same restriction enzyme and has complementary sticky ends. The sticky ends hydrogen-bond, and DNA ligase seals the recombinant molecules. The plasmids are introduced into bacterial cells by transformation and reproduce as the bacteria develop clones of cells. The resulting replication of the inserted foreign genes is called *gene cloning*. Clones carrying the recombinant DNA are identified by the use of selective media (usually for antibiotic resistance).

Bacteriophages can also serve as vectors for introducing recombinant DNA into cells. Recombinant phage DNA is created using restriction enzymes and DNA ligase to insert restriction fragments of foreign DNA into the viral genome. Once inside a cell, the phage DNA replicates itself to form new phage particles, simultaneously cloning the inserted genes.

Sources of Genes for Cloning   The genes used for cloning can be isolated directly from a particular organism by cutting its DNA into thousands of pieces with restriction enzymes and inserting them into plasmids or viral DNA. The collection of the huge number of DNA segments from a genome, derived in this shotgun approach, is called a *genomic library*.

Complementary DNA (cDNA) eliminates the problem of the large introns that may make a eukaryotic gene too large to clone easily. Using the enzyme reverse transcriptase, obtained from retroviruses, mRNA is used as a template to produce cDNA. Since cDNA has the introns already removed, it will produce mRNA that can be more easily translated by bacterial cells, which lack RNA-processing machinery. Bacterial transcription also requires the attachment of a promoter and other transcription signals to the cDNA. A partial genomic library can be produced from the mRNA molecules found in a cell.

Finding the Gene   Finding the bacterial clone that contains the gene of interest within a genomic library may be difficult. If the gene is translated into protein, the presence of the protein can be determined by its activity or its structure (using antibodies). Techniques for detecting the gene itself involve the use of a nucleic acid sequence called a *probe*, which has complementary sequences to segments of the gene. The radioactively labeled probe will hybridize with the gene of interest by base-pairing and thus locate bacterial clones with the gene.

DNA Synthesis and Sequencing   Genes for cloning may be chemically synthesized when the exact nucleotide sequence is known. The sequencing of genes is made possible by the use of restriction enzymes, which cut long DNA sequences into fragments, and gel electrophoresis, which can separate the fragments according to their lengths.

In the Sanger method of DNA sequencing, single-stranded restriction fragments are incubated with radioactively labeled primer and the four nucleotides, which include modified nucleotides (dideoxy forms) that block further synthesis when they are incorporated into a growing DNA strand. The result is four sets of radioactive strands of varying lengths that can be separated by gel electrophoresis. The nucleotide sequence can be determined by the bands produced in the gel.

Thousands of DNA sequences are being collected in computer data banks where they can be analyzed for genetic control elements and similarities among genes of different organisms and automatically translated into amino acid sequences.

Making the Gene Product Differences between prokaryotes and eukaryotes in the mechanisms for transcription and translation present problems in getting bacteria to express eukaryotic genes. Large-scale commercial production of a gene product in bacteria has been accomplished, however, with the aid of genetic tricks such as using plasmid vectors that produce many gene copies per cell; changing the gene's promoter to a highly active one; and attaching the eukaryotic gene to the start of a bacterial gene that is produced normally in large quantities. When bacterial cells are engineered to secrete the protein product, the task of purification is simplified.

Yeast present an excellent vehicle for recombinant DNA technology because of their ease of culture, their ability to take up DNA by transformation, and their capacity to express eukaryotic genes since they are eukaryotic cells. Cultured animal and plant cells must be used in some genetic research and in commercial applications that require synthesis of complex products (such as antibodies).

Applications of Recombinant DNA Technology

Biological Research Recombinant DNA technology has facilitated the molecular study of eukaryotic gene structure and function.

Genes can be produced in large amounts and used as labeled probes to locate similar DNA segments in the same or other genomes, providing information on evolutionary relationships. Complementary DNA probes allow location and study of the natural form of the gene, including its regulatory and noncoding sequences.

DNA probes can also be used to map genes on eukaryotic chromosomes with an *in situ hybridization* technique. Labeled DNA probes base-pair (hybridize) with intact chromosomes on a microscope slide, and autoradiography and chromosome staining show the location of the gene.

Many molecules controlling cell metabolism and development are produced in such small quantities that they cannot be purified and characterized by nor-

mal biochemical techniques. The production of cDNA from mRNA molecules and its cloning in bacteria have provided large enough quantities of these proteins to investigate their structures and functions.

The Human Genome Project Four complementary components of the human genome project are genetic (linkage) mapping, physical mapping, nucleotide sequencing, and similar analysis of other species important in genetic research. The location of genetic markers spaced throughout the chromosomes will facilitate the mapping of other genes. Physical mapping involves determining the order of identifiable fragments of each chromosome. The sequencing of the 3 billion nucleotide pairs may be the most time-intensive part of the project. The analysis of the genomes of other species will allow comparisons with the human genome. Information from the human genome project should contribute to the diagnosis, treatment, and prevention of genetic diseases, and provide insights into basic questions of molecular genetics and evolutionary biology. New techniques developed in basic biological and biochemical research are expected to contribute to the building of the human genetic map.

RFLP, restriction fragment length polymorphisms, provide valuable markers for the creation of the human genetic map. Restriction fragment length refers to the lengths of the DNA fragments, shown by the pattern of bands produced by gel electrophoresis, that result when DNA is cut by a specific restriction enzyme. Fragment length variations or polymorphisms occur as a result of differences in base sequences that may delete or add restriction sites to homologous chromosomes. When these naturally occurring variations in DNA occur frequently in a population, geneticists can use them as reference points along a chromosome. DNA identification tests, based on RFLP markers, have been used in criminal and paternity cases. Comparisons of patterns of inheritance of genetic diseases and RFLP markers have led to the mapping of several disease genes.

In RFLP analysis, DNA extracted from white blood cells is mixed with a restriction enzyme, and the resulting restriction fragments are separated by gel electrophoresis. The fragments are blotted onto a filter, and the bands of interest are identified by the use of a radioactively labeled probe.

Polymerase chain reaction (PCR) is another new technique that is contributing to the human genome project, genetic research, and forensic applications. By incubating DNA with special primers and DNA polymerase, millions of copies of a section of DNA can be rapidly produced. Through the amplification of DNA from single sperm cells, the results of meiotic recombination may be studied, and more closely linked genes and markers may be mapped.

Medicine Diagnosis of genetic diseases and heterozygote carriers of harmful recessive mutations is becoming possible using gene cloning, with the normal gene providing a probe for locating the corresponding gene in the cells being tested. Restriction enzymes and gel electrophoresis are then used to compare the restriction fragment patterns formed by the two segments of DNA. When the linkage of a disease gene with a RFLP marker has been determined, comparisons of blood samples within a family may be used to diagnose the presence of the abnormal allele. Alleles for cystic fibrosis and Huntington's disease have been linked with a RFLP marker and can be detected in this manner.

Genetic engineering may provide the means for correcting genetic disorders in individuals by replacing or supplementing defective genes. New genes would presumably be introduced into a few somatic cells of types that actively reproduce within the body. Thus the gene would be replicated when the cells are reinserted into the individual. The protein product of the introduced normal gene would need to correct the biochemical defect despite the absence of the gene and protein in most other tissues of the body.

Some of the technical problems involved with human gene therapy include how to get the proper control mechanisms to operate on the transferred gene, and when in development and into which tissues the gene should be introduced. An ethical and social question is whether the germ cells should be treated in order to correct defects in future generations. Opponents fear that tampering with human genes will eventually lead to the practice of eugenics, the deliberate effort to control the distribution of human genes.

Traditional vaccines against viral diseases are either particles of a virulent virus that have been inactivated or active virus particles of an attenuated (nonpathogenic) viral strain. These vaccines induce an immune response in the organism during which antibodies specific for the invading pathogen are developed. Recombinant DNA techniques have been used to make large amounts of protein molecules from disease-causing viruses and bacteria. If the protein, called a subunit, triggers an immune response against the pathogen, it can be used as a vaccine. Genetic engineering methods can also be used to directly modify the genome of the pathogen so as to attenuate it. The live vaccinia virus, which is the basis of the smallpox vaccine, has been modified to carry genes that induce immunity to other diseases.

Gene splicing has been used to produce hormones and proteins, such as interferons, in bacteria. Insulin and human growth hormone were the first two polypeptide hormones made by recombinant DNA techniques to be approved for use in the United States. Before 1982, insulin for treating diabetics was obtained from pig and cattle pancreatic tissues. Insulin that is chemically identical to human insulin is now produced in genetically engineered bacteria.

Because growth hormones are species-specific, human growth hormone, used for treating children with hypopituitarism that results in dwarfism, had to be obtained in the small quantities available from human cadavers. A genetically engineered version of human growth hormone was developed in 1979 and approved for use in 1985.

Agriculture Recombinant DNA techniques are providing means to improve the productivity of agricultural plants and animals. Genes for desirable traits can be microinjected into egg cells or early embryos. Vaccines, growth factors, and hormones are produced with new technology. Plant crops with engineered genes may be made resistant to diseases, harsh growth conditions, or herbicides.

The regeneration of plants from single cells growing in tissue culture has made plant cells easier to manipulate than are mammalian cells. The search for genetic vectors for plants cells, however, has been more difficult. A plasmid carried by the bacterium *Agrobacterium tumefaciens* is the best-developed vector.

This bacterium produces crown gall tumors in the plants it infects when the *Ti plasmid* (tumor inducing) integrates a segment of its DNA into the plant cells' chromosomes. Foreign genes are inserted into the Ti plasmid, and *Agrobacterium* with recombinant plasmids are used to infect plant cells growing in culture. When these cells regenerate whole plants, the foreign gene is included in the plant genome.

Only dicotyledons are susceptible to infection by *Agrobacterium*. Other techniques are being developed to transfer foreign genes into important monocots such as corn and wheat. Brief jolts of electricity used in electroporation temporarily open holes in the cell membrane through which naked DNA can enter a cell. DNA can also be injected through microscopic needles into cell nuclei. Challenges still remain in identifying beneficial genes that will improve plant traits, particularly those traits that are polygenic.

Early results of genetic engineering in plants have been positive. Plant strains that carry a bacterial gene for resistance to a herbicide have been developed. Crop yields may be improved by the production of bacterial species, found surrounding plant roots, that are genetically engineered to produce a toxin harmful to crop pests. The production of bacteria with increased nitrogen-fixing potential and the engineering of plants that can fix nitrogen themselves are examples of a valuable potential use of recombinant DNA technology in improving plant productivity.

Safety and Policy Matters

Scientists have worried that there might be dangerous consequences to recombinant DNA manipulations. One particular hazard concerns the production of new pathogens and their release into the environment. Early scientists working in this field developed a set of guidelines to deal with policy and safety issues. The National Institutes of Health (NIH) adopted these recommendations as part of their formal guidelines for biomedical research.

The safety measures were of two types: the first to protect scientists from infection by engineered microbes and to prevent the microbes from accidentally escaping from the laboratory, and the second to develop mutant strains of microbes that could not possibly survive outside the laboratory.

A new genetic engineering industry began to develop, with many of its scientists recruited from academic posts. Critics feared that safety and ethical questions would not be dealt with in the rush to build profitable business enterprises.

As organisms were developed to be introduced into the environment, it became necessary to decide which federal agencies were best suited to evaluate proposals for the release of genetically engineered organisms. The Food and Drug Administration, the Environmental Protection Agency, the U.S. Department of Agriculture, and the N.I.H. Recombinant DNA Advisory Committee are now involved in regulating and guiding the use of this powerful new technology.

STRUCTURE YOUR KNOWLEDGE

1. Fill in the table below on the basic strategies of gene manipulation used in recombinant DNA technology.

2. There are many applications for recombinant DNA techniques in basic research, in agriculture, and in medicine. Describe four or five examples in each of these categories.

3. Make a schematic diagram of the steps required to produce marketable amounts of human insulin.

TEST YOUR KNOWLEDGE

MULTIPLE CHOICE: *Choose the one best answer.*

1. The role of restriction enzymes in recombinant DNA technology is to
   a. provide a vector for the transfer of the recombinant DNA.

Technique or Tool	Brief Description of Technique or Tool	Use in Recombinant DNA Technology
Restriction enzymes		
DNA ligase		
Gel electrophoresis		
Reverse transcriptase		
Labelled probe		
Gene cloning		

b. produce cDNA from mRNA.

c. produce a staggered cut at specific recognition sequences on DNA.

d. reseal "sticky ends" after base-pairing of complementary bases.

2. Molecules are separated by gel electrophoresis on the basis of

a. their charge.

b. their charge and size.

c. their polarity.

d. their shape.

3. Plasmid DNA and bacteriophages

a. are involved in cloning genes.

b. are both vectors for transferring recombinant DNA into cells.

c. can be treated so that they carry foreign genes in their DNA.

d. all of the above

4. The best working definition for gene cloning is

a. life from nonlife.

b. synthetic production of organic molecules.

c. manipulation of natural DNA sequences.

d. harnessing of microbial cell chemistry to produce numerous copies of foreign DNA.

5. Genes for cloning may be chemically synthesized

a. when the exact sequence of nucleotides is known.

b. through the use of restriction enzymes and gel electrophoresis to separate restriction fragments.

c. by the Sanger method.

d. using cDNA.

6. Petroleum-lysing bacteria are being engineered for the removal of oil spills. What is the most realistic danger to the environment?

a. mutations leading to the production of a strain pathogenic to humans

b. extinction of natural microbes due to the competitive advantage of the "Petro-bacterium"

c. destruction of natural oil deposits

d. none of the above

7. You have been contracted to introduce a gene that imparts larval moth resistance to bean plants. Which of the following vectors are you most likely to use?

a. bacteriophage

b. Ti plasmids

c. *Agrobacterium*

d. both b and c

8. An attenuated virus

a. is a live virus that is nonpathogenic.

b. consists of viral particles that have been inactivated.

c. can transfer recombinant DNA to its host cell.

d. will not produce an immune response.

9. Difficulties in getting prokaryotic cells to express eukaryotic genes include the fact that

a. the signals that control gene expression are different.

b. the genetic code differs between the two.

c. prokaryotic cells cannot transcribe introns.

d. all of the above

10. Copy DNA cannot create as complete a gene library as the shotgun approach because

a. it eliminates introns from the genes.

b. a cell produces mRNA for only a small portion of its genes.

c. the shotgun approach produces more restriction fragments.

d. cDNA is not as easily integrated into plasmids or phage genomes.

11. Which of the following is not true of recognition sequences?

a. Modification by methylation of bases within them prevents restriction of bacterial DNA.

b. They are usually symmetrical sequences of 4 to 8 nucleotides.

c. They signal the attachment of RNA polymerase.

d. There is a specific restriction enzyme for each recognition sequence.

12. Recombinant DNA technology contributes to basic research in

a. the evolutionary relationships of organisms.

b. the control of gene expression in eukaryotes.

c. the biochemical analysis of growth factors and molecules produced in small quantities.

d. all of the above.

CHAPTER 12
MEIOSIS AND SEXUAL LIFE CYCLES

Suggested Answers to Structure Your Knowledge

1.

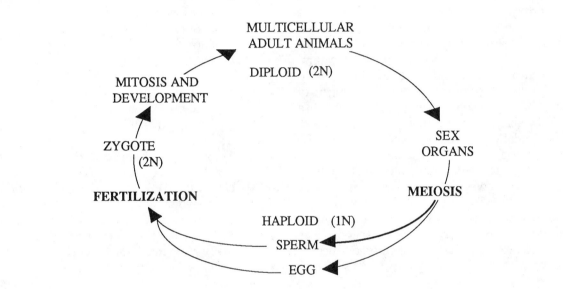

2.

STAGE	KEY EVENTS OF MEIOSIS
Interphase I	Chromosome replication, sister chromatids attached at centromere
Prophase I	Synapsis of homologous pairs, crossing over, spindle forms
Metaphase I	Tetrads align at metaphase plate, independent assortment
Anaphase I	Homologous chromosomes separate and move toward opposite poles
Telophase I	Haploid sets of double-stranded chromosomes reach poles, cytokinesis
Interkinesis	Nuclear membranes may form, no replication of chromosomes
Prophase II	Spindle forms
Metaphase II	Chromosomes align at metaphase plate
Anaphase II	Centromeres of sister chromatids separate, chromosomes move apart
Telophase II	Cytokinesis, four haploid cells with single-stranded chromosomes

Answers to Test Your Knowledge

Multiple Choice:

1.	b	4.	a	7.	c	10.	b
2.	d	5.	d	8.	c	11.	a
3.	d	6.	b	9.	d	12.	b

CHAPTER 13
MENDEL AND THE GENE IDEA

Suggested Answers to Structure Your Knowledge

1. The segregation of gene pairs (alleles) takes place at anaphase I when homologous chromosomes move to opposite poles of the cell. The two cells formed from this division have one-half the number of chromosomes and one copy of each gene. Independent assortment relates to the lining up of synapsed chromosomes at the equatorial plate in a random fashion during metaphase I. Chromosomes initially received from the mother and father may orient in either direction. The number of different chromosome arrangements in gametes is equal to 2^N where N is the haploid number of chromosomes. The number of gametic arrangements of genes in crosses involving more than one gene is equal to 2^n where n is the number of genes (as long as they are on different chromosomes).

3. Generation I
   Genotype of father = Aa
   Genotype of mother = Aa
   Generation II
   Genotype of mate 1 = AA (probably)
   Genotype of mate 2 = Aa
   Generation III
   Genotype of son 4 = Aa

   The genotype of son 3 could be AA or aa. If his wife is AA, then he could be Aa (both his parents are carriers), and the recessive allele would never be expressed in his offspring. Even if he and his wife were both carriers, there would be a 81/256 or 31.6% chance that all four children would be normally pigmented. (3/4 x 3/4 x 3/4 x 3/4)

2.

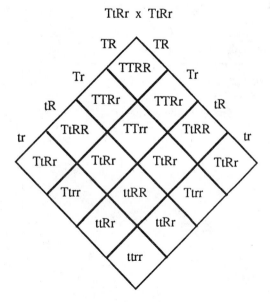

TtRr x TtRr

PHENOTYPE	NUMBER
tall, red	3
tall, pink	6
tall, white	3
short, red	1
short, pink	2
short, white	1

Answers to Test Your Knowledge

Matching:

1. I	4. D	7. L	10. K
2. A	5. N	8. O	11. H
3. F	6. J	9. C	12. M

Multiple Choice:

1. b	4. c	7. c	10. d	13. b	16. d
2. d	5. b	8. d	11. b	14. d	17. a
3. b	6. c	9. b	12. d	15. c	18. c

Genetics Problems:

1.

Mother	Child	Man exonerated if
AB	A	no groups exonerated
O	B	A or O
A	AB	A or O
O	O	AB
B	A	B or O

2. Father's genotype must be Pp since polydactyly is dominant, and he has had one normal child. Mother's genotype is pp. The chance of the next child having normal digits is 1/2 or 50% because the mother can only donate a p allele, and there is a 50% chance that the father will donate a p allele.

3. AB gamete from AABb parent = 1/2 (1 x 1/2)
   ab gamete from AaBb parent = 1/4 (1/2 x 1/2)
   ABc gamete from AABbcc parent = 1/2 (1 x 1/2 x 1)
   ABc gamete from AaBbCc parent = 1/8 (1/2 x 1/2 x 1/2)

4. probability of AAbb offspring = 1/2 (to get AA) x 1/2 (bb) = 1/4
   probability of aaBB offspring = 1/4 (aa) x 1/2 (BB) = 1/8
   probability of AaBbCc = 1 (Aa) x 1/2 (Bb) x 1 (Cc) = 1/2
   probability of aabbcc = 1/4 (aa) x 1/4 (bb) x 1/2 (cc) = 1/32

5. The genotypes of the puppies were 3/8 B-S- (because B and S are dominant alleles, the "-" indicates that the second gene can be either B or b, S or s, and still produce a dominant phenotype), 3/8 B-ss, 1/8 bbS-, and 1/8 bbss. Because recessive traits show up in the offspring, both parents had to be at least heterozygous for both genes. Black occurs in a 6:2 or 3:1 ratio, indicating a heterozygous cross. Solid occurs in a 4:4 or 1:1 ratio, indicating a cross between a heterozygote and homozygous recessive. Parental genotypes were BbSs x Bbss.

6. The F₁ cross produced AaBbCc plants with 3 units of 5 cm added to the base height of 10 cm, to produce 25-cm tall F₁ plants. Of the 64 possible combination of gametes in the F₂, there will be 7 different phenotypic classes, varying from all dominant alleles, 5 dominant and 1 recessive, 4 dominant and 2 recessive, and so on, through all 6 recessive alleles.

7. It is good to decide first what phenotypes are produced by each genotype. M-B- individuals will be black, because both genes are dominant. M-bb individuals are brown. mm-- individuals will all be white because no melanin is produced to be laid down (the B genotype does not matter). All F₁ guinea pigs will be MmBb and black. The normal 9 M-B-, 3 M-bb, 3 mmB-, 1 mmbb phenotypic ratio will be 9 black, 3 brown, and 4 white due to the epistatic effect of mm on the B gene.

8. The 1:1 ratio of the first cross indicates a cross between a heterozygote and a homozygote, Hh x hh, or Hh x HH. The second cross indicates that the hairless hamsters cannot be hh, or only hairless hamsters should be produced. If the hairless hamsters are Hh, then one would expect a 3:1 phenotypic ratio and a 1:2:1 genotypic ratio. The 1:2 ratio indicates that the homozygous-recessive genotype may be lethal, so that embryos that are hh never develop.

9. The 1:2:1 ratio indicates a case of incomplete dominance. Medium-tailed pigs are probably Tt, and the heterozygous cross produced three tt stub-tailed pigs, six Tt medium-tailed pigs, and four TT long-tailed pigs.

10.

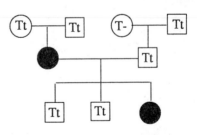

At least one of these grandparents must have been Tt.)

11. You can determine the genotype of the black parent to be CCᶜʰ. The Himalayan parent could be either CʰCʰ or Cʰc. First list the alleles in order of dominance: C, Cᶜʰ, Cʰ, c. To approach this problem you should write down as much of the genotype as you can be sure about for each phenotype involved. The parents were black (C-) and Himalayan (Cʰ-). The offspring genotypes would be black (C-) and Chinchilla (Cᶜʰ-). Right away you

can tell that the black parent must have been CC^{ch} because half of the offspring are Chinchilla; C^{ch} is dominant over C^h and c and could not have been present in the Himalayan parent. The genotype of this parent could be either C^{h}C^h or C^{h}c. A test cross would be necessary to determine this genotype.

You cannot definitely determine the genotype of the parents in the second cross. You can, however, eliminate some genotypes. The black parent could not be CC or CC^{ch}, because both the C and C^{ch} alleles are dominant to C^h, and Himilayan offspring were produced. The Chinchilla parent could not be C^{ch}C^{ch} for the same reason. One or both parents had to have C^h as their second allele.

12. This problem is challenging. First write out the phenotypes possible and the genotypes that could produce them. BB- - and - -SS genotypes are lethal. Bristled flies must be Bbss. Nonbristled flies could be BbSs, bbSs, or bbss. The nonbristled parents must be BbSb. Now determine the probablilty of obtaining these genotypes.

bristled = Bbss = 1/2 x 1/4 = 1/8 or 2/16
nonbristled = BbSs (1/4) + bbSs (1/8) + bbss (1/16) = 7/16

The phenotypic ratio of the F_1 would be 2:7, bristled to nonbristled.

A backcross of Bbss to BbSs would yield a 1:2 ratio of bristled to nonbristled.
bristled = Bbss = 1/2 x 1/2 = 1/4
 nonbristled = BbSs (1/4) + bbSs (1/8) + bbss (1/8) = 4/8

CHAPTER 14
CHROMOSOMAL BASIS OF INHERITANCE

Suggested Answers to Structure Your Knowledge

1.

SYSTEMS OF SEX DETERMINATION					
	X - Y	X - O	Z - W	HAPLO-DIPLOIDY	NO SEX CHROMOSOMES
FEMALE	XX	XX	ZW	Diploid	Both gametes produced by same organism
MALE	XY	XO	ZZ	Haploid	
EXAMPLES	Humans, mammals, fruit flies, dioecious plants	Crickets, roaches, grasshoppers	Birds, some fish, moths, butterflies	Bees, ants, wasps	Worms, snails, monoecious plants

2. Genes that are not linked assort independently, and the ratio of offspring from a testcross with a dihybrid heterozygote should be 1:1:1:1 (AaBb x aabb gives AaBb, Aabb, aaBb, aabb offspring). Genes that are linked and do not cross over should produce a 1:1 ratio (AaBb and aabb). If crossovers occur 50% of the time, then the heterozygote will produce equal quantities of AB, Ab, aB, and ab gametes, and the genotypic ratio of offspring will be 1:1:1:1, the same as for unlinked genes. Crosses with intermediate genes on the chromosome could establish both that the genes A and B are on the same chromosome and that they are a certain distance apart.

3. If the gene for the mutant is sex-linked, you could assume it was on the X chromosome. A cross between mutant female flies and normal males should produce all normal female and mutant male flies.
X^{m}X^m x X$^+$Y produces X$^+$X^m and X^{m}Y offspring

4.

NAME OF DISEASE OR SYNDROME	CHROMOSOMAL ALTERATION INVOLVED	SYMPTOMS OR ASSOCIATED TRAITS
Down syndrome	Trisomy 21	Characteristic facial features, short stature, heart defects, respiratory infections, mental retardation
Klinefelter syndrome	XXY (or XXYY, XXXY, XXXXY, XXXXXY)	Sterility, small testes, enlarged breasts or feminine body contours, normal intelligence or mental retardation
Turner syndrome	Monosomy X	Short stature, sex organs do not mature, no 2ndary sex characteristics, normal intelligence
Metafemale	Trisomy X	Limited fertility, often mentally retarded
Cri du chat syndrome	Deletion in chromosome 5	Mental retardation, small head, unusual facial features, unusual cry

Answers to Test Your Knowledge

Multiple Choice:

1. d	5. d	9. a	13. c	17. b
2. a	6. b	10. a	14. d	18. c
3. b	7. b	11. d	15. a	19. a
4. d	8. b	12. b	16. c	20. a

Genetics Problems:

1. The gene is sex-linked, so a good notation is X^C, X^c, and Y so that you will remember that the Y does not carry the gene. Capital C indicates normal sight. Genotypes are:
   1. X^CY 3. X^CX^C or X^CX^c 5. X^CY 7. X^CX^c
   2. X^CX^c 4. X^CX^c 6. X^cY

2. e   a   c   b   d

3. The genes appear to be linked because the parental types appear most frequently in the offspring. Recombinant offspring represent 10 out of 40 total offspring for a recombination frequency of 25%, indicating that the genes are 25 map units apart.

4. Let X^L represent the dominant allele and X^l represent the recessive lethal allele. The mother is X^LX^l; the father must be X^LY. Male fetuses with the genotype X^lY would be expected one-half of the time and would spontaneously abort. Assuming an equal sex ratio at conception, the ratio of girl to boy children would be 2:1, or 6 girls and 3 boys.

5. Female flies produce v^+bw^+ gametes and males produce either vbw or Ybw gametes. All the F_1 are phenotypically wild-type red-eyed flies. Females are v^+vbw^+bw and males are v^+Ybw^+bw. In the reciprocal cross (vvbwbw x $v^+Ybw^+bw^+$), all F_1 females are v^+vbw^+bw (red-eyed) and males are $vYbw^+bw$ (vermillion-eyed).

An F_2 is produced from the heterozygote red-eyed F_1 that resulted from the first cross. The following Punnett square shows the gametes that each parent can produce and the resulting phenotypes for the F_2.

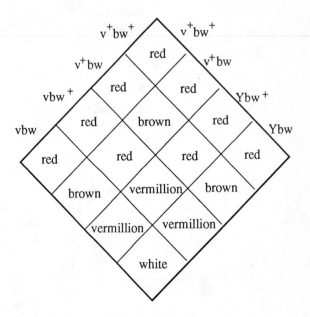

PHENOTYPE	NUMBER
red-eyed female	6
brown-eyed female	2
red-eyed male	3
brown-eyed male	1
vermillion-eyed male	3
white-eyed male	1

6. Females are the heterogametic sex in birds. For female chicks to be black, they must have received a recessive allele from the male parent. If all female chicks are black, the male parent must have been Z^bZ^b. If the male parent was homozygous recessive, then all male offspring will receive a recessive allele, and the female parent would have to be Z^BW to produce all barred males.

For female chicks to be both black and barred, the male parent must have been Z^BZ^b. If the female parent were Z^BW, only barred male chicks would be produced. To get an equal number of black and barred male chicks, the female parent must have been Z^bW.

CHAPTER 15
MOLECULAR INHERITANCE

Suggested Answers to Structure Your Knowledge

1.

Investigator	Organisms Used	Techniques, Experiments	Conclusions
Griffith	S and R strains of bacteria, S caused pneumonia in mice	Inject mice with heat-killed S and live R cells	R cells transformed by genetic material from S cells. Heat-stable factor may not be protein.
Avery	S and R strains of bacteria	Purify chemicals from heat-killed S cells; test for transforming live R cells	Transforming agent was DNA.
Hershey & Chase	Bacteriophage (phage T2) and the bacterium, *E. coli*.	Label DNA, protein of phage; infect *E. coli.*; blend and centrifugeso bacteria in bottom, viral parts in top.	DNA of phage injected into host; proteins stay outside; DNA directed host to replicate phage—DNA hereditary material
Chargaff	DNA from several organisms	Paper chromatography to separate A, T, G, C bases; compare base ratios	DNA species-specific: ratio of bases different between species; but A=T and G=C.
Franklin	DNA	X-ray crystallography of DNA molecules	DNA is in shape of helix
Watson & Crick	DNA	Information from X-ray photo; wire scale models	Double helix with rungs of A-T and G-C, sides of phosphate–sugar chain
Meselson & Stahl	*E. coli.*	*E. coli.* labelled with heavy isotope of N, grow one generation with light N, centrifuge to find density	Semiconservative replication of DNA.

2.

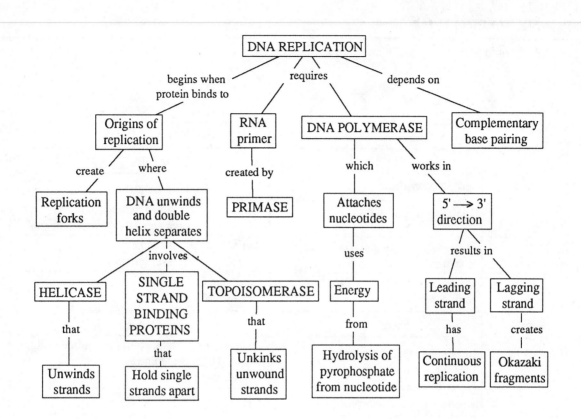

Answers to Test Your Knowledge

Multiple Choice:

1. c	**6.** b	**11.** c	**16.** d	**21.** h	**26.** a
2. a	**7.** a	**12.** d	**17.** d	**22.** g	**27.** b
3. c	**8.** b	**13.** b	**18.** d	**23.** d to c	**28.** c
4. b	**9.** a	**14.** c	**19.** b	**24.** a	
5. d	**10.** d.	**15.** a	**20.** f	**25.** e	

CHAPTER 16
FROM GENE TO PROTEIN

Suggested Answers to Structure Your Knowledge

1.

	TRANSCRIPTION	TRANSLATION
Template	DNA	mRNA
Location	nucleus	ribosomes
Molecules involved	ribonucleotides, DNA and enzymes	amino acids, attached to tRNA, mRNA, ribosomes, enzymes, ATP, GTP and others
Enzymes needed	RNA polymerase, and processing enzymes	aminoacyl–tRNA synthetase, peptidyl transferase, and others
Control-start & stop	promoter sequence on DNA, terminator sequence	initiation factors, AUG codon, release factor
Product	mRNA	protein
Processing involved	RNA processing: introns removed, Poly A and cap added	spontaneous folding, coiling; disulfide bridges; removal of some amino acids; cleaving, or quarternary structure
Energy source	ribonucleoside triphosphate	ATP and GTP

2. The genetic code includes the sequence of nucleotides on DNA that is transcribed into the codons found on mRNA and translated into their corresponding amino acids. There are 64 possible codons created from the four nucleotides used in the triplet code (4^3). Codons are needed only for 20 amino acids and stop codons. Redundancy of the code refers to the fact that several triplets may code for the same amino acid. Often these triplets only differ in the third nucleotide. The wobble hypothesis explains the fact that there are only about 40 different tRNA molecules that pair with the 64 possible codons. The third nucleotide of many tRNAs can form H bonds with more than one base. Because of the redundancy of the genetic code, these wobble tRNAs still place the correct amino acid in position. The genetic code was thought to be universal—each codon coded for exactly the same amino acid no matter which organism it was found in. Recently, some exceptions to this universality have been found in a few ciliates.

3.

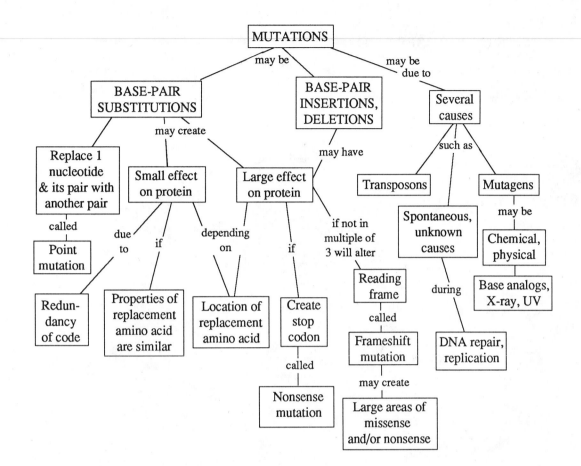

Answers to Test Your Knowledge

Multiple Choice:

1. d	**5.** b	**9.** d	**13.** c	**17.** c
2. a	**6.** d	**10.** b	**14.** d	**18.** c
3. b	**7.** a	**11.** b	**15.** c	**19.** c
4. d	**8.** a	**12.** d	**16.** a	**20.** b

Fill in the Blanks:

DNA	TAC	CAG	T T C	ATG
mRNA codon	AUG	GUC	AAG	U A G
Anticodon	UAC	C A G	UUC	none—release factor binds
Amino acid	methionine	valine	lysine	STOP

CHAPTER 17
THE GENETICS OF VIRUSES AND BACTERIA

Suggested Answers to Structure Your Knowledge

1.

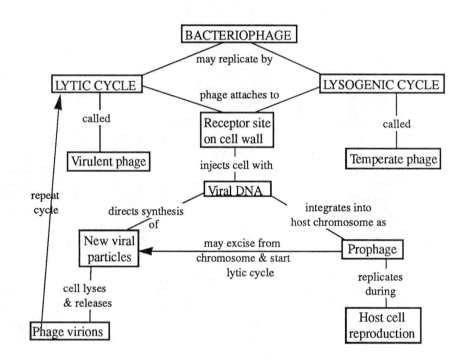

2. A. Promoter
   B. Operator
   C. Structural gene
   D. Operon
   E. Corepressor or
      inducer metabolite
   F. Repressor
   G. RNA polymerase
   H. mRNA
   I. Regulatory gene

3.

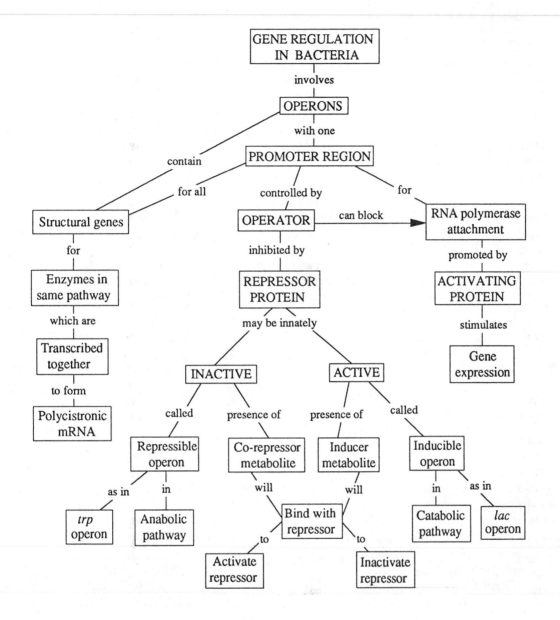

Answers to Test Your Knowledge

Matching:

1. C	**5.** A	**9.** K
2. I	**6.** M	**10.** G
3. H	**7.** L	
4. B	**8.** J	

Matching:

1. C	**5.** F
2. D	**6.** B
3. G	**7.** A
4. E	**8.** H

Multiple Choice:

1. b	**7.** b	**13.** c
2. d	**8.** b	**14.** d
3. a	**9.** c	**15.** b
4. b	**10.** d	**16.** c
5. a	**11.** a	**17.** a
6. c	**12.** d	**18.** c

CHAPTER 18
GENE EXPRESSION AND DEVELOPMENT IN EUKARYOTES

Suggested Answers to Structure Your Knowledge

1.

ORGANIZATION OF EUKARYOTIC DNA				
PHYSICAL STRUCTURE	REGULATORY SEQUENCES	STRUCTURAL SEQUENCES	GENES	
			Coded sequences	Noncoded sequences
DNA packing	Recognition sequences (for genes in same pathway)	Highly repetitive sequences (satellite DNA at centromere)	EXONS	INTRONS
nucleosome 30 nm chromatin fiber	Promoters (bind RNA polymerase)		Multigene families (copies of same or simlar genes; tandem arrays)	Pseudogenes (gene-like sequences in tandem arrays)
looped domain hetero-chromatin	Enhancers (increase transcription)		Gene amplification (many copies of same gene)	Gene diminution (chromosomes lost from cell)
			Development control genes (proto-oncogenes, homeotic genes)	

2. Development involves the complex processes through which a fertilized egg grows into a multicellular organism consisting of differentiated cells organized into tissues and organs. The key to this process is the selective expression of genes from the complete genome present in every cell. The zygote is said to be totipotent, capable of forming all the specialized cells of the adult organism. Determination is the progressive restriction of the developmental potential as the fates of the cells become more limited during development. Even though the cells are still genomically equivalent, long-term changes in gene expression may have irreversibly turned off some genes. Differentiation is the specialization of cells by selective gene expression by which a cell produces its characteristic proteins and develops specialized structures and functions.

3.

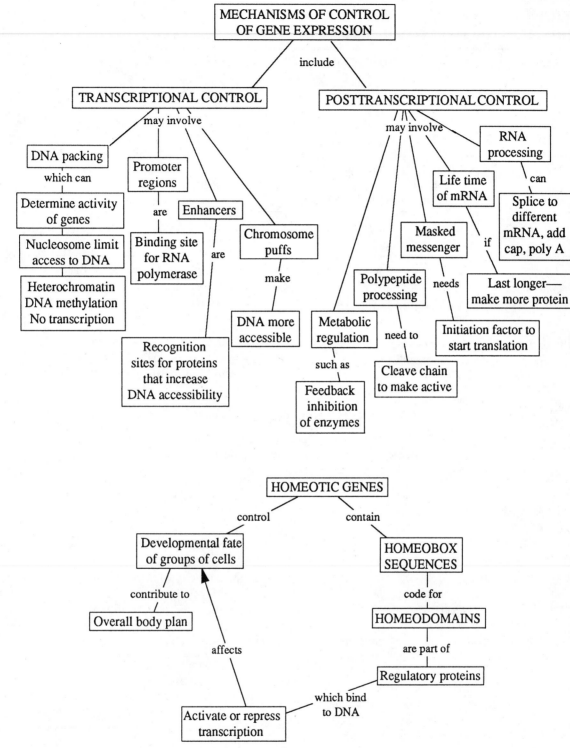

4.

Answers to Test Your Knowledge

Multiple Choice:

1. b	**6.** c	**11.** c	**16.** b
2. a	**7.** d	**12.** c	**17.** d
3. d	**8.** d	**13.** d	**18.** c
4. b	**9.** c	**14.** c	**19.** a
5. a	**10.** c	**15.** a	**20.** d

CHAPTER 19
RECOMBINANT DNA TECHNOLOGY

Suggested Answers to Structure Your Knowledge

1.

Technique or Tool	Brief Description of Technique or Tool	Use in Recombinant DNA Technology
Restriction enzymes	Bacterial enzyme that cuts foreign DNA, leaving "sticky ends" that can base-pair with other restriction fragments	Form restriction fragments that can be used to sequence DNA, and to make recombinant DNA
DNA ligase	Enzyme that forms phosphodiester bonds in DNA molecule	Seal sticky ends of restriction fragments, make recombinant DNA
Gel electro-phoresis	Apply mixture of macromolecules to slab of gel under electric field, molecules separate by of movement due to size and charge	Separate restriction fragments. Bands permit identification or sequencing of DNA or removal of fragments for gene library.
Reverse transcriptase	Retrovirus enzyme that produces DNA from mRNA	Create copy DNA (cDNA) from mRNA, gives smaller gene (RNA processed—no introns) more easily translated by bacteria
Labelled probe	Radioactively labeled single-stranded DNA or mRNA that will base-pair with complementary sequences of DNA	Locate gene in clone of bacteria, identify similar nucleic acid sequences, make cytological map of genes
Gene cloning	Reproduce large quantities of a gene in phage, bacteria, or yeasts; use bacteria or phage as vectors	Produce many copies of a gene for basic research or for large-scale production of gene product

2. Basic Research:
(1) In the Sanger method, techniques using restriction enzymes and gel electrophoresis are used to determine nucleotide sequences in DNA. (2) Radioactively labeled probes are used to map genes on chromosomes. (3) Molecules that control metabolism and development are often present in very small quantities. With cDNA, their genes can be cloned and the molecules produced in large enough quantities to study. (4) cDNA can be used as a probe to locate genes and study their noncoded regions such as introns and regulator sequences. (5) Using probes to locate genes with similar sequences, the evolutionary relationships between genes and between organisms can be studied.

Agricultural Applications:
(1) Vaccines, hormones, and growth factors, which will improve the health or productivity of livestock, can be produced using recombinant DNA technology. (2) The genome of agricultural plants and animals may eventually be directly altered and improved. (3) Improvements in the nitrogen-fixing capacity of bacteria or plants are being developed. (4) Soil bacteria that produce pesticides are being produced. (5) Plant varieties that have genes for resistance to adverse conditions, diseases, and herbicides are being developed.

Medical Applications:
(1) Vaccines are being developed, either through the production of virus proteins to determine if these subunits produce immune responses, or in the development of attenuated viruses. (2) Diagnosis of genetic diseases is possible by probing for a potentially defective gene with a normal gene. (3) Treatment of genetic disorders may be possible through the replacement of a defective gene with a normal one. (4) Insulin, human growth factor, and several products of the immune system have been produced by biotechnology.

3.

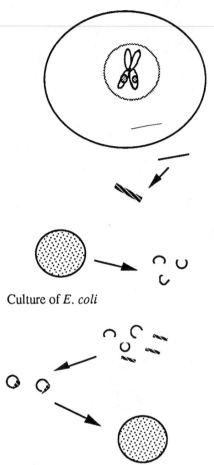

Human chromosome carrying gene for insulin

Nuclear DNA contains introns, would produce nonfunctional insulin if cloned directly

Obtain mRNA for insulin from cytoplasm

Mix mRNA with reverse transcriptase and nucleotides to get cDNA sequence (without introns) for insulin. Attach active promotor. Use restriction enzyme to produce sticky ends.

Culture of *E. coli*

Obtain plasmids with identification markers (antibiotic resistanct). Treat plasmids with restriction enzyme that cuts plasmid and produces sticky ends.

Mix cDNA and plasmids. Add ligase to seal ends.

Introduce recombinant plasmids into bacterial cells.

Bacterial antibody-resistance colonies clone insulin gene and produce insulin.

Answers to Test Your Knowledge

Multiple Choice:

1.	c	**4.**	d	**7.**	d	**10.**	b
2.	b	**5.**	a	**8.**	a	**11.**	c
3.	d	**6.**	c	**9.**	a	**12.**	d

MECHANISMS OF EVOLUTION

THE FAR SIDE   By GARY LARSON

Great moments in evolution

CHAPTER 20

DESCENT WITH MODIFICATION: A DARWINIAN VIEW OF LIFE

FRAMEWORK

This chapter describes Darwin's formulation of evolution—descent with modification from a common ancestor by the mechanism of natural selection, which results in the evolution of species adapted to their environments. The scientific and philosophical climate of Darwin's day was quite inhospitable to the implications of evolution, but most biologists accepted the theory of evolution quite rapidly. Acceptance of natural selection as the mechanism of evolution occurred later, with the incorporation of genetic principles into the modern synthesis of evolution in the 1940s. Evidence for evolution is drawn from biogeography, the fossil record, taxonomy, comparative anatomy, comparative embryology, and molecular biology.

CHAPTER SUMMARY

Evolution refers to all the changes that have occurred in the history of life on Earth. Darwin presented the first convincing case for evolution in his book, *On the Origin of Species by Means of Natural Selection*, published in 1859. Darwin made two major claims: that species had not been specially created in their present forms but had evolved from ancestral species and that natural selection provided the mechanism for evolution.

Pre-Darwinian Views

Darwin's theory was truly radical, for it challenged both the prevailing scientific views and the world view that had been held for centuries in Western culture.

The Scale of Life and Natural Theology The Greek philosopher Plato believed in two worlds, an ideal and eternal real world and the illusory world perceived by the senses. The variations in plant and animal populations were simply imperfect representatives of ideal forms. Plato's philosophy of ideal forms is known as idealism or essentialism.

Aristotle believed that all living forms could be arranged on a scale of nature of increasing complexity. Each group of organisms was fixed, permanent, and did not evolve.

The Judeo-Christian account of creation, the special design of each species during the week in which the Creator formed the universe, embedded the idea of the fixity of species in Western thought. Biology during Darwin's time was dominated by natural theology, the study of nature to reveal the Creator's plan.

One of the goals of natural theology was to classify species to reveal the rungs on the scale of life that God had created. Carl Linnaeus, working in the 18th century, developed both a binomial system for naming organisms according to their genus and species and a hierarchy of classifications for grouping species. *Taxonomy*, the branch of biology that names and classifies organisms, was begun by Linnaeus.

Cuvier, Fossils, and Catastrophism Fossils are remnants or impressions of organisms laid down in rock, usually *sedimentary rocks*, such as sandstone and shale, that form through the compression of layers of sand and mud into superimposed layers called strata. Fossils reveal a succession of flora and fauna (plant and animal life).

Cuvier, the father of *paleontology*, the study of fossils, observed the differences between older fossils and modern life forms and the occurrences of extinctions. Adopting the view of history known as *catastrophism*, he speculated that the differences in fossil strata were the result of local catastrophic events such as floods or drought and not indicative of evolution.

Gradualism in Geology   Gradualism, the idea that profound change is the cumulative result of slow but continuous processes, was used by Hutton in 1795 to explain the geological state of the earth. Lyell, the leading geologist of Darwin's time, extended gradualism to a theory of *uniformitarianism*, that held that the uniform rates of geological processes cause their effects to balance out through time.

Darwin took two ideas from the observations of Hutton and Lyell: Earth must be very old if geological change is slow and gradual, and very slow and subtle processes occurring over long periods of time could effect substantial change.

Lamarck's Theory of Evolution   Lamarck, in 1809, was the first to publish a theory of evolution that explained how life evolves. Lamarck believed that evolution was driven by the tendency toward greater complexity and that, as organisms evolved, they became better adapted to their environments. Lamarck explained the mechanism of evolution with two principles: use or disuse of body parts leads to their development or deterioration, and acquired characteristics can be inherited. In the creationist–essentialist climate of the time, Lamarck's views were dismissed and vilified; under present genetic knowledge, his inheritance of acquired characteristics is sometimes ridiculed. Lamarck's theory, however, presented many key evolutionary ideas: that evolution is the best explanation for the fossil record and the current diversity of life, that Earth is very old, and that adaptation to the environment is the main result of evolution.

On the Origin of Darwinism

The Voyage of the Beagle   Darwin was 22 years old when he sailed from England as the naturalist on the *Beagle*. He spent the voyage collecting thousands of specimens of the fauna and flora of South America, observing the various adaptations of organisms living in very diverse habitats, and making note of the geographic distribution of species. He was particularly struck by the uniqueness of the fauna of the Galapagos Islands.

Through his experiences on the *Beagle* and the reading of Lyell's *Principles of Geology*, Darwin came to doubt the church's orthodox view that the earth was static and had been created only a few thousand years ago.

Darwin Frames His View of Life   Upon learning that the 14 types of finches he had collected on the Galapagos were indeed separate species, Darwin began, in 1837, the first of several notebooks on the origin of species. He began to perceive that the origin of new species was closely related to the process of adaptation to different environments.

In 1844, Darwin wrote a long essay on the origin of species and natural selection but was reluctant to introduce his theory publicly. In 1858, Darwin received Wallace's manuscript describing a theory of natural selection identical to Darwin's. Wallace's paper and extracts of Darwin's unpublished essay of 1844 were jointly presented to the Linnean Society on July 1, 1858. Darwin quickly finished and published *The Origin of Species* the next year. Within a decade, Darwin's book and its defenders had convinced the majority of biologists that evolution was the best explanation for the forms of life on Earth.

The Concepts of Darwinism

The Principle of Common Descent   Darwin's concept of *descent with modification* included the notion that all organisms were related through descent from some unknown ancient prototype and had developed increasing modifications as they adapted to various habitats. Darwin's view of the history of life is analogous to a tree with a common ancestor at the fork of each new branch, and modern species at the tips of the living twigs. Most branches are dead ends; about 99% of all species that have lived are extinct.

Natural Selection and Adaptation   Darwin's book focused on how populations of individual species adapt to their environments through natural selection. Ernst Mayr's description of Darwin's theory of *natural selection* includes the following facts and inferences: (Fact 1) species have great potential fertility; (Fact 2) most population sizes are stable; (Fact 3) natural resources are limited; (Inference 1) since only a fraction of offspring survive, there is a struggle for limited resources; (Fact 4) there is variation between individuals in a population; (Fact 5) much of this variation can be inherited; (Inference 2) individuals whose inherited characteristics fit them best to the environment are likely to leave more offspring; (Inference 3) unequal reproduction leads to gradual change in a population. Natural selection is the differential reproductive success within a population that leads to adaptation to the environment.

Variability arises through chance, but natural selection is the result of definite environmental criteria for reproductive success. The excessive production of offspring sets up the struggle for life; only a small proportion can live to leave offspring of their own. Darwin's ideas were influenced by the writing of Malthus on the potential for the human population to grow much faster than the supply of food and resources.

The *artificial selection* used in the breeding of domesticated plants and animals provided Darwin with evidence that selection among the variations present in a population can lead to substantial changes. He reasoned that natural selection, working over hundreds or thousands of generations, could gradually create the modifications essential for the present day diversity of life. Gradualism is basic to the Darwinian view of evolution.

Some Subtleties of Natural Selection   Natural selection results in the evolution of populations, groups of interbreeding individuals of the same species in a common geographical area. Evolution is measured only as change in the relative proportions of variations in a population over time. Natural selection affects only those traits that are heritable; acquired, nongenetic characteristics cannot evolve. And natural selection is a local and temporal phenomenon, depending on the specific environmental factors present in a region at a given time.

Natural Selection at Work: A Case History   The best known example of natural selection in action is the English peppered moth, *Biston betularia*. Before the Industrial Revolution, the light colored variant of this moth was the one most commonly collected, and dark moths were rare. As industrial pollution darkened the tree trunks in the late 1800s, the light moths began to decrease in number and the dark variants to increase. This change in the color of moths in polluted areas is known as industrial melanism.

Predation by birds increases when the moths contrast with their background. Predation served as the natural selection agent that contributed to the shift in the composition of peppered moth populations in industrial regions in a short period of time.

The Modern Synthesis

Most biologists rapidly accepted evolution but not Darwin's suggestion of natural selection as the mechanism of evolution. Without a theory of genetics, the perpetuation of parental traits in offspring alongside the chance occurrences of variations could not be explained.

Although Mendel was a contemporary of Darwin, his contribution to the theory of inheritance and natural selection was not recognized. When Mendel's work was rediscovered, many geneticists believed that Darwin's reliance on the inheritance of quantitative traits was not explained by the inheritance of discrete traits with which Mendel and later researchers worked.

During the 1920s, the research focus on mutations provided an alternative explanation for evolution as occurring by rapid phenotypic changes caused by mutations. Biologists also resisted Darwin's mechanistic explanations as support of orthogenesis, or goal-oriented evolution, grew.

The emergence of *population genetics* in the 1930s, with its emphasis on quantitative inheritance and genetic variation within populations, reconciled Mendelism with Darwinism.

In the early 1940s a comprehensive theory of evolution, known as the *modern synthesis*, or neo-Darwinism, was developed. This theory, which includes the genetic basis of variation, emphasizes the importance of populations as the units of evolution, the essential role of natural selection, and the gradualness of evolution.

Today, nearly all biologists accept the fact of evolution, although debate still continues on models of how evolution occurs. The rate of evolution and the role of mechanisms other than natural selection are areas in which many evolutionists are challenging the modern synthesis.

Evidence for Evolution

Biogeography   The geographic distribution of species, or *biogeography*, first suggested common descent to Darwin. Islands have endemic species that are related to species on the nearest island or mainland. Widely separated islands having similar environments are more likely to have species taxonomically related to those of the nearest mainland, regardless of environment, than to each other.

The Fossil Record   The succession of fossil forms supports the existence of the major branches of descent that were established with evidence from anatomy, molecular biology, and other sources. Darwin was troubled by the fragmentary fossil record that did not contain many transitional fossils linking modern species to their ancestral forms. The record is still incomplete, although many key links have been found.

Taxonomy   The taxonomy developed by Linnaeus provided a hierarchical organization of groups that suggested the branching genealogy of the tree of life. Most taxonomists organize life forms with the intent of reflecting evolutionary relatedness.

Comparative Anatomy   The anatomical similarities among species grouped in the same taxonomic category provide evidence of common descent. The same skeletal elements make up the forelimbs of all mammals regardless of function or external shape. These forelimbs are *homologous structures*, similar because of their common ancestry. Comparative anatomy illustrates that evolution is a remodeling process in which ancestral structures are modified for new functions.

Vestigial organs are rudimentary structures, of little or no value to the organism, that are historical remnants of ancestral structures. The skeletons of some snakes retain vestigial pelvic and leg bones, once used in their walking ancestors.

Comparative Embryology Comparative embryology shows that closely related organisms have similar stages in their embryonic development. Early embryos of vertebrates look very similar due to the presence of the same structures. They pass through a stage in which they have gill slits on the sides of their throats; these slits develop into gills in fish but into different structures in other vertebrates.

In the late 19th century, many embryologists adopted the view that "ontogeny recapitulates phylogeny," stating that the embryonic development (*ontogeny*) of an organism is a replay of its evolutionary history (*phylogeny*). Although the recapitulation theory is an overstatement, ontogeny does relate to phylogeny and can provide evidence of homology between structures that are very different in their adult forms.

Molecular Biology An organism's DNA reflects its ancestry; closely related species should have a larger proportion of DNA and proteins in common than do more distantly related species. Molecular biologists have measured the degree of similarity in DNA sequences and amino acid sequences in proteins, and have found that the closer two species are taxonomically, the greater the DNA and protein similarities. In *DNA–DNA hybridization*, single-stranded DNA strands extracted from two species are mixed and form hybridized double-stranded DNA. The temperature at which this hybrid DNA "melts" (separates into single strands) depends on the number of hydrogen bonds that formed due to base pairing and reflects the degree of similarity between the base sequences of the DNA of the two species.

Darwin's boldest claim—that all forms of life descended from the earliest organisms and are thus related—is supported through molecular biology. For example, all aerobic species, even though they may be as unrelated as grass and fungi and butterflies, have the respiratory protein, cytochrome *c* (with variations in the amino acid sequences, of course, but with the same essential structure and function). The genetic code, evidently passed along through all branches of evolu-

tion, is also important evidence that all life is related. Evolution is the key to both the unity and diversity of life.

STRUCTURE YOUR KNOWLEDGE

1. Briefly state the main components of Darwin's theory of evolution.

2. Darwin and present-day biologists have drawn on six sources of evidence for evolution. Briefly describe the contributions from each of these areas.

TEST YOUR KNOWLEDGE

MATCHING: *Match the theory or philosophy and its proponent(s) with the following descriptions.*

A. catastrophism	a. Aristotle
B. early theory of the mechanism of evolution	b. Cuvier
C. essentialism	c. evolutionists of this century
D. gradualism	d. Hutton
E. modern synthesis	e. Lamarck
F. natural theology	f. Linnaeus
G. ontogeny recapitulates phylogeny	g. some embryologists
H. scale of nature	h. Plato
I. uniformitarianism	i. Wallace

Theory Proponent

1. ____ ____ discovery of the Creator's plan through the study of His works

2. ____ ____ history of Earth marked by floods or droughts that resulted in extinctions

3. ____ ____ replication of ancestry of organism during its development

4. ____ ____ inheritance of acquired characteristics

5. ____ ____ profound change is the cumulative product of slow but continuous processes

6. ___ ___ fixed species on a continuum from simple to complex

7. ___ ___ evolutionary theory with genetic explanation of natural selection

8. ___ ___ ideal world with perfect forms of which world of senses is imperfect representation

MULTIPLE CHOICE: *Choose the one best answer.*

1. The classification of organisms into hierarchical groups is called
   a. the scale of nature.
   b. taxonomy.
   c. natural theology.
   d. ontogeny.

2. The study of fossils is called
   a. phylogeny.
   b. gradualism.
   c. paleontology.
   d. fossilogy.

3. To Cuvier, the differences in fossils from different strata were evidence for
   a. changes occurring as a result of cumulative but gradual processes.
   b. divine creation.
   c. evolution by punctuated equilibrium.
   d. local catastrophic events such as droughts or floods.

4. Darwin proposed that new species evolve from ancestral forms by
   a. the gradual accumulation of adaptations to changing environments.
   b. the inheritance of acquired adaptations to the environment.
   c. the struggle for limited resources.
   d. the accumulation of mutations.

5. Artificial selection
   a. was used by Darwin as evidence for changes possible with natural selection.
   b. involves the artificial insemination of females of a species.
   c. resulted in the change in the English peppered moth population.
   d. all of the above.

6. The best description of natural selection is
   a. the survival of the fittest.
   b. the struggle for existence.
   c. the reproductive success of the members of a population most fit to the environment.
   d. the overproduction of offspring in environments with limited natural resources.

7. The biogeographic distribution of species
   a. provides evidence for evolution.
   b. provides evidence for natural selection.
   c. shows that many endemic island species are related to species on the nearest mainland.
   d. all of the above

8. The remnants of pelvis and leg bones in a snake
   a. are vestigial structures.
   b. provide support for the fact that ontogeny recapitulates phylogeny.
   c. are homologous structures.
   d. provide evidence for orthogenesis.

9. Which of the following would provide the best information for distinguishing phylogenetic relationships between several very similarly appearing organisms?
   a. the fossil record
   b. homologous structures
   c. comparative embryology
   d. DNA–DNA hybridization

10. Darwinism and Mendelism were reconciled by
    a. the explanation for the inheritance of quantitative traits.
    b. the development of population genetics.
    c. the deciphering of the common genetic code.
    d. biogeography.

CHAPTER 21

HOW POPULATIONS EVOLVE

FRAMEWORK

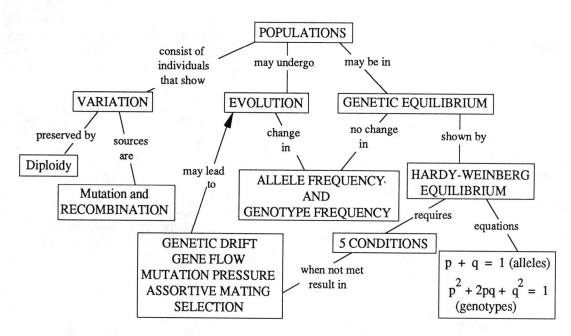

CHAPTER SUMMARY

Although natural selection selects for or against individuals, it is populations that actually evolve. The application of genetics to the theory of natural selection has been the most important post-Darwinian advance in evolutionary biology.

The Genetics of Populations

A *species* is a group of populations that can interbreed in nature. Within the geographic range of a species, populations (localized groups of individuals of that species) may be isolated or overlapping.

The Gene Pool and Microevolution   The *gene pool* is the term for all the genes present in a population at any given time. The pool includes the two alleles for each gene locus for individuals of a diploid species. If all individuals are homozygous for the same allele, the allele is said to be *fixed* in the gene pool. More often, two or more alleles are present, each at a certain frequency in the gene pool. *Microevolution* is a change in the relative frequencies of alleles in a population over time.

The Hardy–Weinberg Theorem   In the absence of selection and other agents of change, the allele frequencies within a population will remain constant from one generation to the next. This stasis is formulated as the *Hardy–Weinberg theorem*. The allele frequency within a population determines the proportion of gametes that will contain that allele. The random combination of gametes will yield offspring with genotypes that reflect and reconstitute the allele frequencies. The frequencies of both alleles and genotypes will remain stable in a population that is in *Hardy–Weinberg equilibrium.*

With the Hardy–Weinberg equation, the frequencies of alleles within a population can be estimated from the genotype frequencies, and vice versa. The letters p and q represent the proportions of the two alleles within a population. The combined frequencies of the alleles must equal 100% of the genes for that locus within a population: $p + q = 1$.

The frequencies of the genotypes in the offspring reflect the frequencies of the alleles and the probability of each combination. According to the rule of multiplication, the probability that two gametes containing the same allele will come together in a zygote is equal to ($p \times p$) or p^2, or ($q \times q$) or q^2. For a p and q allele to combine, the p allele could come from one parent or the other. The frequency of a heterozygous offspring reflects these two possibilities and is equal to $2pq$. The sum of the frequencies of all possible genotypes in the population adds up to one: $p^2 + 2pq + q^2 = 1$.

Allele frequencies can be determined from genotype frequencies. If the frequencies of the homozygous genotype (p^2) and the heterozygous genotype ($2pq$) are known, then the frequency of the p allele can be determined. All the gametes from the homozygotes and one-half the gametes from the heterozygotes will contain p. The frequency of p in the gene pool will equal the frequency of the homozygous dominant genotype and one-half the frequency of heterozygous genotypes. If the frequency of homozygous recessive individuals is known (q^2), then the frequency of q is equal to the square root of q^2 (assuming the population is in Hardy–Weinberg equilibrium for that gene).

As an example, if p = frequency of allele A = 0.7, and q = frequency of allele a = 0.3, then, according to $p^2 + 2pq + q^2 = 1$, AA = 0.49, Aa = 0.42, and aa = 0.09. To determine allele frequencies, p = frequency AA + 1/2 frequency Aa, or 0.7 (0.49 + 0.21), and q = frequency aa + 1/2 frequency Aa, or 0.3 (0.09 + 0.21). Alternately, $q = \sqrt{aa} = 0.3$.

The Hardy–Weinberg equilibrium is maintained only if all of the following five conditions are met: (1) a large population, (2) no migration into or out of the population, (3) no net changes in the gene pool due to mutation, (4) random mating, and (5) equal reproductive success of all genotypes. The Hardy–Weinberg predictions can serve as a baseline for comparison with actual populations, in which these five conditions are almost never met and gene pools are changing.

Causes of Microevolution

Five potential agents of microevolution arise from the five conditions of Hardy–Weinberg equilibrium. Natural selection, resulting in the unequal reproductive success of different genotypes, tends to increase the fitness of a population to its environment; the other four agents are chance events and usually nonadaptive.

Genetic Drift   Chance deviations from expected results are more likely to occur in a small sample, a phenomenon known as *sampling error*. If a population is small, the random drawing of alleles to form the next generation may not represent the allele frequencies in the gene pool due to sampling error. *Genetic drift* is a chance change in the gene pool of a small population; such a change is not related to the fitness of individuals. Genetic drift may play a major role in the microevolution of populations of less than 100 individuals.

The *bottleneck effect* occurs when some disaster or other factor reduces the population size dramatically, and the few surviving individuals are unlikely to represent the original genetic makeup of the population. Genetic drift will remain a factor in the population until it is large enough for chance events to be less significant. A bottleneck usually reduces variability because some alleles are lost from the gene pool.

The genetic drift found when a few individuals colonize a new area is known as the *founder effect*. The small sample size represented by the few colonists is unlikely to be representative of the parent population, and genetic drift will continue to affect the gene pool of the new population until it is large enough that sampling errors will not be a factor.

Gene Flow   The migration of individuals or the transfer of gametes between populations may result in the gain or loss of alleles. This phenomenon is known as *gene flow*. The differences in allele frequencies between populations, which may have developed by natural selection or genetic drift, tend to be reduced by gene flow.

Mutation   The altering of allele frequencies due to mutation is probably of little importance in microevolution due to the very low mutation rates for most gene loci (one mutation in every 10^5 to 10^6 gametes). Mutation is the original source of genetic variation, however; as such, it is central to evolution.

Nonrandom Mating   Individuals tend to mate more often with close neighbors than with more distant population members. The effect of this *inbreeding* is that genotype frequencies may vary from those predicted from Hardy–Weinberg equilibrium. The proportions of homozygotes may increase, and thus more individuals expressing recessive phenotypes will be found. Although inbreeding may change the ratios of genotypes and phenotypes, the allele frequencies within the population will remain the same.

In *assortative mating*, the choice of mates may be due to preference for like individuals. This nonrandom mating may result, as does inbreeding, in fewer heterozygous individuals, but it does not change the allele frequencies in the gene pool.

Natural Selection   For the Hardy–Weinberg equilibrium to be maintained, there must be no differential success in reproduction, a condition that is probably never met. Some individuals are more successful in producing viable, fertile offspring and thus passing their alleles to the next generation. Natural selection is likely to be adaptive; favorable genotypes are maintained in a population. Environmental changes result in selection favoring genotypes adapted to the new conditions.

The Genetic Basis of Variation

The Nature and Extent of Variation Within and Between Populations   Individual variation, the slight differences between individuals as a result of their unique genomes, is the raw material for natural selection. Polygenic traits that vary quantitatively provide much of the heritable variation within a population. Some traits vary categorically as distinct phenotypes and may be determined by a single gene locus. *Polymorphism* is a situation in which two or more distinct forms, or *morphs*, are evident in a population. Human blood groups are an example of a polymorphism.

The extent of genetic variation is evident in the molecular differences found by using biochemical methods such as electrophoresis to compare the protein products of specific gene loci among individuals in a population. In *Drosophila*, any two flies in a population may differ genetically at 25% of their loci.

Geographic variations are regional differences in allele frequencies among the populations of a species. These variations may be due to differing environmental selection factors or simply genetic drift. If an environmental parameter changes gradually across a distance, there may be graded variations within a species, called a *cline*, that parallel the environmental gradient. A cline may also exist along the graded

overlap between neighboring populations. A *step cline* occurs across a relatively short distance.

Sources of Variation   New alleles originate by mutations, most of which occur in somatic cells and cannot be passed on to the next generation.   Mutations that alter a protein enough to affect its function are more often harmful than beneficial. Rarely, however, a mutation may result in an individual who is better adapted to the environment and more reproductively fit, or a mutation already present in the population may be selected for when the environment changes.

Chromosomal mutations also are most often deleterious. Occasionally, a translocation may bring alleles together that are beneficial in combination. A *supergene* is a cluster of genes on a chromosome that have a cooperative function. An inversion of a supergene may preserve it by preventing crossing over with its noninverted homologous section.

Duplications that do not upset the genetic balance within cells may provide an expanded genome with superfluous loci that could eventually take on new functions by mutation.

Most of the genetic variation present in a population is due to the unique recombinations of existing alleles that each individual inherits from the gene pool. Crossing over and independent assortment produce gametes with a great deal of genetic variation, and each zygote has a unique assortment of genes from two parents. Animals and plants depend on sexual recombination for the genetic variation needed for adaptation.

Mutation can be a source of genetic variation, however, for bacteria and microorganisms with very short generation times. A new beneficial mutation can increase in frequency rapidly in a bacterial population that is growing by the asexual expansion of clones.

How Variation Is Preserved   Natural selection selects for favorable genotypes and tends to eliminate others, setting a trend toward genetic uniformity. The diploid character of most eukaryotes maintains genetic variation by the hiding of recessive alleles from selection when present in heterozygotes, perpetuating alleles that could be selected for should the environment change.

Balanced polymorphism occurs when natural selection serves to maintain variation at some gene loci. When individuals heterozygous at a certain gene locus are reproductively more fit, this *heterozygote advantage* tends to maintain two or more alleles in the population. In the case of sickle cell anemia, a lethal recessive allele is maintained in the population in countries with malaria due to the malaria-resistant advantage of the heterozygote.

Hybrid vigor may be seen in plants when two highly inbred varieties are crossbred. The crossbreeding may segregate harmful recessive alleles that were homozygous in the inbred varieties and produce a heterozygote advantage at other loci.

Balanced polymorphism may be maintained in a population that ranges across a patchy environment. In the land snail, *Cepaea nemoralis*, each of the distinctively colored morphs is camouflaged in a particular patch of the habitat. In *frequency-dependent selection*, a morph's reproductive success declines if it becomes too common. The polymorphism of female African swallowtail butterflies that resemble several different species of noxious butterflies increases the effectiveness of this Batesian mimicry. Should any one morph become too common, birds would stop associating that coloration with bad taste.

Is All Variation Adaptive? Some of the diversity seen in populations may be called *neutral variation*, genetic variations that do not result in a selective advantage for some individuals over others. According to a *neutral theory* of molecular evolution, many of the mutations affecting the beta chain of human hemoglobin do not appear to confer a selective advantage or disadvantage. The frequencies of these alleles will not be affected by natural selection.

The large quantity of noncoding DNA in a genome is considered by some scientists as evidence that DNA is inherently "selfish," replicating itself to the greatest extent possible within a species.

There is no consensus on how much variation, if any, is truly neutral, since it cannot be shown that an allele brings no benefits at all to an organism.

Adaptive Evolution

Adaptive evolution is a combination of the chance occurrence of new genetic variations by mutation and sexual recombination and the selection of those chance variations that fit organisms to their environments.

Fitness Darwinian fitness is measured by the relative contribution of an individual to the next generation's gene pool. Success is determined not simply by survival but by the number of offspring produced.

Population geneticists speak in terms of the *relative fitness* of a genotype as its contribution to the next generation as compared to the contributions of other genotypes. The most fecund variants are said to have a relative fitness of 1, whereas the relative fitness of another genotype is measured as the percentage of offspring it produces compared with the most fit vari-

ant. The *selection coefficient* is the difference between the two fitness values, a measure of the selection against the inferior variant. A lethal genotype would have a selection coefficient of 1.0.

The rate at which a deleterious allele declines in a population depends both on the selection coefficient and on whether the allele is dominant or recessive. Harmful recessives are rarely eliminated due to heterozygote protection. Likewise, beneficial recessives increase slowly because they must occur as a homozygote to be selected for. Dominant alleles can be selected for and against much more rapidly.

What Selection Acts On Selection acts on phenotype, the physical traits of an organism, and indirectly adapts a population to its environment by selecting for and maintaining favorable genotypes in the gene pool.

Genotypes may be pleiotropic—having more than one effect—some of which may be positive whereas others may be negative. Many traits are polygenic, influenced by several genes. Selection acts on an organism that is an integrated composite of many phenotypic features. The fitness of any genotype depends on the entire genetic context of the individual. Genes whose functions are related make up a *coadapted gene complex*. The integrated development of an organ requires coadaptation of alleles at many gene loci.

Under the influence of the environment, genotypes determine phenotypes within a range of possibilities, known as the *norm of reaction*. The norm of reaction can be specifically determined, as in blood groups, or broadly variable, as in mental abilities and behavioral patterns.

Modes of Natural Selection The frequency of a trait, especially a quantitative trait determined by many gene loci, may be affected by natural selection in three common ways. *Stabilizing selection* acts against extreme phenotypes and favors more intermediate forms, tending to reduce phenotypic variation in stable environments. *Directional selection* occurs most frequently during periods of environmental change in which individuals that deviate from the average for some phenotypic trait may be favored. *Diversifying selection*, or disruptive selection, occurs when the environment favors individuals on both extremes of a phenotypic range. Balanced polymorphism may result from diversifying selection.

Sexual Selection Sexual dimorphism is the distinction between males and females on the basis of secondary sexual characteristics such as size, plumage, or antlers. In vertebrates, the male is usually the showier sex, and these characteristics may serve to attract females or compete with other males for females. *Sexual selection* is the selection for traits that may not be adaptive to the environment but do enhance reproductive success by increasing the individual's success in attracting a mate.

Does Evolution Fashion Perfect Organisms?

There are at least four reasons why evolution does not produce perfect organisms. First, each species has evolved from a long line of ancestral forms, many of whose structures have been co-opted for new situations. Second, adaptations are often compromises between the needs to do several different things, such as swim and walk, be fast and strong. Third, the evolution that occurs as the result of chance events, such as genetic drift, is not adaptive. Fourth, natural selection can act on only those variations that are available; new genes do not arise when needed.

The Tempo of Microevolution   Natural selection is an agent of change and also an agent of the status quo. Stabilizing selection probably acts most of the time, resisting maladaptive change. Evolution occurs in spurts. When a population is faced with a major change in environment or genome, it may evolve, perhaps to become a new species, or, more commonly, it becomes extinct.

STRUCTURE YOUR KNOWLEDGE

1. For the Hardy–Weinberg equation, explain what p and q are and how they can be used to predict the frequencies of the genotypes of the next generation. Make sure you understand both how to use the Hardy–Weinberg equation and why it works.

2. Create a concept map that organizes your understanding of the possible causes of microevolution.

3. Natural selection tends to work toward genetic unity; the genotypes that are most fit produce the most offspring and increase the frequency of adaptive genotypes in the population. Yet there remains a great deal of variability within the populations of a species. Describe some of the factors that contribute to this genetic variability.

4. You collect 100 samples from a large butterfly population. 50 specimens are dark brown, 20 are speckled, and 30 are white. Coloration in this species of butterfly is controlled by one gene locus: BB individuals are brown, Bb are speckled, and bb are white. What are the allele frequencies for the coloration gene in this population? Is this population in Hardy–Weinberg equilibrium? Explain your answer.

TEST YOUR KNOWLEDGE

MULTIPLE CHOICE: *Choose the one best answer.*

1. The best description of a gene pool is
   a. all the genes that are present in a given species.
   b. the aggregate of all the alleles present in a population at a given time.
   c. the basic unit in which microevolution can occur.
   d. both b and c.

2. According to the Hardy–Weinberg theorem,
   a. the gene pool should remain constant from one generation to the next if five conditions are met.
   b. only natural selection, resulting in unequal reproductive success, will cause evolution.
   c. the square root of the frequency of individuals showing the recessive trait will always equal the frequency of q.
   d. all of the above are correct.

3. If a population has the following genotype frequencies, AA = 0.42, Aa = 0.46, aa = 0.12, what are the allele frequencies?
   a. A = 0.42, a = 0.12
   b. A = 0.88, a = 0.12
   c. A = 0.65, a = 0.35
   d. A = 0.6, a = 0.4

4. In a population with two alleles, B and b, the allele frequency of B is 0.8. What would be the frequency of heterozygotes if the population is in Hardy–Weinberg equilibrium?
   a. 0.8
   b. 0.16
   c. 0.32
   d. 0.64

5. In a population that is in Hardy–Weinberg equilibrium, 16% of the population shows a recessive trait. What percent shows the dominant trait?
   a. 6%
   b. 48%
   c. 60%
   d. 84%

6. Genetic drift is likely to be seen in a population
   a. that has a high migration rate.
   b. that has a high mutation rate.
   c. in which there is assortative mating.
   d. that is very small.

7. Cheetahs are believed to have experienced a bottleneck effect, which would have resulted in
   a. reduced genetic variability.
   b. many alleles that are fixed in their population.

c. a small population that is then subject to genetic drift.

d. all of the above.

8. Gene flow usually results in
   a. populations that are better adapted to the environment.
   b. an increase in sampling error in the formation of the next generation.
   c. a gain or loss of alleles caused by migration.
   d. nonassortative matings.

9. The existence of two distinctly colored forms in a species is known as
   a. geographic variation.
   b. sexual selection.
   c. heterozygote advantage.
   d. polymorphism.

10. Assortative mating will most likely result in
    a. a change in allele frequency.
    b. an increase in the gene loci that are homozygous.
    c. sexual selection.
    d. neutral variations.

11. Variations in a species existing over a large temperature range may
    a. result in stabilizing selection.
    b. be a cline if the variations parallel the temperature change.
    c. result in directional selection, favoring the milder climate.
    d. increase gene flow.

12. Mutations are rarely the cause of microevolution because
    a. they are most often harmful and do not get passed on.
    b. even if they are not harmful, they may be masked in the diploid condition and not able to be selected for.
    c. they occur very rarely.
    d. of all of the above.

13. In a study of population of field mice, you find that 48% of the mice have a coat color that indicates that they are heterozygous for a particular gene. What would be the frequency of the dominant allele in this population?
    a. 0.4
    b. 0.5
    c. 0.7
    d. you cannot estimate allele frequency from this information.

14. In a random sample of a population of Shorthorn cattle, 73 animals were red ($C^R C^R$), 63 were roan ($C^R C^W$—a mixture of red and white), and 13 were white ($C^W C^W$). Estimate the allele frequencies of C^R and C^W and determine whether the population is in Hardy–Weinberg equilibrium.
    a. $C^R = 0.7$, $C^W = 0.3$; the genotypic ratio is what would be predicted from these frequencies and the population is in equilibrium.
    b. $C^R = 1.04$, $C^W = 0.44$; the allele frequencies add up to greater than 1 and the population is not in equilibrium.
    c. $C^R = 0.64$, $C^W = 0.36$; because the population is large and a random sample was chosen, the population is in equilibrium.
    d. you cannot estimate allele frequency from this information.

15. The rate at which a harmful allele declines in a population depends on
    a. whether the allele is dominant or recessive.
    b. the selection coefficient for the genotypes in which it occurs.
    c. the frequency of the allele in the population.
    d. all of the above.

16. The phenotype that a genotype determines may vary due to
    a. the occurrence of pleiotropic genes.
    b. the effect of the environment.
    c. a very narrow norm of reaction.
    d. coadapted gene complexes.

17. Disruptive selection may result in
    a. balanced polymorphism.
    b. one phenotype gradually disappearing from the population.
    c. sexual dimorphism.
    d. a high selection coefficient.

18. Sexual selection will
    a. result in individuals better adapted to the environment.
    b. increase assortative mating.
    c. select for traits that enhance an individual's chances of mating.
    d. result in stabilizing selection.

19. Which of these is *not* a reason why evolution does not produce perfect organisms?
    a. Much of evolution may occur as the result of chance.
    b. Adaptations are often compromises between different needs.
    c. Selection works on the structures available at the time.
    d. Unless conditions are always changing, equilibrium sets in and evolution does not occur.

20. The greatest source of genetic variation in most populations is from

    a. mutations.
    b. recombination.
    c. selection.
    d. polymorphism.

21. A supergene is a
    a. coadapted gene complex located close together on a chromosome.
    b. pleiotropic gene.
    c. polygenic complex.
    d. gene found on a polytene chromosome.

22. A population of a plant lives in an area that is becoming more arid. The average surface area of leaves has been decreasing over the generations. This is an example of
    a. stabilizing selection.
    b. directional selection.

    c. disruptive selection.
    d. heterozygote superiority.

23. If individuals with a particular genotype have a relative fitness of 0.7, they
    a. have a selection coefficient of 0.7.
    b. are carrying a lethal gene.
    c. leave 70% as many offspring as do individuals with the most fecund genotype.
    d. are reproductively less fit than genotypes with a relative fitness of 0.3.

24. Neutral variations
    a. are not acted on by natural selection.
    b. may be present in the large amount of DNA that does not get expressed.
    c. appear not to influence reproductive success.
    d. all of the above

CHAPTER 22

THE ORIGIN OF SPECIES

FRAMEWORK

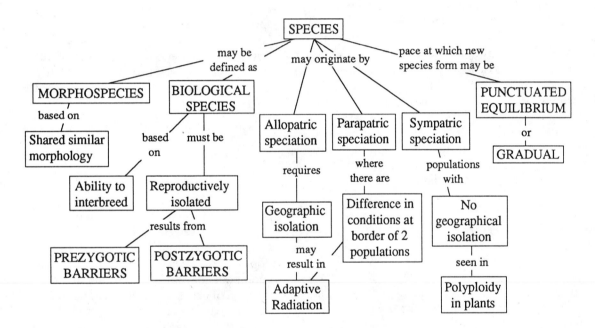

CHAPTER SUMMARY

The origin of species, or *speciation*, is the basis of the evolution of biological diversity. *Anagenesis*, or *phyletic evolution*, involves the transformation of an entire population into a different enough form that it is renamed a new species. In *cladogenesis*, or branching evolution, a new species arises from a parent species that continues to exist. Cladogenesis is both the more common pattern of evolution and the process that increases biological diversity. Evolutionary theory attempts to determine the mechanisms by which new species originate.

The Species Problem

Taxonomists often find that their scientific classification of local species corresponds to the folk taxonomy of the region.

Two Concepts of Species *Species* are most often characterized by their physical form or morphology. Such groups, determined on the basis of anatomical features, are known as *morphospecies*. The concept of *biological species*, developed by Mayr in 1942, goes beyond the physical differences between species and considers reproductive and thus genetic isolation to be the basis for separating species.

A biological species consists of a population or group of populations of individuals that have the potential to interbreed with each other in nature and produce fertile offspring. A biological species is the largest unit in which gene flow is possible. Members of a species are said to be *conspecific*.

Limitations of the Biological Species Concept

The biological species concept does not work for species that are completely asexual, such as prokaryotes and some protists and fungi. It is also impossible to group extinct species on the criterion of interbreeding. Geographically isolated populations may or may not be able to interbreed. Even in some sexually reproducing, geographically neighboring species, it may be difficult to apply the biological species concept. Phenotypically distinct populations that are separated geographically but presumably capable of interbreeding are sometimes called subspecies. The gene flow among subspecies may be so circuitous or slight that it is difficult to decide whether some subspecies should be designated as separate species. Perhaps instances in which the biological species concept is difficult to apply are in the midst of the gradual evolution of populations into new species. Most often, the application of the concepts of morphospecies and biological species results in the classification of the same groups as species. Reproductive isolation preserves the physical distinctions between species.

Reproductive Barriers

Any intrinsic mechanism that prevents two species from producing fertile hybrids is a reproductive barrier that serves to preserve the genetic integrity of a species. *Prezygotic barriers* function before the formation of a zygote by preventing the mating between species or the successful formation of a zygote. Should a hybrid zygote form, *postzygotic barriers* prevent it from developing into a fertile adult.

Prezygotic Barriers

Prezygotic barries include *ecological isolation*, in which two species may live in the same area but occupy different habitats; *temporal isolation*, in which two species breed at different times; and *behavioral isolation*, in which courtship rituals and physical or chemical signals attract mates of the proper species. Pheromones are volatile chemical compounds that may serve as species-specific sexual attractants.

Anatomical incompatibility may provide a *mechanical isolation* for closely related species. Mechanical barriers may also be present in flowering plants in which the floral anatomy is adapted to a specific pollinator.

Gametic Isolation

Should other prezygotic barriers fail, *gametic isolation* will usually prevent interbreeding between species because the gametes will not fuse to form a zygote. Gamete recognition may depend on molecular receptors on the egg cell; similar molecular recognition mechanisms are involved in pollen discrimination by flowers.

Postzygotic Barriers

Postzygotic barriers may include *hybrid inviability*, in which a hybrid zygote fails to survive embryonic development due to genetic incompatibility; *hybrid sterility*, in which a viable hybrid individual is sterile, often due to the inability to produce normal gametes in meiosis; or *hybrid breakdown*, in which the hybrids are viable and fertile but their offspring are defective or sterile.

Introgression

When fertile hybrids occasionally successfully mate with one of their parent species, genes may pass between species in a process called *introgression*. This small amount of gene transplantation may increase the reservoir of genetic variation present in a species without threatening its integrity as a distinct species.

The Biogeography of Speciation

Allopatric speciation occurs when the gene pool of a population is segregated geographically from other populations and follows its own evolutionary course as a result of selection, genetic drift, and mutation. If gene flow is very slight at the boundary between two populations, *parapatric* speciation may occur when the gene pools of both populations diverge without the dilution of genes from their neighbors. *Sympatric* speciation may occur when a subpopulation becomes reproductively isolated in the midst of its parent population.

Allopatric Speciation

Geological changes can cause sympatric populations to become allopatric. The extent of the geographical barrier necessary to maintain separation depends on the ability of the organisms to disperse.

The geographic isolation of a small population at the fringe of the parent population's range may result in speciation because of three factors. First, the gene pool of the peripheral isolate probably represents the extremes of any clines in the population. If the isolated population is small, the founder effect also may have produced a population whose gene pool differs initially from the parent pool. Second, genetic drift in the small gene pool may cause phenotypic divergence from the parent population by chance. Third, selection pressures are likely to be different, and possibly more severe, on the fringe of a population's range, pushing the peripheral isolate's gene pool in a different direction than the parent population. These factors, however, do not guarantee speciation. Most pioneer populations probably become extinct before they change enough to become new species.

Allopatric speciation occurs much faster in small populations, perhaps within only hundreds to thousands of generations.

Allopatric speciation may occur on island chains, where small founding populations may diverge in isolation. A single ancestral species of finch probably gave rise to the 13 species of finches now found on the Galapagos through the process of isolation and recolonization of islands that may have presented varying habitats. *Adaptive radiation* is the formation of numerous species from an ancestral species that has been confronted with a new and diverse habitat.

Parapatric Speciation   Parapatric populations occupy separate ranges that abut along a common border. This border may represent a region of change for some important environmental factor, such that the gene pools of the two populations have different selective pressures. The two gene pools may diverge enough that the limited breeding at the border may not maintain genetic continuity between the populations. Hybrids produced at the border may be less well adapted to conditions on either side, and postzygotic barriers may facilitate speciation. Parapatric speciation is theoretically possible, although no specific instances can be cited.

Sympatric Speciation   In sympatric speciation, a radical change in the genome of a subpopulation may result in its reproductive isolation in the midst of the parent population. Such events are more common in plants than in animals. In *autopolyploidy*, a single species may double its chromosome number to the tetraploid ($4N$) state through the fusion of diploid gametes. Tetraploids can fertilize themselves or mate with other tetraploids but cannot mate with diploids from the parent population. This postzygotic barrier results in reproductive isolation in just one generation. De Vries documented this form of speciation in the evening primrose.

Polyploid species may arise more commonly through *allopolyploidy*. An interspecific hybrid plant is usually sterile due to difficulties in the meiotic production of gametes. One of two mechanisms may result in the production of a fertile polyploid from a sterile hybrid clone. A mitotic nondisjunction in the reproductive tissue would allow the hybrid to produce gametes with complete sets of chromosomes, which could fuse with gametes from the same plant or other hybrid tetraploids and create a new species that is reproductively isolated from both parent species. Or an initial diploid gamete from one plant may fuse with a normal haploid gamete of another, creating a sterile triploid hybrid. If a later meiotic nondisjunction in the asexual clone produces a triploid gamete that then unites with a normal haploid gamete of the second species, a fertile tetraploid hybrid would result. The chromosome number of an allopolyploid is equal to the sum of the chromosome numbers of the parent species.

Speciation of polyploids, especially allopolyploids, has been frequent and important in plant evolution. Many of our agricultural plants are polyploids. Plant geneticists are now hybridizing plants and using chemicals to induce nondisjunction in order to create new, specially adapted polyploid species.

Genetic Mechanisms of Speciation

Speciation by Divergence   Reproductive barriers may arise coincidently as two populations diverge as they adapt to different environments.

When allopatric populations become sympatric again, either in nature or experimentally in the laboratory, they may not be reproductively isolated and may interbreed as one species, or they may have developed intrinsic reproductive barriers that maintain them as separate species. The populations could also interbreed on a limited scale, and the hybrids may be less fit than individuals of the parent populations.

Populations that are not quite reproductively isolated are called *semispecies*. One hypothesis holds that the association of semispecies will reinforce prezygotic reproductive barriers, because the individuals with these barriers will not mate outside their population and will leave more offspring than individuals that interbreed and produce inviable or sterile hybrids. Population geneticists debate the importance of this reinforcement of reproductive isolation in the speciation process.

In evolution by divergence, natural selection does not necessarily favor reproductive barriers; genetic isolation may be a secondary consequence of the diverging adaptations to different environments. Postzygotic barriers that affect the viability of hybrids may result from pleiotropic effects of important genes controlling development. As populations diverge in their adaptation to specific environments, prezygotic ecological barriers can also evolve.

Sexual selection, which enhances an organism's reproductive success with its own species, may lead indirectly to reproductive barriers. The *recognition concept of species* emphasizes that natural selection would act on characteristics that maximize successful mating within a species, as opposed to driving the development of reproductive barriers in an isolated population that may be speciating.

Speciation by Peak Shifts   According to Sewell Wright's "adaptive landscape" metaphor, *adaptive peaks* occur when a population's gene pool is at an equilibrium that maximizes fitness, and valleys separating the peaks occur when a change in allele frequencies has reduced the average fitness of individuals. Natural selection will push the population back to the former peak or on to an alternative adaptive peak. When environmental conditions change, the adaptive landscape changes.

Peak shifts, associated with nonadaptive changes in the gene pool, can be initiated by a founder effect or a bottleneck in a population and the accompanying genetic drift. The adaptive landscape is redefined by a change in location or a major change in karyotype, as in the establishment of polyploidy in a plant. Adaptive evolution can push a destabilized gene pool to a new adaptive peak.

How Much Genetic Change is Required for Speciation?   Substantial genetic change is not required for speciation. Reproductive isolation may occur by a change at a single gene locus, especially in a coadapted gene complex, or by the cumulative divergence at many gene loci.

Gradual and Punctuated Interpretations of Speciation

The traditional evolutionary concept of the origin of species involves the gradual divergence of populations by microevolution, with each newly formed species continuing to evolve over long periods of time. The fossil record, however, provides few cases of gradually transitioning forms. Rather, new forms appear rather suddenly, persist unchanged for a long time, and then disappear.

In 1972 Eldredge and Gould proposed the theory of *punctuated equilibrium*, in which long periods of stasis are punctuated by episodes of relatively rapid change and speciation. Speciation both by polyploidy in plants and by small allopatric populations may occur fairly rapidly. In geological time, a few thousand years for a species to evolve is small compared to the few millions of years a successful species may remain in existence.

The degree to which a species changes after its origin may be very little if it remains in a stable environment. Once selection has resulted in a complex of coadapted genes, new changes in the genome will tend to be disruptive. Long periods of stasis may result from the tendency of natural selection to maintain a population at an adaptive peak.

Some gradualists maintain, however, that fossils report stasis only in external anatomy and that changes in internal anatomy, physiology, and behavior go unrecorded. Even the reports of long periods of stasis is debatable. After analyzing a particularly complete set of trilobite fossils, Sheldon reported a gradual change in aspects of their morphology. He disagreed with paleontologists who had assigned the youngest and oldest fossils in each evolutionary lineage to a different species. Many additional studies of fossil lineages will be needed to determine the roles of gradual and punctuated tempos in the origin of species.

STRUCTURE YOUR KNOWLEDGE

1. How are speciation and microevolution different?

2. What is probably the key event in the origin of a species? How might this event occur?

3. Compare the gradual and punctuated equilibrium theories of evolution. Which theory seems to have the most evidence supporting it? Describe this evidence.

TEST YOUR KNOWLEDGE

MULTIPLE CHOICE: *Choose the one best answer.*

1. Most new species probably have arisen by
   a. anagenesis.
   b. cladogenesis.
   c. phyletic evolution.
   d. both a and c.

2. Which of the following is *not* a type of intrinsic reproductive isolation?
   a. mechanical isolation
   b. behavioral isolation
   c. geographical isolation
   d. gametic isolation

3. The individuals placed in a morphospecies and a biological species may differ because
   a. organisms that may look different may be able to interbreed.
   b. organisms that do not breed in nature may do so in captivity.

c. hybrids formed from the mating of two morphospecies are sterile.

d. anatomical features do not provide a reliable basis for grouping organisms.

4. Gene flow can occur within all of the following groups of organisms except
   a. a population.
   b. a species.
   c. a genus.
   d. a subspecies.

5. For which of the following is the biological species concept least appropriate?
   a. plants
   b. prokaryotes
   c. two subspecies that are geographically isolated
   d. field biologists

6. Subspecies are usually
   a. separated geographically.
   b. phenotypically distinct.
   c. assumed to be capable of interbreeding.
   d. all of the above.

7. If two species are able to breed but produce sterile hybrids, their genetic integrity is maintained by
   a. gametic isolation.
   b. a prezygotic barrier.
   c. hybrid inviability.
   d. a postzygotic barrier.

8. Introgression occurs when
   a. a hybrid successfully breeds with an individual of a parent species.
   b. hybrids successfully breed with each other.
   c. individuals from two species successfully breed.
   d. all of the above take place.

9. Speciation is most likely to occur in allopatric populations when
   a. the splinter population is small.
   b. a peripheral isolate is exposed to more severe selection pressures at the boundary of the population's range.
   c. the flow of genes across the border between the two populations does not mediate the differing selective pressures to which the populations are exposed.
   d. both a and b.

10. According to one hypothesis, reproductive isolation may be reinforced
    a. by postzygotic barriers when semispecies breed and produce less fit hybrids, thus reinforcing the development of prezygotic reproductive barriers.

b. when two populations evolve in different environments.

c. when allopatric populations become sympatric again and they have developed intrinsic reproductive barriers.

d. when hybrids are capable of breeding with members of the parent's populations.

11. Sexual selection may lead indirectly to reproductive barriers because
    a. isolated populations are exposed to different selection pressures.
    b. natural selection could act on characteristics that maximize successful mating within a species.
    c. hybrids may be reproductively more fit.
    d. prezygotic barriers are more likely to evolve before postzygotic barriers.

12. Adaptive radiation may occur when
    a. a small population is reunited with its parent population.
    b. numerous invasions and allopatric speciations take place on islands that provide diversity of environmental opportunities and problems.
    c. a segment of the population develops a reproductive isolating mechanism.
    d. all of the above

13. According to Wright's evolutionary metaphor,
    a. a population has only one adaptive peak and is maintained there by natural selection.
    b. a small population that drifts into a valley surrounding an adaptive peak will either become extinct or evolve into a new species.
    c. an environmental change may create a new adaptive landscape and natural selection can push a destabilized gene pool to a new adaptive peak.
    d. all of the above

14. The punctuated equilibrium theory maintains that
    a. long periods of stasis are punctuated by episodes of relatively rapid speciation and change.
    b. microevolution is the driving force of speciation.
    c. most rapid speciation events involve polyploidy in plants.
    d. evolution occurs rapidly when the adaptive landscape changes.

15. The act of speciation is most dependent on
    a. natural selection.
    b. allopatric populations.
    c. chance events and reproductive isolation.
    d. changing environments.

16. Sheldon's anaylsis of huge numbers of trilobite fossils provides support for

    a. the punctuated equilibrium model of evolution.

    b. the gradual tempo of the evolution of new species.

    c. the need for additional exhaustive studies of fossil lineages.

    d. both b and c.

CHAPTER 23

MACROEVOLUTION

FRAMEWORK

This chapter considers the major events and evolutionary trends that have led to the biological diversity of today. A goal of systematics is to determine the phylogenetic history of species. The branch of systematics called taxonomy names and classifies species according to their presumed evolutionary relationships. Phylogenetic trees are constructed from evidence gathered from the fossil record and from anatomical and molecular homologies.

Mechanisms for macroevolution include the gradual modification of preadapted structures for new functions, alterations in regulatory genes that result in major morphological changes, evolutionary trends resulting from species selection, and adaptive radiations as new adaptive zones appear when evolutionary novelties develop or mass extinctions occur. Continental drift and other major geological events have shaped the direction of macroevolution. The relationship between microevolution and macroevolution and the role of natural selection and chance in macroevolution are controversial topics in contemporary evolutionary theory.

CHAPTER SUMMARY

The term *macroevolution* refers to major events and evolutionary trends in the history of life. When biologists study macroevolution, they consider large diversifications of taxonomic groups, the origin of novel biological designs, major extinctions, and the mechanisms that may have produced these evolutionary developments.

The Record of the Rocks

Paleobiologists reconstruct evolutionary history by studying the succession of organisms found in the fossil record. Paleontologists focus on animal fossils; paleobotanists study the fossils of plants.

Fossils Fossils are preserved impressions or remnants of past organisms. Sometimes enough organic material is retained in the fossil that biochemical analysis and electron microscopic study of cells can be done. Hard parts of animals, such as bones, teeth, or shells, may remain as fossils. Petrification, the replacement of tissues with dissolved minerals, may turn the fossil to stone. Molds of organisms, left when they were covered by mud or sand, are a common type of fossil.

The Geological Time Scale Sedimentary rocks, formed from the compression of deposits of mud or sand, are the richest source of fossils. Layers, or strata, of rock form during the varying periods of sedimentation within bodies of water. The order in which fossils appear in the strata of sedimentary rocks indicates the relative ages of the fossils.

Index fossils, such as shells of widespread animals, are used to correlate strata from different locations and allow geologists to develop a composite picture of a consistent sequence of geological periods. There are four geological eras—the Precambrian, Paleozoic, Mesozoic, and Cenozoic—that are delineated by major transitions in the fossils found in the rocks. The eras are subdivided into periods and epochs, which are associated with particular evolutionary developments.

Radioactive dating is often used to determine the age of rocks and fossils. During an organism's lifetime, it accumulates radioactive isotopes of certain elements in proportions equal to the relative abundances of the isotopes in the environment. After the organism dies, the isotopes begin to decay at a fixed rate, known as the *half-life*, or the number of years it takes for one-half of the radioactive isotopes present in a specimen to decay. Carbon-14 dating is used to determine the age of relatively young fossils; potassium-40, with a half-life of 1.3 billion years, can be used to date rocks hundreds of millions of years old.

During an organism's life, only *L*-amino acids are synthesized, but following death, *L*-amino acids are slowly converted to *D*-amino acids. The proportion of *L*- and *D*-amino acids can be used to date some fossils.

The formation of a fossil is an unlikely occurrence. The imperfection of the fossil record is understandable considering that a large number of species that lived probably left no fossils, most fossils that form are destroyed, and only a fraction of existing fossils have been found. Although more transitional fossils have been found since Darwin's day, there is still a scarcity of fossils that show a progression of changes from ancestral to present-day forms. Some paleontologists interpret the discontinuities in the fossil record as evidence for the punctuated equilibrium model of evolution.

Systematics: Tracing Phylogeny

Phylogeny is the evolutionary history of a species. A phylogenetic tree is a diagram of the proposed evolutionary relationships of various groups. *Systematics* is the branch of biology concerned with the diversity of life and its phylogenetic history.

Taxonomy   Taxonomy involves the identification and classification of species. Linnaeus developed a system of taxonomy that assigned to each species a two-part Latin name—a *binomial* consisting of a genus and species name. Linnaeus also developed a hierarchy of classifications that organizes similar groups into more general categories, proceeding from species to *genus, family, order, class, phylum,* and finally to *kingdom*.

Taxonomy describes diagnostic characteristics for closely related organisms and assigns names to new species. Species are ordered into the *taxa*, or units, of the increasingly broader taxonomic categories. Both words of the binomial are italicized, and the names for all taxa at the genus level and higher are capitalized.

Only species exist in nature as biologically identifiable groups, connected by interbreeding and reproductively isolated from other groups. The assignment of taxa to each broader category is often a subjective enterprise, depending on the distinctions the taxonomist deems important. A goal of classification is to reflect the evolutionary relationships of species. Each

taxon should be *monophyletic*, meaning that it should contain only species that are derived from a single ancestor. A *polyphyletic* taxon, such as the plant kingdom, would include groups having different ancestry, whereas a *paraphyletic* taxon, such as the reptilian class, excludes species that share a common ancestor with other species in the taxon.

Sorting Homology from Analogy   Species are generally classified into higher taxa based on similarities in morphology and other characteristics. *Homology* is likeness based on a shared ancestry; *analogy* is similarity due to evolutionary convergence. In *convergent evolution*, unrelated species develop similar features because they have similar ecological roles and natural selection has chosen for analogous adaptations.

Phylogenetic trees are built on homologous similarities. In general, the more homology between two species, the more closely they are related. The issue can be confused when adaptation leads to large differences and convergence creates misleadingly similiar structures.

Molecular Systematics   A powerful taxonomic tool is the comparison of sequences of amino acids in proteins and nucleotides in DNA as evidence of shared ancestry. The protein cytochrome *c* is found in all aerobic organisms. Comparisons of the amino acid sequences of this protein from many species show that differences increase as the groups are more taxonomically distant. Phylogeny based on cytochrome *c* is consistent with comparative anatomy and fossil evidence.

DNA–DNA hybridization is a technique that can compare single-stranded DNA of two species by determining the extent of their base-pairing and, thus, the homology between the DNA sequences of the species. Evidence from this technique generally supports phylogenetic relationships established by other methods and has been used to settle some old taxonomic disputes.

Restriction mapping of DNA has been used to compare segments of DNA from different species. If two species have diverged greatly, their collections of restriction fragments will not match closely. Restriction mapping of mitochondrial DNA, which mutates about ten times faster than does nuclear DNA, provides phylogenetic information for closely related species.

The actual sequencing of DNA from two species provides the most accurate information on genetic divergence. The sequencing of ribosomal RNA, whose genes change slowly compared to other DNA, can be used to determine early phylogenetic relationships.

Some proteins seem to change or evolve at consistent rates. Thus, the number of amino acid substitutions in homologous proteins is proportional to the time elapsed from when two species branched off from a common ancestor. DNA comparisons can also be used as molecular clocks to date phylogenetic branchings. Dates obtained by this method are generally consistent with the fossil record and may be more closely correlated with the time that species have been separated than are morphological differences.

The consistent rate of protein change and the rate of DNA divergence indicate that the accumulation of neutral mutations may change the genome as a whole more than do changes brought about by selection. Disagreement about the extent of neutral mutations has caused some evolutionists to rely on molecular systematics to determine the sequence of branches in phylogeny but not the actual dates for the origin of taxa.

Schools of Taxonomy   Phylogenetic trees indicate the relative time of origin of different taxa and the degree of divergence or difference that develops between branches. *Phenetics* is a school of taxonomy that determines taxa strictly on the basis of measurable similarities and differences for as many anatomical characteristics as possible. Computers are used to make quantitative comparisons between species and then groups of species, producing a dichotomously branched tree called a phenogram.

Cladistics is a taxonomic school that classifies organisms according to the order in time that clades, or evolutionary branches, arise. Phylogeny is diagrammed on a cladogram, a series of dichotomous forks, each of which is defined by novel homologies for the species on that branch. A primitive character, or *plesiomorphic character*, is common for all the species to be grouped, and derived characters, or *apomorphic characters*, are identified as homologies that evolved after a branch point and define that branch.

In the cladistic approach, birds are closer relatives of crocodiles than crocodiles are of lizards and snakes. Birds and crocodiles share apomorphic characters not found in snakes and lizards. According to cladistics, the class Aves does not exist because birds are found within the clade of reptiles. The degree of morphological difference between branches is not a consideration in cladistics, only the time and sequence of evolutionary origin.

Classical evolutionary taxonomy, which predates both phenetics and cladistics, considers both the homology of structures and the sequence of branching. In cases of conflict between these two characteristics, a subjective decision is made. According to classical taxonomy, even though birds may share a close phylogenetic branch with crocodiles, they are assigned to their own class on the basis of the major adaptive divergence that has resulted from the ability to fly.

Phenetics and cladistics produce different classification schemes because the evolutionary rate of morphological change is not constant. Taxonomic disagreements arise from the need for compromise between a classification system that accurately reflects genealogies and one that provides a useful organization for the diversity of species.

Mechanisms of Macroevolution

Origins of Evolutionary Novelties   Evolutionary novelties that define higher taxa may evolve by the gradual modification of existing structures for new functions. *Preaptation* is the term for structures that evolved and functioned in one setting and are co-opted for a new function. Feathers and wings may have developed as prey-capturing structures that were then shaped into structures adapted for flight. Along with the Darwinian tradition of gradualism, preaptation provides a mechanism by which novel designs arise slowly through many small changes in current structures that become adapted for another function.

Slight changes in the function of regulatory genes may create major changes in development and adult morphology. Since each regulatory gene may influence hundreds of structural genes, evolutionary novelties could arise fairly rapidly due to a change in a regulatory gene.

Allometric growth, the different rates of growth in various parts of the body, results in the final shape of the organism. A minor genetic alteration that affects allometric growth could have a major morphological effect.

Novel organisms could also be produced by genetic changes that affect the timing of development. *Paedomorphosis* is the retention in the adult of juvenile traits of ancestral organisms. The continuation of growth of the human brain for several years longer than the chimpanzee brain is a prolonging of a juvenile process that has greatly influenced human characteristics.

What Produces Evolutionary Trends?   Evolutionary trends rarely occur in the fossil record as a sequence of gradually changing intermediate forms. *Equus* is not the direct result of trends of increasing size, reduction in the number of toes, and changes in dentition occurring by phyletic evolution from the much smaller ancestor *Hyracotherium* to the modern horse of today. Several adaptive radiations resulted in many different species that appeared in the fossil record, remained fairly unchanged, and then became extinct.

Evolutionary trends that are evident in the fossil record are most often the result of punctuated equilibrium, in which new species change by increments as they branch from ancestral ones.

According to Steven Stanley's model of *species selection*, an evolutionary trend is analogous to a trend in a population produced by natural selection when the best-adapted individuals are most successful reproductively. The species that live the longest before extinction and speciate most (produce the most offspring) determine the direction of the trend. Evolutionary trends are ultimately dictated by environmental conditions; if conditions change, an evolutionary trend may end or change.

Continental Drift and Macroevolution Biogeography, the geographic distribution of species, is correlated with the geological history of the earth. The continents rest on great plates of crust and upper mantle that float on the molten mantle. These plates shift and move, creating earthquakes, volcanoes, and mountains in regions where plates abut.

Large-scale continental drift brought all the land masses together into a supercontinent named *Pangaea* about 250 million years ago, near the end of the Paleozoic era. This tremendous change undoubtably had a great environmental impact as shorelines were eliminated, oceans got deeper and shallow coastal areas were drained, and the climate of the land mass changed. Species that had evolved in isolation came together and competed, many species became extinct, and new opportunities for remaining species became available.

About 180 million years ago, during the early Mesozoic era, Pangaea broke up and the continents drifted apart, creating a huge geographic isolation event. The current biogeography reflects this separation.

Punctuations in the History of Biological Diversity Major adaptive radiations have occurred during the early history of some taxa when the evolution of a novel characteristic opened a new *adaptive zone*, or way of life with unexploited opportunities. Nearly all the animal phyla that exist today evolved during the first 10 to 20 million years of the Cambrian, the first period of the Paleozoic era. The origin of shells and skeletons in a few taxa opened an adaptive zone by making more complex designs possible and changing predator-prey relationships.

An empty adaptive zone can be exploited only if appropriate evolutionary novelties arise; novelties that do arise cannot enable organisms to move into adaptive zones that do not exist or are filled. Mammals existed at least 75 million years before their first major adaptive radiation, which has been linked to the ecological void left by the extinction of the dinosaurs. Mass extinctions are often followed by new adaptive radiations.

A species may become extinct due to a change in its physical or biological environment. An evolutionary change in one species may impact on other species in the community. Extinction, inevitable in a changing world, usually occurs at a rate of between 2.0 and 4.6 families per million years. During periods of major environmental change, mass extinctions may occur.

About a dozen mass extinctions have been found in the fossil record. The Permian extinctions, occurring at the boundary between the Paleozoic and Mesozoic eras (about 250 million years ago), claimed over 90% of the species of marine animals. This extinction coincided with the formation of Pangaea and may be related to the changes associated with that event.

The Cretaceous extinction, which marks the boundary between the Mesozoic and Cenozoic eras about 65 million years ago, claimed over one-half the marine species, many families of terrestrial plants and animals, and the dinosaurs. The climate was cooling during that time, and shallow seas were receding from continental lowlands.

Sunlight may have been blocked as a result of increased volcanic activity. There is also evidence that an asteroid or comet collided with Earth during the Cretaceous extinctions. Separating Mesozoic and Cenozoic sediments is a thin layer of clay enriched in iridium, an element rare on earth but common in meteorites. This layer may have been the fallout from a huge cloud of dust that was created when an asteroid hit Earth. This cloud, similar to what is projected in a nuclear winter, would have blocked light and severely affected weather for several months.

The linking of the evidence for an asteroid and the mass extinctions occurring at this time does not indicate cause and effect. Many scientists believe that changes in climate due to continental drift and other processes were sufficient to account for the mass extinctions that took place.

Is a New Synthesis Necessary?

The modern synthesis, which has dominated evolutionary theory for 50 years, combines contributions from paleontology, biogeography, systematics, and population genetics, along with recent discoveries in molecular biology, to reaffirm the Darwinian view of life. This paradigm maintains that the gradual accumulations of many small changes occurring over vast periods of time can result in large-scale evolutionary changes. Microevolution, changes in gene frequencies in populations, is sufficient to explain most macroevolution, and natural selection is seen as the major cause of evolution at all levels.

A current evolutionary debate concerns both the rates of evolution (gradualism versus punctuated equilibrium) and the relative contribution of microevolution to macroevolution. Some scientists favor a hierarchical theory, which maintains that events that lead to speciation and episodes of macroevolution may have little to do with adaptation.

According to the hierarchical theory, the beginnings of most new species result from chance geographic isolation, genetic accidents, and genetic drift in small, isolated populations. Macroevolution related to continental drift and mass extinctions is believed to have affected biological diversity as much as has the gradual adaptation that results from selection. And most evolutionary trends occur not by phyletic transitions caused by microevolution, but by species selection—the differential survival and speciation of separate species that change little after they come into existence.

The modern synthesis does not claim that evolution is always gradual or that chance events do not change gene pools. The major difference between the modern synthesis and the hierarchical theory is the relative importance of the different evolutionary mechanisms of natural selection and chance events, and the connection between microevolution and macroevolution. Evolutionists from both schools, however, agree that natural selection is the mechanism of adaptation that fine-tunes a population to its environment.

STRUCTURE YOUR KNOWLEDGE

1. Organize a concept map that describes the major activities and objectives of taxonomy.

2. Compare microevolution and macroevolution. To what extent are these two related?

TEST YOUR KNOWLEDGE

MULTIPLE CHOICE: *Choose the one best answer.*

1. The richest source of fossils is found
   a. where roadways cut through rock layers.
   b. along gorges.
   c. within sedimentary rock strata.
   d. encased in volcanic rocks.

2. Which of the following is least likely to leave a fossil?
   a. a soft-bodied land organism such as a slug
   b. a marine organism with a shell such as a mussel
   c. a vascular plant embedded in layers of mud
   d. a fresh-water snake

3. Index fossils are fossils of
   a. unique organisms that are used to determine the relative rates of evolution in different areas.
   b. widespread organisms that allow geologists to correlate strata of rocks from different locations.
   c. transitional forms that link major evolutionary groups.
   d. extinct organisms that mark the separation of different eras.

4. The half-life of Carbon-14 is 5600 years. A fossil that has 1/8 the normal proportion of C-14 to C-12 is probably
   a. 2800 years old.
   b. 11,200 years old.
   c. 16,800 years old.
   d. 22,400 years old.

5. Which of the following is *not* a reason for an incomplete fossil record?
   a. Fossils only form when organisms are buried in sand or mud.
   b. Erosion and other processes may destroy fossils.
   c. The large majority of fossils are not found.
   d. Some of the transitional forms expected by evolutionists may not exist due to a punctuated equilibrium pace of evolution.

6. The relatively youngest fossils
   a. will have the greatest amount of D amino acids.
   b. will be found in the highest strata.
   c. will have the greatest amount of DNA–DNA hybridization.
   d. all of the above.

7. In the binomial *Homo Sapiens*,
   a. *Homo* is the name of the genus.
   b. *Sapiens* should not be capitalized.
   c. the words are italicized because they are Latin.
   d. All of the above are true.

8. Related families are grouped into the next highest taxon called a
   a. class.
   b. order.
   c. phylum.
   d. genus.

9. The only biologically identifiable group in nature is the
   a. kingdom.
   b. phylum.
   c. genus.
   d. species.

10. When two structures are homologous, they
    a. indicate that the species have evolved from a common ancestor.
    b. share common embryological development patterns.

c. still may be adapted for different functions in the two species.

d. all of the above

11. DNA–DNA hybridization compares the genomes of two species by
    a. determining the temperature at which hybrid DNA separates.
    b. determining the extent of base pairing between single-stranded DNA of two species.
    c. establishing the degree of DNA homology and thus the degree of relatedness between two species.
    d. all of the above

12. Convergent evolution may result
    a. when species have similar ecological roles.
    b. when homologous structures are adapted for different functions.
    c. as a result of adaptive radiation.
    d. when species are widely separated geographically.

13. *Lystrasaurus*, a fossil dinosaur, is found on most continents. This distribution is best explained by
    a. the relative ease of finding dinosaur bones.
    b. divergent evolution.
    c. continental drift.
    d. dispersal capabilities of ancient reptiles.

14. The development of feathers in the ancestors of birds is an example of
    a. convergent evolution.
    b. divergent evolution.
    c. preaptation.
    d. adaptive radiation.

15. Shared derived characters are
    a. used to characterize all the species on a branch of a cladogram.
    b. common for all groups on a major line of a cladogram.
    c. homologous structures that develop during adaptive radiation.
    d. used to indicate the degree of morphological differences between clades.

16. Many hypotheses have been suggested for the extinction of the dinosaurs. The explanation that probably accounts for most of the decline in species diversity at the end of the Mesozoic is
    a. the collision of an asteroid with Earth.
    b. mammalian competition with poorly adapted reptiles.
    c. ecological shifts due to climatic change.
    d. overpopulation of the huge reptiles.

17. Allometric growth
    a. is the uneven growth of different body parts.

b. involves paedomorphosis.
c. results in an evolutionary trend of increasing body size.
d. results in the phyletic evolution of a change in body size.

18. Punctuated equilibrium is a part of the hierarchical theory because
    a. it emphasizes the role of chance in speciation events before natural selection has a chance to develop adaptations.
    b. it emphasizes phyletic evolution of preadapted structures.
    c. it emphasizes continental drift and mass extinctions as major factors of macroevolution.
    d. it recognizes the central role of natural selection in microevolution.

TRUE OR FALSE: *Indicate T or F, and then correct the false statements.*

1. ____ A monophyletic taxon includes only species that share a common ancestor.

2. ____ The more the sequences of amino acids in homologous proteins vary, the more recently the two species have diverged.

3. ____ Phylogenetic trees determined on the basis of similar structures may be inaccurate when adaptive radiations have created large differences or convergent evolution has created misleading similarities.

4. ____ Phenetics is the school of taxonomy that is concerned with the order in which new groups branch over time on the phylogenetic tree.

5. ____ The retention in the adult of juvenile traits of ancestors is called paedomorphosis.

6. ____ A slight change in the function of a regulator gene may result in a major morphological change because there are so many regulator genes present in a genome.

7. ____ An adaptive zone is the region in which a species can successfully live.

8. ____ The layer of clay enriched in iridium between the sediments of the Mesozoic and Cenozoic eras is indicative of a nuclear winter.

9. ____ According to species selection, the best adapted species will be selected for by the environment.

10. ____ According to the modern synthesis, natural selection is seen as the major cause of evolution at all levels.

CHAPTER 20
DESCENT WITH MODIFICATION

Suggested Answers to Structure Your Knowledge

1. The two major components of Darwin's evolutionary theory are that all of life has descended from a common ancestral form and that the modification of that form has been the result of natural selection of the offspring best adapted to the environment.

2. (a) The distribution of species geographically (biogeography) supports evolution because of the similarities among species in close proximity and the differences in species that are separated by great distances, even if their habitats are similar.
(b) The fossil record documents the changes that have occurred in species throughout time and occasionally provides transitional forms that link different phylogenetic groups.
(c) The taxonomic groupings of species into a hierarchy of related groups points to the evolution of species from common ancestors.
(d) Comparative anatomy illustrates the relatedness of groups based on their similar structures. Homologous structures and vestigial structures provide evidence of evolution.

(e) Comparative embryology illustrates the commonalities of developmental patterns in related groups. It also helps to identify homology of structures that may appear different in adult form.
(f) The similarities in the nucleotide sequences of DNA and protein products among related species indicates descent from a common ancestor.

Answers to Test Your Knowledge

Matching:

1.	F	f	5.	D	d
2.	A	b	6.	H	a
3.	G	g	7.	E	c
4.	B	e	8.	C	h

Multiple Choice:

1.	b	5.	a	9.	d
2.	c	6.	c	10.	b
3.	d	7.	d		
4.	a	8.	a		

CHAPTER 21
HOW POPULATIONS EVOLVE

Suggested Answers to Structure Your Knowledge

1. In the Hardy–Weinberg equation, p and q refer to the frequencies of two alleles (A and a) in the gene pool. The frequency or proportion of gametes containing the A allele is equal to p, and the frequency of gametes containing the a allele is equal to q. The proportion of offspring resulting from gametes containing two A alleles is ($p \times p$) or p^2. Likewise, aa offspring will equal q^2. Aa individuals can be formed with the A allele coming from the egg or sperm while the a allele is contributed by the sperm or egg. Thus the frequency of Aa offspring is equal to $2pq$.

2.

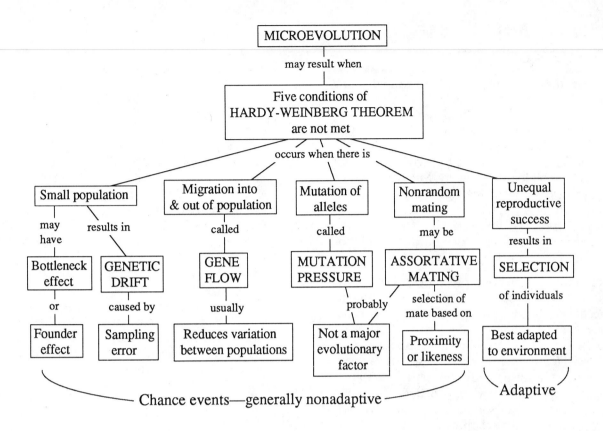

3. Genetic variation is retained within a population by three general mechanisms: diploidy, heterozygote advantage, and balanced polymorphism. Diploidy masks recessive alleles from selection when they occur in the heterozygote. Thus, less adaptive or even harmful alleles are maintained in the gene pool and are available should selection pressures change. In situations in which there is heterozygote advantage or balanced polymorphism, two or more alleles are simultaneously selected for and retained within the gene pool.

4. The allele frequencies estimated from this sample are B = 0.6 and b = 0.4. Assuming this sample truly represents the individuals in the larger population, then we must say that the population is not in Hardy-Weinberg equilibrium. If it were, we would expect to find 36% BB, 48% Bb, and 16% bb individuals. The larger numbers of both homozygous genotypes indicates that assortative mating may be affecting the genotypic frequencies in the population. The allele frequencies may still remain constant from one generation to the next, so microevolution may not be occurring. Other possible explanations for these data are incorrect sampling techniques or the action of some selection agent on speckled butterflies that reduces the numbers of heterozygotes in the population.

Answers to Test Your Knowledge

Multiple Choice:

1. d	5. d	9. d	13. d	17. a	21. a
2. a	6. d	10. b	14. a	18. c	22. b
3. c	7. d	11. b	15. d	19. d	23. c
4. c	8. c	12. d	16. b	20. b	24. d

CHAPTER 22
THE ORIGIN OF SPECIES

Suggested Answers to Structure Your Knowledge

1. Speciation is the process by which a new, reproductively isolated species evolves from its predecessor. It is part of the process of evolution and the increase in biological diversity. Microevolution is the process by which changes occur within the gene pool of a population, as a result of either chance events or natural selection. If the makeup of the gene pool changes enough, microevolution may lead to speciation.

2. Reproductive isolation is probably the key event in the origin of a species. Reproductive isolating mechanisms may develop as the result of chance events, and then natural selection may work on the newly isolated gene pool to develop a population adapted for its environment. These chance events may occur because of the geographic separation of a small population (in which genetic drift may play a large role) or chance mutations (as in polyploidy in plants). Sexual selection, by which natural selection reinforces mechanisms that maximize successful mating within a species, may lead indirectly to reproductive barriers between speciating populations.

3. In the gradual tempo theory of evolution, small changes accumulate within populations as a result of chance events and natural selection, and these may lead (particularly as environmental conditions change) to the gradual evolution of new life forms. The punctuated equilibrium theory holds that evolution occurs in spurts of relatively rapid change inserted within large periods of stasis. The fossil record seems to indicate that new species appear fairly rapidly and change most early in their existence, and then remain relatively unchanged during their duration on earth. The lack of transitional forms, intermediate between species or between higher taxonomic groups, is further evidence for the punctuated pace of evolution. Additional studies of fossil lineages should help distinguish between the gradual or punctuated tempos in the evolutionary origin of species.

Answers to Test Your Knowledge

Multiple Choice:

1. b	5. b	9. d	13. c
2. c	6. d	10. a	14. a
3. a	7. d	11. b	15. c
4. c	8. a	12. b	16. d

CHAPTER 23
MACROEVOLUTION

Suggested Answers to Structure Your Knowledge

1.

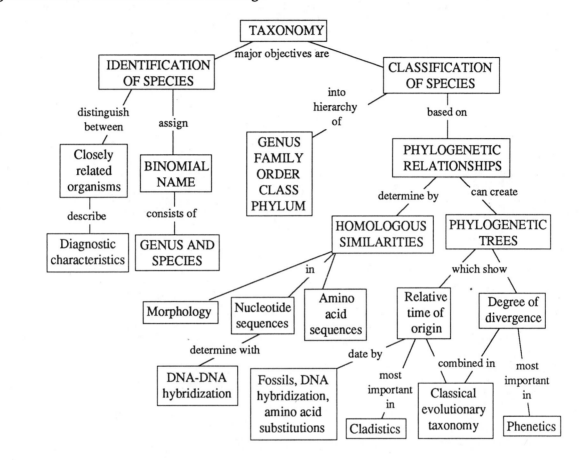

2. Microevolution consists of the changes in gene frequencies that occur within a population over the generations due both to chance events, such as genetic drift, gene flow, and mutation pressure, and to differential reproductive success as a result of natural selection. New species may form as the result of microevolution or as the result of geographical separations or mutations. Macroevolution includes the major events and trends that have occurred in the history of life as taxa higher than species evolve. Macroevolution can result from the gradual modification of preadapted structures for new functions through natural selection (as in microevolution) or, more likely, from chance-related mechanisms such as major geological changes resulting in geographic rearrangements of species, alterations in regulatory genes causing relatively rapid morphological changes, and species selection. Microevolution and macroevolution both involve natural selection and the element of chance. Natural selection appears to play a larger role in microevolution, while chance may play the largest role in macroevolution.

Answers to Test Your Knowledge

Multiple Choice:

1. c	5. a	9. d	13. c	17. a
2. a	6. b	10. d	14. c	18. a
3. b	7. d	11. d	15. a	
4. c	8. b	12. a	16. c	

True or False:

1. True
2. False, change the *more* to the *less* they vary
3. True
4. False, change *phenetics* to *cladistics*
5. True
6. False, change to: because *each regulator gene may control many structural genes.*
7. False, change to: is *an unexploited niche or way of life.*
8. False, change to: indicative of *an asteroid having crashed into the earth.*
9. False, the species that exist the longest and speciate the most often will determine the trend of evolution.
10. True

THE EVOLUTIONARY HISTORY
OF BIOLOGICAL DIVERSITY

Ed and Barbara are visited by the insects of the
Amazon Basin.

CHAPTER 24

EARLY EARTH AND THE ORIGIN OF LIFE

FRAMEWORK

This chapter describes the formation of Earth and a scenario for the chemical evolution of life between 4.1 and 3.5 billion years ago. Conditions on the primitive Earth are thought to have favored the spontaneous formation of organic monomers, the linking of these monomers into polymers, the grouping of aggregates of organic molecules into droplets called protobionts, which became capable of metabolism and reproduction, and the development of self-replicating genetic information capable of directing metabolism and reproduction.

From this proposed beginning, an incredible biological diversity has evolved. This heterogeneous group of organisms is now classified into five kingdoms: Monera, Protista, Plantae, Fungi, and Animalia.

CHAPTER SUMMARY

Life on Earth has been evolving for 3.5 billion years. Geological events have altered the course of biological evolution, and life has changed the earth. The fossil record documents, however incompletely, the development of much of the diversity of life, a development that has often been episodic. The origin of life on the young planet Earth, however, remains unrecorded and a matter of speculation.

Formation of the Earth

Most astronomers believe that all matter was once concentrated in a giant mass that blew apart with a "big bang" 10 to 20 billion years ago. Our sun formed about 5 billion years ago when most of the swirling matter in the cloud of dust that formed our solar system condensed in the center. Earth and the rest of the planets formed about 4.6 billion years ago as gravity attracted the remaining dust and ice to a few kernels.

Geologists believe that Earth went through a molten period, when its components sorted into layers of different densities. Nickle and iron sank to the core, less dense material formed a mantle, and the least dense material solidified into a thin crust. The continents are borne on plates of crust that float on the mantle.

Scientists speculate that Earth's first atmosphere of hot hydrogen gas was added to by gases from volcanoes and other vents. This early atmosphere consisted mostly of water vapor, carbon monoxide and dioxide (CO, CO_2), nitrogen (N_2), methane (CH), and ammonia (NH_3). The first seas were formed by torrential rains; lightning, volcanic activity, and UV radiation were quite intense.

The Antiquity of Life

During Darwin's time, no fossils prior to the Cambrian Period had been found. Now, Precambrian fossils provide evidence of animals dating back 700 million years, and a succession of microorganisms, most of which were prokaryotes, spanning nearly 3 billion years. Fossils that appear to be represent a diverse prokaryote fauna have been discovered in a rock formation called the Fig Tree Chert, which is 3.4 billion years old. Fossils resembling spherical and filamentous prokaryotes have been found in rocks called *stromatolites* that are 3.5 billion years old. Stromatolites are banded domes of sediment that form around the jellylike coats of microbes. It is possible that the earliest life forms may have emerged as long as 4 billion years ago.

The Origin of Life

Between 4.1 billion years ago, when Earth's crust began to solidify, and 3.5 billion years ago, when records of stromatolites are found, life began. Most biologists believe that life developed on Earth from nonliving materials that became ordered into molecular aggregates capable of self-replication and metabolism.

According to one hypothesis, a chemical evolution in four stages produced the first organisms: (1) the abiotic (non-living) synthesis and accumulation of small organic molecules or monomers; (2) the joining of organic monomers into polymers; (3) the aggregation of molecules into droplets, called *protobionts*; and (4) the origin of heredity.

Abiotic Synthesis of Organic Monomers   Oparin and Haldane independently postulated that primitive Earth conditions, in particular the reducing atmosphere and the lightning and intense ultraviolet radiation that penetrated the early atmosphere, favored the synthesis of organic compounds from inorganic precursors available in the atmosphere and seas.

In 1953, Miller and Urey tested the Oparin–Haldane hypothesis with an apparatus that simulated early Earth conditions. After a week of applying sparks (lightning) to a warmed flask of water (the primeval sea) in an atmosphere of H_2O, H_2, CH_4, and NH_3, Miller and Urey found a variety of amino acids and other organic compounds in the flask.

Laboratory replications of the early earth have been able to produce all 20 common amino acids, several sugars, lipids, the purine and pyrimidine bases of DNA and RNA, and even ATP. The abiotic synthesis and accumulation of organic monomers could have been a natural process on primeval Earth.

Abiotic Synthesis of Polymers   The abiotic synthesis of polymers may have occurred when dissolved organic monomers splashed onto hot rocks. Fox has created polypeptides he calls proteinoids by dripping dilute solutions of organic monomers onto hot sand or rock in the laboratory.

Clay may have been an important substratum for polymerization because of its ability to concentrate amino acids and other organic monomers when they bind to charged sites on the clay particles. Metal atoms, such as iron and zinc, present at some of the binding sites may have functioned as catalysts facilitating the dehydration reactions that link monomers.

Formation of Protobionts   Protobionts, aggregates of abiotically produced organic molecules that exhibit some of the properties associated with life, probably preceded living cells. Protobionts may have developed such capabilities as metabolism, excitability, and the ability to maintain an internal chemical environment different from the surroundings.

Laboratory experiments indicate that protobionts could have formed spontaneously. Proteinoids, when mixed with cool water, self-assemble into tiny droplets called *microspheres*, which are coated by a selectively permeable membrane across which a membrane potential may develop. Protobionts can discharge this voltage in nervelike fashion.

When the organic ingredients include certain lipids, droplets called *liposomes*, surrounded by a lipid bilayer, may spontaneously form. Oparin has made *coacervates*, which are colloidal droplets that form when a solution of polypeptides, nucleic acids, and polysaccharides is shaken. If enzymes are included, the coacervates are capable of absorbing substrates, catalyzing reactions, and releasing the products.

The Origin of Genetic Information   The last step necessary for the evolution of life was the origin of genetic information. Successful protobionts would need not only to grow and divide, but also to develop a mechanism for replicating their successful characteristics, for creating instructions for making their key molecules. The hereditary pathway of DNA→RNA→protein was too complex to evolve all at once. One hypothesis maintains that the first genes were short strands of RNA, shown in the laboratory to be capable of self-replication without the aid of enzymes. Cech's recent discovery of RNA catalysts, called *ribozymes*, which remove introns from RNA and catalyze RNA synthesis in modern cells, indicates that RNA molecules may have been autocatalytic and self-replicating in the prebiotic world.

Natural selection has been observed operating on populations of RNA molecules in test tubes. Early populations of RNA molecules may have evolved as those molecules with the most stable three-dimensional conformation and greatest autocatalytic activity within a particular environment successfully competed for monomers and generated families of similar sequences.

RNA-directed protein synthesis may have begun with the weak binding of specific amino acids to bases along the RNA molecules. Perhaps with a catalyst such as zinc, the amino acids could link together to form a short polypeptide. This polypeptide may have then behaved as an enzyme to help the RNA molecule replicate.

When these early RNA and polypeptide molecules became packaged into protobionts, molecular cooperation could become more efficient due to the concentration of components and the potential for the protobiont to evolve as a unit. If primitive enzymes that developed the ability to extract energy from an organic fuel were contained within a membrane, that energy would be made available for the other reactions within the protobiont.

This four-step scenario results in a hypothetical antecedent of a cell, in which an aggregate of molecules selectively incorporates monomers from its surroundings and uses enzymes produced by genes to make polymers and to carry out chemical reactions. The protobiont could grow and split, and distribute copies of its genes to its offspring. Errors in the copying of RNA would lead to variations, some of which may have metabolic or genetic advantages. DNA is a more stable genetic molecule, and apparently replaced RNA as the carrier of genetic information.

Laboratory simulations cannot establish that this sequence actually occurred in the evolution of life, but they have shown that some of the key steps are possible. Alternate explanations include an idea hypothesized by Crick, called *panspermia*, that some organic compounds may have reached the early earth in meterorites and comets. Modern meterorites have been found to contain amino acids and other organic compounds.

Via whatever route, at some point membrane-enclosed compartments capable of metabolism and genetic replication evolved past the grey border separating them from true cells; by 3.5 billion years ago, prokaryotes were already flourishing.

The Kingdoms of Life

The kingdom is the most inclusive taxonomic category. Intuitively and historically, humans have divided the diversity of life into two kingdoms—plants and animals. The study of microbial life often confused the two-group classification. The plant kingdom was stretched to make room for bacteria, unicellular eukaryotes with chloroplasts, and fungi, whereas unicellular creatures that move and ingest food were placed in the animal kingdom.

Whittaker's proposal for a five-kingdom system, made in 1969, has slowly gained favor among biologists. The five kingdoms are Monera, Protista, Plantae, Fungi, and Animalia. The prokaryotes are set apart from the eukaryotes and placed in the kingdom Monera. The kingdoms Plantae, Fungi, and Animalia are multicellular eukaryotes, defined by characteristics of nutrition, structure, life style, and life cycle. Plants are autotrophic organisms; animals are heterotrophic organisms that ingest and digest their food; and fungi are absorptive heterotrophic organisms. Protists are unicellular eukaryotes, or simple multicellular organisms that descended from them.

STRUCTURE YOUR KNOWLEDGE

1. What do scientists think the primitive earth was like? How could life possibly evolve in such an inhospitable environment?

2. Describe the significance of the four steps that are proposed as a possible route for the evolution of protobionts.

TEST YOUR KNOWLEDGE

MATCHING: *Match the scientist with the theory or experiment.*

1. ____	produced organic compounds in a lab apparatus	A. Fox
2. ____	discovered RNA catalysts called ribozymes	B. Oparin and Haldane
3. ____	developed the five-kingdom system	C. Crick
4. ____	claimed that conditions of primitive Earth favored synthesis of organic compounds	D. Cech
5. ____	advanced panspermia theory that meterorites and comets carried organic compounds to Earth	E. Whittaker
6. ____	produced protobionts called coacervates in the laboratory	F. Miller and Urey

MULTIPLE CHOICE: *Choose the one best answer.*

1. According to the "big bang" theory,
   a. the expanding universe blew apart from a concentrated giant mass between 10 and 20 billion years ago.
   b. our solar system was created from a giant explosion about 5 billion years ago.
   c. a huge meteorite reached Earth 4.6 billion years ago, melting its matter, which then sorted into layers of varying density.
   d. most galaxies were formed 20 billion years ago and have already burned out.

2. The primitive atmosphere of Earth may have favored the synthesis of organic molecules because
   a. it was highly oxidative.
   b. it was reducing and had energy sources in the form of lightning and UV radiation.
   c. it had a great deal of methane and organic fuels.
   d. it had plenty of water vapor, carbon, and nitrogen, and was highly oxidative.

3. Life on Earth is thought to have begun
   a. 700 million years ago.
   b. 3 billion years ago.
   c. between 3.5 and 4.1 billion years ago.
   d. 5 billion years ago.

4. Stromatolites are
   a. the earliest fossils.
   b. similar to the layered mats formed by cyanobacteria.
   c. banded domes of sediment formed around motile microbes.
   d. all of the above

5. Proteinoids are
   a. droplets of molecules covered with a selectively permeable membrane.
   b. colloidal droplets formed from polypeptides, nucleic acids, and polysaccharides.
   c. polypeptides formed in the laboratory by dripping organic monomers onto hot rocks.
   d. organic molecules associated with clay crystals.

6. A proposed hypothesis for the origin of genetic information is that
   a. early DNA molecules coded for RNA, which then catalyzed the production of proteins.
   b. early polypeptides became associated with RNA bases, and a catalyst such as zinc could have linked the bases into RNA molecules.
   c. short RNA strands were capable of self-replication and evolved by the natural selection of molecules that were most stable and autocatalytic.
   d. as protobionts grew and split, they distributed copies of their DNA molecules to their offspring.

7. Evidence that protobionts may have formed spontaneously comes from
   a. the discovery of ribozymes, showing that prebiotic RNA molecules may have been autocatalytic.
   b. the laboratory synthesis of microspheres, liposomes, and coacervates.
   c. the fossil record found in the Fig Tree Chert.
   d. the abiotic synthesis of polymers.

8. In the five-kingdom system of classification,
   a. prokaryotes are placed in the kingdom Monera.
   b. the fungi include absorptive heterotrophs.
   c. single-celled or simple multicellular eukaryotes are placed in the Protista.
   d. all of the above are correct.

9. If you examined one of the early Precambrian fossils in detail you would be likely to observe impressions of structures resembling
   a. cell walls.
   b. nuclear membranes.
   c. mitochondria.
   d. xylem cells.

10. Polymer formation from organic monomers may have been facilitated by
    a. the electrification of monomers by lightning.
    b. proteinoid formation by pH changes.
    c. amino acid concentration on clay particles.
    d. sedimentary accumulation on stromatolites.

11. The earliest life forms may have exhibited all of the following characteristics *except*
    a. nerve-like action potentials.
    b. sexual reproduction.
    c. membranes formed from lipid bilayers.
    d. metabolism of chemical compounds.

12. To place a newly identified species into one of the five kingdoms, the most basic information you would need to acquire is
    a. the way in which it acquires food.
    b. the absence or presence of a nuclear membrane.
    c. whether the organism is multicellular.
    d. all of the above.

CHAPTER 25

PROKARYOTES AND THE ORIGINS OF METABOLIC DIVERSITY

FRAMEWORK

This chapter describes the bacteria and cyanobacteria of Kingdom Monera. It reviews the morphology of the prokaryotic cell and presents an overview of the groups archaebacteria and eubacteria. Every mode of nutrition and most metabolic pathways that are found today evolved within the prokaryotes. Prokaryotes exist in all conceivable habitats and in various symbiotic relationships with other organisms. These most numerous and diverse of all organisms are essential to the chemical cycles necessary to maintain life on Earth.

CHAPTER SUMMARY

The history of prokaryotes spans at least 3.5 billion years. The kingdom Monera contains *bacteria*, which include the *cyanobacteria* (formerly blue-green algae), and is characterized by the prokaryotic cell. Prokaryotes evolved alone for the first 2 billion years of life and still outnumber all eukaryotes combined. They flourish in all habitats, including ones that are too harsh for any other forms of life. The collective impact of these microscopic organisms on Earth is huge.

Prokaryotic Form and Function

Morphology of Prokaryotes Prokaryotes exist primarily as single cells, although a few species may form aggregates of cells that stick together after dividing, and some cyanobacteria exhibit a simple multicellular form. The three most common shapes of prokaryotes are spherical (cocci), rod (bacilli) and spiral (spirilla). Prokaryotic cells are usually 1–10μm in diameter, one-tenth the size of eukaryotic cells.

The Prokaryotic Genome The haploid genome of prokaryotes contains only 1/1000 as much DNA as a eukaryote. The circular double-stranded DNA chromosome has very little associated protein and is concentrated in a *nucleoid region*. Smaller rings of DNA, called plasmids, may carry genes for antibiotic resistance, metabolism of unusual nutrients, or other functions. They replicate independently of the genophore, or bacterial chromosome, and may be transferred between bacteria during conjugation.

Prokaryotic ribosomes are smaller than eukaryotic ribosomes and differ in their protein and RNA content. Some antibiotics can bind to prokaryotic ribosomes and block protein synthesis without inhibiting eukaryotic ribosomes.

Membranous Organization Extensive internal membrane systems are not found in prokaryotic cells, although there may be some infoldings of the plasma membrane that function in respiration and thylakoids in cyanobacteria that function in photosynthesis.

The Cell Surface Nearly all prokaryotes have cell walls composed of *peptidoglycan*, a matrix composed of polymers of sugars cross-linked by short polypeptides. Many antibiotics, such as penicillin, inhibit the cross-links in peptidoglycan and prevent wall formation. The *gram stain* is an important tool for identifying bacteria as *gram-positive* (bacteria with simpler walls containing a thick layer of peptidoglycan) or *gram-negative* (bacteria with more complex walls and an outer *lipopolysaccharide* membrane). Pathogenic gram-negative bacteria are often more harmful because their lipopolysaccharides may be toxic, and the outer membrane protects them from the defenses of their hosts and from antibiotics.

Many prokaryotes secrete a sticky *capsule* outside the cell wall that serves as protection and as glue for adhering to a substratum or each other. Bacteria may also attach by means of surface appendages called *pili*. Pili may be specialized for transfer of DNA during conjugation.

Motility of Prokaryotes   Some motile bacteria secrete slime and glide; spirochaetes have *axial filaments* and move in a corkscrew fashion; and other bacteria are equipped with flagella, either scattered over the cell surface or concentrated at one or both ends of the cell. Many motile bacteria exhibit *taxis*, an oriented movement in response to chemical, light, magnetic, or other stimuli. Alternating runs and tumbles may result in the movement of a bacterium in a beneficial direction along a chemical gradient.

Reproduction and Growth of Prokaryotes   Prokaryotes reproduce by binary fission, producing a colony of progeny. The geometric growth of bacterial colonies usually stops at some point due to the exhaustion of nutrients or the toxic accumulation of wastes.

Mutations are the major source of genetic variation in prokaryotes. Since generation times are less than a few hours, a favorable mutation can be passed rapidly to a large number of progeny. Despite the lack of meiosis and sexual cycles, some genetic recombination occurs by conjugation and viral transduction.

Metabolic Diversity   Nutrition refers to how an organism obtains both energy and the carbon it uses for synthesizing organic compounds. Prokaryotes may be phototrophs, using light for energy, or chemotrophs, obtaining energy from chemicals. If CO_2 is the only carbon source, organisms are called autotrophs; when organic nutrients are needed, organisms are called heterotrophs.

Thus, there are four major nutritional categories of bacteria: (1) *photoautotrophs*, cyanobacteria, and other photosynthetic prokaryotes that use light energy and CO_2 to synthesize organic compounds; (2) *photoheterotrophs*, which use light to generate ATP but obtain carbon in organic form; (3) *chemoautotrophs*, which obtain energy by oxidizing inorganic substances (such as H_2S, NH_3, or Fe^{2+}) and need only CO_2 as a carbon source; and (4) *chemoheterotrophs*, which use organic molecules as both an energy and carbon source.

The majority of bacteria are chemoheterotrophs, including *saprophytes*, which decompose and absorb nutrients from dead organic matter, and *parasites*, which absorb nutrients from living hosts. Some fastidious bacteria require all 20 amino acids, vitamins, and other organic compounds in order to grow, whereas others may flourish with glucose as their only organic nutrient. There is such diversity of chemoheterotrophs that almost any organic molecule can serve as food for some species. The few synthetic organic compounds that cannot be broken down by any bacteria are called *nonbiodegradable*.

Obligate aerobes need oxygen for cellular respiration; *facultative anaerobes* can use oxygen but also can grow in anaerobic conditions using fermentation; *obligate anaerobes* cannot use—and are poisoned by—oxygen.

The Diversity of Prokaryotes

The Status of Prokaryotic Taxonomy   Molecular systematics, using comparisons of amino acid sequences and base sequences in DNA and RNA, provide the best approach to the phylogenetic classification of this immense and heterogeneous kingdom that diversified so long ago. These comparisons show that the prokaryotes split into at least two divergent lineages very early in the history of life: the *archaebacteria* and the *eubacteria*.

Archaebacteria   The archaebacteria include only a few surviving genera that inhabit extreme environments, perhaps resembling habitats on early Earth. Their cell walls lack peptidoglycan; their plasma membranes contain a unique lipid composition; and their RNA polymerase and a ribosome protein are very different, resembling those of eukaryotes. The archaebacteria presently consist of three subgroups: the methanogens, the extreme halophiles, and the thermoacidophiles.

Methanogens have a unique energy metabolism in which H_2 is used to reduce CO_2 to methane (CH_4). These strict anaerobes live in swamps and marshes, bubbling up marsh gas or methane. They are important decomposers in sewage treatment and have been used to convert manure to methane on some farms. Methanogens in the guts of cattle and other herbivores contribute to the animals' nutrition.

The extreme halophiles (salt-loving) live in saline places such as the Great Salt Lake and the Dead Sea. They use a simple mechanism of photophosphorylation in which bacteriorhodopsin, the pink pigment built into the plasma membrane, absorbs light and uses the energy to pump H^+ out of the cell. The proton gradient drives the synthesis of ATP.

Thermoacidophiles need an environment that is both hot and acidic. They may be found oxidizing sulfur for energy in the hot sulfur springs in Yellowstone National Park.

Eubacteria   Within the group eubacteria, the cyanobacteria represent a photoautotrophic mode of nutrition. Commonly known as blue-green algae, these prokaryotes have a plantlike mechanism of photosynthesis in which they use chlorophyll *a* and two photosystems to transfer electrons from water to NADP+, releasing O_2. The components of the light reaction are built into thylakoid membranes.

Some filamentous cyanobacteria are capable of *nitrogen fixation*, the assimilation of atmospheric nitrogen into organic compounds. *Heterocysts* are specialized cells with enzymes that reduce N_2 to NH_3 (ammonia).

Other photoautotrophs include green sulfur and purple sulfur bacteria. They have only one photosystem and reduce NADP+ with electrons from H_2S rather than H_2O. These bacteria commonly are found in anaerobic habitats, such as sediments at the bottom of bodies of water.

The genus *Pseudomonas*, found in almost all aquatic and soil habitats, includes the most versatile chemoheterotrophs. Species of this group are capable of utilizing a large range of organic nutrients, including some pesticides and synthetic compounds.

Spirochaetes are spiral-shaped saprophytic or parasitic bacteria that move in a corkscrew fashion. Both syphilis and Lyme disease are caused by members of this group.

Endospore-forming bacteria can survive long periods of harsh conditions—including dessication, heat, cold, and exposure to most poisons—when they are protected within their resistant, thick-walled endospores. Microbiologists use an autoclave to kill endospores and sterilize laboratory media and glassware. Canned foods must be heated to temperatures high enough to kill the endospores of bacteria, including those of *Clostridium botulinum*, the bacterium that produces the lethal botulism toxin.

Enteric bacteria inhabit the intestinal tracts of animals. Many are harmless permanent residents, but a few may cause diseases such as the *Salmonella* species that cause typhoid fever and food poisoning.

Rickettsias are small parasitic bacteria, including ones that cause Rocky Mountain spotted fever and typhus. Chlamydias are similar to rickettsias. Nongonococcal urethritis (NGU), a sexually transmitted disease, is caused by a chlamydia.

Mycoplasmas may be the smallest of all cells. They are the only prokaryotes that lack cell walls.

Actinomycetes form branching chains that resemble fungi. Tuberculosis and leprosy are caused by actinomycetes, although most species are free-living soil organisms. Cultured species of *Streptomyces* are used to produce commercial antibiotics.

Myxobacteria form gliding colonies that move through the soil on a secreted slime. Under adverse conditions, the cells congregate to form an erect fruiting body in which durable spores are formed.

The Importance of Prokaryotes

Prokaryotes and Chemical Cycles Prokaryotes are indispensible in the *chemical cycles* through which chemical elements are recycled between the biological and physical worlds. Bacteria, along with fungi, are *decomposers* of dead organisms and of the waste products of living ones, and thus return carbon, nitrogen and other elements to the environment for assimilation into new living forms. Autotrophic bacteria bring carbon from CO_2 into the food chain and release O_2 to the atmosphere, and certain bacteria—the only organisms capable of fixing atmospheric nitrogen—supply plants with the nitrogen they need to make proteins.

Symbiotic Bacteria Symbiosis is an ecological relationship involving direct contact between organisms of two different species. Usually the smaller organism, or *symbiont*, lives within or on the larger organism, the *host*. In *mutualism*, both host and symbiont benefit. *Nodules* on the roots of legumes house symbiotic bacteria that fix nitrogen for the host. In *commensalism*, the symbiont benefits and the host is neither harmed nor helped. Many of the bacteria found in the human body are commensal, although some may be mutualistic. In *parasitism*, the symbiont, now called a parasite, benefits at the expense of the host.

Bacteria and Disease About one-half of all human diseases are caused by pathogenic bacteria that manage to invade the body, resist internal defenses, and grow enough to harm the host. *Opportunistic* bacteria are normal inhabitants of the human body that cause illness when the body's defenses are weakened.

Koch, the first to connect particular diseases to specific bacteria, suggested four criteria for establishing this connection. Called *Koch's postulates*, they include the following: (1) find the same pathogen in each diseased individual studied; (2) isolate and grow the pathogen in a pure culture; (3) induce the disease in experimental animals with the cultured pathogen; (4) isolate the same pathogen from the experimental animal after it develops the disease.

Pathogenic bacteria commonly cause disease by producing toxins. *Exotoxins*, among the most potent poisons known, are proteins secreted by the bacterium that can induce various symptoms. *Endotoxins*, which are components of the outer membrane of certain gram-negative bacteria, all produce fever and aches in the host.

In the past century, improved hygiene, sanitation, and the development of antibiotics have decreased the incidence and severity of bacterial disease, reduced infant mortality, and extended life expectancy in developed countries.

Putting Bacteria to Work   The diverse metabolic capabilities of prokaryotes have been used to digest organic wastes, produce chemical products, make vitamins and antibiotics, and produce food products such as yogurt and cheese. Research using *Escherichia coli* and other bacteria has expanded our understanding of molecular biology, and recombinant DNA techniques using bacteria develop both basic knowledge and practical applications.

The Origins of Metabolic Diversity

All types of nutrition and most metabolic pathways evolved in prokaryotes before there were eukaryotes. Geological evidence about conditions of early Earth, combined with molecular systematics and the energy metabolism found in existing groups, indicates a possible sequence of events in metabolic evolution.

Nutrition of the Earliest Prokaryotes   The first prokaryotes probably were chemoheterotrophs that absorbed abiotically synthesized organic compounds. ATP must have been an important energy molecule early in life's history, based on its universal role in energy exchange in all modern organisms. Selection may have resulted in the gradual evolution of glycolysis, the metabolic pathway that breaks down organic molecules and generates ATP by substrate phosphorylation. As the only metabolic pathway common to all modern organisms, glycolysis must have developed relatively early in metabolic evolution.

Glycolysis and fermentation do not require molecular oxygen. The archaebacteria and other obligate anaerobes that use fermentation are believed to have forms of nutrition most like that of the original prokaryotes that developed on anaerobic Earth.

The Origin of Electron Transport Chains   The excretion of organic acids formed by fermentation would have acidified the environments of early prokaryotes. Transmembrane proton pumps, driven by ATP, may have helped to regulate internal pH. An electron transport chain that coupled the oxidation of organic acids to the transport of H^+ out of the cell would have conserved ATP, and an excess of H^+ so produced could then diffuse back, reverse the proton pump, and thus generate ATP. This type of anaerobic respiration persists in some modern bacteria. The basic chemiosmotic mechanism that uses proton gradients to transfer energy from redox reactions to ATP synthesis is common to all cells.

The Origin of Photosynthesis   The development of pigments and photosystems, co-opting components of electron-transport chains, may have allowed some prokaryotes to use light energy to drive electrons from H_2S to $NADP^+$ and thus generate reducing power to fix CO_2 and make their own organic molecules. The nutrition of the anaerobic green and purple sulfur bacteria probably is most like that of the early photosynthetic prokaryotes.

The first cyanobacteria evolved a two-photosystem mechanism that was able to use water as a source of electrons and hydrogen for fixing CO_2 and began to release O_2 as a byproduct of their photosynthesis.

The Oxygen Revolution and the Origins of Respiration   Cyanobacteria evolved at least 2.5 billion years ago, contributing to the fossil stromatolites that have been found all over the world. Banded iron formations are found in marine sediments from that same time. The oxygen evolved by the cyanobacteria probably precipitated out dissolved iron ions as iron oxide. When the dissolved iron was finally exhausted, oxygen began to accumulate in the seas and bubble out into the atmosphere. Oxidized iron is found in terrestrial rocks from about 2 billion years ago.

The change to a more oxidizing atmosphere probably caused the extinction of many bacteria, while others survived in anaerobic habitats where they are still found today. Some bacteria evolved antioxidant mechanisms that protected them from rising oxygen levels. Some photosynthetic prokaryotes, however, began using the oxidizing power of O_2 to pull electrons from organic molecules down existing transport chains, leading to the development of aerobic respiration by the co-opting of transport chains from photosynthesis. Several bacterial lines gave up photosynthesis, and their electron transport chains became adapted to function exclusively in aerobic respiration.

Thus, on ancient Earth, every type of nutrition and energy metabolism evolved within the prokaryotes.

STRUCTURE YOUR KNOWLEDGE

1. Fill in the following chart with a brief description of the characteristics of prokaryotic cells.

2. Create a concept map that organizes the various modes of nutrition (along with their associated terminology) found in the kingdom Monera.

CHARACTERISTICS OF THE PROKARYOTIC CELL

Characteristic	Description
Cell shape	
Genome	
Internal membranes	
Cell surface	
Forms of motility	
Reproduction and growth	

3. List the five or six stages that may have occurred in the early evolution of nutrition and metabolism in the prokaryotes.

4. Describe four positive ways in which prokaryotes impact our lives and the world around us.

TEST YOUR KNOWLEDGE

MULTIPLE CHOICE: *Choose the one best answer.*

1. Which of the following is *not* true of plasmids?
   a. They replicate independently of the main chromosome.
   b. They may carry genes for antibiotic resistance or special enzymes.
   c. They are essential for the existence of bacterial cells.
   d. They may be transferred between bacteria during conjugation.

2. Gram-positive bacteria
   a. have peptidoglycan in their cell walls, whereas gram-negative bacteria do not.
   b. have an outer membrane around their cell walls.
   c. have lipopolysaccharides in their cell walls and thus may be less pathogenic than gram-negative bacteria.
   d. have simpler, thick peptidoglycan cell walls.

3. Because prokaryotes have ribosomes different from those of eukaryotes,
   a. some selective antibiotics can block protein synthesis of bacteria without harming the eukaryotic host.
   b. it is believed that eukaryotes did not evolve from prokaryotes.
   c. protein synthesis can occur at the same time as transcription.
   d. their rates of protein synthesis are much slower.

4. Many prokaryotes secrete a sticky capsule outside the cell wall that
   a. allows them to glide through a slime layer.
   b. serves as protection and glue for adherence.
   c. reacts with the gram stain.
   d. is used for attaching cells during conjugation.

5. Chemoautotrophs
   a. are photosynthetic.
   b. use organic molecules for an energy and carbon source.
   c. oxidize inorganic substances for energy and use CO_2 as a carbon source.
   d. use light to generate ATP but need organic molecules for a carbon source.

6. The major source of genetic variation in prokaryotes is
   a. binary fission.
   b. mutation.

c. conjugation.

d. plasmid exchange.

7. A symbiotic relationship in which the symbiont benefits and the host is neither harmed nor helped is called

a. saprocism.

b. opportunism.

c. mutualism.

d. commensalism.

8. According to Koch's postulates, a specific bacteria can be linked to a disease if

a. the same pathogen is isolated from each diseased individual studied.

b. the cultured pathogen causes the disease when introduced to experimental animals.

c. the same pathogen can be isolated from the experimental animal after it develops the disease.

d. all of the above are done.

9. The first form of nutrition to evolve was probably that of

a. photoautotrophs that used light energy to reduce CO_2 with electrons from H_2S.

b. chemoheterotrophs that used abiotically made organic compounds.

c. anaerobic chemoautotrophs.

d. photoheterotrophs that used light for energy and abiotically made organic compounds for a carbon source.

10. Evidence of the early evolution of cyanobacteria comes from

a. stromatolites.

b. banded iron formations in marine sediments.

c. oxidized iron layers in terrestrial rocks.

d. all of the above.

11. Glycolysis

a. is the only metabolic pathway common to all modern organisms and may have been the first metabolic pathway to evolve.

b. breaks down organic molecules and generates ATP by creating a proton gradient that reverses transmembrane proton pumps.

c. uses the electron transport chain to produce ATP.

d. All of the above are correct.

12. Aerobic respiration may have originated when

a. some bacteria evolved antioxidant mechanisms to protect them from rising oxygen levels.

b. photosynthetic bacteria produced an excess of organic molecules.

c. some bacteria used modified electron transport chains from photosynthesis to pass electrons from organic molecules to O_2.

d. some bacteria gave up photosynthesis and reverted to chemoheterotrophic nutrition.

FILL IN THE BLANKS

1. _____ the name for rod-shaped bacteria

2. _____ region in which the prokaryotic chromosome is found

3. _____ important laboratory technique for identifying bacteria

4. _____ surface appendages of bacteria used for adherence to substrate

5. _____ an oriented movement in response to light or chemical stimuli

6. _____ nutrition in which organic molecules are used for both energy and carbon source

7. _____ type of nutrition of purple and green sulfur bacteria

8. _____ organism that can use oxygen or grow anaerobically

9. _____ organisms that decompose and absorb nutrients from dead organic matter

10. _____ proteins that are secreted by bacteria and are potent poisons

11. _____ thickenings on roots that house nitrogen-fixing bacteria

12. _____ bacterial group that lacks peptidoglycan in the cell walls and is found in harsh habitats

MATCHING: *Match the bacterial group with its characteristics.*

1. _____ grow in hot and acidic habitats

    A. cyano bacteria

2. _____ versatile chemo-heterotrophs found in water and soil

    B. halophiles

3. _____ bacteria found in animal intestinal tracts

    C. actino-mycetes

4. _____ form branching docolonies; used to produce antibiotics

    D. thermaci-philes

5. _____ spiral-shaped bacteria; one causes syphilis

    E. *Pseudomonas*

6. _____ photosynthetic bacteria, use chlorophyll *a* and evolve O_2

    F. myxo-bacteria

7. _____ gliding colonies that erect fruiting bodies

    G. methano-gens

8. _____ anaerobes that produce marsh gas

    H. spirochaetes

9. _____ smallest bacteria; lack cell walls

    I. rickettsias

10. _____ bacteria that can survive heat, cold, and dessication in resistant stage; one causes botulism sulfur bacteria

    J. green and purple

11. _____ small parasitic bacteria; one causes Rocky Mountain spotted fever

    K. endospore-forming bacteria

12. _____ photoautotrophic bacteria that use H_2S rather than H_2O to reduce $NADP^+$

    L. enteric bacteria

    M. myco-plasmas

CHAPTER 26

PROTISTS AND THE ORIGIN OF EUKARYOTES

FRAMEWORK

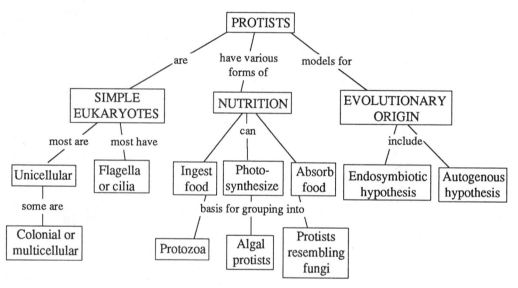

CHAPTER SUMMARY

The kingdom Protista consists of eukaryotic unicellular organisms, with a few colonial and multicellular organisms. Ancient protists, probably the first eukaryotes to evolve from prokaryotes, were the ancestors of modern protists, plants, fungi, and animals. The eukaryotic characteristics of a true nucleus, mitochondria, chloroplasts, ER, Golgi apparatus, 9 + 2 flagella and cilia, mitosis, and meiosis evolved in the protists.

Characteristics of Protists

Protists abound almost anywhere there is water—they occupy freshwater, marine, and moist terrestrial habitats, or live symbiotically within their hosts' bodies.

They form an important part of the *plankton*, the drifting community of organisms found in bodies of water. Some forms attach to or creep along the substrate.

Nearly all protists are aerobic and use mitochondria for their respiration. Their nutrition can be photoautotrophic, heterotrophic, or both. Most protists have flagella or cilia at some point during their lifetime. These structures are called *undulipodia* by some biologists to distinguish them from prokaryotic flagella.

Mitosis occurs in most phyla of protists, but the process may have variations. All protists can reproduce asexually; some also have sexual reproduction. Many protists form resistant *cysts* to withstand harsh conditions.

These organsims are exceedingly complex at the cellular level. A single cell must be capable of performing all the basic functions that can be divided among the specialized cells of plants and animals.

The Boundaries of Kingdom Protista

Whittaker originally assigned unicellular eukaryotes to the kingdom Protista. Because multicellularity evolved several times within the protists, giving rise to plants, animals, fungi, and some organisms that lack the distinctive traits of those three groups, there is disagreement over which groups to include in this kingdom. Some biologists have advocated the use of the name Protista and the inclusion of some phyla of multicellular organisms that seem to be closely related to unicellular eukaryotes. The classification scheme used by this book moves the multicellular algae and some funguslike phyla into the kingdom Protista.

The morphology and life styles of the protists are diverse. An informal scheme for surveying the diversity of protists (but one that does not reflect evolutionary relationships) involves the use of three general categories: (1) protozoa (animallike), (2) algal protists (plantlike), and (3) protists resembling fungi.

Protozoa

Protists that ingest food are informally grouped as protozoans. Six phyla are briefly described in the text.

Rhizopoda The *amoebas* are naked or shelled simple protists that move and feed with *pseudopodia*. The cytoskeleton functions in the formation of these cellular extensions. Nontypical forms of mitosis occur in the asexual reproduction of this group. Most amoebas are found freeliving in freshwater, marine, or soil habitats. Some are parasites, such as *Entamoeba histolytica*, which causes amoebic dysentery in humans.

Actinopoda The slender projections called axopods, which radiate from these protists, help these planktonic organisms remain buoyant and capture microscopic food organisms. *Heliozoa* are freshwater actinopods. *Radiolarians* are primarily marine and may have delicate silica shells.

Foraminifera *Forams* are marine organisms known for their porous, calcareous shells. Cytoplasmic strands, extending through the pores, function in swimming, shell formation, and feeding. Some forams derive food from symbiotic algae. Foram fossils are useful for dating sedimentary rocks.

Apicomplexa These parasites of animals have complex life cycles that often include several host species. Infections are spread by tiny infectious cells called sporozoites (the group was formerly called sporozoa). Malaria, caused by the apicomplexan *Plasmodium* that is carried by the anopheles mosquito, affects over two hundred million people a year.

Zoomastigina These protozoa are characterized by whiplike flagella used for motility. *Zoomastigotes* are heterotrophic organisms that may be freeliving or symbiotic. Mutualistic flagellates that live in the gut of termites digest the cellulose eaten by the host. *Trypanosoma*, a parasitic zoomastigote, causes African sleeping sickness.

Ciliophora These protists are characterized by the cilia they use to move and feed. Most *ciliates* are complex unicellular organisms found in fresh water. Ciliates have two types of nuclei—a large *macronucleus* that controls everyday functions of the cell and splits during binary fission (asexual reproduction) and many small *micronuclei*, which are exchanged in the sexual process called *conjugation*.

Algal Protists

The protists of the algal phyla are mainly photosynthetic. Except for cyanobacteria (blue-green algae), all organisms generically called algae are included in the kingdom Protista. Some biologists place the green, red, and brown algae in the plant kingdom.

All photosynthetic protists have chlorophyll *a*, as do the cyanobacteria and plants. The accessory pigments vary among the algal protists and are used—along with cell-wall chemistry, stored food products, and chloroplast structure—as classification characteristics. The text describes seven of the algal phyla.

Dinoflagellata *Dinoflagellates* make up a large proportion of the phytoplankton that forms the basis of most marine food chains. Blooms of dinoflagellates are responsible for the often harmful red tides in coastal waters. The photosynthesis of symbiotic dinoflagellates, living in small coral cnidarians, is the main food source for coral reef communities. Some dinoflagellates are parasitic or carnivorous heterotrophs.

Dinoflagellates have brownish plastids containing chlorophylls *a* and *c* and a mixture of pigments, including peridinin, which is unique to this phylum. Food is stored as starch. The beating of two flagella in perpendicular grooves in the outer cellulose plates of the cell produces a characteristic spinning movement. The division of the dinoflagellate nucleus during asexual reproduction is unusual. Sexual reproduction does not occur.

Chrysophyta The color of *golden algae* results from yellow and brown carotenoid pigments, present in addition to chlorophylls *a* and *c*. Carbohydrates are stored in the form of laminarin. There are many colonial forms of these flagellated, freshwater plankton.

Bacillariophyta Diatoms are common unicellular plankton with unique glasslike shells. Massive quantities of fossilized shells make up diatomaceous earth. Although diatoms have the same pigments as golden algae, most *phycologists* (specialists in algae) classify them in a separate phylum. Food reserves are stored in the form of an oil.

Euglenophyta The best known member of this phylum is *Euglena*, a common green flagellate found in pond water. Euglenophytes have chlorophyll *b* in addition to *a* and store the polysaccharide paramylon. *Euglena* photosynthesizes in the light but can ingest food particles in the dark.

Chlorophyta The chloroplasts of *green algae* resemble those of plants, and botanists believe that plants evolved from chlorophytes. Most species of green algae live in freshwater, although many are marine. Various unicellular forms may be planktonic, inhabitants of damp soil, or symbionts in invertebrates or in fungi—forming associations known as *lichens*. Colonial species may be filamentous. Some large multicellular forms resemble true plants but are more closely related to the green algae. The *thallus*, or body of an alga, lacks true roots, stems, and leaves.

Three separate evolutionary trends within the green algae have led to increasingly complex flagellated colonies, as in *Volvox*; multinucleated filaments, as in *Bryopsis*; and definite multicellular thalli, as seen in *Ulva*.

The life cycles of most green algae include sexual (with biflagellated gametes) and asexual stages. *Conjugating algae*, such as *Spirogyra*, produce amoeboid gametes.

In the life cycle of the unicellular flagellated *Chlamydomonas*, the haploid mature organism reproduces asexually by dividing mitotically to form four zoospores, which grow into mature haploid cells. Under harsh conditions, the cell may produce many haploid gametes, which pair off and fuse with similar-looking gametes (*isogamy*) of opposite mating strains. The resulting diploid zygote secretes a protective coat. Following dormancy, the zygote undergoes meiosis, forming four haploid cells, which grow into mature individuals. Some species of green algae exhibit *oogamy*, in which a flagellated sperm fertilizes a nonmotile egg.

Some multicellular green algae have an *alternation of generations*, in which a haploid multicellular individual (called the *gametophyte*) mitotically produces haploid gametes that fuse to form a diploid zygote. The zygote develops into the multicellular *sporophyte*, which produces haploid cells called zoospores by meiosis. Zoospores grow into the haploid gametophyte generation. In *Ulva*, the two generations are *isomorphic*.

Phaeophyta The mostly marine *brown algae* are the largest and most complex protists. They contain the pigment fucoxanthin in addition to chlorophylls *a* and *c*, and produce laminarin, the same storage compound produced by the golden algae. The brown and golden algae are closely related. Most common seaweeds are brown or red algae.

The gelatinous algin of the cell walls protects the thallus in the intertidal zones. The giant seaweeds known as kelps are brown algae that fasten to the sea floor by holdfasts. Their gas-filled bladders support long leaflike blades, and their tubular cells transport photosynthetic products down the stemlike stipe.

In most brown algae, there is an alternation of generations. The gametophyte and sporophyte may be isomorphic or *heteromorphic*, as in *Laminaria*. Kelps and seaweeds are both ecologically and commercially important.

Rhodophyta The color of *red algae* is due to the accessory pigment phycoerythrin, which may be present in varying amounts to maximize photosynthesis at the wavelengths of light reaching the depth of water in which the algae are found. Floridean starch is the storage product, and the cell walls consist of gelatinous materials in addition to cellulose. Red algae are multicellular, often with filamentous, delicately branched thalli. The base of the thallus may be differentiated as a holdfast. The *coralline algae* are red algae with walls encrusted with calcium carbonate.

All red algae reproduce sexually, and they often have an alternation of generations. Unlike other algal protists, they have no flagellated cells in their life cycles.

Protists Resembling Fungi

Although the slime molds resemble fungi in appearance and life style, their cellular organization, reproduction, and life cycles are different enough to suggest that they are most closely related to protists. General consensus places the following two phyla with the protists, although they are still studied by *mycologists* (biologists who study fungi). Water molds are also classified as protists.

Myxomycota Plasmodial slime molds are heterotrophic, engulfing food particles by phagocytosis as they grow through leaf litter or rotting logs. A coenocytic amoeboid mass called a *plasmodium* is the feeding stage. A sexual stage develops erect stalks that produce resistant spores by meiosis. When the spores germinate, they fuse to form diploid zygotes in which the nucleus repeatedly divides without cytoplasmic divisions.

Arasiomycota *Cellular slime molds* consist of solitary cells during the feeding stage of the life cycle. In the absence of food, the cells congregate into a mass resembling a plasmodial slime mold, except that it is not coenocytic. Since the cellular slime molds are haploid, the fruiting bodies produce resistant spores asexually by mitosis. In sexual reproduction, two amoebas fuse to form a zygote, which develops into a giant cell enclosed in a protective wall. Following meiosis and mitosis, haploid amoebas are released.

Oomycota The water molds, white rusts, and downy mildews of this phylum resemble fungi in appearance and nutrition. Oomycetes, however, have cell walls made of cellulose, a predominant diploid stage in their life cycles, and flagellated cells—all characteristics that are different from those of true fungi.

The large egg cell of a water mold is fertilized by a smaller unflagellated sperm. Zygotes may become dormant within resistant walls and release flagellated zoospores upon germination. The saprophytic water molds are important decomposers in aquatic ecosystems. White rusts and downy mildews can be destructive parasites of land plants.

The Origin of Eukaryotes

The Antiquity of Eukaryotes The oldest reputed fossils of eukaryotes, found in rocks 1.5 billion years old, are *acritarchs*, Precambrian structures that resemble ruptured coats of cysts similar to those of algal protists. Almost all eukaryotes require oxygen. The cyanobacteria probably had produced enough oxygen to accumulate in the atmosphere during the two billion years of prokaryotic history predating that time.

Models of Eukaryotic Origins There are two major theories on the origin of the eukaryotic cell. According to the *autogenous hypothesis*, eukaryotic cells are believed to have evolved by the specialization of internal membranes derived from invaginations of the plasma membrane.

In the *endosymbiotic hypothesis*, developed by Margulis, eukaryotic cells evolved from symbiotic combinations of prokaryotic cells. Chloroplasts are thought to be descendants of photosynthetic prokaryotes that became endosymbionts within larger cells. Mitochondria may have developed from aerobic heterotrophic bacteria that were incorporated into larger cells.

Chloroplasts appear to have had at least three separate origins, as evidenced by three lineages seen in algal protists. In the red line, the chloroplasts of red algae show similarities to the thylakoid arrangement and pigment composition of cyanobacteria. The chloroplasts of green algae, the green line, are more similar to a eubacterium named *Prochlorothrix*, which is the only know prokaryote with chlorophyll *b* as well as *a*. The thylakoids of the brown line generally are in stacks of three, and chlorophyll *c* and fucoxanthin are accessory pigments. A modern prokaryote that resembles a possible ancestor of the prokaryotic symbiont of the brown plastids has not been found.

Various similarities, including size, membrane enzymes and transport systems, circular DNA molecules not associated with proteins, and the process of division, exist between eubacteria and the chloroplasts and mitochondria of eukaryotes. Chloroplasts and mitochondria contain the ribosomes and equipment needed to transcribe and translate their DNA into proteins. The ribosomes of chloroplasts and mitochondria resemble prokaryotic ribosomes. These similarities, as well as the present-day existence of endosymbiotic relationships, are cited as evidence of the endosymbiotic hypothesis. Molecular systematics, comparing base sequences of ribosomal RNA, also suggest eubacterial origins for chloroplasts and mitochondria.

Even though chloroplasts and mitochondria are not genetically autonomous, the billions of years of coevolution and the existence of transposons (jumping genes) may have allowed the host cell to develop nuclear control over its symbionts.

The autogenous and endosymbiotic models need not be mutually exclusive. The chloroplasts and mitochondria may have originated as endosymbionts, whereas the components of the endomembrane system may have developed as modifications of the plasma membrane. A theory for the development of the eukaryotic cell also must explain the evolution of the 9 + 2 flagella and cilia and the processes of mitosis and meiosis, which also rely on microtubules.

The Origins of Multicellularity

Multicellularity evolved in the kingdom Protista many times, creating the ancestors of plants, fungi, and animals. Colonial aggregations of cells probably became more interdependent, leading to the specialization of cells and the division of labor. Some cells in the colony may have lost their flagella and eventually become specialized for other functions, such as reproduction.

STRUCTURE YOUR KNOWLEDGE

1. (a) What are the criteria or characteristics of the organisms placed in the kingdom Protista? (b) Why is there a controversy over which groups belong in this kingdom?

2. Describe the two hypotheses for the origin of eukaryotic cells. What evidence is given to support the endosymbiotic model?

3. Draw a concept map to show the various types of life cycles found in the protists.

TEST YOUR KNOWLEDGE

MATCHING: *Match the protistan phyla with their descriptions.*

1. ___ Naked and shelled amoebas   **A.** Acrasiomycota

2. ___ Photosynthetic or heterotrophic flagellates   **B.** Actinopoda

3. ___ Heliozoa, radiolarians, siliceous skeletons   **C.** Apicomplexa

4. ___ Plasmodial slime molds, coenocytic   **D.** Bacillariophyta

5. ___ Marine, with calcareous porous shells   **E.** Chlorophyta

6. ___ Green algae, probable ancestors of plants   **F.** Chrysophyta

7. ___ Brown algae, large and complex, seaweeds   **G.** Ciliophora

8. ___ Golden algae, flagellated, freshwater plankton   **H.** Dinoflagellata

9. ___ Diatoms, two-piece shells of silica   **I.** Euglenophyta

10. ___ Red algae, pigment phycoerythrin   **J.** Foraminifera

11. ___ Complex, unicellular, ciliated, macro- and micronuclei   **K.** Myxomycota

12. ___ Water molds, white rusts, downy mildews   **L.** Oomycota

13. ___ Whiplike flagella, may be free-living, mutualistic, parasitic   **M.** Phaeophyta

14. ___ Parasites, complex life cycles, sporozoites; malaria   **N.** Rhizopoda

15. ___ Whirling movement, flagella in grooves in shell, cause red tides   **O.** Rhodophyta

16. ___ Cellular slime molds, unicellular feeding stage, aggregate to form asexual fruiting body   **P.** Zoomastigina

MULTIPLE CHOICE: *Choose the one best answer.*

1. The term *undulipodia* is used to refer to
   a. protists that use an amoeboid type of movement.
   b. the drifting community of organisms found in aquatic environments.
   c. the eukaryotic 9 + 2 form of flagella and cilia.
   d. the multicellular forms included in the kingdom Protista.

2. The slime molds and multicellular algae are included in the kingdom Protista because
   a. they appear to be more closely related to unicellular eukaryotes.
   b. they lack important characteristics of the fungi and plants.
   c. the classification system is trying to reflect phylogenetic relationships.
   d. of all of the above.

3. The category of protozoa includes the protists that
   a. are motile.
   b. ingest their food.
   c. do not have chlorophyll.
   d. are all of the above.

4. Genetic variation is generated in the ciliates when
   a. a micronucleus replicates its genome many times and becomes a macronucleus.
   b. isogametes, in the form of zoospores, fuse.
   c. micronuclei are exchanged in conjugation.
   d. oogamous sexual reproduction occurs.

5. Phytoplankton
   a. form the basis of most marine food chains.
   b. include the dinoflagellates and diatoms.
   c. are the main food source for coral reef communities.
   d. All of the above are true.

6. Mycologists are specialists in the study of
   a. algae.
   b. fungi.

c. protists.

d. phylogeny.

7. The chlorophyta are believed to be the ancestors of plants because
   a. they are the only multicellular algal protists.
   b. they do not have flagellated gametes.
   c. of similarities in chloroplasts and pigment composition.
   d. they exhibit an alternation of generations as do plants.

8. Which of the following protist phyla is incorrectly paired with its accessory pigments?
   a. Rhodophyta—phycoerythrin
   b. Phaeophyta—fucoxanthin and chlorophyll *c*
   c. Dinoflagellata—fucoxanthin and chlorophyll *c*
   d. Euglenophyta—chlorophyll *b*

9. The oldest reputed eukaryotic fossils
   a. date from 1.5 billion years ago.
   b. resemble the ruptured coats of algal cysts.
   c. are acritarchs.
   d. All of the above are correct.

10. According to the endosymbiont theory,
    a. multicellularity evolved when primitive cells incorporated prokaryotic cells that then took on specialized functions.
    b. the symbiotic associations found in lichens resulted from the incorporation of algal protists into the ancestors of fungi.
    c. the chloroplasts and mitochondria of eukaryotic cells originated as prokaryotic endosymbionts.
    d. the infoldings and specializations of the plasma membrane led to the evolution of eukaryotic cells.

11. Examples of mutualistic symbiotic relationships include
    a. termites and zoomastigotes.
    b. dinoflagellates living in small coral cnidarians.
    c. chlorophytes and fungi in lichens.
    d. all of the above.

12. The Precambrian fossil evidence of protists, in the form of cysts called acritarchs, indicates that the atmosphere at that time
    a. was fairly high in oxygen because most protists are aerobic.
    b. was still low in oxygen because most protists are anaerobic.
    c. had a high sulfur content due to chemoautotrophic protists.
    d. cannot be determined because protists are strictly aquatic.

CHAPTER 27

PLANTS AND THE COLONIZATION OF LAND

FRAMEWORK

This chapter details the evolution of plants and their adaptations to the terrestrial habitat. The kingdom Plantae includes multicellular, photoautotrophic eukaryotes that develop from embryos retained in a protective jacket of cells. Most plants exhibit an alternation of generations, in which the sporophyte is the most conspicuous stage in all divisions except the bryophytes. The development of a cuticle and jacketed reproductive organs, vascular tissue, seeds and pollen, and flowers are linked to the major periods of plant evolution in which mosses, ferns, conifers, and flowering plants appeared and radiated.

CHAPTER SUMMARY

For the first 3 billion years, life was confined to the seas, ponds, lakes, and streams. Plants began the movement of life onto land about 400 million years ago, and the evolutionary history of the plant kingdom involves the increasing adaptation to changing terrestrial conditions.

Introduction to the Plant Kingdom

General Characteristics of Plants   Plants are multicellular, photoautotrophic eukaryotes. Nearly all plants are terrestrial, although some have returned to water during their evolution. Adaptations to life on land include a waxy *cuticle* that prevents desiccation and *stomata* or pores to allow for the exchange of gases through the cuticle. Plant chloroplasts contain chlorophylls *a* and *b* and a variety of carotenoids. The cell walls of plants consist of cellulose, and starch is the food storage compound.

A Generalized View of Plant Reproduction and Life Cycles   Nearly all plants reproduce sexually; most can also propagate asexually. Gametes are produced and the egg is fertilized and develops into an embryo within *gametangia*—multicellular organs with a protective jacket of sterile cells.

The life cycle of plants involves the alternation of generations between the haploid gametophyte and the diploid sporophyte. In all extant plants, these generations are *heteromorphic*. In all but the bryophytes, the sporophyte is the more conspicuous stage. Evolutionary trends include the reduction of the haploid generation with the dominance of the diploid and adaptations to terrestrial life such as the replacement of flagellated sperm by pollen.

Some Highlights of Plant Evolution   Four major periods of plant evolution are linked to the evolution of structures that opened new adaptive zones on land. The first period involved the origin of plants, probably from green algae, during the Silurian period, which ended 400 million years ago. Two distinct groups of early plants developed, both showing the terrestrial adaptations of a cuticle and jacketed reproductive organs. One group, which included the ancestors of all plants except the bryophytes, evolved *vascular* tissue—specialized cells joined into tubes for transport throughout the plant. The nonvascular plants were the ancestors of the mosses.

The early adaptive radiation of vascular plants took place during the early Devonian period, the second period of plant evolution.

About 360 million years ago, the origin of the first vascular plants with *seeds*, or embryos enclosed with a store of food inside a protective coat, marked the third major period of plant evolution. Early seed plants gave rise to *gymnosperms* (naked seed plants), such as the conifers. Some seedless plants, such as the ferns, dominated the landscape along with gymnosperms for over 200 million years.

The emergence of flowering plants about 130 million years ago during the early Cretaceous period started the fourth major evolutionary episode. The majority of contemporary plants are *angiosperms*, or flowering plants, which bear seeds in protective chambers called ovaries.

Classification of Plants   Most botantist use the term *division* in place of *phylum* for the major taxonomic groups within the plant kingdom. This textbook recognizes ten divisions within the kingdom Plantae. Alternative classification schemes divide plants into two divisions: the Bryophyta (mosses) and the Tracheophyta (vascular plants).

The Move Onto Land

The Case for Green Algae as the Ancestors of Plants   Almost all botanists agree that plants evolved from green algae, most of which share the following characteristics with plants: (1) chlorophyll *a*, and the accessory pigments chlorophyll *b* and beta-carotene; (2) thylakoid membranes stacked into grana; (3) cell walls of cellulose; (4) starch as carbohydrate reserve; (5) cell plate formation (in some green algae) in cytokinesis.

Filamentous chlorophytes along the edges of bodies of water may have been the first green algae to colonize the land. Fluctuations in water levels may have resulted in selection for species that could survive exposed periods. The development of waxy cuticles and jacketed reproductive organs opened the adaptive zone of land. Bryophytes and vascular plants either had separate origins from green algae or diverged early in their evolution from an algal ancestor.

Division Bryophyta   This division has three classes: mosses, liverworts and hornworts. *Bryophytes* have cuticles and protective gametangia, but they are restricted to damp, shady places due to both their lack of vascular tissue and their flagellated sperm. Sperm are produced in *antheridia*. One egg is produced and fertilized within the *archegonium*, where the zygote develops into an embryo.

Mosses grow as water-absorbing mats of many plants in a tight pack. Each gametophyte plant attaches to the soil with elongated cells called rhizoids. Photosynthesis occurs in small leaflike appendages. The embryo grows into the sporophyte generation while still attached within the archegonium and depends on the gametophyte for water and nutrients. At the tip of the sporophyte stalk, a *sporangium* produces haploid spores by meiosis. Germinating spores grow into a small, green, threadlike protonema from which the gametophyte plant develops.

Most *liverworts* have bodies that are divided into lobes. Their life cycle is similar to that of mosses, although they also reproduce asexually by the formation of gemmae that grow in cups on the surface of the gametophyte. *Hornworts* resemble liverworts, but their sporophytes are elongated capsules that grow like horns from the flat gametophyte.

Mosses are a separate phylogenetic branch and are represented in the fossil record from about 350 million years ago. Vascular plants were already established on land by that time.

Terrestrial Adaptations of Vascular Plants   During the evolution of vascular plants, there was an increasing differentiation of an underground root system to absorb water and minerals and an aerial shoot system of stems and leaves to receive light for photosynthesis. *Lignin*, a hard material embedded in the cellulose of plant cell walls, was an important adaptation to provide support for the aerial parts of the plant. The vascular system of *xylem* and *phloem* provided the conducting vessels to connect roots and leaves. Empty xylem cells carry water and minerals up from the roots; their lignified walls also provide support. The living cells of phloem distribute organic nutrients throughout the plant. In some vascular plants, seeds and pollen were additional terrestrial adaptations.

The Earliest Vascular Plants   Fossils of vascular plants are present in the sedimentary rocks of the late Silurian and early Devonian periods. *Cooksonia* is the oldest plant that has been discovered. *Zosterophyllum* and *Rhynia* are characteristic genera of the diversity of early Devonian species that are the probable ancestors, respectively, of the division Lycophyta and the other divisions of vascular plants.

Seedless Vascular Plants

Four extant divisions have retained the primitive seedless condition.

Division Psilophyta   *Psilotum*, known by the common name *whiskfern*, is one of two genera of relatively simple, primitive vascular plants in this division. Rhizomes bear tiny rhizoids, and the scales emerging from the dichotomously branching stems lack vascular tissue and thus are not true leaves. Sporangia along the stems release haploid spores. The nonphotosynthetic gametophyte depends on symbiotic fungi.

Division Lycophyta   *Lycopods* were a major part of the landscape during the Carboniferous period (340 to

280 million years ago). The division Lycophyta split into two evolutionary lines: one that led to huge woody trees and a second that remained small and herbaceous. The giant lycopods thrived for millions of years but became extinct when the Carboniferous swamps dried up. The small lycopods are represented today by the genera *Lycopodium* and *Selaginella*, commonly called club mosses or ground pines. Many tropical species grow on trees as *epiphytes*.

In the club moss, the spores are borne on *sporophylls*—leaves specialized for reproduction. Spores germinate and grow into inconspicuous, subterranean gametophytes, which depend on symbiotic fungi for nutrition. These gametophytes produce both archegonia and antheridia. *Lycopodium* is said to be *homosporous* because it produces only one kind of spore, which produces bisexual gametophytes. *Selaginella* is *heterosporous*; it produces *megaspores* that develop into female gametophytes bearing archegonia and *microspores* that develop into male gametophytes with antheridia that make flagellated sperm.

Division Sphenophyta   This division originated in the Devonian radiation and produced giant plants during the Carboniferous period. Today, only the genus *Equisetum* is found. Commonly called horsetails, these homosporous plants grow in damp locations. The gametophyte is minute but freeliving and photosynthetic.

Division Pterophyta   Ferns were also found in the great forests of the Carboniferous period and are the most numerous of seedless plants in the modern flora. Fern leaves, commonly called fronds, are much larger than are those of lycopods. The leaves of lycopods probably evolved as emergences from the stem containing a single strand of vascular tissue and are thus called *microphylls*. Fern leaves, called *megaphylls*, have branching systems of veins. One theory holds that megaphylls evolved by the formation of webbing between close-growing separate branches.

Some fern leaves are specialized sporophylls with sporangia, arranged into clusters called sori, on their undersides. Most ferns are homosporous, although the archegonia and antheridia on the small, photosynthetic gametophyte mature at different times. Sperm swim to fertilize the egg in neighboring archegonia, and the young sporophyte grows out from the archegonium.

The Coal Forests   The seedless plants of the Carboniferous forests left behind extensive beds of coal. Dead plants did not completely decay in the stagnant swamp waters, and great accumulations of peat developed. When the sea later covered the swamps and marine

sediments piled on top, heat and pressure converted the peat to coal.

Terrestrial Adaptations of Seed Plants

Three life-cycle modifications contributed to the success of seed plants on land: (1) the further-reduced gametophyte generation was retained within the sporophyte plant; (2) pollination replaced the swimming of sperm to egg; (3) the seed, with embryo surrounded by a food supply within a seed coat, maintained dormancy through harsh conditions and functioned in dispersal.

Gymnosperms

Of the two groups of seed plants, gymnosperms appear earlier in the fossil record. Three of the four divisions are relatively small: Cycadophyta, which includes the palmlike cycads; Ginkgophyta, with the deciduous, ornamental ginkgo tree; and Gnetophyta, which includes the bizarre *Welwitschia* as one of its three genera.

Division Coniferophyta   The name *conifer* comes from the reproductive structure, the cone. Most conifers are evergreens, with needle-shaped leaves covered with a thick cuticle. Coniferous trees are among the tallest, largest, and oldest living organisms.

The pine tree is a heterosporous sporophyte. Pollen cones consist of many tiny sporophylls that bear sporangia. Meiosis gives rise to microspores that develop into pollen grains—the immature male gametophyte. Scales of the ovulate cone hold ovules, each of which contains a sporangium, called a nucellus, within a protective integument. A megaspore mother cell undergoes meiosis, and one of the resulting megaspores undergoes repeated divisions to produce a female gametophyte in which a few archegonia develop.

Pollination occurs when a pollen grain is drawn through the micropyle, an opening in the integument around the nucellus. A pollen tube grows and digests its way through the nucellus. Fertilization occurs when one of the two sperm nuclei entering an egg cell joins with the egg nucleus to produce a zygote. The zygote develops into a sporophyte embryo with a rudimentary root and several embryonic leaves, or cotyledons. The remaining female gametophyte tissue nourishes the embryo, and a seed coat develops from the integuments of the ovule. The production and maturation of seeds takes three years.

The History of Gymnosperms   Gymnosperms probably developed from a group of plants that had evolved seeds by the end of the Devonian Period. The divisions of gymnosperms arose during the Carboniferous and early Permian period. As the Permian progressed, continental interiors became warmer and drier. Conifers and the cycads replaced the lycopods, horsetails, and ferns that dominated the Carboniferous swamps. The end of the Permian marked the boundary between the Paleozoic and Mesozoic eras. The Mesozoic is sometimes called the age of dinosaurs, and conifers and the great palmlike cycads supported these giant reptiles. When the climate became cooler at the end of the Mesozoic, the dinosaurs and many plants became extinct, but some gymnosperms, particularly conifers, persisted.

Angiosperms

Angiosperms, or flowering plants, are the most diverse and widespread of modern flora. The division Anthophyta is divided into two classes: Monocotyledones and Dicotyledones.

Most angiosperms rely on insects or other animals for transferring pollen to female sex organs, an advance over the more random wind pollination of gymnoperms. Gymnosperms transport water through *tracheids*, relatively primitive, tapered cells that also function in mechanical support. In angiosperms, shorter, wider *vessel elements* are arranged end-to-end to form more specialized tubes for water transport. *Fibers* have thick lignified walls that add support to the xylem of angiosperms.

The Flower The *flower*, the reproductive structure of an angiosperm, is a compressed shoot with four whorls of modified leaves: *sepals, petals, stamens,* and *carpels*. Sepals are usually green, whereas petals are brightly colored in most flowers pollinated by insects. A stamen consists of a stalk, called a *filament*, and an *anther*, where pollen is produced. The carpel has a sticky *stigma*, which receives pollen, and a *style*, which leads to the *ovary*. The ovary contains ovules that develop into seeds.

Four evolutionary trends in angiosperms include: (1) reduction of the number of floral parts, (2) fusion of floral parts (compound carpels are common; *pistil* refers to single or fused carpels), (3) change from radial to bilateral symmetry, and (4) lowering the position of the ovary to below petals and sepals for better protection.

The Fruit A *fruit* is a ripened ovary, which functions in the protection and dispersal of seeds. Simple fruits develop from a single ovary; aggregate fruits come from several ovaries of the same flower; and multiple fruits, such as a pineapple, develop from several sepa-

rate flowers. Fruits may be modified in various ways to disperse seeds.

Life Cycle of an Angiosperm Angiosperms are heterosporous. Immature male gametophytes, consisting of two haploid nuclei, are *pollen grains*, which develop within the anthers. *Ovules* contain the female gametophyte, which consists of an *embryo sac* with seven haploid cells (one of which contains two nuclei).

Most flowers have some mechanism to ensure *cross-pollination*, the transfer of pollen from one plant to another. A pollen grain germinates on the stigma and extends a pollen tube down the style to the ovule, where it releases two sperm nuclei into the embryo sac. In a process called *double fertilization*, one sperm unites with the egg to form the zygote, and the other sperm nucleus fuses with the large cell containing two nuclei in the center of the embryo sac. The *endosperm*, which develops from the triploid (3N) nucleus, serves as a food reserve for the embryo. The embryo consists of a rudimentary root and one (in monocots) or two (in dicots) seed leaves, called *cotyledons*. The seed is a mature ovule, containing an embryo, endosperm, and seed coat derived from the integuments.

The Rise of Angiosperms Angiosperms appear rather suddenly in the fossil record about 120 million years ago in the early Cretaceous period, with no transitional links to ancestors. By the end of the Cretaceous, 65 million years ago, angiosperms had radiated and become the dominant plants, as they are today. Nearly all paleobotanists agree that the angiosperms must have evolved from some group of gymnosperms, although the lack of transitional forms obscures their ancestry. The abruptness of their appearance may be due to an imperfect fossil record or perhaps to the punctuated nature of evolution.

Angiosperms rose to prominence during the Cretaceous, another period of climatic change and extinctions. Dinosaurs and many cycads and conifers disappeared and were replaced by mammals and flowering plants. The end of the Cretaceous marks the boundary between the Mesozoic and Cenozoic Eras.

Relationships between Angiosperms and Animals Animals influenced the evolution of plants and vice versa. Predation of animals on plants may have provided the selective pressure for plants to keep spores and vulnerable gametophytes attached to themselves. As flowers and fruits evolved, some predators became beneficial as pollinators and seed dispersers. The *coevolution* of angiosperms and their pollinators is seen in the diversity of flowers, whose color, fragrance, and shape are usually matched to the pollinator's sense of sight and smell, or its particular morphology. As

Characteristics \ Plant Groups / Common Names	Bryophyta	Lycophyta Sphenophyta	Pterophyta	Conifero-phyta	Anthophyta
Cuticle, stomata	X	X	X	X	X

seeds mature, the fruit softens and increases in sugar content, attracting bird and mammal seed dispersers.

All of our fruit and vegetable crops are angiosperms. Grains are grass fruits; their endosperm is the main food source for most people and their domesticated animals. Humans have intervened in plant evolution by selective breeding. Agriculture is a unique case of coevolution between plants and animals.

STRUCTURE YOUR KNOWLEDGE

1. The evolution of plants shows a trend of increasing adaptations to a terrestrial habitat. In the table above, list the many plant characteristics that contribute to survival on land, and check which of the major plant groups have those characteristics.

2. The table on the following page will help you to organize plant evolution within a geological time frame. Plot the major events (origin, dominance, and decline) for vascular plants, seed plants, bryophytes, lycopods, sphenopsids, ferns, gymnosperms, conifers, cycads, and angiosperms. The evolution of major animal groups is shown.

TEST YOUR KNOWLEDGE

MULTIPLE CHOICE: *Choose the one best answer.*

1. Adaptations for terrestrial life seen in all plants are
   a. chlorophylls *a* and *b*.
   b. cell walls of cellulose and vascular tissue.
   c. a waxy cuticle and stomata.
   d. pollen and gametangia.

2. Plants are thought to have originated from chlorophytes
   a. over 400 million years ago.
   b. during the Carboniferous period.
   c. when the climate became warmer and drier at the end of the Paleozoic era.
   d. as the ancient atmosphere became oxygen-rich.

3. Plants are believed to have evolved from green algae because both groups share the following characteristics:
   a. gametangia as sex organs and alternation of generations.
   b. starch as carbohydrate reserve and cytokinesis by cleavage furrow.
   c. thylakoid membranes stacked into grana, chlorophylls *a* and *b*, and cellulose in cells walls.
   d. all of the above

4. Which of the following lack true roots and leaves?
   a. bryophytes
   b. bryophytes and whiskferns (*Psilotum*)
   c. bryophytes, whiskferns, and lycopods
   d. bryophytes, whiskferns, lycopods, and pterophytes

5. Megaphylls
   a. are gametophyte plants that develop from megaspores.
   b. are leaves specialized for reproduction.
   c. are leaves with branching vascular systems.
   d. are large leaves.

6. Bryophytes differ from the other plant groups because
   a. their gametophyte generation is dominant.
   b. they are lacking cuticle and lignin.
   c. they have flagellated sperm.
   d. of all of the above.

7. Pollination in a pine occurs when
   a. a pollen grain is drawn through the micropyle.
   b. a sperm nucleus fuses with an egg in an archegonium.
   c. two sperm nuclei enter the nucellus.
   d. pollen grains reach the ovulate cone.

8. Which of the following incorrectly pairs a sporophyte embryo with its food source?
   a. pine embryo — female gametophyte tissue in nucellus
   b. corn embryo — 3N endosperm tissue in seed
   c. moss embryo — archegonium and gametophyte plant
   d. fern embryo — female sporophyte surrounding archegonium

9. A difference between seedless vascular plants and the plants with seeds is that
   a. the gametophyte generation is conspicuous in the seedless plants, whereas the sporophyte is dominant in the seeded plants.
   b. the spore is the agent of dispersal in the first, whereas the seed functions in dispersal in the second.
   c. the gametophyte is photoautotrophic in the seedless plants but dependent on the sporophyte generation in the other group.
   d. the embryo is unprotected in the seedless plants but retained within the female reproductive structure in the other group.

10. As a result of coevolution,
    a. two organisms mutually influence each other's evolution.
    b. a flower may have nectar guides that lead bees to its nectaries.
    c. the red color of ripe fruits may attract birds or mammals that disperse the seeds.
    d. all of the above

11. Which of the following does *not* function in support in angiosperms?
    a. cuticle.
    b. fibers.
    c. xylem.
    d. lignin.

12. If you could take a time machine back to the Carboniferous period, which of the following scenarios would you most likely confront?
    a. creeping mats of low growing bryophytes
    b. fields of tall grasses swaying in the wind
    c. swamps dominated by large lycopods, sphenopsids, and ferns
    d. forests of naked-seed trees filling the air with pollen

TRUE OR FALSE: *Indicate T or F, and then correct the false statements.*

1. _____ All photoautotrophic, multicellular eukaryotes are plants.

2. _____ Heteromorphic plants produce male and female gametophytes.

3. _____ Bryophytes are terrestrial nonvascular plants.

4. _____ The club mosses and horsetails are naked seed plants.

5. _____ A sporangium produces spores, no matter what group it is found in.

6. _____ A seed consists of an embryo, nutritive material, and a protective coat.

7. _____ Tracheids are wide, specialized cells arranged end-to-end for water transport and are found in angiosperms.

8. _____ A stamen consists of a filament and anther in which pollen is produced.

9. _____ The female gametophyte in angiosperms consists of haploid cells in which a few archegonia develop.

10. _____ The male gametophyte in angiosperms consists of a pollen grain.

CHAPTER 28

FUNGI

FRAMEWORK

This chapter describes the morphology, life cycles, and the ecological and economic importance of the kingdom Fungi. The four divisions of fungi are established on the basis of variations in sexual reproduction. The lichens are symbiotic complexes of fungi and algae. Fungi play an important ecological role, both as decomposers and by their association with plant roots in mycorrhizae.

CHAPTER SUMMARY

Characteristics of Fungi

The fungi were once classified with plants. These eukaryotic, mostly multicellular organisms are now placed in their own kingdom.

Nutrition Fungi are heterotrophs that obtain their nutrients by *absorption* of the small organic molecules formed when they secrete acids and enzymes into their surrounding food media. Fungi may be saprophytes that feed on nonliving organic material as *decomposers*, parasites that absorb nutrients from living hosts, or mutualistic symbionts who feed on but also benefit their hosts.

Structure Most fungi are composed of filaments called *hyphae*, which form a netlike mass called a *mycellium*. Hyphae may be divided into cells by *septa*, which usually have pores through which cell organelles and nutrients can pass. Cell walls are composed of chitin. Hyphae also may be *aseptate* and *coenocytic*, consisting of a continuous cytoplasmic mass with many nuclei. Parasitic fungi may penetrate their host cells or tissues with modified hyphae called *haustoria*.

The filamentous hyphae grow rapidly and provide a large surface area for absorption. Although fungi are nonmotile, they rapidly enter into new food territory by the extension of their hyphae.

During mitosis, the spindle forms within the nucleus, which later divides after the chromosomes separate. Nuclei of the mycelia are haploid. Some septate fungi are *dikaryons* with two nuclei per cell. No flagellated cells are found in the life cycles of the organisms included as fungi in this text's classification scheme.

Reproduction Fungi produce huge quantities of asexual spores when conditions are favorable. Sexual reproduction may occur, usually by conjugation, when environmental conditions change. Hyphae of opposite mating strains may join and their haploid nuclei come together. The hyphae may exist as a dikaryon for a while before syngamy. After a diploid zygote forms, it undergoes meiosis to return to the haploid condition characteristic of most fungi.

Diversity of Fungi

Fungi may be classified into four divisions, which are based primarily on variations in sexual reproduction, and the lichens.

Division Zygomycota Zygomycetes are mostly terrestrial fungi that produce asexual spores in *sporangia* at the tips of aerial hyphae. Zygosporangia, formed in sexual reproduction, can remain dormant during harsh conditions.

Rhizopus stolonifer, the black bread mold, is a common zygomycete that spreads horizontal hyphae across its food substrate, produces rhizoids for anchorage and absorption, and erects hyphae with bulbous black sporangia containing hundreds of haploid spores. Sexual reproduction involves the conjugation of hyphal extensions from neighboring mycelia of opposite mating strains. The fused gametangia develop a tough protective coat, and the zygosporangium may remain dormant until conditions are favorable. *Pilobolus* is a dung fungus that can bend toward bright light to shoot its sporangia onto grass, which is eaten by grazing animals who scatter the spores in feces.

Division Ascomycota The ascomycetes or sac fungi, ranging from the unicellular yeasts to the cup fungi, produce their sexual spores in sacs called *asci*, which may be packed together to form *ascocarps* or "fruiting" structures. The hyphae of multicellular forms are septate. Ascomycetes produce chains of asexual spores, called *conidia*, at the tips of specialized hyphae.

Ascospores are haploid spores that are produced by sexual reproduction. In some ascomycetes, the eight ascospores, formed from a zygote by meiosis and a subsequent mitotic division, are lined up in the ascus in the order in which they were formed, allowing geneticists to study the genetic recombination, crossing over and independent assortment of chromosomes during meiosis.

Many unicellular yeasts produce sexual spores in an ascuslike structure. Division Ascomycota includes many important decomposers and destructive plant parasites, such as powdery mildews and the fungus that causes Dutch elm disease.

Division Basidiomycota The basidiomycetes, which include the mushrooms, shelf fungi, puffballs, and stinkhorns, produce a club-shaped *basidium*, which represents a short diploid stage in the life cycle. A large mass of dikaryotic, septate hyphae extends under the ground from the reproductive structure, called a *basidiocarp*, of a mushroom. Syngamy and meiosis occur in cells at the tip of hyphae lining the gills of a mushroom, forming basidia that produce four haploid basidiospores. A fairy ring of mushrooms may be produced at the perimeter of the expanding underground mycelium.

Division Deuteromycota The large group of fungi that seem to reproduce only asexually are lumped into the *imperfect fungi*. When a sexual stage is discovered in the life cycle of one of these, the fungus is reclassified into the appropriate group. Many deuteromycetes, including *Penicillium*, are probably related to the ascomycetes, as suggested by their formation of conidia.

Lichens Lichens are associations of millions of algal cells in a lattice of fungal hyphae. The fungus is usually an ascomycete, and the algae are usually unicellular or filamentous green algae or cyanobacteria. Lichens reproduce asexually either as fragments or by tiny airborn clumps called *soredia*. The algal and fungal components can also reproduce independently, sexually or asexually.

The symbiotic relationship between the alga and fungus may be mutual, or perhaps parasitic. The alga provides the fungus with food; it is debated whether the alga benefits from the fungus. These associations are given genus and species names, usually after their fungal member. Naturalists categorize lichens according to their morphology. Some vividly colored species are used to produce dyes.

Lichens can grow on bare rock and can withstand desiccation and great cold. In the tundra, herds of caribou and reindeer graze on lichens. Lichens cannot, however, tolerate air pollution, as they take up most of their minerals from the air in the form of compounds dissolved in rain.

Ecological and Commercial Importance of Fungi

Fungi as Decomposers Fungi and bacteria are the principal decomposers of organic matter that make possible the essential recycling of chemical elements between living organisms and their abiotic surroundings.

Mycorrhizae Mycorrhizae are common and very important mutualistic associations of plant roots and fungi, in which the fungal hyphae provide the plant with minerals absorbed from the soil. Basidiomycetes are the fungi most often found in mycorrhizae.

Commercial Uses of Fungi Fungi have commercial uses as food (mushroooms, truffles, and flavoring in cheese), yeast (*Saccharomyces cerevisiae*) used in baking and brewing, and sources of antibiotics. In 1928, Fleming discovered that a strain of the fungus *Penicillium* killed bacteria, and Florey purified the antibiotic penicillin a decade later.

Fungi as Spoilers Fungi are also spoilers of food, clothing, and wood used in buildings and boats. Their ubiquitous spores may land on and decompose our food as well as the organic litter in our ecosystem.

Pathogenic Fungi Pathogenic fungi cause athlete's foot, ringworm, vaginal yeast infections, and lung infections in humans. Fungal diseases of plants are common. Dutch elm disease is caused by an ascomycete. Another ascomycete forms ergots on rye that can cause various serious symptoms when accidentally milled into flour. Lysergic acid, the raw material of LSD, is one of the toxins in the ergots.

Evolution of Fungi

Fungi are not related to plants or animals; they appear to have had a separate origin from protists. The phylogeny of fungi is uncertain. Ascomycota may have evolved from Zygomycota and later given rise to Basidiomycota. The ancestors of Zygomycota may have been red algae or conjugating green algae.

All major groups of fungi had evolved by the end of the Carboniferous period, 300 million years ago. The oldest fossils of fungi date back 450–500 million years. The first terrestrial vascular plant fossils have petrified mycorrhizae, indicating that plants and fungi moved onto land together.

STRUCTURE YOUR KNOWLEDGE

1. Fill in the following table that summarizes the characteristics of the major groups of fungi.

2. The kingdom Fungi contains members with saprophytic, parasitic, and mutualistic symbiotic modes of nutrition. How do these types of nutrition relate to the ecological and economic importance of this group?

TEST YOUR KNOWLEDGE

FILL IN THE BLANKS

1. _____ division between cells in fungal hyphae

2. _____ hyphal cells with two nuclei

3. _____ component of cell walls in most fungi

4. _____ asexual spores produced in chains at ends of hyphae

5. _____ club-shaped reproductive structure found in mushrooms

6. _____ sacs that contain sexual spores in cup fungi

7. _____ mutualistic associations between plant roots and fungi

8. _____ tough, protective zygote produced by zygomycetes

9. _____ type of hyphae with many nuclei

10. _____ hyphae of parasitic fungi that grows into cells or tissues

Division	Common Members	Morphology	Asexual Reproduction	Sexual Reproduction	Miscellaneous Information
Zygomycota					
Ascomycota					
Basidio-mycota					
Deutero-mycota					
Lichens					

MULTIPLE CHOICE: *Choose the one best answer.*

1. The major difference between fungi and plants is that fungi
   a. have an absorptive form of nutrition.
   b. do not have a cell wall.
   c. are not eukaryotic.
   d. are multinucleate but not multicellular.

2. Fungal mitosis
   a. produces spores in dikaryotic cells.
   b. does not involve the formation of a spindle.
   c. takes place within the nucleus.
   d. all of the above

3. A fungus that is both a parasite and a saprophyte is one that
   a. digests only the nonliving portions of its host's body.
   b. lives off the sap within its host's body.
   c. first lives as a parasite but then consumes the host after it dies.
   d. lives as a mutualistic symbiont on its host.

4. The Deuteromycota differ from all the other fungi in that
   a. their hyphae are coenocytic.
   b. they have flagellated cells.
   c. they have cell walls of cellulose and are predominantly diploid.
   d. they have no known sexual stage.

5. The group that is believed to have evolved first is the
   a. Ascomycota.
   b. Basidiomycota.
   c. Deuteromycota.
   d. Zygomycota.

6. Basidiomycetes differ from all other fungi in that they
   a. have dikaryotic cells.
   b. have no asexual reproductive stage.
   c. form symbiotic relationships with algae.
   d. are coenocytic.

7. In the Ascomycota,
   a. sexual reproduction occurs by conjugation.
   b. spores line up in a sac in the order they were formed by meiosis.
   c. asexual spores form in sporangia on erect hyphae.
   d. the vast majority of spores formed are asexual.

8. Lichens are symbiotic associations that
   a. usually involve an ascomycete and a green alga or cyanobacterium.
   b. can reproduce asexually by forming soredia.
   c. are able to colonize bare rocks and cold habitats.
   d. are all of the above.

CHAPTER 29

INVERTEBRATES AND THE ORIGINS OF ANIMAL DIVERSITY

FRAMEWORK

Animals are multicellular eukaryotic heterotrophs whose mode of nutrition is ingestion. Characteristics of animals include sexual reproduction; dominant diploid stage; specialized cells, tissues and organs; embryonic development with a blastula stage; and the presence of muscles and nerves.

This chapter surveys the characteristics and representatives of the major animal phyla. Fossil evidence for the kingdom Animalia's origin and phylogeny is lacking. Comparative anatomy and embryology are used to construct phylogenetic trees. Phyla are organized into several broad groupings based on general organization, symmetry, presence of a coelom, and embryology. The sequence at the bottom of the page outlines these divisions.

CHAPTER SUMMARY

The evolution of multicellular organisms that eat other organisms opened a new adaptive zone as early animal life in Precambrian seas radiated and populated first the seas and eventually the land. Animals are grouped into about 35 phyla, with more than a million recognized species and perhaps another million not yet

identified. All but one subphylum of these are *invertebrates*, animals that lack backbones.

Characteristics of Animals

Animals are multicellular eukaryotes that use *ingestion* as their mode of nutrition. (The use of absorption by parasites probably evolved secondarily.) Carbohydrates are stored as glycogen. Animal cells lack walls, and desmosomes, gap junctions, and tight junctions may connect adjacent cells. The cells of an animal are specialized in structure and function; in all but the simplest animals, cells are organized into tissues, tissues into organs, and organs are grouped into systems. Muscles and nerves are unique to animals and are necessary for their active behavior.

The diploid stage is dominant, and reproduction is primarily sexual, with a flagellated sperm fertilizing a larger, nonmotile egg. The zygote undergoes a series of divisions, called *cleavage*, and usually passes through a *blastula* (hollow ball of cells) stage during embryonic development. Some animal life cycles include a *larva*— a free-living, sexually immature form. *Metamorphosis* transforms a larva into a sexually mature adult.

The greatest number of animal phyla live in the stable environment of the oceans. Many animals live in freshwater. Terrestrial habitats have been extensively exploited by only a few animal phyla, notably the vertebrates and arthropods.

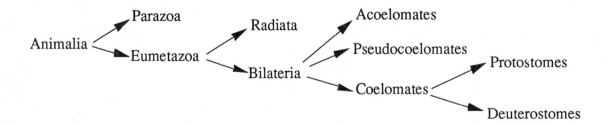

Clues to Animal Phylogeny

Because of the lack of transitional fossils linking phyla to their predecessors, animal phylogeny depends on comparative anatomy and embryology.

Major Branches of the Animal Kingdom   Many zoologists maintain that the animal kingdom had its origin from one group of protists. The sponges of the phylum Porifera are given a separate subkingdom, *Parazoa*. All other phyla are grouped into the subkingdom *Eumetozoa*.

The eumetozoa is divided into two groups, partly on the basis of body symmetry. Branch *Radiata*, containing the hydras, jellyfishes and their relatives, has *radial symmetry*. The *Bilateria* includes animals with *bilateral symmetry*, having distinct head and tail ends, and left and right sides. Bilateral symmetry is associated with *cephalization*, the concentration of sensory organs in the head end, which is an adaptation for unidirectional movement. The radial symmetry of a sessile or planktonic animal allows it to confront its environment equally on all sides.

Development and Body Plan   An early embryo of an animal in the branch Bilateria usually develops three layers, called the *germ layers*. The *ectoderm* develops into the outer body covering and, in some phyla, into the central nervous system. The *endoderm* lines the primitive gut, or *archenteron*, and gives rise to the lining of the digestive tract and associated organs. The muscles and most other organs arise from the middle layer, the *mesoderm*.

Animals of the Bilateria that have solid bodies are called *acoelomates* and include the flatworms (Platyhelminthes) and a few other phyla. A fluid-filled cavity separating the digestive tract and the outer body wall is found in the other bilateria animals. In *pseudocoelomates*, including the rotifers (Phylum Rotifera) and roundworms (Phylum Nematoda), the cavity is not completely lined by mesoderm and is called a *pseudocoelom*. *Coelomates* have a true *coelom*, a body cavity completely lined by mesoderm. A fluid-filled body cavity cushions internal organs, allows organs to grow and move independently of the outer body wall, and also functions as a hydrostatic skeleton in soft-bodied coelomates.

The Protostome–Deuterostome Dichotomy   Differences in embryology and coelom formation divide the coelomates into two evolutionary lines. The *protostome* line includes annelids, mollusks and arthropods. The *deuterostomes* include echinoderms and chordates.

Most protostomes have *spiral cleavage*, in which the planes of cell division are oblique to the polar axis of the egg and newly formed cells fit in the grooves between cells of adjacent tiers. The *determinate cleavage* of some protostomes sets the developmental fate of each embryonic cell very early. Many deuterostomes exhibit *radial cleavage*, in which parallel and perpendicular cleavage planes result in aligned tiers of cells in the dividing zygote. *Indeterminate cleavage* in deuterostomes means that early embryonic cells retain the capacity to develop into a complete embryo.

The *blastopore* is the opening of the developing archenteron. In typical protostomes, the blastopore develops into the mouth, and a second opening forms at the end of the archenteron to produce an anus. In the deuterostomes, the second opening develops into the mouth, whereas the blastopore becomes the anus.

In protostomes, the coelom forms from splits within solid masses of mesoderm, called *schizocoelous* development. In deuterostomes, the mesoderm begins as lateral outpocketings from the archenteron, called *enterocoelous* development.

Parazoa (Phylum Porifera)

Sponges are sessile, mostly marine animals. Water is drawn through pores in the body wall of this saclike animal, into a central cavity, the *spongocoel*, from where it flows out through the *osculum*. Sponges are filter-feeders, collecting food particles from the water circulating through their bodies.

In the *mesohyl*, or gelatinous matrix, between the two body-wall layers are *amoebocytes*. These cells take up food from the choanocytes, flagellated collar cells lining the inside cavity that phagocytize food particles, and carry nutrients to the epidermal cells. Amoebocytes also form skeletal fibers, which may be spicules made of calcium carbonate or silicate, or flexible fibers composed of a protein called spongin.

Most sponges are *hermaphrodites*, producing both eggs and sperm, which differentiate from amoebocytes. Sperm, released into the spongocoel and carried out through the osculum, cross-fertilize eggs retained in the mesohyl of neighboring sponges. Swimming larvae disperse through the osculum.

Sponges are capable of extensive *regeneration*, replacing damaged body parts and reproducing asexually from fragments. Sponges lack organs and have relatively unspecialized cells. Ancestors of sponges may have been choanoflagellates, collared protists that resemble choanocytes.

Radiata

Phylum Cnidaria   The phylum Cnidaria, formerly called Coelenterata, includes hydras, jellyfishes, sea anemones, and corals. These mostly marine animals have radial symmetry and a simple anatomy consisting of a sac with a central *gastrovascular cavity* and a single opening serving as both mouth and anus. *Polyps* are sessile, cylindrical forms, whereas *medusae* are flattened, mouth-down polyps that move by passive drifting and weak contractions of the bell. Both these body forms occur in the life cycles of some cnidarians.

Cnidarians use their ring of tentacles, armed with *cnidocytes* or stinging cells, to capture prey. Cells of the epidermis and gastrodermis have bundles of microfilaments arranged into contractile fibers. A nerve net coordinates the contraction of cells against the hydrostatic skeleton of the gastrovascular cavity, resulting in movement.

The class *Hydrozoa* includes animals that alternate between polyp and medusa forms, as seen in the colonial *Obelia*. Sexual reproduction occurs in the medusa stage, whereas the asexual polyp stage is usually more conspicuous. The common, freshwater *Hydra* exists only in polyp form, reproducing asexually by *budding*, or sexually when environmental conditions deteriorate.

In the class *Scyphozoa*, the medusa stage is more prevalent. The sessile polyp stage often does not occur in the jellyfishes of the open ocean. The class Anthozoa includes sea anemones and most corals. They occur only as polyps.

Phylum Ctenophora   The comb jellies of the phylum Ctenophora are the largest animals that use cilia for locomotion. Nerves running from a sensory organ to combs of cilia coordinate movement.

Acoelomates

Phylum Platyhelminthes   The flatworms of the phylum Platyhelminthes are soft-bodied, ribbonlike animals that lack a body cavity. They develop from a three-layered embryo, have distinct organs, and have a gastrovascular cavity with only one opening.

The class *Turbellaria* includes freshwater *planarians* and other, mostly marine, free-living flatworms. Their branching gastrovascular cavity functions for both digestion and circulation. Gas exchange and diffusion of nitrogenous wastes occurs across the body wall. Ciliated flame cells function in osmoregulation. Planarians move using cilia to glide or body undulations to swim. Eyespots on its head detect light, and auricles function for smell. A pair of ganglia (rudimentary brain) leads into a ladderlike nervous system. Planarians reproduce asexually by regeneration or sexually by copulation between hermaphroditic worms.

Flukes are placed in the class *Trematoda*. Adapted to live as parasites in other animals, flukes have a tough outer covering, suckers, and extensive reproductive glands. The life cycles of flukes are complex, usually including asexual and sexual stages, and intermediate hosts in which larvae develop.

The tapeworms of the class *Cestoda* are parasites, mostly of vertebrates. Tapeworms consist of a scolex with suckers and hooks for attaching to the host's intestinal lining, and a ribbon of proglottids packed with reproductive organs. Absorption of predigested food from the host eliminates the need for a digestive system. The life cycles of tapeworms may include intermediate hosts.

Phylum Nemertea   Most proboscis worms in the phylum Nemertea are marine. Their long, retractable *proboscis* is used to explore the environment and capture prey. These worms probably evolved from flatworms but show two new anatomical features: a digestive tube with separate mouth and anus and a circulatory system of vessels through which blood flows.

Pseudocoelomates

Phylum Rotifera   The phylum Rotifera includes the rotifers, which are smaller than many protists but have a pseudocoelom, a complete digestive tract, and other organ systems. A crown of cilia draws water and microscopic food organisms into the mouth. Some species of rotifers reproduce by *parthenogenesis*; female offspring develop from unfertilized eggs. In other species, another type of egg parthenogenetically develops into degenerate males, which produce sperm that fertilize eggs. The resulting resistant zygotes survive harsh conditions in a dormant state.

Phylum Nematoda   The round worms in the phylum Nematoda are among the most abundant of all animals, found inhabiting water, soil, and the bodies of plants and animals. These cylindrical worms are covered with tough cuticles, have complete digestive tracts, and have a thrashing movement produced by contraction of their longitudinal muscles. Reproduction is exclusively sexual; fertilization is internal, and most zygotes form resistant cells. Numerous nematode species are ecologically important in decomposition and nutrient cycling. The developmental history of every cell of *Caenorhabditis elegans*, an important research organism, has been traced.

Many nematodes are serious agricultural pests and animal parasites. Humans are host to 50 species, including some pinworms, hookworms, and *Trichinella spiralis*, the nematode that causes trichinosis.

Protostomes

Phylum Mollusca The mollusks, in the phylum Mollusca, are mostly marine, soft-bodied animals, many of which are protected by a shell. The molluscan body plan has three main parts: a muscular *foot* used for movement, a *visceral mass* containing the internal organs, and a *mantle* that covers the visceral mass and may secrete a shell. A muscular tongue with a rasping *radula* is used for feeding by many mollusks. Most mollusks have separate sexes.

Some marine mollusks have a life cycle that includes a ciliated larva called the *trochophore*, also found in annelids. Mollusks do not share the trait of segmentation with annelids, and many zoologists believe that mollusks and annelids had separate origins from flatworms. Of the seven molluscan classes, four are discussed in the text.

Chitons, in the class *Polyplacophora*, are oval marine animals with shells that are divided into eight dorsal plates. Chitons are found clinging tightly to rocks in the intertidal zone.

Gastropoda is the largest molluscan class. Most members are marine, although there are many freshwater species, and some snails and slugs are terrestrial. Land snails lack gills; their vascularized mantle cavity functions as a lung. Most gastropods have single spiraled shells. Many gastropods have distinct heads with eyes at the tips of tentacles. A distinctive feature of this class is *torsion*, the embryonic asymmetric growth that results in a U-turn of the digestive tract, with the anus ending up near the mouth.

The clams, oysters, mussels, and scallops of the class *Bivalvia* have the two halves of their shells hinged at the mid-dorsal line and closed by powerful adductor muscles. Bivalves have no heads. These filter-feeders are fairly sedentary. Water flows into and out of the mantle cavity through siphons, and food particles are trapped in the mucus that coats the gills and then swept to the mouth by cilia.

The active squids and octopuses of the class *Cephalopoda* are rapidly moving carnivores. The mouth at the center of the foot has beaklike jaws to crush prey. The foot is drawn out to form several long tentacles. The shell is either reduced and internal (squids) or absent (octopuses). The giant squid is the largest of all invertebrates.

Cephalopods are the only mollusks with *closed circulatory systems*. They have well-developed nervous systems, sense organs, and complex brains—important features for an active predator. The ancestors of octopuses and squids were probably shelled predacious mollusks that eventually reduced their shells. Shelled *ammonites* were the dominant invertebrate predators until their extinction at the end of the Cretaceous period. The chambered nautilus has survived.

Phylum Annelida The phylum Annelida includes segmented worms that are found in most marine, freshwater, and soil habitats. The earthworm, a well-known example, has a complete digestive system with specialized regions and a closed circulatory system with some blood vessels functioning as pumping hearts. Respiration occurs across the moist skin. Septa partition the coelom into segments, in which are found a pair of excretory metanephridia and fused ganglia along the ventral nerve cords. The brain is a pair of cerebral ganglia. Earthworms contain both male and female reproductive organs and exchange sperm during copulation.

The noncompressible coelomic fluid, enclosed by septa within the body segments, serves as a hydrostatic skeleton; circular and longitudinal muscles contract alternately to extend the body and to pull it forward while the setae grip the substrate. Segmentation may have first evolved as an adaptation for this type of burrowing or creeping movement.

The class *Oligochaeta* includes the earthworms and other aquatic species. The mostly marine segmented worms in the class *Polychaeta* have parapodia on each segment that function in gas exchange and locomotion (aided by stiff setae). Many marine polychaetes are tube-dwellers. Most of the leeches in the class *Hirudinea* inhabit fresh water. Many feed on small invertebrates, whereas others are parasites and temporarily attach to animals, slit or digest a hole through the skin, and suck the blood of their host.

Phylum Arthropoda Considering species diversity, distribution, and sheer numbers, the phylum Arthropoda represents the most successful phylum of animals. Characteristics of arthropods include jointed appendages and a *cuticle* that completely covers the body as an *exoskeleton* composed of chitin and protein. To grow, an arthropod must *molt*, shedding its old exoskeleton and secreting a larger one. Arthropods have extensive cephalization and well-developed sensory organs.

A heart pumps hemolymph through an open circulatory system consisting of short arteries and a network of sinuses known as the hemocoel. Respiration may occur through gills that extend from the body in most aquatic species, or tracheal systems of branching internal ducts in terrestrial arthropods.

The segmented arthropods probably evolved from annelids or from a segmented protostome that was a common ancestor of both phyla. Early arthropods may have resembled onychophorans, a phylum of worms with unjointed walking appendages. There are Cambrian fossils of jointed-legged animals that resemble segmented worms.

The *trilobites*, with their pronounced segmentation and uniform appendages, were common early arthropods throughout the Paleozoic era. The large eurypterids, or sea scorpions, were marine predators that belonged to the arthropod subphylum called *chelicerates*. The chelicerate body is divided into a cephalothorax and abdomen, with the most anterior appendages modified as pincers or fangs, called *chelicerae*. The horseshoe crab is the one surviving marine chelicerate. Most modern species are in the terrestrial class Arachnida.

Another evolutionary line produced several groups that had jawlike *mandibles*, one or two pairs of sensory *antennae*, and *compound eyes*. Crustaceans, with two pairs of antennae and branched appendages, are an aquatic subphylum that is believed to have evolved in the ocean. The subphylum Uniramia, including the classes of insects, centipedes, and millipedes, contains organisms with one pair of antennae and unbranched (uniramous) appendages. They are believed to have evolved on land.

The exoskeleton of early marine arthropods, which probably functioned in protection and anchorage for muscles, helped preadapt arthropods for terrestrial life by providing the needed support and protection from water loss. During the early Devonian period, both chelicerates and uniramians spread onto land. The oldest evidence of terrestrial animals are burrows of millipedelike arthropods about 450 million years old.

The five groups of extant arthropods covered by the text are the chelicerate class *Arachnida*, the subphylum *Crustacea*, and the uniramian classes *Chilopoda, Diplopoda,* and *Insecta*.

Scorpions, spiders, ticks, and mites belong to the class Arachnida. The cephalothorax has six pairs of appendages: two pairs of chelicerae, a pair of pedipalps that usually functions in feeding, and four pairs of walking legs. In most spiders, *book lungs* are used in gas exchange. Many spiders spin silk webs for capturing prey, which they attack with poison-filled fanglike chelicerae.

The class Crustacea has remained mostly in marine and freshwater habitats. Crustaceans have two pairs of antennae, three or more pairs of mouthpart appendages, walking legs on the thorax, and appendages on the abdomen. The pincers in lobsters and crayfish are modified walking legs. Larger crustaceans have gills. In their open type of circulatory system, a heart pumps hemolymph through arteries into sinuses that bathe the organs. Nitrogenous wastes pass by diffusion through thin areas of the cuticle, and a pair of glands regulates salt balance in the hemolymph. Sexes usually are separate. A swimming larval stage occurs in most aquatic crustaceans.

Lobsters, crayfish, crabs, and shrimp are relatively large crustaceans called decapods. The dorsal side of the cephalothorax is covered by the carapace. Isopods are mostly small marine crustaceans but also include the terrestrial sow bugs, which are limited to damp habitats. The small, numerous copepods are important members of the plankton communities that form the foundation of marine and freshwater food chains. Barnacles are sessile crustaceans that feed by using their feet to sweep food toward their mouths.

The millipedes of the class Diplopoda are wormlike, segmented vegetarians with two pairs of walking legs per segment. The centipedes of the class Chilopoda are terrestrial carnivores with appendages modified as mouthparts and poison claws for paralyzing prey. Each segment of the trunk has one pair of legs.

The class Insecta is divided into about 26 orders and has more species than all other forms of life combined. The study of insects, called *entomology*, is a vast field. The oldest insect fossils date back to the Devonian period, but a major diversification occurred in the Carboniferous and Permian periods with the evolution of flight. Another important radiation occurred as insects developed coevolutionary relationships with flowering plants during the Cretaceous period.

Many insects have one or two pairs of wings emerging from the dorsal thorax, which may have evolved as extensions of the cuticle that helped the insect body absorb heat. Insects flap their wings by the action of muscles that change the shape of the thoracic cuticle. Dragonflies were among the first winged insects. Later modifications of wings included overlapping or hooking the two pairs together during flight and hardening of the anterior wings as a protective cover.

A generalized insect body has three regions: a head, a thorax, and an abdomen. The head, consisting of several fused segments, has one pair of antennae, a pair of compound eyes, and several pairs of mouthparts modified for various types of ingestion. The thorax bears wings and three pairs of walking legs.

The digestive tract has several specialized regions. The circulatory system is open; the heart pumps hemolymph through an artery into the hemocoel. *Malpighian tubules* are outpocketings of the gut that function as excretory organs, removing nitrogenous wastes from the hemolymph. Gas exchange is accomplished by a *tracheal system* of chitin-lined tubes that ramify throughout the body and open to the outside through spiracles.

The nervous system consists of a pair of ventral nerve cords with segmental ganglia and a dorsal brain in the head formed from several fused ganglia. Insects are capable of complex behavior.

In *incomplete metamorphosis*, the young are smaller versions of the adult and pass through several molts. In *complete metamorphosis*, the larvae, known as grubs or caterpillars, look entirely different from the adult. Larvae eat and grow; adults mate and reproduce.

Reproduction is almost always sexual; fertilization is usually internal.

Insects have a huge impact on humans as pollinators of food crops, vectors of diseases, and competitors for food.

The Lophophorate Animals

The three phyla of *lophophorate animals, Phoronida, Bryozoa* and *Brachiopoda*, are difficult to assign as protostomes or deuterostomes on the basis of their embryonic development. One hypothesis holds that the lophophorate animals and deuterostomes branched from a common coelomate ancestor. All three phyla have a *lophophore*, a horseshoe-shaped or circular fold bearing ciliated tentacles, which surrounds the mouth and functions in filter-feeding. The U-shaped digestive tract and lack of a distinct head may have evolved separately in each group as adaptations to a sessile life.

Phoronids are marine worms that live buried in sand in chitin tubes with their lophophore extended from the tube for feeding. Bryozoans are tiny animals that live in colonies that resemble mosses and may be encased in a hard exoskeleton. The marine brachiopods, or lamp shells, resemble bivalves. They attach to the substratum by a stalk and open their shell to allow water to flow through the lophophore. *Lingula* is a living brachiopod genus that has changed little in 400 million years.

Deuterostomes

The echinoderms and chordates are grouped together as deuterostomes based on common embryological traits.

Phylum Echinodermata The phylum Echinodermata includes sessile or sedentary marine animals with radial symmetry. *Echinoderms* have a thin skin covering an endoskeleton of hard calcareous plates. A *water vascular system* with a network of hydraulic canals controls extensions called tube feet that function in locomotion, feeding and gas exchange. Sexual reproduction usually involves separate sexes and external fertilization. Bilateral larvae metamorphosize into radial adults.

The sea stars of the class *Asteroidea* have five arms radiating from a central disc. Sea stars use tube feet to creep slowly and to open bivalves for food. They evert their stomach through their mouth and slip it into a slightly opened shell, where digestive juices begin to digest the bivalve. Sea stars are capable of regeneration from fragments.

Brittle stars, of the class *Ophiuroidea*, have smaller central discs and longer arms than do sea stars. They move by lashing their arms.

The sea urchins and sand dollars of the class *Echinoidea* have no arms but do have five rows of tube feet by which they move slowly. The mouth is ringed by complex jawlike structures used to eat seaweeds and other foods.

Sea lilies, in the class *Crinoidea*, live attached by stalks to the substratum and extend their arms upward from around the mouth. Their form has changed little in 500 million years of evolution.

The class *Holothuroidea* includes the sea cucumbers. These elongated animals with five rows of tube feet bear little resemblance to other echinoderms.

Phylum Chordata The phylum Chordata contains two subphyla of invertebrates plus the subphylum Vertebrata. Echinoderms and chordates share the developmental characteristics of deuterostomes, but they have existed as separate phyla for at least 500 million years.

The Origins of Animal Diversity

Several hypothetical ancestors have been suggested for the origin of animals from the protists. One hypothesis is that animals developed from a coenocytic ancestor, perhaps a ciliate, that became divided into many cells. Early animals may have resembled bilateral acoel worms, an order of flatworms that lack digestive cavities and take food in through a pharynx that leads into a solid coenocytic mass. Another theory suggests a colonial flagellate ancestor, perhaps involving a ball of flagellated cells similar to a present day choanoflagellate colony. Cell layers may have developed from internal cell proliferation or invagination. Another suggested hypothetical ancestor resembles *Trichoplax adhaerens*, a contemporary tiny, flat marine creature with a ciliated epidermis. Lacking a gut, this animal hunches up to form a temporary digestive cavity when it crawls over a food particle. The first archenteron may have originated in this fashion.

There is no fossil record of transitional forms between protists and the first animals. The early diversification of animals is also a phylogenetic mystery. The oldest known fossils come from the end of the Precambrian era, in a span from about 700 to 590 million years ago known as the *Ediacaran period*. These soft-bodied creatures resembled cnidarian medusae, some colonial cnidarians, and annelids.

A second radiation occurred in the Cambrian period at the beginning of the Paleozoic era about 590 million years ago. The evolution of shells and hard skeletons may have opened new adaptive zones. Nearly all modern phyla, along with many extinct phyla, are present in Cambrian fossils.

The Cambrian appearance of so many complex animals is one of the great mysteries of biological history. Some paleobiologists see the Cambrian radiation as a continuation of the diversification of body forms seen in the Precambrian worms and jellyfish. Seilacher interprets the Ediacaran fossils as discontinuous with the Cambrian fauna. In his view, the two groups represent different solutions to problems associated with increased size. The Ediacaran animals were flat and foliate, exposing most cells to the environment. The Cambrian fauna and their descendents were thicker, with extensive internal surfaces for exchange with the environment. The Cambrian radiation may represent new designs that evolved from burrowing cylindrical survivors of a mass extinction.

STRUCTURE YOUR KNOWLEDGE

1. The diverse phyla of the kingdom Animalia are structured into broad groups based on a series of general characteristics (symmetry, coelom, and so on). Create a concept map or diagram that shows these divisions, the criteria used to determine them, and the major phyla included in each group. Include common examples for each taxon.

2. Why are the origin and phylogeny of the kingdom Animalia a mystery? Describe some of the hypotheses concerning animal evolution.

3. Complete the following table showing the increase in complexity of the structures used by members of each phylum for accomplishing the basic life functions.

	Digestion	Excretion	Respiration	Nervous control	Skeleton	Locomotion
Porifera						
Cnidaria						
Platyhelminthes						
Rotifera						
Nematoda						
Mollusca						
Annelida						
Arthropoda						
Echinodermata						
Chordata invertebrate phyla						

TEST YOUR KNOWLEDGE

MATCHING: *Match the following organisms with their classes and phyla. Answers may be used more than once or not at all.*

Class	Phylum	Organism
1._____	_____	jellyfish
2._____	_____	crayfish
3._____	_____	snail
4._____	_____	leech
5._____	_____	planarian
6._____	_____	cricket
7._____	_____	scallop
8._____	_____	tick
9._____	_____	sea urchin
10._____	_____	hydra

Classes

A. Arachnida
B. Bivalvia
C. Cestoda
D. Crustacea
E. Echinoidea
F. Gastropoda
G. Hirudinea
H. Hydrozoa
I. Insecta
J. Oligochaeta
K. Ophiuroidea
L. Scyphozoa
M. Turbellaria

Phyla

a. Annelida
b. Arthropoda
c. Cnidaria
d. Echinodermata
e. Mollusca
f. Nematoda
g. Nemertea
h. Platyhelminthes
i. Porifera
j. Rotifera

MULTIPLE CHOICE: *Choose the one best answer.*

1. Invertebrates include
   a. all animals except for the phylum Vertebrata.
   b. all animals without backbones.
   c. only animals that use hydrostatic skeletons.
   d. the Radiata and Protostomia, but not the Deuterostomes.

2. The greatest number of animal phyla are found

   a. in the ocean, where environmental conditions are most stable.
   b. in freshwater habitats, where organisms do not face problems of salt balance.
   c. on land, where gas exchange and locomotion are easiest.
   d. in the tropical rain forests, where temperatures are warm and rainfall is abundant.

3. An insect larva
   a. is a sexually immature organism that may or may not resemble the adult.
   b. is transformed into an adult by metamorphosis.
   c. is specialized for eating and growth.
   d. is all of the above.

4. Sponges differ from the rest of the Animalia because
   a. they are completely sessile.
   b. they have radial symmetry and are filter feeders.
   c. their simple body structure has no tissues or organs.
   d. they have an unusual feeding structure called a lophophore and their embryology differs from both protostomes and deuterostomes.

5. Cephalization
   a. is the development of bilateral symmetry.
   b. is associated with motile animals that concentrate sensory organs in a head region.
   c. is a diagnostic characteristic of deuterostomes.
   d. is common in radially symmetrical animals.

6. A true coelom
   a. is found in deuterostomes.
   b. is a fluid-filled cavity completely lined by mesoderm.
   c. may be used as a hydrostatic skeleton by soft-bodied coelomates.
   d. all of the above

7. Which of the following is descriptive of protostomes?
   a. radial and determinate cleavage, blastopore becomes mouth
   b. spiral and indeterminate cleavage, coelom forms as split in solid mass of mesoderm
   c. spiral and determinate cleavage, blastopore becomes mouth, schizocoelous development
   d. radial and determinate cleavage, enterocoelous development, blastopore becomes anus

8. Hermaphrodites
   a. contain male and female sex organs but usually cross fertilize.

  b. include sponges, earthworms, and most insects.
  c. are characteristically found in rotifers.
  d. develop by parthenogenesis from unfertilized eggs.

9. A gastrovascular cavity
  a. functions in both digestion and circulation.
  b. has only one opening that serves as mouth and anus.
  c. is found in the phyla Cnidaria and Platyhelminthes.
  d. all of the above

10. Which of the following is *not* true of cnidarians?
  a. An alternation of medusa and polyp stage is common in the class Hydrozoa.
  b. They use a ring of tentacles armed with stinging cells to capture prey.
  c. They include hydras, jellyfishes, sea anemones and barnacles.
  d. They have a nerve net that coordinates contraction of microfilaments for movement.

11. Which of the following combinations of phylum and characteristic is incorrect?
  a. Nemertea — proboscis worm, first complete digestive tract
  b. Rotifera — parthenogenesis, crown of cilia, microscopic animals
  c. Nematoda — gastrovascular cavity, tough cuticle, ubiquitous
  d. Annelida — segmentation, closed circulation, hydrostatic skeleton

12. Which of the following is either not an excretory structure or is incorrectly matched with its class?
  a. metanephridia — Oligochaeta
  b. Malpighian tubules — Echinoidea
  c. flame cells — Turbellaria
  d. thin region of cuticle — Crustacea

13. The major characteristics of the phylum Mollusca are
  a. radial symmetry, visceral mass, hydrostatic skeleton.
  b. a foot, mantle, visceral mass.
  c. shell, mantle, radula.
  d. tube feet, visceral mass, mantle.

14. Torsion
  a. is embryonic asymmetric growth that results in a U-shaped digestive tract in gastropods.
  b. is characteristic of mollusks.

  c. is responsible for the spiral growth of bivalve shells.
  d. is all of the above.

15. Bivalves differ from other mollusks in that
  a. they are predaceous.
  b. they have no heads and are filter feeders.
  c. they have shells.
  d. they have open circulatory systems.

16. The oldest fossil evidence of terrestrial animals are
  a. trilobites.
  b. insect castes.
  c. burrows of millipedelike organisms.
  d. eurypterids.

17. The exoskeleton of arthropods
  a. functions in protection and anchorage for muscles.
  b. is composed of chitin and cellulose.
  c. is absent in millipedes and centipedes.
  d. expands at the joints when the arthropod grows.

18. Which of the following is true of the subphylum Uniramia?
  a. The horseshoe crab is the one surviving marine member of this group.
  b. It contains the primarily aquatic crustaceans.
  c. It contains insects, centipedes, and millipedes, characterized by their unbranched appendages.
  d. It is characterized by jawlike mandibles, antennae, compound eyes, and includes all arthropods except the chelicerates.

19. Echinoderms are characterized by
  a. a calcareous exoskeleton.
  b. bilateral larvae; radially symmetrical adults with rows of tube feet used for movement and feeding.
  c. a water vascular system that serves both digestive and circulatory functions.
  d. all of the above.

20. The oldest known animals
  a. were soft-bodied creatures from the Ediacaran Period, over 590 million year ago.
  b. were from the Cambrian Period at the beginning of the Paleozoic Era.
  c. were planuloids—swimming, cigar-shaped, solid masses of cells.
  d. were coenocytic ciliates resembling acoel worms.

CHAPTER 30

THE VERTEBRATE GENEALOGY

FRAMEWORK

This chapter focuses on the origin and characteristics of the subphylum Vertebrata and its classes: Agnatha, Placodermi, Chondrichthyes, Osteichthyes, Amphibia, Reptilia, Aves, and Mammalia. The fossil evidence and speculation concerning human ancestry is described.

CHAPTER SUMMARY

Phylum Chordata

There are three subphyla of *chordates*, two without backbones—the *cephalochordates* and the *urochordates*—and one with backbones—the *vertebrates*.

Chordate Characteristics There are four distinguishing anatomical features present in the embryo stage of chordates: (1) a flexible rod called a *notochord* running the length of the animal, (2) a *dorsal, hollow nerve cord* that develops into the brain and spinal cord of the central nervous system, (3) *pharyngeal slits* that open from the anterior region of the digestive tube to the outside, and (4) a *postanal tail* containing a skeleton and muscles and extending beyond the anus.

Chordates Without Backbones The lancelet is a tiny marine animal in the subphylum Cephalochordata. It traps food particles through a mucous net secreted across the pharyngeal slits. The lancelet swims by coordinated contractions of serial muscle segments that flex the notochord from side to side. Blocks of mesoderm called *somites*, which flank the notochord of chordate embryos, develop into these muscle segments.

Tunicates are sessile, filter-feeding, marine animals in the subphylum Urochordata. Water moves through the saclike, tunic-covered animal from the incurrent siphon, through pharyngeal slits, to the excurrent siphon. The larval traits of notochord, nerve cord, and tail are lost in the adult animal.

The Origin of Vertebrates

Fossils resembling cephalochordates have been dated at 550 million years old; vertebrates appear 50 million years later. The fossil record does not show the origin of vertebrates from invertebrate ancestors. Protovertebrates may have been urochordatelike larval forms that became sexually mature and reproduced before undergoing metamorphosis. Mutations in regulatory genes could vary the rate of development, resulting in sexual maturity within a larval form, a phenomenon known as *paedogenesis*. If the larval-like forms became more active swimmers and foragers, natural selection would have favored the development of a head with sensory organs and a brain.

Vertebrate Characteristics

Cephalization, the development of sense organs and a brain, and *vertebrae*, a series of skeletal units enclosing the nerve cord, are the most fundamental differences setting off vertebrate from invertebrate chordates. The skull enclosing the brain and the vertebral column are made of bone and/or cartilage, which is secreted by living cells and can grow with the animal. Many vertebrates also have ribs and a breastbone and an appendicular skeleton supporting two pairs of limbs.

Vertebrates have a closed circulatory system with a ventral, two-to-four-chambered heart pumping blood through arteries to capillaries and back through veins. The blood, containing hemoglobin in red blood cells, is oxygenated through the skin, gills or lungs. Kidneys remove waste products as the blood passes through them. Reproduction is almost always sexual with separate male and female sexes.

Four classes of the subphylum Vertebrata are called *tetrapods* because they have two pairs of limbs. The other three extant classes are commonly called fishes.

Class Agnatha

The oldest vertebrate fossils are jawless creatures called *agnathans*, including the fishlike *ostracoderms* that have an armor of bony plates. First traces of agnathans are found in the Cambrian strata, and a tremendous radiation occurred by the late Silurian, about 400 million years ago. Agnathans were generally small, lacked paired fins, and probably filtered food from the mud or water. Most disappeared during the Devonian period. The jawless fishes today are represented by lampreys, which are blood-sucking parasites as adults, and marine hagfishes, which are scavengers.

Class Placodermi

A group of armored fishes called *placoderms* replaced the agnathans during the late Silurian and early Devonian periods. These larger fish had paired fins that enhanced swimming and jaws that evolved from the skeletal supports of the anterior gill slits. Both these traits facilitated a predatory lifestyle. The remaining gill slits, no longer needed for filter-feeding, maintained their function in gas exchange.

The Devonian period is known as the age of fishes. The placoderms and acanthodians (another class of jawed fishes) radiated in both fresh and salt water, and the sharks and bony fishes first evolved. The placoderms and acanthodians disappeared almost completely by the Carboniferous period, 350 million years ago.

Class Chondrichthyes

Sharks and rays, the cartilaginous fishes, have skeletons made of cartilage, well-developed jaws, and paired fins. Sharks swim to maintain buoyancy and to move water into their mouths and past their gills. The largest sharks filter-feed on plankton, although most are carnivores. Shark teeth evolved from the jagged scales that cover the skin. A spiral valve in the intestine increases surface area within the relatively short digestive tract.

Sharks have keen senses: sharp vision, nostrils, and regions in the head that can detect electric fields generated by muscle contractions of nearby fish. The *lateral line system* is a row of microscopic organs, running along the sides of the shark body, that are sensitive to water pressure changes. Sharks also have a pair of auditory organs.

Following internal fertilization, *oviparous* species of sharks lay eggs that hatch outside the mother. *Ovoviviparous* species retain the fertilized eggs until they hatch within the uterus. A few species are *viviparous*; the developing young are nourished by a placenta until born. The *cloaca* is the common chamber for the openings of the reproductive tract, excretory system, and digestive tract.

Class Osteichthyes

The skeleton of most bony fishes has a hard matrix of calcium phosphate. The skin is often covered by flattened bony scales. Bony fishes have a lateral line system and glands that secrete mucus to reduce drag. The movements of muscles in the gill chambers and the *operculum*, or flap over the gill chamber, draw water across the gills. Bony fishes also have a *swim bladder*, an air sac that controls buoyancy.

The flexible fins of bony fish are better for steering and propulsion than are the stiffer fins of sharks. The common fusiform body shape of fish, sharks, and aquatic mammals—an example of convergent evolution—reduces drag while swimming. Reproduction may vary, but most species of fish are oviparous, and fertilization is external.

Bony fishes probably originated in freshwater, and the swim bladder was modified from lungs that were used to augment gills. Three subclasses evolved: the *ray-finned fishes*, including most of the familiar fish, many of which spread to the seas during their evolution; the *lobe-finned fishes*; and the *lungfishes*, both of which remained in freshwater and continued to use their lungs. Lobefins and some lungfishes had muscular pectoral and pelvic fins that were supported by extensions of the skeleton and may have enabled the fish to "walk" along the bottom or, occasionally, on land. Three present-day genera of lungfishes are found in the Southern Hemisphere. The coelocanth is the only living species of lobe-finned fish.

Class Amphibia

Early Amphibians   Lobe-finned fishes with lungs, living in freshwater habitats subject to drought, may have been the ancestors of tetrapods. The amphibians were the first vertebrates on land; the oldest fossils date back to the late Devonian, about 350 million years ago. Amphibians have always been predators, eating the insects and other invertebrates that preceded them onto land. The Carboniferous period is known as the Age of Amphibians, during which a diversity of new forms and sizes evolved. Amphibians declined during the late Carboniferous, and by the Mesozoic era, the survivors resembled modern amphibians.

Modern Amphibians   There are three extant orders of amphibians: *urodeles* or salamanders, *anurans* or frogs, and *apodans* or caecilians. Salamanders, which may be aquatic or terrestrial, swim or walk with a back-and-forth bending of the body. Anurans are more specialized than urodeles for moving on land, as seen by the powerful hindlegs of frogs used for hopping. Protective adaptations of frogs include camouflage coloring and distasteful or poisonous skin mucus. Tropical apodans are burrowing, legless, nearly blind amphibians.

Some frogs undergo a metamorphosis from a tadpole, an aquatic herbivorous larval form with gills and a tail, to the carnivorous terrestrial adult form with legs and lungs. Many amphibians do not have an aquatic tadpole stage. Some species are entirely aquatic or terrestrial. Paedogenesis occurs in some groups of salamanders; the mudpuppy retains gills and other larval features.

Amphibians are most abundant in damp habitats. They carry out much of their gas exchange across their moist skin, even when they have lungs.

In mating, the male frog grasps the female, but fertilization is external. Oviparous amphibians generally lay their eggs in aquatic or moist environments. Ovoviviparous or viviparous species are known, and some species show varying amounts of parental care. During breeding season, frogs may communicate with vocalizations and undergo migrations to specific breeding sites.

Class Reptilia

Reptiles, including the lizards, snakes, turtles, and crocodilians, exhibit adaptations for living on land not seen in the amphibians.

Reptilian Characteristics   The skin is covered with keratinized waterproof scales, and lungs are the main vehicle for gas exchange. Most species lay eggs covered by protective shells on land. The embryo develops within a fluid-filled amniotic sac; the evolution of the *amniote egg* allowed vertebrates to complete their life cycles on land.

Although reptiles do not use their metabolism to control body temperature, they can regulate their temperature through behavioral adaptations. Reptiles are *ectothermic*; they absorb external heat rather than generating their own, and thus have much lower caloric needs than a mammal has.

The Age of Reptiles   The oldest reptilian fossils date from 300 million years ago from the upper Carboniferous period. The *cotylosaurs*, or *"stem reptiles,"* were the small insect-eating, lizardlike ancestors of the various reptilian orders that arose in two major adaptive radiations. During the Permian period, the first radiation gave rise to terrestrial predators called *synapsids*, from which one lineage led to the *therapsids*. Mammals are believed to have evolved from the therapsid line. Some stem reptiles returned to water, leading to the plesiosaurs and ichthyosaurs. The *thecodonts* also arose during the Permian radiation and, after surviving the mass extinctions of the Permian period, served as the ancestor to the dinosaurs, crocodilians, and birds.

The second great radiation occurred about 200 million years ago during the late Triassic period, when the diverse dinosaurs evolved from thecodonts. The flying pterosaurs, with wings formed from a membrane of skin stretched between an elongated finger and the body, and the terrestrial dinosaurs were the dominant vertebrates on Earth for millions of years. They included the largest animals ever to live on land. Some scientists postulate that dinosaurs were active, agile, and *endothermic*—able to use metabolic heat to warm the body.

During the Cretaceous period at the end of the Mesozoic era, the climate became cooler and more variable. Probably within 5 to 10 million years, the dinosaurs declined and became extinct. A few reptilian groups survived and are successful today.

Reptiles of Today   The three largest orders of extant reptiles are: *Squamata*, lizards and snakes; *Chelonia*, turtles; and *Crocodilia*, alligators and crocodiles.

Lizards, the most numerous and diverse reptiles, are relatively small. Snakes, apparently descended from lizards that adopted a burrowing lifestyle, are known for their absence of limbs. Snakes are carnivorous, with adaptations for locating (vibration-sensing and heat-detecting and olfactory organs), killing (sharp teeth, often with toxins), and swallowing (loosely articulated jaws) their prey.

Turtles evolved from stem reptiles during the Mesozoic era and have changed little since then. Crocodiles and alligators are among the largest living reptiles.

Class Aves

Birds evolved from dinosaurs during the reptilian radiation of the Mesozoic era. The amniote egg and scales on the legs of birds are two reptilian traits birds display.

Characteristics of Birds   The characteristics of birds relate to their ability to fly. Their bones are honeycombed—strong but light. They are toothless; food is ground in the gizzard. The keratinized beak has assumed a great variety of shapes associated with different diets. Birds are endothermic; feathers and a fat layer help retain metabolic heat. The four-chambered heart segregates oxygenated and oxygen-poor blood, which facilitates a high metabolic rate.

Birds have excellent vision. Their relatively large brains provide for visual processing, motor coordination, and more complex behavior. Fertilization is internal. Eggs are laid and kept warm during development by the female and/or male bird.

Aerodynamic wings are flapped by large pectoral muscles attached to a keel on the sternum. Feathers, made of keratin and evolved from reptilian scales, are extremely light and strong and shape the wing into an airfoil.

The Origin of Birds   Many dinosaurs, including some from the diverse *theropods*, built nests, cared for their young, and were most likely endothermic. It has been argued that birds are living dinosaurs and should be regarded as reptiles.

Fossils of *Archaeopteryx lithographica*, an ancient feathered creature with clawed forelimbs, teeth, and long tail, are found from the Jurassic Period, 150 million years ago. This weak flyer was probably not a direct ancestor of modern birds. A fossil dating from 125 million years ago seems to represent a later stage of avian evolution. The evolution of a strong flying ability enhanced hunting, foraging, escape from predators, and migration to favorable habitats.

Modern Birds   The ostrich, kiwi, and emu are examples of flightless birds, called ratites, that lack a keeled breastbone and large flight muscles. Carinate birds have a sternal keel to which attach large breast muscles. The passeriforms, or perching birds, include 60 percent of living bird species.

Class Mammalia

Mammalian Characteristics   Mammals have hair, made of keratin, which helps to insulate these endothermic animals. Their active metabolism is provided for by an efficient respiratory system that uses a *diaphragm* to help ventilate the lungs and a four-chambered heart that separates oxygenated and oxygen-poor blood. Milk, produced in mammary glands, is used to nourish mammalian young. Fertilization is internal, and the embryo develops in the uterus. In placental mammals, a *placenta*, formed from the uterine lining and embryonic membranes, nourishes the developing embryo.

Mammals have relatively large brains and are capable of learning. Extended parental care of the young provides them time to learn complex skills.

Mammalian teeth come in a diverse assortment of shapes and sizes that are specialized for eating a variety of foods.

Evolution of Mammals   The oldest fossils of mammals date back to the Triassic period, 190 million years ago. The reptilian therapsids were mammalian ancestors. Small Mesozoic mammals, which were probably nocturnal and insectivorous, coexisted with dinosaurs. Mammals underwent an extensive radiation in the adaptive zones created by the extinction of the dinosaurs and the rise of flowering plants toward the end of the late Cretaceous period. There are three major mammal groups today: *monotremes*, egg-laying mammals; *marsupials*, mammals with pouches; and *placental mammals*.

Monotremes   The platypus and spiny anteaters (echidnas) are the only egg-laying mammals today. The egg is reptilian in structure, but monotremes have hair and produce milk for their young. Monotremes seem to have descended from a very early mammalian branch.

Marsupials   Marsupials, including opossums, kangaroos, and the koala, complete their embryonic development in a maternal pouch called a *marsupium*. Born very early, the neonate fixes its mouth to a teat and completes its development while nursing.

Australian marsupials have radiated and filled the niches occupied by placental mammals in other parts of the world. Marsupials apparently originated in what is now North America, spreading southward. After the breakup of Pangaea, South America and Australia became island continents where the marsupials diversified in isolation from placental mammals. Geological history helps to explain the evolutionary history and distribution of marsupials.

Placental Mammals   A placental mammal completes its development attached to a placenta within the uterus. Placentals and marsupials may have diverged from a common ancestor about 80 to 100 million years ago. The major orders of placental mammals radiated during the late Cretaceous and early Tertiary, about 70 to 45 million years ago.

There appear to be four main evolutionary lines of

placental mammals. The orders Chiroptera and Insectivora, containing the bats and shrews respectively, resemble early mammals. A second branch from a lineage of medium-sized herbivores gave rise to the lagomorphs (rabbits), perissodactyls (odd-toed ungulates such as horses and rhinos), artiodactyls (even-toed ungulates such as deer and swine), sirenians (sea cows), proboscideans (elephants), and cetaceans (porpoises and whales).

The order Carnivora, a third evolutionary line that developed during the early Cenozoic era, includes cats, dogs, raccoons, and the pinnipeds (seals, sea lions, and walruses). The most extensive radiation resulted in the primate–rodent complex. The order Rodentia includes rats, squirrels and beavers. The order Primates includes monkeys, apes and humans.

The Human Ancestry

Evolutionary Trends in Primates   The first primates probably were small arboreal animals, descended from insectivores late in the Cretaceous. Characteristics that were shaped by this arboreal ancestry include limber shoulder joints that permit *brachiation* (swinging from one hold to another), dextrous hands, eye–hand coordination, forward-facing and close-together eyes to provide depth perception, and parental care.

Modern Primates   Two subgroups of modern primates are *prosimians* and *anthropoids*. Prosimians, such as lemurs, lorises, and tarsiers, probably resemble early arboreal primates. The anthropoids include monkeys, apes, and humans. The first fossils of anthropoids are monkeylike primates that probably evolved from prosimian stock about 40 million years ago in Africa or Asia. Somehow these ancestral monkeys reached South America, and the strictly arboreal New World monkeys have been evolving separately from Old World monkeys for millions of years.

There are four genera of apes: gibbons, orangutans, gorillas, and chimpanzees. Modern apes generally are larger than monkeys, with relatively long legs, short arms, and no tails.

The Emergence of Humankind   Paleoanthropology is the study of human origins and evolution. Chimpanzees and humans represent two divergent branches from a common less specialized ancestor. Human evolution has not occurred as phyletic change within an unbranched hominid line; there have been times when several different species coexisted. Different human features have evolved at different rates, known as *mosaic evolution*. Thus, bipedalism or erect posture evolved before an enlarged brain developed.

The oldest known ape fossils are of a cat-sized tree-dweller that lived about 35 million years ago. About 25 million years ago, during the Miocene epoch, ape descendants diversified and spread into Eurasia. About 20 million years ago, the Himalayan range arose, the climate became drier, and the forests contracted. Some of the Miocene apes living on the edge of the forest may have begun foraging for food on the savanna. *Ramapithecus* is an anthropoid known from fossils 8 to 14 million years old. Most anthropologists believe that humans and apes diverged four to five million years ago.

In 1924, a British anthropologist found a skull of an early human that he called *Australopithecus africanus*. It appears that *Australopithecus* was an upright hominid, with humanlike hands and teeth but a small brain, who foraged on the African savannas for nearly two million years, beginning about three million years ago.

In 1974, a petite *Australopithecus* skeleton was discovered in the Afar region of Ethiopia. Lucy, as this fossil was named, and similar fossils have been designated as a separate species, *Australopithecus afarensis*. There is debate whether Lucy represents the most ancient of the australopithecine group and the common ancestor of two hominid branches (one leading to *A. africanus* and one leading to *Homo*, our genus); or whether *A. afarensis* had diverged from the lineage leading to *Homo*.

Australopithecus walked erect for over 1 million years with no enlargement of the brain. Larger-brained fossils date back about 2 million years. Simple stone tools sometimes have been found with these fossils. Some paleoanthropologists place these fossils in the genus *Homo*, calling them *Homo habilis*, or "handy man," whereas other scientists retain them in *Australopithecus*.

Homo habilis and *Australopithecus* coexisted for nearly 1 million years. One theory is that these two were distinct lines of hominids, one of which was an evolutionary deadend, while the other, *Homo habilis*, was on the line to modern humans.

Homo erectus, the first hominid to migrate out of Africa into Europe and Asia, is represented by fossils, known as Java Man and Peking Man, which range in age from 1.5 million years to 300,000 years. They had a greater height and brain capacity than *H. habilis* and survived in the colder climates of the north, living in shelters, building fires, wearing animal skins, and using more elaborate stone tools.

The oldest fossils classified as *Homo sapiens* are the Neanderthals, who lived from about 130,000 to 30,000 years ago. They had heavier brow ridges, less pronounced chins and larger brains than modern humans. They were skilled tool-makers and left evidence of burials and other rituals.

The oldest fully modern *Homo sapiens* fossils date back 90,000 years. Neanderthals may have overlapped modern humans in Eurasia by about 30,000 years. Neanderthals may have become extinct without genetically contributing to modern *Homo sapiens*. Molecular genetics has shown that the mitochondrial DNA of today's human population is very uniform. Some scientists speculate that the modern human gene pool is traceable to a single female, probably living in Africa between 200,000 and 400,000 years ago. Debate concerning the evolution of modern humans continues, hopefully to be fueled with additional fossil and molecular evidence.

Evolution of the erect stance required remodeling of the foot, pelvis and vertebral column. Prolonging the period of growth of the brain after birth made possible the enlargement of the human brain. An extended period of parental care helps to make possible the basis of culture—the transmission of knowledge accumulated over the generations. Language, spoken and written, is the major means for this transmission.

Cultural evolution has included several stages: the communal efforts of nomads who hunted and gathered food on African grasslands, the development of agriculture in Eurasia and the Americas about 10,000 to 15,000 years ago, and the Industrial Revolution, which began in the eighteenth century. Since then, the human population and new technology have been growing exponentially. The biological evolution of life has been linked with environmental changes. Cultural evolution has made *Homo sapiens* a new force in the history of life—a species that no longer needs to adapt to its environment but simply changes the environment to meet its needs. As agents of rapid environmental change, we are altering the planet faster than many species can adapt. The next crisis in the history of life may be the habitat destruction, pollution, resource decline, and climatic changes resulting from the actions of *Homo sapiens*.

STRUCTURE YOUR KNOWLEDGE

1. Place the major events in the evolution of Phylum Chordata on the geological timetable on the next page.

2. Describe several examples from vertebrate evolution that illustrate the common evolutionary theme that new adaptations usually evolve from preexisting structures.

3. How did various vertebrate groups meet the challenges of a terrestrial habitat?

TEST YOUR KNOWLEDGE

FILL IN THE BLANKS

1. _____ Chordate subphylum of sessile, filter-feeding marine animals

2. _____ subphylum that includes lancelet

3. _____ blocks of mesoderm along notochord that develop into muscles

4. _____ jawless fish with armor of bony plates

5. _____ flap over the gills of bony fishes

6. _____ adaptation that allows reptiles to reproduce on land

7. _____ lizardlike ancestors of the various reptilian orders

8. _____ group from which birds descended

9. _____ egg-laying mammals

10. _____ order in which snakes are placed

11. _____ structure that helps mammals ventilate their lungs

12. _____ mammals that walk on the tips of the toes

13. _____ genus in which the fossil Lucy is placed

14. _____ subgroup that includes monkeys, apes, and humans

15. _____ major means for the transmission of human culture

MULTIPLE CHOICE: *Choose the one best answer.*

1. Pharyngeal slits appear to have functioned first as
   a. filter-feeding devices.
   b. gill slits for respiration.
   c. components of the jaw.
   d. portions of the inner ear.

2. Which of the following characteristics does *not* differentiate the vertebrates from the invertebrate chordates?

Era	Period	Age (millions of years ago)	Major Animal Events	Plants Events
CENOZOIC	Neogene			
		5		
		24		
	Paleogene			Dominance of angiosperms
		65		
Climatic Change—Cooler				Flowering plants appear
MESOZOIC	Cretaceous	144		
	Jurassic	213		
	Triassic			Conifers and cycads dominant
		248		
Climatic Change—Warmer and Drier				
PALEOZOIC	Permian	286		Dominance of lycopods, sphenopsids, ferns; great coal forests; first seed plants appear
	Carboniferous	360		Radiation of vascular plants
	Devonian	408		
	Silurian	438		Origin of vascular plants
	Ordovician	505		Marine algae abundant
	Cambrian	590		

a. cephalization
b. vertebrae
c. postanal tail
d. ventral heart and closed circulatory system

3. Modern agnathans are represented by
   a. hagfishes and lampreys.
   b. sharks and rays.
   c. the lobe-fin coelocanth.
   d. lungfishes.

4. Jaws first occurred in the class
   a. Agnatha.
   b. Chondrichthyes.
   c. Osteichthyes.
   d. Placodermi.

5. Which of the following is incorrectly paired with its gas exchange mechanism?

a. amphibians—skin and lungs
b. lobed-finned fishes—gills and lungs
c. reptiles—lungs
d. bony fishes—swim bladder and gills

6. Urodeles include
   a. limbless, burrowing amphibians.
   b. frogs that have a tadpole stage.
   c. frogs and toads that may or may not have a tadpole stage.
   d. salamanders.

7. Reptiles have lower caloric needs than do mammals of comparable size because they
   a. are ectotherms.
   b. have waterproof scales.
   c. have an amniote egg.
   d. move by the bending back and forth of their vertebral column.

8. Which of the following best describes the earliest mammals?
   a. large, herbivorous
   b. large, carnivorous
   c. small, insectivorous
   d. small, herbivorous

9. Oviparity is a reproductive strategy that
   a. allows mammals to bear well-developed young.
   b. is evidence of the relatedness of birds and reptiles.
   c. is the only means by which sharks reproduce.
   d. is a necessity for all flying vertebrates.

10. In Australia, marsupials fill the niches that placental mammals fill in other parts of the world because
    a. they are better adapted and have outcompeted placental mammals.
    b. their offspring complete their development attached to a teat in a marsupium.
    c. after Pangaea broke up, they diversified in isolation from placental mammals.

    d. they evolved from monotremes that migrated to Australia about 12 million years ago.

11. Which of the following characteristics of primates is not associated with an arboreal existence?
    a. shoulder joints that permit brachiation
    b. upright posture
    c. eyes close together on the face to improve depth perception
    d. parental care

12. Which of the following is *not* true of Neanderthals?
    a. They were the first to be classified as *Homo sapiens*.
    b. They walked erect for over a million years with no substantial enlargement of the brain.
    c. They disappeared about 30,000 years ago, and it is not known whether they contributed to the modern human gene pool.
    d. They were skilled tool-makers and had a culture that included burials and other rituals.

CHAPTER 24
EARLY EARTH AND THE ORIGIN OF LIFE

Suggested Answers to Structure Your Knowledge

1. Primitive Earth is believed to have had seas, volcanoes, and large amounts of ultraviolet radiation passing through the thin atmosphere, which probably consisted of H_2O, CO, CO_2, N_2, CH_4, and NH_3. Life was able to evolve in this environment because the reducing (electron-adding) atmosphere and availability of energy from lightning and UV radiation would facilitate the abiotic synthesis of organic molecules, and the hot lava rocks may have aided the polymerization of these monomers. The availability of methane or organic fuels may have provided an energy source for early protobionts.

2. (1) The abiotic synthesis of organic molecules provided the amino acids, simple sugars, and nucleotides that served as the building blocks of proteins, energy molecules, and precursors for the information molecules of RNA and DNA.
   (2) In order for life to develop, organic monomers had to be linked into polymers that have emergent properties deriving from their structural organization. Early polypeptides may have functioned as simple enzymes, lipids formed bounding membranes, and RNA molecules provided self-repli-cating structures capable of ordering amino acids into polypeptides.
   (3) The aggregation of molecules into discrete packets, separated from the surroundings by selectively permeable membranes, allowed for the concentration of critical molecules and the emerging properties of metabolism, excitability, and primitive reproduction.
   (4) The development of a precise mechanism for the recording and replication of genetic material allowed for the reproduction and evolution of successful aggregations of organic molecules.

Answers to Test Your Knowledge

Matching:

1. F	3. E	5. C
2. D	4. B	6. A

Multiple Choice:

1. a	4. d	7. b	10. c
2. b	5. c	8. d	11. b
3. c	6. c	9. a	12. d

CHAPTER 25
PROKARYOTES AND THE ORIGINS OF METABOLIC DIVERSITY

Suggested Answers to Structure Your Knowledge

1.

CHARACTERISTICS OF THE PROKARYOTIC CELL

Characteristic	Description
Cell shape	May be spherical (cocci), rod (bacilli), or spiral (spirilla).
Genome	Circular DNA molecule with little associated protein; 1/1000 as much DNA as eukaryote, concentrated in nucleoid region. May have plasmids—circular DNA with genes for antibiotic resistance or special metabolic enzymes.
Internal membranes	Few internal membrane systems, some infoldings of plasma membrane may be used in respiration; thylakoid membranes in cyanobacteria for photosynthesis.
Cell surface	Cell wall made of peptidoglycan, may have outer lipopolysaccharide membrane (gram-negative bacteria); sticky capsule for adherence, pili for attachment or conjugation
Forms of motility	May glide in secreted slime or "swim" with flagella that are either scattered over cell or concentrated at ends of cell, or use axial filament in twisting movement. May use alternating runs and tumbles, can result in taxis toward or away from stimulus.
Reproduction and growth	Divide by binary fission, rapidly grow into colony of progeny. Geometric growth slowed by accumulation of toxic wastes or depletion of nutrients. Some genetic exchange during conjugation. Most genetic variation is a result of mutations.

2.

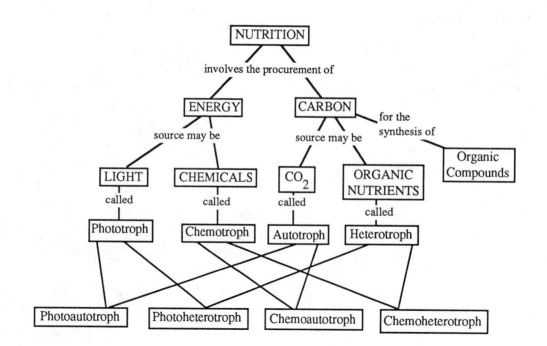

3. • Breakdown of abiotically synthesized organic compounds for energy led to *glycolysis* and formation of *ATP* by *substrate level phosphorylation*.

   • *Fermentation* and excretion of acids led to development of *proton pumps* (using ATP) to maintain pH of cell.

   • *Electron transport chain* linked oxidation of organic compounds with pumping of H⁺ out of cell. Efficient electron transport chain could have formed proton gradient so that H⁺ diffused back in, reversing proton pump and generating ATP (*chemiosmosis—anaerobic respiration*).

   • Use of light energy to drive electrons from H_2S to NADP⁺, co-opting electron transport chain, to generate reducing power to fix CO_2 (*photosynthesis*).

   • Use of water as source of electrons and hydrogen for fixing CO_2 (*photosynthesis evolving oxygen*).

   • Use of O_2 to pull electrons from organic molecules down existing electron transport chain (*aerobic respiration*).

4. Decomposers—recycle global nutrients, used for sewage treatment, clean up oil spills

   Nitrogen Fixers—provide nitrogen to the soil in nodules on legume roots

   Symbionts—enteric bacteria, some produce vitamins

Biotechnology—used for production of antibiotics, various useful products

Answers to Test Your Knowledge

Multiple Choice:

1. c	4. b	7. d	10. d
2. d	5. c	8. d	11. a
3. a	6. b	9. b	12. c

Fill in the Blanks:

1. bacilli
2. nucleoid region
3. gram stain
4. pili
5. taxis
6. chemoheterotroph
7. photoautotroph
8. facultative anaerobe
9. saprophytes
10. exotoxins
11. nodules
12. archaebacteria

Matching:

1. D	4. C	7. F	10. K
2. E	5. H	8. G	11. I
3. L	6. A	9. M	12. J

CHAPTER 26
PROTISTS AND THE ORIGIN OF EUKARYOTES

Suggested Answers to Structure Your Knowledge

1. (a) The kingdom Protista includes unicellular—and a few colonial and multicellular—eukaryotic organisms. Nearly all protists use aerobic respiration and have flagella or cilia at some point in their life. They all reproduce asexually, and some have sexual reproduction. They are found in water or moist habitats (including inside the bodies of hosts), and many form cysts to withstand harsh conditions. Protists may be photoautotrophic or heterotrophic or both.

   (b) There is controversy over the boundaries of the kingdom Protista. Originally Whittaker assigned only unicellular eukaryotes to this kingdom. Some colonial and multicellular eukaryotes, however, lack the distinctive traits of either fungi, plants, or animals, and appear to be more closely related to unicellular protists than to these multicellular kingdoms. Margulis and others have advocated an expanded kingdom Protista to include these multicellular descendents of unicellular eukaryotes.

2. According to the autogenous hypothesis, eukaryotic cells are thought to have evolved internal membrane structures from the specialization of membranes invaginated from the plasma membrane.

   The endosymbiotic hypothesis claims that at least the mitochondria and chloroplasts of eukaryotic cells developed from photosynthetic and aerobic heterotrophic prokaryotes that became incorporated into larger cells. The lines of evidence for this model include the following similarities: the chloroplasts of red algae to cyanobacteria; the chloroplasts and pigments of green algae to the bacterium *Prochlorothrix*; the size, membrane enzymes and transport systems, division process, circular DNA molecule, ribosomal RNA sequences, and ribosomes of chloroplasts and mitochondria to those found in eubacteria. The existence of endosymbiotic relationships in modern organisms and the recent discovery of transposons, which may have transferred endosymbiont DNA to the host cell nucleus, provide additional evidence for this model.

3.

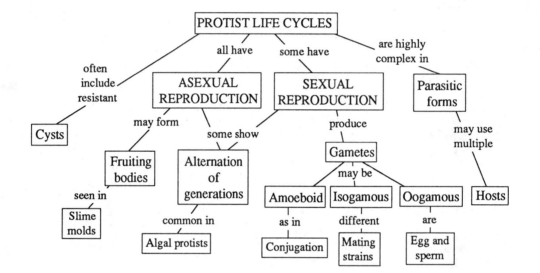

Answers to Test Your Knowledge

Matching:

1.	N	5.	J	9.	D	13.	P
2.	I	6.	E	10.	O	14.	C
3.	B	7.	M	11.	G	15.	H
4.	K	8.	F	12.	L	16.	A

Multiple Choice:

1.	c	4.	c	7.	c	10.	c
2.	d	5.	d	8.	c	11.	d
3.	b	6.	b	9.	d	12.	a

CHAPTER 27
PLANTS AND THE COLONIZATION OF LAND

Suggested Answers to Structure Your Knowledge

1.

Characteristics \ Plant Groups / Common Names	Bryophyta — Mosses, liverworts	Lycophyta Sphenophyta — Club mosses, horsetails	Pterophyta — Ferns	Coniferophyta — Conifers, pines	Anthophyta — Flowering plants
Cuticle, stomata	X	X	X	X	X
Protected embryo, jacketed reproductive organs	X	X	X	X	X
Vascular tissue for transport and support		X	X	X	X
Protected gametophyte retained on sporophyte plant				X	X
Nonswimming sperm, pollen				X	X
Fibers for support in vascular tissue				X	X
Seed protects and encloses embryo with food source				X	X
Flowers and fruits, animal pollinators, protected seeds					X

2.

Era	Period	Age (millions of years ago)	Major Plants Events	Animal Events
CENOZOIC	Neogene		Dominance of angiosperms	Major radiation and dominance of mammals, birds, pollinating insects
CENOZOIC	Paleogene	— 65 —	Dominance of angiosperms	Major radiation and dominance of mammals, birds, pollinating insects
	Climatic Change—Cooler			
MESOZOIC	Cretaceous	144	Flowering plants (angiosperms) appear	Dinosaurs become extinct
MESOZOIC	Jurassic	213	Conifers and cycads dominant	Dinosaurs dominant
MESOZOIC	Triassic	— 248 —		
	Climatic Change—Warmer and Drier			
PALEOZOIC	Permian	286	Dominance of lycopods, sphenopsids, ferns; great coal forests; first seed plants appear; fossil record of mosses	Mass extinction of marine invertebrates, radiation of reptiles
PALEOZOIC	Carboniferous	360	Dominance of lycopods, sphenopsids, ferns; great coal forests; first seed plants appear; fossil record of mosses	Amphibians dominant
PALEOZOIC	Devonian	408	Adaptive radiation of vascular plants	Fish diversify
PALEOZOIC	Silurian	438	Origin of vascular plants	
PALEOZOIC	Ordovician	505	Marine algae abundant	First vertebrates
PALEOZOIC	Cambrian	— 590 —		Origin of most invertebrates

Answers to Test Your Knowledge

Multiple Choice:

1.	c	5.	c	9.	b
2.	a	6.	a	10.	d
3.	c	7.	a	11.	a
4.	b	8.	d	12.	c

True or False:

1. F—add *or algae.*
2. F—change *heteromorphic* to *heterosporous,* or change *produce . . .* to *have gametophyte and sporophyte that are morphologically distinct.*

3. T
4. F—change *naked seed* to *seedless,* or change *club mosses and horsetails* to *gymnosperms.*
5. T
6. T
7. F—change *tracheids* to *vessel elements*
8. T
9. F—change *angiosperm* to *gymnosperm,* or change *haploid cells . . .* to *an embryo sac with seven haploid cells.*
10. T

CHAPTER 28
FUNGI

Suggested Answers to Structure Your Knowledge

1.

Division	Common Members	Morphology	Asexual Reproduction	Sexual Reproduction	Miscellaneous Information
Zygomycota	black bread mold, *Pilobolus*	coenocytic hyphae, rhizoids, horizontal aerial hyphae	spores in sporangia at tips of aerial hyphae	resistant zygospores, conjugation	terrestrial
Ascomycota	sac fungi, yeasts, cup fungi	septate hyphae or unicellular	conidia, wind-dispersed spores	dikaryotic hyphae, ascus, ascospores, ascocarps	genetic studies, decomposers, yeasts, in most lichens
Basidio-mycota	club fungi, mushrooms, puffballs, shelf fungi	dikaryotic hyphae, mycelia, basidiocarp	no asexual stage	basidium with basidiospores	fairy rings, mycorrhizae
Deutero-mycota	fungi imperfecti, *Penicillum*		many form conidia	no sexual stage	many produce antibiotics
Lichens	lichens	foliose, crustose, fruticose	soredia or fragments	fungus or alga may sexually reproduce	symbiotic association, harsh habitats

2. The role of fungi as saprophytes and decomposers of organic material is central to the recycling of chemicals between living organisms and their physical environment. Decomposers also prevent the extensive accumulation of litter, feces, and dead organisms. Parasitic fungi have economic and health effects on humans. Plant pathogens cause extensive loss of food crops and trees (Dutch elm disease, powdery mildews, ergots on rye). Fungal pathogens of humans can cause annoying to serious problems. An example of a mutualistic symbiont is the mycorrhizae found on the roots of most plants. These associations greatly benefit the plant.

Answers to Test Your Knowledge

Fill in the Blanks:

1. septum	5. basidium	9. coenocytic
2. dikaryon	6. asci	10. haustoria
3. chitin	7. mycorrhizae	
4. conidia	8. zygospore	

Multiple Choice:

1. a	3. c	5. d	7. b
2. c	4. d	6. b	8. d

CHAPTER 29
INVERTEBRATES AND THE ORIGINS OF ANIMAL DIVERSITY

Suggested Answers to Structure Your Knowledge

1.

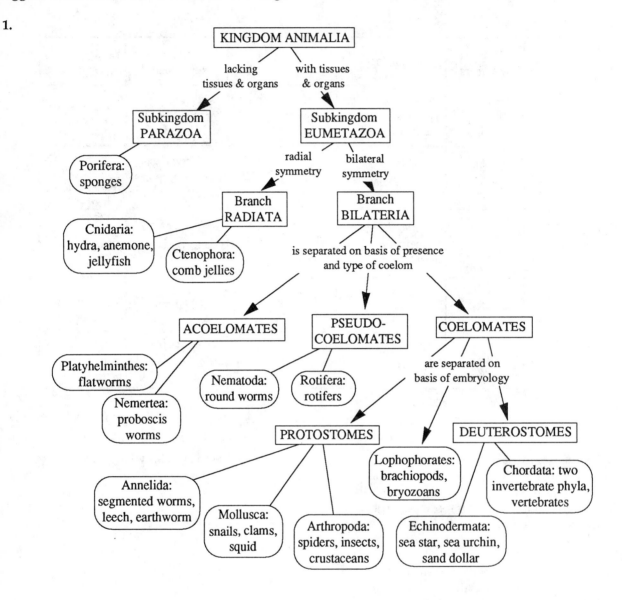

2. Transitional forms between protists and the first animals and between the various animal phyla that would shed light on the origin and phylogeny of animals have not been found in the fossil record. Suggested hypothetical ancestors of the eumetozoa include a colonial flagellate, a planuloid, an acoel flatworm, or a coenocytic ciliate. The fossil record begins with soft-bodied medusae and annelidlike worms from the Ediacaran period of the Precambrian era. The fossils of the next period include nearly all modern phyla. The Cambrian appearance of so many complex animals remains unexplained.

3.

	Digestion	Excretion	Respiration	Nervous control	Skeleton	Locomotion
Porifera	choanocytes, amoebocytes	diffusion	diffusion	none	spicules or spongin	none
Cnidaria	gastrovascular cavity	↓		nerve net	hydrostatic	contractile fibers
Platyhelminthes	↓	diffusion, flame cells		ganglia, nerve cord	↓	cilia, muscles, glide or swim
Rotifera	complete digestive teact	↓			hydrostatic, cuticle	muscles, tapered "foot"
Nematoda			↓		hydrostatic, cuticle	longitudinal muscles
Mollusca		"kidney," nephridia	gills, diffusion	↓	external shell, hydrostatic	muscular foot, jet propulsion
Annelida		metanephridia	diffusion, skin blood vessels	ganglia, ventral nerve cords	hydrostatic, compartments	setae, muscle sets
Arthropoda		Malphigian tubules	gills, tracheal system	↓	exoskeleton of chitin	jointed legs, wings
Echinodermata		diffusion	skin gills	radial nerve system	calcareous endoskeleton	tube feet, water vascular system
Chordata (invertebrate phyla)	↓		pharyngeal gill slits	dorsal, hollow nerve cord	notochord	muscles, swim or sessile

Answers to Test Your Knowledge

Matching:

1.	L	c	5.	M	h	9.	E	d
2.	D	b	6.	I	b	10.	H	c
3.	F	e	7.	B	e			
4.	G	a	8.	A	b			

Multiple Choice:

1.	b	5.	b	9.	d	13.	b	17.	a
2.	a	6.	d	10.	c	14.	a	18.	c
3.	d	7.	c	11.	c	15.	b	19.	b
4.	c	8.	a	12.	b	16.	c	20.	a

CHAPTER 30
THE VERTEBRATE GENEALOGY

Suggested Answers to Structure Your Knowledge

1.

Era	Period	Age (millions of years ago)	Major Animal Events	Plants Events
CENOZOIC	Neogene		*Homo sapiens* *Homo erectus, Homo habilis* *Australopithecus*	
		5		
	Paleogene	24	Origin of apes	Dominance of angiosperms
		65	Major radiation and dominance of mammals, birds, pollinating insects	
	Climatic Change—Cooler			
MESOZOIC	Cretaceous	144	Dinosaurs become extinct Primates probably appear	Flowering plants appear
	Jurassic	213	Dinosaurs dominant Birds appear	
	Triassic		Second radiation of reptiles first dinosaurs, mammals, and birds	Conifers and cycads dominant
		248		
	Climatic Change--Warmer and Drier			
PALEOZOIC	Permian	286	First reptilian radiation Origin of most modern insect orders	Dominance of lycopods, sphenopsids, ferns; great coal forests; first seed plants appear
	Carboniferous	360	Amphibians dominant Reptiles appear	
	Devonian	408	Age of Fishes	Radiation of vascular plants
	Silurian	438	Diversity of jawless fish	Origin of vascular plants
	Ordovician	505	First vertebrates	Marine algae abundant
	Cambrian	590	First invertebrate Chordate Origin of most invertebrates	

2. The following are examples of new adaptations that evolved from preexisting structures: (1) use of pharyngeal slits, through which water flowed for filter feeding, for gas exchange, development of gills; (2) transformation of lungs, developed in bony fish in stagnant waters for gas exchange, to an air bladder functioning in buoyancy; (3) development of fleshy fins, supported by skeletal elements, into limbs for terrestrial animals; (4) use of feathers, probably functioning for insulation in endothermy, for flight; (5) use of dextrous hands, important for an arboreal life, to manipulate tools.

3. Amphibians remained in damp habitats, burrowing in mud during droughts; some secrete foamy protection for eggs laid on land. Reptiles developed scaly, waterproof skin, an amniote egg that provided an aquatic environment for developing embryo, and behavioral adaptations to modulate changing temperatures. All terrestrial groups are tetrapods, using limbs for locomotion on land.

Answers to Test Your Knowledge

Fill in the Blanks:

1. Urochordates
2. Cephalochordates
3. somites
4. ostracoderms
5. operculum
6. amniote egg
7. cotylosaurs, stem reptiles
8. archosaurs, dinosaurs
9. monotremes
10. Squamata
11. diaphragm
12. ungulates
13. *Australopithecus*
14. anthropoid
15. language

Multiple Choice:

1. a
2. c
3. a
4. d
5. d
6. d
7. a
8. c
9. b
10. c
11. b
12. b

PLANTS: FORM AND FUNCTION

THE FAR SIDE By GARY LARSON

Feb. 22, 1946: Botanists create the first artificial flower.

CHAPTER 31

ANATOMY OF A PLANT

FRAMEWORK

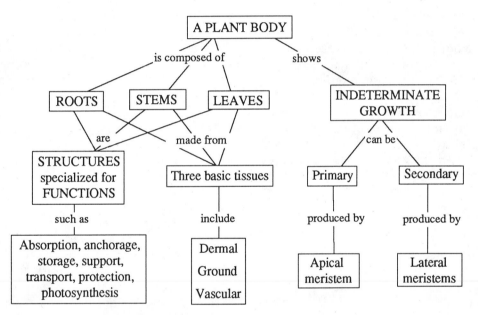

A rather large new vocabulary is needed to name the specialized cells and structures in a study of plant anatomy. Focus your attention on how the structure of the parts of a plant integrates into the functioning of the entire living organism.

CHAPTER SUMMARY

Botany is the study of plants. The most recently evolved and currently dominant group of plants is the angiosperms, characterized by flowers and fruits. This group is split into two classes: *monocots*, which have one embryonic seed leaf or cotyledon, and *dicots*, which have two. Plants show adaptations to their terrestrial environment in both evolutionary and physiological time scales. The anatomy of plants also illustrates the correlation between structure and function.

The Parts of a Flowering Plant

Plants have an underground *root system* for obtaining water and minerals from the soil and an aerial *shoot system*, consisting of stems, leaves, and flowers, for absorbing light and carbon dioxide for photosynthesis. Vascular tissue transports materials between the interdependent roots and shoots. *Xylem* carries water and dissolved minerals upward, whereas *phloem* conducts food made in the leaves to nonphotosynthetic plant parts and stored food from the root to regions of active growth.

The Root System   The structure of roots is adapted for anchorage, absorption, conduction, and storage of food. A *taproot* system is found commonly in dicots, whereas monocots generally have *fibrous root* systems. Most absorption of water and minerals occurs through the tiny *root hairs* that are clustered near the root tips. *Adventitious* roots arise from stems or leaves and may serve as props to help support stems.

The Shoot System   The stem consists of *nodes*, at which leaves are attached, and segments between nodes, called *internodes*. An *axillary bud* is found in the angle between a leaf and the stem. A *terminal bud*, consisting of developing leaves, nodes and internodes, is found at the tip of a shoot. The terminal bud may exhibit *apical dominance* and inhibit the growth of axillary buds. Axillary buds may develop into flowers or *vegetative* branches with their own terminal buds, leaves, and axillary buds.

Modifications of stems include stolons, horizontal stems that grow along the ground; rhizomes, horizontal stems growing underground that may end in enlarged tubers; and bulbs, vertical, underground shoots functioning in food storage.

Leaves, the main photosynthetic organs of most plants, usually consist of a flattened blade and a *petiole*, or stalk. Most monocots lack petioles. Monocot leaves usually have parallel major veins, whereas dicot leaves have networks of branched veins. Leaves emerge from the stem in an alternate, opposite, or whorled arrangement, and they may be simple or compound.

Plant Cells and Tissues

How Plant Cells Grow   The growth of roots and stems is largely a result of cell growth—an irreversible increase in cell size. About 90% of the growth of a plant cell is due to the uptake of water by the vacuole. The cell wall loosens and the cell swells, usually in one direction. Unlike animal cell growth, which requires the synthesis of cytoplasm, this economical means of cell elongation allows for the rapid growth of shoots and roots. When a plant cell stops growing, the wall becomes thicker and more rigid.

Types of Plant Cells   *Parenchyma* cells, relatively unspecialized cells that lack secondary walls and have large central vacuoles, function in photosynthesis and food storage. *Collenchyma* cells also may lack secondary walls but have thickened primary walls and can function in support. Strands or cylinders of collenchyma cells elongate along with the young parts of the plant.

Schlerenchyma cells have thick secondary walls strengthened with lignin. These supporting elements often lose their protoplasts (the living portion of a plant cell) at maturity. *Fibers* are long, tapered schlerenchyma cells that usually occur in bundles. *Sclereids* are shorter and irregular in shape.

After the cells of xylem form secondary walls and mature, the protoplast disintegrates, leaving behind a conduit through which water flows. The secondary walls may be deposited in spiral or ring patterns, or they may have scattered *pits*, or thinner regions consisting only of primary walls. *Tracheids* are long, thin, tapered xylem cells with lignin-strengthened walls. *Vessel elements* are wider, shorter, thinner-walled, and probably more efficient water conductors. Gymnosperms have only tracheids, whereas angiosperms have both tracheids and vessel elements, the latter probably having evolved from the former in ancient flowering plants.

Phloem consists of chains of cells called *sieve-tube members*, which remain alive but lack nuclei, ribosomes, and vacuoles at functional maturity. Fluid flows through pores in the *sieve plates* of the end walls between cells. The nucleus and ribosomes of a *companion cell*, which is connected to a sieve-tube member by numerous plasmodesmata, may serve both cells.

Plant Tissues   *Simple tissues* are groups of cells of a single cell type, such as parenchyma or sclerenchyma. *Complex tissues*, such as xylem or phloem, are composed of more than one type of cell.

Tissue Systems   Plants consist of three tissue systems. The *dermal tissue system*, or *epidermis*, is a single layer of cells that covers and protects the young parts of the plant. Root hairs are extentions of epidermal cells that increase the surface area for absorption. The epidermis of leaves and some stems is covered with a *cuticle*, a waxy coating that prevents excess water loss.

The *vascular tissue system* consists of xylem and phloem and functions in transport and support.

The *ground tissue system*, making up the bulk of a young plant, is predominately parenchyma. Ground tissue functions in photosynthesis, support, and storage.

Primary Growth

Indeterminate Growth   Most plants exhibit *indeterminate growth*: they continue to grow as long as they live. Animals have *determinate growth*: they stop growing after reaching a certain size. Plants may be *annuals*, which complete their life cycle in one growing season or year; *biennials*, which have a life cycle spanning two years; or *perennials*, which live many years.

Plants have tissues called *meristems* that remain

embryonic and divide to form new cells at the growing points of the plant. *Apical meristems*, located at the tips of roots and in the buds of shoots, produce *primary growth*, resulting in the elongation of roots and shoots and the formation of the primary tissues.

Primary Growth of Roots   The meristem of the root tip is protected by a *root cap*, which secretes a polysaccharide slime to lubricate the growth route. The *zone of cell division* includes the apical meristem and the primary meristems of the *protoderm*, *procambium*, and *ground meristem*. These meristems divide and give rise to the cells of the three tissue systems of the root. In the *zone of cell elongation*, the cells lengthen to more than ten times their original size, which pushes the root tip through the soil. Cells specialize in structure and function in the *zone of cell differentiation*.

The protoderm gives rise to the single cell layer of the epidermis. The procambium forms a central vascular cylinder, or *stele*. In most dicots, xylem cells radiate from the center of the stele in spokes, with phloem in between. The stele of a monocot may have *pith*, a central core of parenchyma cells, inside the xylem and phloem.

The ground meristem produces the ground tissue system, filling the *cortex* or region of the root between stele and epidermis. A one-cell-thick *endodermis* forms the boundary between cortex and stele and regulates the passage of materials into the stele. *Lateral roots*, or *secondary roots*, may develop from the *pericycle*, the outermost layer of the stele.

Primary Growth of Shoots   In the shoot, the dome-shaped mass of the apical meristem in the terminal bud forms the primary meristems: protoderm, procambium, and ground meristem. Leaf primordia form as tiny bulges on the flanks of the apical dome, and axillary buds develop from clumps of meristematic cells left at the bases of the leaf primordia. Although most of the elongation of dicot shoots occurs at the tip, the shoots of monocots may grow at each node, where meristematic cells continue to divide.

Vascular tissue runs through the stem in *vascular bundles*. In dicots, the vascular bundles may be arranged in a ring, with pith internal and cortex external to the ring. The pith and cortex are connected by pith rays. Xylem is located internal to the phloem in the vascular bundles. In most monocot stems, the vascular bundles are scattered throughout the ground tissue, and no definite pith region exists.

The leaf is covered by the waxy-coated, tightly interlocking cells of the epidermis. *Stomata*, tiny pores flanked by *guard cells*, permit both gas exchange and *transpiration*, the evaporation of water from the leaf.

The *mesophyll* consists of parenchyma ground tissue cells containing chloroplasts. In most dicot leaves, the columnar palisade mesophyll is located above the spongy mesophyll, which has loosely packed, irregularly shaped cells surrounding many air spaces.

A *leaf trace*, branching from a vascular bundle at a node, continues into the petiole as a vein, which divides repeatedly within the blade of the leaf, providing support and vascular tissue to the photosynthetic mesophyll.

Secondary Growth

Secondary growth results in an increase in diameter as new cells are produced by two *lateral meristems*, the *vascular cambium* and *cork cambium*. Most monocots do not have secondary growth.

Secondary Growth of Stems   A band of parenchyma cells between the xylem and phloem of each vascular bundle and extending laterally into the ground tissue forms the vascular cambium, a continuous cylinder of meristematic cells. Cells produced internally to the vascular cambium differentiate into *secondary xylem*, whereas externally produced cells become *secondary phloem*. Wood is the accumulated secondary xylem cells, which have thick, lignified walls. *Rays* of parenchyma cells are produced by the vascular cambium and function in storage and lateral transport through the secondary xylem. Annual growth rings result from the seasonal cycle of xylem production.

The epidermis splits off during secondary growth and is replaced by new protective tissues produced by the cork cambium. Cork cells, with waxy walls impregnated with suberin, are produced externally, and parenchyma tissue called phelloderm is produced internally. The protective coat formed by these three layers (cork, cork cambium, phelloderm) is called the *periderm*. As secondary growth continually splits the periderm, a new cork cambium develops, eventually forming from parenchyma cells in the secondary phloem. Older secondary phloem, outside the cork cambium, also helps protect the stem as part of the bark. *Lenticels* are spongy regions in the bark through which gas exchange occurs. In older trees, the *heartwood* consists of older secondary xylem cells that have become clogged with resins and metabolic by-products. The *sapwood* consists of younger secondary xylem, vascular cambium, younger secondary phloem, and cork cambium.

Secondary Growth of Roots   The two lateral meristems also function in the secondary growth of roots. Vascular cambium produces xylem internal and phloem external to itself, and a cork cambium forms from the pericycle and produces the periderm.

STRUCTURE YOUR KNOWLEDGE

1. In the following table, list the processes or functions of a plant and the specialized cells, tissues, or structures that perform those functions within the roots, stems and/or leaves. Examples of functions would be anchorage, support, transport, and so on.

2. Compare and contrast primary and secondary growth.

TEST YOUR KNOWLEDGE

MATCHING: *Match the plant structure with its description.*

1. _____ schlerenchyma
2. _____ collenchyma
3. _____ tracheid
4. _____ fibers
5. _____ stele
6. _____ pericycle
7. _____ mesophyll
8. _____ periderm

A. chains of cells that transport food materials

B. tapered xylem cells with lignin in cell walls

C. parenchyma cells with chloroplasts in leaves

D. protective coat of woody stems and roots

E. bundles of long schlerenchyma cells

F. supporting cells with thickened primary walls

G. parenchyma cells inside vascular ring in stem

H. layer from which lateral roots originate

FUNCTION	CELLS, TISSUES, OR STRUCTURES that perform FUNCTION in		
	ROOT	STEM	LEAF

9. _____ endodermis

10. _____ pith

I. central vascular cylinder of root

J. unspecialized cells functioning in food storage or photosynthesis

K. cell layer in root regulating movement into central vascular cylinder

L. supporting cells with thick secondary walls

MULTIPLE CHOICE: *Choose the one best answer.*

1. Monocots differ from dicots in all of the following ways *except*
   a. flower parts in multiples of fours or fives rather than threes.
   b. leaves with parallel veins rather than branching venation.
   c. usually no secondary growth.
   d. stem with scattered vascular bundles rather than bundles in a ring.

2. Which of the following is incorrectly paired with its function?
   a. adventitious roots—branching root system found in monocots
   b. tap root—large vertical root found in dicots
   c. root hairs—absorption of water and dissolved minerals
   d. root cap—protect root as it pushes through soil

3. Axillary buds
   a. may not develop due to apical dominance of terminal bud.
   b. form at nodes in the angle where leaves join stem.
   c. have their own apical meristems and leaf primordia.
   d. all of the above are correct.

4. A leaf trace is
   a. a petiole.
   b. the outline of the vascular bundles in a leaf.
   c. a branch from a vascular bundle at a node that extends into a leaf.
   d. a tiny bulge on the flank of the apical dome that grows into a leaf.

5. Ground meristem

   a. produces the root system.
   b. produces the gound tissue system.
   c. produces secondary growth.
   d. is meristematic tissue found at the nodes in monocots.

6. The zone of cell elongation
   a. is responsible for pushing a root through the soil.
   b. comes between the zone of cell division and the zone of cell enlargement.
   c. produces the protoderm, procambium, and ground meristem tissues.
   d. does all of the above.

7. Perennials differ from annuals in that they
   a. have secondary growth and are woody.
   b. persist through many growing seasons.
   c. have indeterminate growth.
   d. all of the above are correct.

8. The function of stomata is to
   a. allow transport of water vapor into the leaf.
   b. allow transport of sugars out of the leaf.
   c. allow gas exchange between leaf and the atmosphere.
   d. open and close the guard cells.

9. Lateral meristems
   a. are responsible for secondary growth.
   b. usually are not present in dicots.
   c. produce protoderm, procambium, and ground meristem tissues.
   d. produce branch roots and lateral buds.

10. Sieve-tube members
    a. are responsible for lateral transport through secondary xylem.
    b. control the activities of phloem cells that have no nuclei or ribosomes.
    c. are transport cells in which fluid passes through sieve plates in the end walls between cells.
    d. have spiral thickenings that allow the cell to grow and function in support.

FILL IN THE BLANKS: *Starting from the outside, place the letter of the tissues in the order in which they are located in a young woody tree trunk.*

___ ___ ___ ___ ___ ___ ___ ___

A. primary phloem
B. secondary phloem
C. primary xylem
D. secondary xylem
E. pith

F. cork cambium
G. vascular cambium
H. cork cells
I. parenchyma tissue called phelloderm

CHAPTER 32

TRANSPORT IN PLANTS

FRAMEWORK

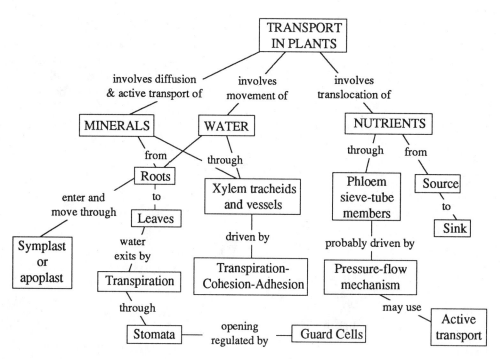

CHAPTER SUMMARY

Transport in plants involves the absorption of water and solutes by individual cells, the movement of substances from cell to cell, and the long-distance transport of sap in xylem and phloem.

Absorption of Water and Minerals by Roots

The soil solution soaks into the hydrophilic walls of epidermal cells and along the matrix of walls into the root cortex, exposing a large surface area of plasma membrane for the uptake of water and minerals.

Active Accumulation of Mineral Ions Minerals may enter cortex cells by diffusion down their concentration gradients or by active transport against a gradient involving specific carrier proteins in the plasma membrane and the expenditure of ATP. The *proton pump* found in plant cell membranes may influence the transport of many different minerals simultaneously. This pump uses ATP to move hydrogen ions (H⁺) out of the cell, generating a voltage or membrane potential across the membrane, which helps drive charged minerals across the plasma membrane.

Entry of Water and Minerals into Xylem Water and minerals may cross the root cortex through the *symplast*, the living connection of cytoplasm extending through plasmodesmata from cell to cell. Or water and minerals may cross through the *apoplast*, the extracellular pathway along the matrix of cell walls. The ring of suberin around each endodermal cell, called the *Casparian strip*, prevents water from the apoplast from entering the stele, and also keeps ions that have accumulated in the stele from leaking back out. The water and minerals must cross the selective plasma membrane of an endodermal cell to move into the stele.

Transfer cells within the stele selectively pump minerals out of their cytoplasm and into their walls. The minerals enter the nonliving xylem tracheids and vessel elements and move again through the apoplast system. Transfer cells, with infoldings of the cell wall and membranes that increase their surface area, are found where there is extensive traffic of solutes between symplast and apoplast.

Ascent of Xylem Sap

Through *transpiration*, plants lose a tremendous amount of water that must be replaced by water transported up from the roots.

Water Potential Water *potential*, abbreviated by Ψ (psi), is a useful measurement for predicting the direction that water will move from one plant part to another. The pressure unit used for water potential is usually *megapascals* (mPa), equal to 10 bars and about the same as ten atmospheres of pressure.

The water potential of pure water in an open container is defined as zero. Water will flow from a region of higher water potential to one of lower water potential. The addition of solutes lowers water potential; a solution thus has a negative water potential. The addition of pressure increases water potential.

A plant cell bathed in a more concentrated solution (hyperosmotic) will lose water by osmosis because the solution has a lower Ψ. In pure water, the cell has the lower Ψ. Water will enter the cell until enough turgor pressure builds up to bring $\Psi = 0$ in the cell, and net movement of water will stop. Differences in water potential affect movement of water both between individual plant cells and between different regions of a plant in long-distance transport.

Root Pressure As minerals are actively pumped into the stele, the water potential of the stele is lowered. Water flows in by osmosis, resulting in *root pressure*, which forces fluid up the xylem. Root pressure may cause *guttation*, the exudation of water droplets through specialized structures in the leaves when more water is forced up the xylem than is transpired by the plant. In most plants, root pressure is not a major mechanism in the movement of xylem sap.

The Transpiration–Cohesion–Adhesion Theory The transpiration–cohesion–adhesion theory is the most widely accepted explanation of the ascent of xylem sap. Transpiration provides the pull on the sap in the xylem. Water moves along a gradient of decreasing water potential from xylem to neighboring cells to air spaces to the drier air outside the leaf.

The transpirational pull on xylem sap is transmitted from the leaves to the root tips by the cohesiveness of water that results from the hydrogen bonds between molecules. The adhesion of water molecules to the hydrophilic walls of the narrow xylem elements and tracheids also helps the upward movement against gravity.

The upward transpirational pull on the cohesive sap creates tension within the xylem, further decreasing water potential and providing for the passive flow of water from the soil, across the cortex, and into the stele.

Cavitation, or a break in the chain of water molecules by the formation of a water vapor pocket in a xylem vessel, breaks the transpirational pull, and the vessel cannot function in transport.

The Control of Transpiration

The Photosynthesis–Transpiration Compromise A plant's tremendous requirement for water is partly a function of making food by photosynthesis. To obtain sufficient CO_2 for photosynthesis, leaves must exchange gases through the stomata, exposing the large surface area of the cells surrounding the air spaces from which water may evaporate. Transpiration is certainly not all bad; it diverts water from the environment through the plant at no cost, making water available and transferring minerals and other substances to the leaves. Transpiration also provides evaporative cooling.

A common ratio of transpiration-to-photosynthesis is 600 grams of water transpired for each gram of CO_2 that is incorporated into carbohydrate. Plants that use the C_4 pathway for photosynthesis have ratios of 300 or less.

How Stomata Open and Close The regulation of the size of stomatal openings can control the rate of transpiration. Each stoma is flanked by a pair of kidney- or dumbbell-shaped guard cells, suspended by subsidiary cells. When the kidney-shaped guard cells of dicots become turgid and swell, their radially oriented microfibrils cause them to buckle outward and increase the size of the gap between them. When they become flaccid, they sag and close the space between them. The thick-walled ends of the dumbbell-shaped monocot guard cells swell when they become turgid, pulling the thinner mid-regions of the cells apart from each other. The reversible active absorption and loss of potassium ions (K^+) by the guard cells result in changes in the water potential of the guard cells, leading to the movement of water and the resulting increase or decrease of turgor pressure. The movement of K^+ across the guard cell membrane is probably coupled to the generation of membrane potentials by proton pumps.

Stomata are usually open during the day and closed at night. The opening of stomata at dawn is related to at least three factors. Light stimulates guard cells to accumulate K^+, perhaps triggered by the illumination of blue-light receptors on the tonoplast of guard cell vacuoles, which activate the proton pumps of the plasma membrane. Light also drives photosynthesis in the guard cell chloroplasts, making ATP available for the active transport of ions. Second, CO_2 is depleted within air spaces of the leaf when photosynthesis begins in the mesophyll—another factor in stomatal opening. The third factor is a daily rhythm of opening and closing that is endogenous to guard cells. Cycles that have intervals of approximately 24 hours are called *circadian rhythms*.

Environmental stress can cause stomata to close during the day. During a water deficiency, the guard cells lose turgor. A hormone called abscisic acid, produced by mesophyll cells in response to a lack of water, signals guard cells to close stomata. High temperatures induce closing, probably by increasing cellular respiration, which raises CO_2 concentrations in the air spaces in the leaf.

Leaf Adaptations that Reduce Transpiration
Xerophytes are plants adapted to arid climates. Their leaves may be small and thick, limiting water loss by reducing surface area relative to volume. Cuticles are thick, and the stomata, concentrated on the lower leaf surface, may be located in depressions. Some desert plants lose their leaves in the driest months.

The succulent plants of the family Crassulaceae assimilate CO_2 by a pathway known as CAM, for crassulacean acid metabolism. CO_2 is assimilated into organic acids during the night and then released for photosynthesis during the day. Thus, the stomata are open at night and can close during the day when transpiration would be greatest.

Transport in Phloem

The transport of food in the plant through phloem is called *translocation*. Phloem consists of parenchyma cells, fiber cells, sieve-tube members, and companion cells. Sap flows between sieve-tube members through porous sieve plates. Phloem sap may have a sucrose concentration as high as 30% and may also contain minerals, amino acids, and hormones.

Source-to-Sink Transport   The sieve tubes of phloem carry food from a sugar *source*, where it is being produced by photosynthesis or the breakdown of starch, to a sugar *sink*, an organ that consumes or stores sugar. When a storage organ, such as a tuber, is storing carbohydrates during the summer, it is a sugar sink; when its starch is broken down to sugar in the spring, it is a sugar source. The direction of transport in any one sieve tube depends on the locations of the source and sink connected by that tube.

Phloem Loading and Unloading   Research on sugar beets and corn has shown that sucrose is loaded by active transport into phloem at the source. Sugar in the leaf appears to move through the symplast of the mesophyll cells. Mesophyll cells bordering veins apparently secrete sugar into the apoplast, from which it is actively accumulated into sieve-tube members, probably being moved through companion cells. This energy-requiring process is believed to be driven indirectly by the proton pumps of plant membranes. In other plants, sugars and nutrients may be transported through the symplast and may not be actively loaded into sieve-tube members. Sucrose may be pumped by active transport out of sieve tubes at the sink end.

Pressure Flow   The rapid movement of phloem sap may be due to a pressure-flow mechanism. The high solute concentration at the source lowers the water potential, and the resulting movement of water into the sieve tube produces hydrostatic pressure. At the sink end, the osmotic loss of water following the exodus of sucrose results in a lower pressure. The difference in these pressures causes water to flow from source to sink, transporting sugar. This model, although difficult to test, is the best explanation for the flow of sap in phloem.

STRUCTURE YOUR KNOWLEDGE

1. Describe the models that explain the movement of xylem and phloem sap throughout the plant.

TEST YOUR KNOWLEDGE

MULTIPLE CHOICE: *Choose the one best answer.*

1. Which of the following is *not* a component of the symplast?
   a. sieve-tube members
   b. xylem tracheids
   c. endodermal cells
   d. transfer cells

2. Proton pumps in the plasma membranes of plant cells may
   a. generate a membrane potential that helps drive charged minerals into the cell through their specific carriers.
   b. be coupled to the movement of K+ into guard cells.
   c. indirectly drive the accumulation of sucrose in sieve-tube members.
   d. be involved in all of the above.

3. The Casparian strip prevents water and minerals from entering the stele through the
   a. plasmodesmata.
   b. endodermal cells.
   c. symplast.
   d. apoplast.

4. The water potential of a plant cell
   a. is equal to 0 when the cell is in pure water and is turgid.
   b. is greater than that of air.
   c. is equal to –0.23 mPa.
   d. becomes greater when K+ ions are actively moved into the cell.

5. Guttation results from
   a. the pressure-flow of sap through phloem.
   b. a water-vapor break in the column of xylem sap.
   c. root pressure causing water to flow up through xylem faster than it can be lost by transpiration.
   d. a higher water potential of the leaves than that of the roots.

6. Which of these is not a major factor in the movement of xylem sap?
   a. transpiration
   b. active transport using a proton pump
   c. adhesion
   d. cohesion

7. Adhesion is a result of
   a. hydrogen bonding between water molecules.
   b. the pull on the water column as water evaporates from the surface of mesophyll cells.
   c. tension within the xylem caused by a lowered water potential.
   d. attraction of water molecules to hydrophilic walls of narrow xylem tubes.

8. A plant with a low transpiration-to-photosynthesis ratio
   a. would lose less water through transpiration for each gram of CO_2 fixed.
   b. could be a C_4 plant.
   c. could be a CAM plant.
   d. all of the above could be correct.

9. Which of these is *not* a factor in the opening of stomata?
   a. a decrease in the turgor of guard cells
   b. depletion of CO_2 in the air spaces of the leaf
   c. stimulation of proton pumps that result in the movement of K+ into the guard cells
   d. circadian rhythm of guard-cell opening

10. Phloem sap moves
    a. from a sugar source to a sugar sink
    b. primarily through the symplast.
    c. by a pressure-flow mechanism.
    d. All of the above are correct.

11. Hydrophytes, plants that are adapted to live in aquatic habitats, are most likely to show which of the following morphologies?
    a. very thick cuticle
    b. leaves reduced to spines
    c. stomata located on the top of the leaves
    d. no vascular tissues

12. Your favorite spider plant is wilting. What is the most likely cause and remedy for its declining condition?
    a. water potential is too low: apply sugar water
    b. the stomata won't open; no remedy available
    c. plasmolysis of its cells; water the plant
    d. cavitation; perform a phloem-cell bypass

CHAPTER 33

PLANT NUTRITION

FRAMEWORK

The nutritional requirements of plants, including essential macronutrients and micronutrients, are discussed in this chapter. Carbon dioxide enters the plant through the leaves, but the other inorganic raw materials, including water and minerals, must be absorbed through the roots. The fertility of soil is influenced by its texture and composition. Nitrogen assimilation by plants is made possible by the nitrogen fixation of bacteria. Mycorrhizae are important associations between fungi and the roots of plants that increase mineral and water absorption.

CHAPTER SUMMARY

Plants, as photosynthetic autotrophs, can make all of their own organic compounds but must obtain carbon dioxide, water, and a variety of minerals or inorganic ions to do so. Plants also need oxygen for cellular respiration. Roots, shoots, and leaves are structurally adapted to obtain essential nutrients from the soil and air.

Nutritional Requirements of Plants

Chemical Composition of Plants Aristotle thought that soil provided the substance for plant growth. In the seventeenth century, van Helmont concluded from his experiment with a willow grown in a tub that the tree had developed mainly from the water that was added. Hales, a physiologist of the eighteenth century, thought that plants were nourished mostly by air. All three of these ideas are partly correct.

Minerals, although essential, make only a small contribution to the mass of a plant. Water supplies most of the hydrogen and some of the oxygen incorporated into organic compounds, but 90% of the water absorbed is lost by transpiration, and most of the water retained is used for growth by cell elongation. By weight, CO_2 is the source of the bulk of organic material of a plant.

Organic substances, most of which are carbohydrates (including cellulose), make up 95% of the dry weight of plants. Thus, carbon, oxygen and hydrogen are the most abundant elements. Nitrogen, sulfur, and phosphorus—also ingredients of organic compounds—are relatively abundant.

Essential Nutrients Essential nutrients are those required for the plant to complete its life cycle from a seed to an adult that produces more seeds. *Hydroponic culture* has been used to determine which of the mineral elements found in plants are essential nutrients. Sixteen elements have been identified as essential in all plants; a few others are necessary for certain plant groups.

Macronutrients, those elements required by plants in relatively large amounts, include the six major elements of organic compounds, as well as calcium, potassium and magnesium. Calcium combines with pectins in the middle lamella to glue plant cells together. Potassium is the major solute for osmotic regulation and an activator of several enzymes. Magnesium is a basic component of chlorophyll and a cofactor of several enzymes.

Seven elements have been identified as *micronutrients*, which are needed by plants in very small amounts. They function in the plant mainly as cofactors of enzymatic reactions. Sodium is a micronutrient for C_4 plants. The slightest trace of an element in a seed or as a contaminant of the glassware, water, or other chemicals used in hydroponic culture may satisfy the needs of the plant, clouding the identification of that element as a micronutrient.

Mineral Deficiencies The symptoms of a mineral deficiency may depend on the functions of that nutrient. A magnesium deficiency results in less chlorophyll and causes *chlorosis*, or yellowing of the leaves. A deficiency of calcium, needed for cell wall synthesis, retards the growth of roots and shoots. The mobility of a nutrient within the plant also affects the symptoms of a deficiency. A mobile nutrient will move to young, growing tissues, and a deficiency will show up first in older parts of the plant.

Symptoms of a mineral deficiency may be distinctive enough for the cause to be diagnosed by a plant physiologist or farmer. Soil and plant anaylsis can confirm a specific deficiency. Nitrogen, potassium, and phosphorus deficiencies are most common.

Soil

Plants that can grow well in a particular region are adapted to the texture and mineral content of that soil.

Texture and Composition of Soils The formation of soil begins with the weathering of rock and accelerates when lichens, fungi, bacteria, and roots secrete acids and when roots expand in fissures, thus breaking rocks into smaller pieces. *Topsoil* is a mixture of decomposed rock, living organisms, and *humus* (decomposing organic matter). Several other distinct soil layers, or *horizons*, are found under the topsoil layer.

Topsoil can vary in the size of its particles from coarse sand to fine clay. *Loams*, made up of a mixture of sand, silt, and clay, are the most fertile soils, having enough fine particles to provide a large surface area for retaining water and minerals but enough coarse particles to provide air spaces with oxygen for respiring roots.

The activities of the numerous soil inhabitants, such as bacteria, fungi, algae, other protoctists, insects, worms, nematodes, and plant roots, affect the physical and chemical properties of soil.

Humus builds a crumbly soil that retains water, provides good aeration of roots, and supplies mineral nutrients.

Availability of Soil Water and Minerals Water, containing dissolved minerals, is available to roots from small spaces in soil where it is bound to hydrophilic soil particles. Many positively charged minerals, such as K^+, Ca^{2+}, Mg^{2+}, adhere to the negatively charged surfaces of finely divided clay particles. Negatively charged minerals, such as nitrate (NO_3^-), phosphate ($H_2PO_4^-$), and sulfate (SO_4^{2-}), tend to leach away more quickly. Roots release acids that facilitate *cation exchange*, in which hydrogen ions displace positively charged mineral ions from the clay particles, making the ions available for absorption. The continuous growth of roots exposes them to new sources of mineral nutrients.

Soil Management Without good soil management, agriculture can greatly reduce the fertility of soil that has built up over centuries. Agriculture diverts essential elements from the chemical cycles when crops are harvested, thus depleting the mineral content of the soil.

Historically, farmers used manure to fertilize their crops. Today in developed nations, commercially produced fertilizers, usually containing nitrogen, phosphorous and potassium, are used. Manure, fishmeal and compost are called organic fertilizers because they contain organic material in the process of decomposing. These materials are slowly decomposed to inorganic nutrients, in the same form supplied by commercial fertilizers. A disadvantage of commercial fertilizers is that they may be rapidly leached from the soil, polluting streams and lakes.

The acidity of the soil affects cation exchange and can alter the chemical form of minerals and thus their ability to be absorbed by the plant. Managing the pH of soil is an important aspect of maintaining fertility.

Irrigation can make farming possible in arid regions, but it places a huge drain on water resources and raises the salinity of the soil. New methods of irrigation and new varieties of plants that can tolerate less water or more salinity may reduce some of these problems.

Many thousands of acres of farmland are lost to water and wind erosion each year. Agricultural use of cover crops, windbreaks, and terracing can minimize erosion.

Nitrogen Assimilation by Plants

Nitrogen usually limits plant growth and yields. To be absorbed by plants, nitrogen must be converted to nitrate or ammonium by the action of nitrogen-fixing soil bacteria or microbes that decompose humus.

Nitrogen Fixation Bacteria capable of *nitrogen fixation* contain *nitrogenase*, an enzyme that reduces N_2 by adding H^+ and electrons to form ammonia (NH_3). In the soil solution, ammonia forms ammonium (NH_4^+), which plants can absorb. Bacteria that oxidize NH_4^+ produce nitrate (NO_3^-), the form of nitrogen most readily absorbed by roots. Within plant leaves, the nitrate is reduced back to ammonium, which is then incorporated into amino acids.

Symbiotic Nitrogen Fixation Plants of the legume family have root swellings, called *nodules*, composed of plant cells containing nitrogen-fixing bacteria, in a form called *bacteroids*. The symbiotic relationship between these bacteria of the genus *Rhizobium* and legumes is illustrated by their cooperative synthesis of *leghemoglobin*. This oxygen-binding protein both releases oxygen for the respiration needed to supply energy for nitrogen fixation and keeps the free oxygen concentration low in the nodules to prevent inhibition of the enzyme nitrogenase.

The root nodules use most of the symbiotically fixed nitrogen to make amino acids, which are then transported throughout the plant. Excess ammonium may be secreted into the soil, increasing the fertility of the soil for nonlegumes, which are grown in rotation with the legumes. Rice farmers culture a water fern that has symbiotic cyanobacteria that fix nitrogen, improving the fertility of rice paddies.

Improving the Protein Yield of Crops   Protein deficiency is the most common form of human malnutrition. Agricultural research attempts to improve the quality and quantity of proteins in crops. Varieties of corn, wheat, and rice have been developed that are enriched in protein, but they require the addition of large quantities of expensive nitrogen fertilizer. Incorporating mutant strains of *Rhizobium*, which do not shut off production of nitrogenase when fixed nitrogen accumulates in the nodules, and developing legumes and bacteria that use less energy to fix nitrogen could improve the productivity of symbiotic nitrogen fixation and increase protein yields in crops.

The tools of genetic engineering are being used to create varieties of *Rhizobium* that can infect nonlegumes and to transplant the genes for nitrogen fixation into other bacteria and, possibly, directly into plant genomes.

Some Nutritional Adaptations of Plants

Parasitic Plants   Parasitic plants, such as the mistletoe, may produce haustoria that invade a host plant and siphon sap from its vascular tissue. *Epiphytes* are plants that grow on the surface of another plant but do not rely on it for nourishment.

Carnivorous Plants   Carnivorous plants, living in acid bogs or other poor soils, obtain nitrogen and minerals by killing and digesting insects. The Venus flytrap, pitcher plants, and sundews have evolved insect traps formed from modified leaves.

Mycorrhizae   Many plants have modified roots called *mycorrhizae*, which are symbiotic associations between the roots and fungi. The hyphal extensions of the fungi provide a large surface area for the absorption

of minerals and water, which are shared with the plant. Photosynthetic products from the plant nourish the fungus. Mycorrhizae are common on plants growing in poor soils. The manipulation of mycorrhizae may have important agricultural applications. The fossil record shows that the earliest land plants had mycorrhizae.

STRUCTURE YOUR KNOWLEDGE

1. Develop a concept map that organizes your understanding of the basic nutritional requirements of plants.

2. List the key properties of a fertile soil.

3. Explain how nitrogen assimilation by plants depends on bacteria and microbes. Why is the nitrogen assimilation of plants so important in agriculture?

TEST YOUR KNOWLEDGE

MULTIPLE CHOICE: *Choose the one best answer.*

1. The inorganic compound that contributes most of the mass to a plant's organic matter is
   a. H_2O.
   b. CO_2.
   c. NO_3^-.
   d. O_2.

2. The effects of mineral deficiencies involving fairly mobile nutrients will first be observed in
   a. older portions of the plant.
   b. new leaves and shoots.
   c. the root system.
   d. the color of the leaves.

3. Most macronutrients are
   a. cofactors in enzymes.
   b. readily available from air and water.
   c. components of organic compounds.
   d. determined by hydroponic culture.

4. The most fertile type of soil is
   a. sand because its large particles allow room for air spaces.
   b. loam, which has a mixture of fine and coarse particles.
   c. clay, because the fine particles provide much surface area to which minerals and water adhere.

d. topsoil—a mixture of decomposed rock, living organisms, and humus.

5. Chlorosis is
   a. a symptom of a mineral deficiency indicated by yellowing leaves due to decreased chlorophyll production.
   b. the uptake of the micronutrient chlorine by a plant.
   c. the formation of chlorophyll by a plant.
   d. a contamination of glassware in hydroponic culture.

6. Negatively charged minerals
   a. are released from clay particles by cation exchange.
   b. are reduced by cation exchange before they can be absorbed.
   c. are leached away by the action of rainwater more easily than positively charged minerals.
   d. are bound when roots release acids into the soil.

7. Nitrogenase
   a. is an enzyme that reduces atmospheric nitrogen to ammonia.
   b. is found in *Rhizobium* and other nitrogen-fixing bacteria.

c. is inhibited by high concentrations of oxygen.
d. is all of the above.

8. Epiphytes
   a. have haustoria for anchoring to their host plants.
   b. are symbiotic relationships between roots and fungi.
   c. live in poor soil and digest insects to obtain nitrogen.
   d. grow on other plants but do not obtain nutrients from their hosts.

9. Which of the following is not an erosion-control measure?
   a. terracing
   b. irrigation
   c. cover crops
   d. windbreaks such as hedge rows

10. Some consider organic fertilizers to be superior to chemical fertilizers because they
    a. are more natural.
    b. provide nutrients in the forms most readily absorbed by plants.
    c. are less likely to be lost to runoff and release their nutrients over a longer period of time.
    d. are easier to mass produce and transport.

CHAPTER 34

PLANT REPRODUCTION

FRAMEWORK

This chapter describes the sexual and asexual reproduction of flowering plants. The flower, with leaves modified for sexual reproduction, produces the haploid gametophyte stages of the life cycle: microspores in the anther develop into pollen grains, and a megaspore in the ovule produces an embryo sac. Pollination and double fertilization of egg and polar nuclei are followed by the development of a seed with a quiescent embryo and endosperm, protected in a seed coat and housed within a fruit. Seed dormancy is broken following proper environmental cues and the imbibition of water.

Vegetative propagation allows successful plants to clone themselves. Agriculture makes extensive use of this type of plant reproduction, including cuttings, grafts, and test-tube cloning.

CHAPTER SUMMARY

Adaptations in reproduction were a key to the spread of plants into a variety of terrestrial habitats. In the conifers and angiosperms, pollen dispersed by wind or animals has replaced flagellated sperm, and zygotes develop into embryos protected within seeds.

Sexual Reproduction of Flowering Plants

The Flower   The flower, specialized for sexual reproduction, is a compressed shoot with four whorls of modified leaves. *Sepals*—the outermost, usually green leaves—enclose and protect a floral bud before it opens. *Petals* are generally brightly colored and may attract pollinators. Sepals and petals are sterile floral parts; gametes develop on stamens and carpels.

A *stamen* consists of a stalk, the *filament*, and a terminal *anther*, with chambers in which pollen develops. Pollen grains contain male gametes in the form of sperm nuclei. A *carpel* consists of a sticky *stigma* at the top of a slender neck or *style*, which leads to an *ovary* in which one or more *ovules* develop. An egg, the female gamete, develops in the ovule. Two or more carpels may be fused to form an ovary with more than one chamber. *Pistil* is a term sometimes used for single or fused carpels.

A *complete flower* has sepals, petals, stamens and carpels. *Incomplete flowers* are missing one or more parts. A *perfect* flower has both stamens and carpels; an *imperfect* flower is missing one of these two. Imperfect flowers may be either staminate or carpellate. A *monoecious* plant species has both staminate and carpellate flowers on the same plant; a *dioecious* species has these flowers on separate plants.

Alternation of Generations: A Review   Plants exhibit an alternation of generations between haploid and diploid generations. The diploid plant, the *sporophyte*, produces haploid spores by meiosis. The haploid multicellular male or female *gametophyte* produces gametes by mitosis. Zygotes grow into new sporophyte plants.

In angiosperms, the gametophytes are reduced in size and remain dependent of the sporophyte plant. The flower is part of the sporophyte generation. Although stamens and carpels are referred to as male and female sex organs, they actually produce spores, not gametes. The spores develop into tiny gametophytes while retained within the flower. A pollen grain is the male gametophyte. The embryo sac that develops within an ovule is the female gametophyte. These gametophytes produce sperm nuclei and eggs.

Development of Pollen   Diploid cells called microsporocytes undergo meiosis to form four haploid *microspores*. The nucleus of a microspore divides once by mitosis to produce a generative nucleus and a tube nucleus. The microspore wall thickens into the durable coat of the pollen grain, with an elaborate pattern characteristic of the plant species. The generative nucleus divides to form two *sperm nuclei*.

Development of Ovules Ovules, which contain the female gametophytes, form within the ovary. A megasporocyte in the sporangium of each ovule grows and undergoes meiosis to form four haploid *megaspores*, only one of which usually survives. This megaspore divides by mitosis, forming the female gametophyte, called the *embryo sac*, which consists typically of eight nuclei contained in seven cells. An egg cell is between two cells at one end; three cells are at the other end; and two nuclei, called polar nuclei, are found in the large central cell. Protective layers called *integuments* form from sporophyte tissue around the embryo sac. The *micropyle* is an opening through the integuments.

Pollination and Fertilization *Pollination* is the arrival onto the stigma of pollen. The majority of angiosperms have mechanisms that prevent self-pollination. The pollen grain germinates and grows a tube down the style and through the micropyle, where it releases its two sperm nuclei within the embryo sac. By *double fertilization*, one sperm fertilizes the egg to form the zygote and the other combines with the polar nuclei to form a triploid nucleus.

The Seed The triploid nucleus divides to form the *endosperm*, a multicellular mass rich in nutrients, which are provided to the developing embryo and stored for use when the seed germinates.

The first mitotic division divides the zygote into a basal cell and a terminal cell. The basal cell divides to produce a thread of cells, called the suspensor, that anchors the embryo and transfers nutrients to it. The terminal cell divides to form a spherical proembryo, on which the cotyledons begin to form as bumps. The embryo elongates to form the torpedo stage with apical meristems at the apexes of the embryonic shoot and the root, where the suspensor attaches.

The root-shoot polarity of the embryo is determined by the first cytoplasmic division of the zygote. Egg organelles and chemicals are unevenly distributed in the cytoplasm, so the two cells differ in cytoplasmic composition. The cytoplasmic environment and position of an embryonic cell influence the selective expression of genes during cellular differentiation.

During maturation, the seed dehydrates to a water content about 5% to 15% of its weight. The embryo becomes dormant until the seed germinates. The embryo and its food supply, the endosperm or enlarged cotyledons, are enclosed in a *seed coat* formed from the ovule integuments.

In a dicot seed such as a bean, the embryo is an elongate embryonic axis, attached to fleshy cotyledons. The axis below the cotyledonary attachment is called the *hypocotyl*; it terminates in the *radicle*, or embryonic root. The upper axis is the *epicotyl*; it terminates as a *plumule*, a shoot tip with a pair of leaves. In some dicots, the cotyledons remain thin and absorb nutrients from the endosperm when the seed germinates.

A monocot seed, such as a corn kernel, has a single thin cotyledon, called a scutellum, which absorbs nutrients from the endosperm during germination. A sheath called a *coleorhiza* covers the root, and a *coleoptile* encloses the embryonic shoot.

Development of the Fruit The ovary of the flower ripens into a *fruit*. Other floral parts may contribute to what we commonly call a fruit. Hormonal changes following pollination cause the ovary to grow tremendously, its thickened wall becoming the *pericarp* of the fruit. Fruit usually does not set if a flower has not been pollinated. Some parthenocarpic plants, such as bananas, do produce fruit without fertilization.

A *simple fruit* develops from a single ovary. An *aggregate fruit* results from a single flower that has several carpels. A *multiple fruit*, such as a pineapple, develops from an *inflorescence*, a group of tightly clustered flowers. Fruits usually ripen as the seeds are completing their development. Ripening of fleshy fruits may include softening of the pulp, a change in color, and an increase in sugar content. Fruits are adapted to disperse seeds, enlisting the aid of either wind or animals.

Humans have selectively bred edible fruits. The cereal grains—the wind-dispersed fruits of grasses, which have dry pericarps—are staple foods for humans.

Seed Germination At germination, the plant resumes the growth and development that was suspended when the seed matured.

Dormancy is an adaptation that increases the chances that the seed will germinate when and where the embryo has a good chance of surviving. The specific cues for breaking dormancy vary with the environment, including heavy rain in deserts, intense heat in areas of frequent fires, and cold in areas with harsh winters. Some seeds must have their seed coats weakened by abrasion or chemicals before they will germinate. The viability of a dormant seed may vary from a few days to decades or longer.

Imbibition, the absorption of water by the dry seed, causes the seed to expand, rupture its coat, and begin a series of metabolic changes. Stored compounds are digested by enzymes, and nutrients are sent to growing regions. Soon after hydration in barley or cereal seeds, the thin outer layer of the endosperm, called the aleurone, begins making a-amylase and other enzymes that digest starch stored in the endosperm.

The radicle emerges from the seed first, followed by the shoot tip. In many dicots, a hook that forms in the hypocotyl is pushed up through the ground, pulling the delicate shoot and cotyledons behind it. Light stimulates the straightening of the hook, and the first foliage leaves begin photosynthesis.

Light seems to be the main cue that the seedling has broken ground. An *etiolated* seedling, grown in the dark, continues to extend its hypocotyl until it exhausts its food reserves. In peas, a hook forms in the epicotyl, lifting the shoot tip out of the soil while the pea cotyledons remain in the ground. In monocots, the coleoptile pushes through the soil, and the shoot tip is protected as it grows up through the tubular sheath.

Germination of a plant seed is a critical and fragile stage in the life cycle. Only a small fraction of seedlings survive to produce seeds themselves.

Asexual Reproduction

Natural Mechanisms of Vegetative Reproduction
Many plant species can clone themselves through asexual or *vegetative reproduction*—an extension of the indeterminate growth of plants, in which meristematic tissues can grow indefinitely, and parenchyma cells can divide and differentiate into specialized cells. A common type of vegetative reproduction is *fragmentation*, the separation of a plant into parts that then form whole plants. Some plants, such as dandelions, can produce seeds asexually, a process called *apomixis*.

Vegetative Propagation in Agriculture   Humans have developed many artifical methods of vegetative propagation of trees, crops, and ornamental plants. New plants develop from shoot or stem cuttings when a *callus*, or mass of dividing cells, forms at the cut end of the shoot, and adventitious roots develop from the callus.

Twigs or buds of one plant can be grafted onto a related plant. The plant that provides the root system is called the stock, and the twig graft is called the scion. Grafting combines the best qualities of different plants.

In test-tube cloning, whole plants can develop from pieces of tissue, called explants, or even from single parenchyma cells, grown on artificial medium. A single plant can be cloned into thousands of plants by subdividing the undifferentiated calluses as they grow in tissue culture. The techniques of genetic engineering are used to insert foreign DNA into plant cells, which

are then used to grow complete plants by test-tube culture.

Protoplast fusion is being coupled with tissue culture to create new plant varieties. A protoplast is the living part of a plant cell contained within the plasma membrane. They can be screened for agriculturally beneficial mutations and then cultured. In some cases, two protoplasts from different plant species can be fused and cultured.

Genetic variability in crops and orchards has been consciously eliminated so that plants grow at the same rate, fruits ripen in unison, and yields at harvest time are predictable. Vegetative propagation is used to clone exceptional plants, and self-pollinating varieties are used when possible. *Monoculture*, the cultivation of a single plant variety on large areas of land, produces a fragile ecosystem, in which there is little genetic variability and little adaptability. "Gene banks" have been created in which plant breeders maintain seeds of many different plant varieties, so that new varieties can be developed if current ones fail.

A Comparison of Sexual and Asexual Reproduction in Plants   Sexual reproduction in plants generates variation in a population, an advantage when the environment changes. Sexual reproduction also produces seeds, a means of dispersal to new locations and dormancy during harsh conditions. By asexual reproduction, plants well-suited to a stable environment can clone exact copies, and the progeny of vegetative propagation are usually not as frail as seedlings. Both modes of reproduction are useful in the adaptation of plant populations to their environments.

STRUCTURE YOUR KNOWLEDGE

1. Create a diagram or concept map that sketches out the major events in the life cycle of angiosperms.

2. List the advantages and disadvantages of sexual and asexual reproduction in plants.

3. Identify the flower parts in the diagram on the following page.

TEST YOUR KNOWLEDGE

FILL IN THE BLANKS

1. _____ flower part modified as male reproductive structure

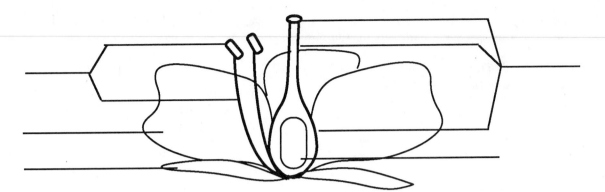

2. _____ generation that produces spores by meiosis

3. _____ species with male and female flowers on same plant

4. _____ female gametophyte of angio-sperms

5. _____ embryonic root

6. _____ embryonic axis above attachment of cotyledon

7. _____ protects dicot shoot as it breaks through the soil

8. _____ plants that produce fruit without fertilization

9. _____ fruit formed from a flower with several separate carpels

10. _____ mass of dividing cells at cut end of a shoot

MULTIPLE CHOICE: *Choose the one best answer.*

1. A flower on a dioecious plant would be
   a. complete.
   b. perfect.
   c. imperfect.
   d. asexual.

2. Which of the following structures is haploid?
   a. pollen grain
   b. ovule
   c. endosperm
   d. microsporocyte

3. The terminal cell in a zygote
   a. develops into the shoot apex of the embryo.
   b. forms the suspensor that anchors the embryo and transfers nutrients.
   c. develops into the endosperm when fertilized by a sperm nucleus.
   d. divides to form the proembryo.

4. A unique feature of fertilization in angiosperms is that
   a. it is double; one sperm fertilizes the egg, one combines with a cell in the embryo sac and forms the endosperm.
   b. cross fertilization occurs.
   c. pollen may be transferred by wind.
   d. a chemical attractant guides the sperm toward the egg.

5. The endosperm
   a. may be absorbed by the cotyledons in the seeds of dicots.
   b. is a triploid tissue.
   c. is digested by enzymes in monocot seeds following hydration.
   d. is all of the above.

6. In an angiosperm, meiotic cell division produces
   a. microspores.
   b. sperm nuclei.
   c. megasporocytes.
   d. all of the above.

7. Which structure protects a monocot shoot as it breaks through the soil?
   a. hypocotyl hook
   b. epicotyl hook
   c. coleoptile
   d. coleorhiza

8. Which of the following is a form of vegetative reproduction?
   a. apomixis
   b. grafting
   c. parthenocarpic fruit production
   d. both a and b

9. A disadvantage of monoculture is that
   a. the whole crop ripens at one time.
   b. since there is no genetic variability, the whole crop could be annihilated by a change in conditions, a new pest, or disease.
   c. it predominately uses vegetative propagation.
   d. all of the above.

10. Protoplast fusion
    a. is used to develop gene banks to maintain genetic variability.
    b. is the method of test-tube cloning.
    c. can be used to form new plant species.
    d. occurs within a callus.

11. In the plant embryo, the suspensor
    a. directs mitotic divisions in the early cells.
    b. connects the early root and shoot apexes.
    c. gives rise to the cotyledons.
    d. is analogous to the umbilical cord in mammals.

12. Flower parts have evolved from modified
    a. leaves.
    b. branches.
    c. sporangia.
    d. sporophytes.

CHAPTER 35

CONTROL SYSTEMS IN PLANTS

FRAMEWORK

This chapter describes the current understanding of the complex systems in plants that control growth, development, movement, flowering, and senescence, as plants respond and adapt to their environments. The functions and interactions of the plant hormones—auxin, cytokinins, gibberellins, abscisic acid, and ethylene—are detailed. Plant movements in response to environmental stimuli include phototropism, gravitropism, thigmotropism, and turgor movements. The biological clock of plants controls circadian rhythms, such as stomatal opening and sleep movements, and may use the phytochrome system to time night length in the photoperiodic control of flowering.

CHAPTER SUMMARY

Plants are able to sense and respond adaptively to many aspects of their environments, generally by altering their patterns of growth and development. Their intricate control systems are the product of the evolutionary history of the interactions between plants and their environments.

The Search for a Plant Hormone

Hormones, chemical signals that coordinate various components of an organism, are produced by one part of the body and translocated to other parts, where they trigger responses in target cells and tissues.

The growth of a shoot toward light is called positive *phototropism*. A coleoptile, enclosing the shoot of a grass seedling, grows straight when in the dark but bends toward the light when illuminated from one side. The

cells on the darker side elongate faster and produce the bending.

Darwin and his son discovered that a grass seedling would not bend toward light if its tip were removed or covered by an opaque cap. They speculated that some signal must be transmitted from the tip down to the elongating region of the coleoptile. Boysen and Jensen demonstrated that the signal was a mobile substance, capable of transmission through a block of gelatin separating a tip from the rest of the coleoptile.

In 1926, Went placed coleoptile tips on blocks of agar to extract the chemical messenger. Decapitated coleoptiles kept in the dark elongated and bent away from the side on which one of these agar blocks was placed. Went concluded that the chemical produced in the tip, which he called auxin, promoted growth as it passed down the coleoptile and was in a higher concentration on the side away from the light.

Functions of Plant Hormones

Five classes of plant hormones have been identified: *auxin, cytokinins, gibberellins, abscisic acid,* and *ethylene.* These hormones affect cell division, elongation, and differentiation. They have a variety of effects, depending on the site of action, the developmental stage of the plant, and relative hormone concentrations. Hormones are effective in very small concentrations; their signal must be amplified in the cell in some way. They may act by affecting the expression of genes, the activity of enzymes, or the properties of membranes.

Auxin Auxin refers to any substance, including synthetic compounds, that stimulates elongation of developing stems or coleoptiles. The natural auxin extracted from plants is indoleacetic acid (IAA).

Auxin is primarily synthesized in the apical meristem of a shoot. Within a range of concentrations, auxin stimulates elongation of cells in young stems or coleoptiles. Above a certain concentration, auxin inhibits growth, probably because it induces the synthesis of ethylene, which inhibits cell elongation. Auxin from the shoot reaches root cells, which are more sensitive to auxin and elongate in response to low concentrations. Auxin illustrates that the same hormone has varying effects on different target cells and, at different concentrations, can have different effects on the same target cells.

Auxin appears to be transported through parenchyma tissue in one direction, from the shoot tip to base. Such polar transport requires energy and probably involves auxin carriers at the basal ends of cells.

According to the acid-growth hypothesis, auxin loosens cell walls by stimulating proton pumps. The resulting lower pH in the cell wall activates enzymes that break cross-links between cellulose microfibrils. The turgor pressure of the cell then exceeds the restraining wall pressure, and the cell elongates.

Auxin also affects cell division in the vascular cambium, differentiation of secondary xylem, formation of adventitious roots at the cut bases of stems, and development of leaf traces. Auxin produced by developing seeds promotes growth of fruit; synthetic auxins may induce parthenocarpic fruit development. The herbicide 2,4-D, a synthetic auxin, is used to disrupt the normal balance of plant growth in "broad leaf" dicot weeds.

Cytokinins   The active ingredients of the coconut milk and degraded DNA that were found to induce plant cell growth in tissue culture were modified forms of adenine. These compounds were named cytokinins, for their ability to stimulate cytokinesis. Cytokinins are produced in actively growing roots, embryos, and fruits. Acting along with auxin, they stimulate cell division and affect differentiation.

When an explant, or piece of tissue, is grown in tissue culture in the absence of cytokinins, the cells grow large but do not divide. If cytokinins and auxin are added, the cells divide. The ratio of the two hormones controls differentiation of the cells: equal concentrations produce undifferentiated cells, more cytokinin results in shoot buds, whereas more auxin leads to root formation.

Cytokinins stimulate RNA and protein synthesis, but the mechanisms of their actions are not known.

The control of apical dominance involves an interaction between auxin, transported down from the terminal bud, which restrains axillary bud development, and cytokinins, transported up from the roots, which stimulate bud growth. This control of lateral branching may coordinate the growth of shoot and root systems. The interaction between auxin and cytokinins works the opposite way in the development of lateral roots.

Both hormones probably regulate growth indirectly by changing the concentration of ethylene.

Cytokinins, which may stimulate protein synthesis, mobilize nutrients, and promote stomatal opening, can retard aging of some plant organs.

Gibberellins   In the 1930s, Japanese scientists determined that the fungus causing "foolish seedling disease" was secreting a chemical, given the name gibberellin, that was producing the hyperelongation of rice stems. Over 70 different gibberellins have now been identified, many occurring naturally in plants.

Gibberellins, which are produced by roots and young leaves near the shoot apex, stimulate growth in both leaves and stem. The application of different concentrations of gibberellin to dwarf plants has demonstrated a positive correlation between growth and the concentration of hormone added. Experimental results such as these can be used in a bioassay to determine the hormone concentration in a sample of unknown concentration. *Bolting*, the growth of an elongated floral stalk, is caused by a surge of gibberellins. In some plants, both auxin and gibberellins contribute to fruit growth. The commercial spraying of these two hormones will induce parthenocarpic fruit development, important in the production of Thompson seedless grapes.

The release of gibberellins from the embryo signals the seed to break dormancy. The germination of grain is triggered when gibberellins stimulate, by way of a second messenger such as Ca^{2+}, the synthesis of messenger RNA that codes for digestive enzymes that break down stored nutrients. Gibberellins are also involved in the breaking of dormancy of apical buds in spring.

Abscisic Acid   The hormone abscisic acid (ABA), which is produced in the bud, slows growth, inhibits cell division in the vascular cambium, and induces leaf primordia to develop into scales that protect the dormant bud during winter.

Abscisic acid may act as a growth inhibitor of the embryo when the seed becomes dormant. ABA must be removed or inactivated, or the ratio of gibberellins to ABA must increase, for dormancy to be broken in some seeds. Abscisic acid also acts to help the plant cope with adverse conditions. In a water-stressed plant, ABA, accumulating in leaves, causes stomata to close.

Ethylene   Plants produce the gas ethylene, which acts as a hormone to promote fruit ripening, to inhibit growth in roots and axillary buds and to contribute to the aging, or *senescence*, of parts of the plant. Ethylene initiates or hastens the degradation of cell walls and the decrease in chlorophyll content that is associated with fruit ripening. Many commercial fruits are ripened in huge containers perfused with ethylene gas.

Deciduous leaf loss protects against winter dehydration. Before leaves are abscised in the autumn, many of their compounds are stored in the stem, awaiting recycling to new leaves. The leaf stops making chlorophyll; fall colors result from a combination of pigments that had been concealed by chlorophyll and new pigments that are made during autumn.

Shortening days and cooler temperatures are the stimuli for leaf abscission. A change in the balance of auxin and ethylene initiates changes in the abscission layer located near the base of the petiole, including the production of enzymes that hydrolyze polysaccharides in cell walls. A layer of cork forms a protective covering on the twig's side of the abscission layer.

Some Unanswered Questions About Plant Hormones

There are so many unanswered questions about the internal chemical signals of plants that some plant physiologists think it is premature to call these growth regulators hormones. Some scientists suggest that responses to growth regulators may be more a product of increased sensitivity of cells to present regulators than the cells' reaction to messages from other parts of the plant.

Plant Movements

Tropisms

Tropisms are growth responses in which a plant organ curves toward or away from a stimulus, as a result of differential cell elongation rates on opposite sides of the organ. These movements may be in response to stimuli such as light (phototropism), gravity (gravitropism), or touch (thigmotropism).

In *phototropism*, as previously discussed, cells on the darker side of a stem elongate faster in response to a larger concentration of auxin moving down from the shoot tip. In a mechanism not yet understood, light causes auxin to migrate laterally across the tip toward the dark side. The photoreceptor is believed to be a yellow pigment that may also be involved in stomatal opening.

Roots exhibit positive *gravitropism*, whereas shoots show negative gravitropism. The settling of *statoliths*, plastids containing dense starch grains, somehow causes hormones to accumulate on the low side of the plant organ. In stems, a higher concentration of auxin and gibberellins cause the elongation of cells on the lower side. In roots, IAA accumulates on the lower side, inhibiting growth, so that the root bends downward. Ca^{2+} appears to be involved in gravitropism, perhaps affecting IAA transport.

Most climbing plants have tendrils that coil around supports. The directional growth in response to touch is *thigmotropism*. The stunting of growth in height and increase in girth of plants that are exposed to wind or mechanical stimulation is called *thigmomorphogenesis*.

Turgor Movements

Turgor movements are reversible movements caused by changes in turgor pressure in response to stimuli.

The folding of the leaves of the sensitive plant *Mimosa* after being touched is due to the rapid loss of turgor by cells in specialized motor organs called pulvini, located at the joints of the leaf. Evidence shows that the motor cells lose potassium when stimulated, resulting in osmotic water loss. The message travels through the plant from the point of stimulation, perhaps as the result of chemical messengers and electrical impulses, called *action potentials*. These electrical messages may be used in plants as a form of internal communication.

Many members of the legume family exhibit *sleep movements*, the daily opening and folding of leaves, caused by changes in the turgor pressure of motor cells in pulvini. Massive migration of potassium ions from one side of the pulvinus to the other may lead to the reversible osmosis in the motor cells.

Circadian Rhythms and the Biological Clock

Many physiological processes in animals and plants fluctuate with the time of day and are controlled by biological clocks—internal oscillators that keep accurate time. A *circadian rhythm* is a physiological cycle with about a 24-hour frequency. These rhythms persist, even when the organism is sheltered from environmental cues. Research indicates that the oscillator for circadian rhythms is endogenous, although the clock is set (entrained) to a precise 24-hour period by daily environmental signals.

Free-running periods, determined in the absence of environmental cues, vary from 21 to 27 hours. The biological clocks in free-running periods still keep perfect time, but they are not synchronized with the outside world. Internal clocks are reset when an organism is returned to normal environment cues or crosses time zones.

The nature of the oscillator is a mystery. The biological clock ticks most likely at the cellular level, perhaps in membranes or protein-synthesizing machinery. The mechanism for circadian rhythms is unaffected by temperature.

Photoperiodism

Seasonal events in the life cycles of plants usually are cued by photoperiod, the relative lengths of night and day. A physiological response to day length is called *photoperiodism*.

Photoperiodic Control of Flowering Garner and Allard discovered that an exceptionally tall variety of tobacco plant flowered only when the day length was 14 hours or shorter. They termed Maryland Mammoth a *short-day plant*, because it needed a light period shorter than a critical length to flower. *Long-day plants* flower when days are longer than a certain number of hours; and *day-neutral plants* are unaffected by photoperiod in their flowering.

In the 1940s, researchers found that night length, not day length, controls flowering and other photoperiod responses. If the nighttime portion of the photoperiod is interrupted by even a few minutes of light, a short-day (long-night) plant such as the cocklebur will not flower. Photoperiodic responses thus depend on a critical night length, the maximum (long-day plants) or minimum (short-day plants) number of hours of uninterrupted darkness required for flowering.

Some plants bloom after a single exposure to the required photoperiod. Some will respond to photoperiod only after exposure to some other environmental stimulus. The need for pretreatment with cold before flowering is called *vernalization*.

Leaves detect the photoperiod. In some species, exposure of a single leaf to the proper photoperiod will induce flowering. A hormone, named florigen, is believed to be the flowering signal that travels from leaves to buds. The signal has yet to be identified; it may be a mixture of several hormones, or even the absence of inhibitors.

Phytochrome Red light is the most effective in interrupting night length. A brief exposure to red light breaks a dark period of sufficient length and prevents short-day plants from flowering, whereas a flash of red light during a dark period longer than the critical length will induce flowering in a long-day plant. A subsequent flash of light from the far-red part of the spectrum negates the effect of the red light.

The photoreceptor responsible for the reversible effects of red and far-red light is the pigment *phytochrome*, which alternates between two forms, one of which absorbs red light, whereas the other absorbs far-red light. These two variations of phytochrome are said to be photo reversible. The P_r to P_{fr} interconversion controls various events in the life of the plant.

Plants synthesize P_r, which is converted to P_{fr} when the phytochrome is illuminated in sunlight. P_{fr} triggers many plant responses to light, such as the breaking of seed dormancy.

Role of the Biological Clock in Photoperiodism
In darkness, the P_{fr} molecules gradually revert to P_r. At sunrise, the P_{fr} level suddenly increases. Nightlength is measured by the biological clock. The role of phytochrome may be to synchronize the clock by signalling when the sun sets and rises. Plants may tell the season by using the clock to measure photoperiod.

Internal and external stimuli influence the complex interactions that control plant growth and development. Seasonal events and flowering are influenced by genes, hormones, an endogenous clock, photoreceptors, and the length of day.

STRUCTURE YOUR KNOWLEDGE

1. Individual plant hormones and combinations of hormones have various effects on the growth and development of plants, depending on their relative concentrations and the target cells. The table on the following page lists the five identified classes of plant hormones and some of the effects they may have. Check which hormones play a role in each process.

2. Plants exhibit tropisms and turgor movements. Create a concept map that organizes your understanding of these plant movements.

3. Develop a concept map to illustrate your understanding of the photoperiodic control of flowering. Do not forget to include the role of the biological clock.

TEST YOUR KNOWLEDGE

TRUE or FALSE: *Indicate T or F and then correct the false statements.*

1. ____ Application of 2,4-D, a synthetic auxin, disrupts the growth of "broad-leaf" dicots.

2. ____ Shoots exhibit positive phototropism and negative gravitropism.

3. ____ Gibberellins, synthesized in the root, counteract apical dominance.

4. ____ Bolting occurs when ethylene stimulates the degradation of cell walls and decreases chlorophyll content.

5. ____ IAA is a synthetic auxin.

Function ↓ Hormone ➤	Auxin	Cytokinins	Gibberellins	Abscisic Acid	Ethylene
Promote stem elongation					
Promote root growth					
Promote fruit development					
Promote fruit ripening					
Promote seed and bud germination					
Promote cell division & differentiation					
Delay senescence					
Contribute to senescence; abscission					
Inhibit growth					
Promote bud and seed dormancy					

6. ____ ABA induces the synthesis of enzymes that hydrolyze polysaccharides in cell walls of the abscission layer.

7. ____ Action potentials are electrical impulses that may be used as internal communication in plants.

8. ____ Thigmotropism is involved in the rapid movements of *Mimosa* leaves.

9. ____ A physiological response to day or night length is called a circadian rhythm.

10. ____ Vernalization is the need for pretreatment with cold before flowering.

MULTIPLE CHOICE: *Choose the one best answer.*

1. The body form of plants within a species may vary more than that of animals within the same species because
   a. growth in animals is indeterminate.
   b. plants respond adaptively to their environments by altering their patterns of growth and development.
   c. plant growth and development is governed by many hormones.
   d. of all of the above.

2. A concentration of auxin that is too low to stimulate growth in a shoot
   a. may induce the synthesis of ethylene, which inhibits growth.
   b. may still be high enough to break dormancy in a seed.
   c. may be high enough to stimulate cell elongation in the root.
   d. may still cause bolting.

3. According to the acid-growth hypothesis,
   a. auxin stimulates membrane proton pumps.
   b. a lowered pH activates enzymes in the wall that break cross-links between cellulose microfibrils.
   c. the turgor pressure of the cell exceeds the lowered restraining wall pressure, and the cell swells and takes up water.
   d. all of the above are involved in cell elongation.

4. Which of the following situations would stimulate the development of axillary buds?
   a. a large quantity of auxin traveling down from the shoot and a small amount of cytokinin produced by the roots
   b. a small amount of auxin traveling down from the shoot and a large amount of cytokinin traveling up from the roots
   c. an equal ratio of cytokinins to auxin
   d. the absence of cytokinins caused by the removal of the terminal bud

5. The growth inhibitor in seeds is usually
   a. abscisic acid.
   b. ethylene.
   c. gibberellin
   d. a small amount of ABA combined with a larger ratio of gibberellins.

6. A circadian rhythm
   a. is controlled by an internal oscillator.
   b. is a physiological cycle of approximately a 24-hour frequency.

c. involves a biological clock that is set by daily environmental signals.

d. involves all of the above.

7. A bioassay

a. uses chemical means to measure the activity of biological compounds.

b. measures the response of living systems to determine the concentration of a biologically active chemical compound.

c. determines the components present in a mixture of biologically active compounds.

d. measures the rate of growth of dwarf plants.

8. The sleep movements of legumes

a. are an example of a phototropism.

b. are an example of a gravitropism.

c. result from changing turgor pressures in motor cells in pulvini.

d. result from the production of action potentials in pulvini.

9. A flash of far-red light during a critical-length dark period

a. will induce flowering in a long-day plant.

b. will induce flowering in a short-day plant.

c. will not influence flowering.

d. will increase the P_{fr} level suddenly.

10. Evidence indicates that

a. a hormone that travels from leaves to buds induces flowering.

b. flowering is induced by phytochrome levels in the leaves.

c. short-day plants need vernalization before photoperiod affects them.

d. a biological clock in the bud determines photoperiod.

CHAPTER 31
ANATOMY OF A PLANT

Suggested Answers to Structure Your Knowledge

1.

FUNCTION	CELLS, TISSUES, OR STRUCTURES THAT PERFORM FUNCTION in		
	ROOT	STEM	LEAF
Anchorage	Tap or fibrous root system		
Absorption	Root hairs on epidermis		Some carnivorous plants
Storage	Parenchyma cells of pith and cortex	Parenchyma cells, may have rhizomes with tubers	Some modified leaves of bulbs
Support	Adventitious roots	Collenchyma, sclerenchyma xylem (wood)	Veins, xylem
Protection	Epidermis, root cap, periderm of woody roots	Epidermis with cuticle, periderm, secondary phloem	Epidermis with cuticle
Transport	Stele with xylem, phloem	Vascular bundles or ring of xylem and phloem	Veins (parallel or branching)
Photosynthesis		Some parenchyma cells with chloroplasts	Mesophyll with chloroplasts, (pallisade and spongy)
Growth	Apical meristem (primary), vascular and cork cambiums (secondary growth)	Apical meristem (primary), vascular and cork cambiums (secondary growth)	Develop from leaf primordia
Gas exchange	Through epidermis, air tubes in swamp trees	Lenticles	Stomata with guard cells

2. Primary growth results in growth at the tips of roots and stems, whereas secondary growth produces greater girth. The apical meristems produce new cells that form the primary meristems: protoderm, procambium, and ground meristem. These cells elongate and differentiate into the dermal, vascular, and ground tissue systems. Secondary growth involves lateral meristems: a vascular cambium, which produces new xylem and phloem cells, and a cork cambium, which produces cells that form the periderm.

Answers to Test Your Knowledge

Matching:

1.	L	**5.**	I	**9.**	K
2.	F	**6.**	H	**10.**	G
3.	B	**7.**	C		
4.	E	**8.**	D		

Multiple Choice:

1.	a	**5.**	b	**9.**	a
2.	a	**6.**	a	**10.**	c
3.	d	**7.**	b		
4.	c	**8.**	c		

Fill in the Blanks:

H̲ F̲ I̲   A̲ B̲ G̲ D̲ C̲ E̲

CHAPTER 32
TRANSPORT IN PLANTS

Suggested Answers to Structure Your Knowledge

1. The transpiration–cohesion–adhesion theory explains the ascent of xylem sap. The lower water potential of the air surrounding the leaves compared to that of the cortex cells of the root and the soil solution results in the passive movement of water down its water potential gradient in a continuous column of water in xylem vessels. The cohesion of water molecules transmits the pull resulting from transpiration throughout the column, and the adhesion of water to the hydrophilic walls of the xylem vessels aids the flow. The tension created within the xylem by the upward pull also helps the passive flow of water.

   A pressure-flow mechanism explains the movement of phloem sap. The active accumulation of sugars in sieve-tube members greatly decreases the water potential at the source end and results in an inflow of water. The removal of sugar from the sink end of a phloem tube results in the osmotic loss of water from the cell. The difference in hydrostatic pressure between the source and sink end of the phloem tube causes the flow of water with the accompanying transport of sugar.

Answers to Test Your Knowledge

1.	b	5.	c	9.	a
2.	d	6.	b	10.	d
3.	d	7.	d	11.	c
4.	a	8.	d	12.	c

CHAPTER 33
PLANT NUTRITION

Suggested Answers to Structure Your Knowledge

1.

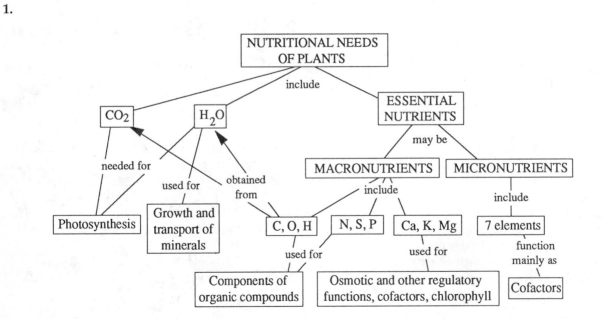

2. The key properties of a fertile soil include:
   - *texture*: mixture of sand, silt and clay so that air spaces containing oxygen are provided as well as sufficient surface area for the binding of water and minerals
   - *humus*: prevents clay from packing so soil retains water and has air spaces; provides reservoir of mineral nutrients
   - *minerals*: adequate supply of nitrogen, phosphorus and potassium as well as other minerals
   - *pH*: proper pH level so that minerals are in a form that can be absorbed and cation exchange can take place

3. Nitrogen must be in the form of nitrate or ammonium in order to be absorbed by plants. Nitrogen-fixing bacteria (using the enzyme nitrogenase) reduce N_2 to form ammonia (NH_3), which converts to ammonium (NH_4^+) in the soil solution. Plants can absorb ammonium, and other bacteria in the soil oxidize the ammonium into nitrate, the form in which plants acquire most of their nitrogen. Microbes decomposing humus also contribute to the ammonium content of the soil. Legumes often have root nodules that house nitrogen-fixing bacteria of the genus *Rhizobium*. Nitrogen is essential to plant growth because it is needed for protein formation. Productive plant growth and high protein content of crops are important in agriculture.

Answers to Test Your Knowledge

Multiple Choice:

1.	b	**5.**	a	**9.**	b
2.	a	**6.**	c	**10.**	c
3.	c	**7.**	d		
4.	b	**8.**	d		

CHAPTER 34
PLANT REPRODUCTION

Suggested Answers to Structure Your Knowledge

1. All structures, except the two circled with the dotted line, belong to the sporophyte generation.

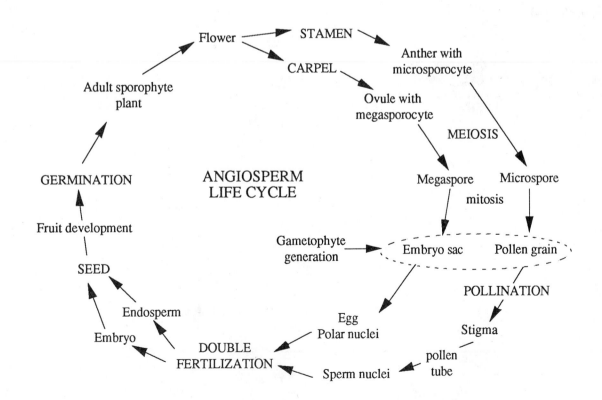

ANGIOSPERM LIFE CYCLE

2. *Sexual:* Advantages include increased genetic variability, which provides the potential to adapt to changing conditions, and the dispersal and dormancy capabilities provided by seeds. Disadvantages are that the seedling stage is vulnerable and that genetic recombination may separate adaptive traits.

Asexual: Advantages include the hardiness of vegetative propagation and the maintenance of well-adapted plants in an environment. Disadvantages relate to the advantages of sexual reproduction: there is no genetic variability from which to choose should conditions change, and there are no seeds for dispersal.

3.

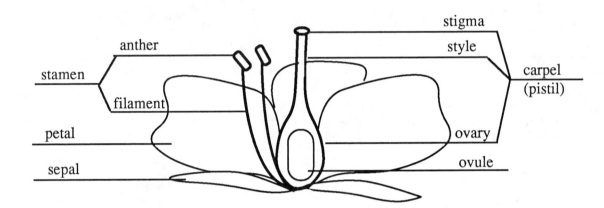

Answers to Test Your Knowledge

Fill in the Blanks:

1. stamen
2. sporophyte
3. monoecious
4. embryo sac
5. radicle
6. epicotyl
7. hypocotyl hook
8. parthenocarpic
9. aggregate
10. callus

Multiple Choice:

1.	c	5.	d	9.	b
2.	a	6.	a	10.	c
3.	d	7.	c	11.	d
4.	a	8.	d	12.	a

CHAPTER 35
CONTROL SYSTEMS IN PLANTS

Suggested Answers To Structure Your Knowledge

1.

Function ▼ Hormone ➤	Auxin	Cytokinins	Gibberellins	Abscisic Acid	Ethylene
Promote stem elongation	√		√		
Promote root growth	√	√	√		
Promote fruit development	√		√		
Promote fruit ripening					√
Promote seed and bud germination		√	√		
Promote cell division & differentiation	√	√	√		
Delay senescence		√			
Contribute to senescence; abscission					√
Inhibit growth				√	√
Promote bud and seed dormancy				√	

2.

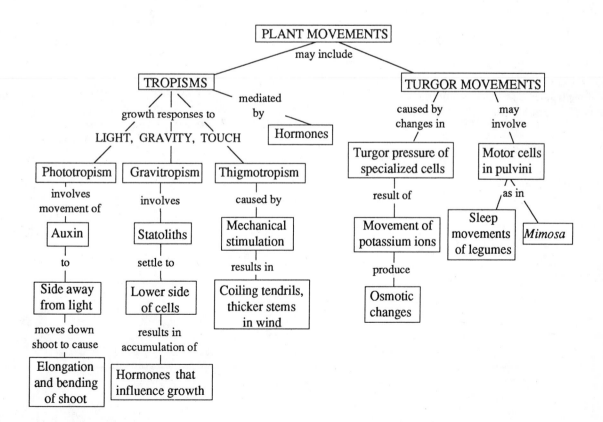

3.

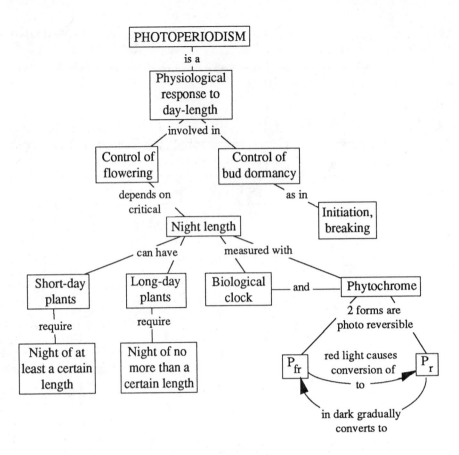

Answers To Test Your Knowledge

True or False:

1. true
2. true
3. false: change *gibberellins* to *cytokinins*
4. false: change *bolting* to *fruit ripening;* or change to read *Bolting is gibberellin-induced elongation of a flower stalk.*
5. false: change *synthetic* to *naturally occurring*
6. false: change *ABA* to *ethylene*
7. true

8. false: change *thigmotropism* to *turgor movements;* or change to read *Thigmotropism is the change in the growth of a plant in response to touch.*
9. false: change *circadian rhythm* to *photoperiodism*
10. true

Multiple Choice:

1.	b	**5.**	a	**9.**	c
2.	c	**6.**	d	**10.**	a
3.	d	**7.**	b		
4.	b	**8.**	c		

ANIMALS: FORM AND FUNCTION

Testing whether or not animals "kiss."

CHAPTER 36

INTRODUCTION TO ANIMAL STRUCTURE AND PHYSIOLOGY

FRAMEWORK

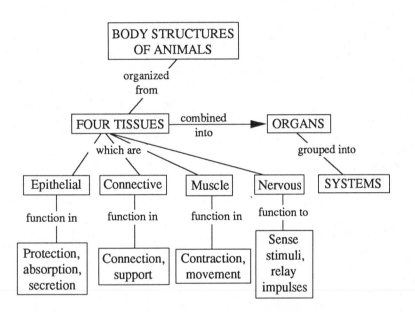

CHAPTER SUMMARY

Almost all plants and animals need an internal transport system, and terrestrial organisms must support themselves against gravity. Networks of tubes and rigid structures have evolved in both these kingdoms. Although fundamentally different in origin, anatomy, and physiology, both groups illustrate the general themes of the correlation of structure and function and the capacity of life to adapt to the environment, both by short-term physiological adjustment and long-term evolution.

Levels of Structural Organization

A hierarchy of structural order characterizes life. Unicellular protozoa are able to carry out all of their life functions without integration beyond the cellular level. Multicellular organisms have specialized cells grouped into tissues. In most animals, tissues are combined into structural and functional units called organs, and various organs may cooperate in organ systems.

Animal Tissues Tissues are collections of cells with a common structure and function. Histologists—biologists who study tissues—classify tissues into four categories.

Epithelial tissue lines the outer and inner surfaces of the body in sheets of tightly packed cells that function in protection, absorption, and secretion. Cells at the base of an epithelium are attached to a *basement membrane*. Epithelia are classified on the basis of the number of cell layers and cell shape. A *simple epithelium* has one layer, whereas a *stratified epithelium* has multiple layers of cells. The cells at the free surface may be *squamous* (scalelike), *cuboidal* (boxlike), or *columnar* (pillarlike).

Epithelia may be specialized for absorption or secretion. *Mucous membranes* are found in the gut, where mucus lubricates and protects the gut lining, and in the air passages, where mucus traps particles, which are removed as this ciliated epithelium moves the mucous layer upward.

Connective tissue connects and supports other tissues and is characterized by having relatively few cells suspended in an extracellular *matrix* of fibers, which may be embedded in a liquid, jellylike or solid ground substance.

Loose connective tissue attaches epithelia to underlying tissues and holds organs in place. Its loosely woven fibers are of three types. *Collagenous fibers* are made of collagen, the most abundant animal protein. Three collagen molecules coil to form a fibril, and several fibrils make up a collagenous fiber, a structure with great tensile strength. *Elastic fibers*, made of the protein elastin, provide loose connective tissue with resilience. Branched *reticular fibers* form a tightly woven connection with adjacent tissues.

The cells scattered in the fibrous mesh of loose connective tissue include *fibroblasts*, which secrete the protein of the extracellular fibers, and *macrophages*, amoeboid cells that engulf bacteria and cellular debris by phagocytosis.

Adipose tissue is a special form of loose connective tissue with adipose cells held in the matrix. Each adipose cell contains a large fat droplet. This tissue pads and insulates the body and stores fuel reserves.

Fibrous connective tissue, with its dense arrangement of parallel collagenous fibers, is found in *tendons*, which attach muscles to bones, and in *ligaments*, which join bones at joints.

Cartilage is composed of collagenous fibers embedded in a rubbery ground substance called *chondrin*, which is secreted by *chondrocytes* found in scattered lacunae, or spaces, in the ground substance. Cartilage is a strong but somewhat flexible support material, making up the skeleton of sharks and vertebrate embryos. We retain cartilage in the nose, ears, intervertebral discs, trachea, and ends of some bones.

Bone is a mineralized connective tissue formed by *osteocytes* that deposit a matrix of collagen and calcium phosphate, which hardens into hydroxyapatite. *Haversian systems* consist of concentric layers of matrix deposited around a central canal containing blood vessels and nerves. Osteocytes are located in lacunae within the matrix and are connected to one another by cellular extensions called canaliculi. In long bones, the hard outer region is compact bone built of repeating Haver-

sian systems, whereas the interior is a spongy bone tissue, called marrow. Red marrow, near the ends of long bones, manufactures blood cells.

Blood is like a connective tissue in that it has a liquid extracellular matrix called plasma, which contains water, salts, and dissolved proteins. Erythrocytes (red blood cells) carry oxygen, leukocytes (white blood cells) function in defense, and platelets are involved in the clotting of blood.

Muscle tissue consists of long, contractile cells that are packed with microfilaments of actin and myosin. Three types of muscle tissues are found in vertebrate bodies. *Skeletal muscle*—also called striated muscle because of its striped appearance, caused by the arrangement of overlapping filaments—is responsible for the voluntary movements of the body. *Cardiac muscle*, forming the wall of the heart, is also striated; the ends of its branching cells are joined by intercalated discs that rapidly relay electrical impulses. *Visceral muscle*, also called smooth muscle, is composed of spindle-shaped cells lacking striations. Visceral muscle is found in the walls of the digestive tract, arteries, and other internal organs. Skeletal muscle is the only muscle under conscious control.

Nervous tissue senses stimuli and transmits signals. The *neuron*, or nerve cell, consists of a cell body and two or more nerve processes that conduct impulses toward (dendrites) and away from (axons) the cell body.

Organs and Organ Systems In all but the sponges and cnidarians, tissues are organized into specialized units of function called *organs*. Organs often consist of a layered arrangement of tissues. Many vertebrate organs are suspended by *mesenteries* in fluid-filled body cavities. Mammals have a *thoracic cavity* separated by a muscular diaphragm from an *abdominal cavity*.

Groups of organs are integrated into *systems*, which perform the major functions required for life. Ten organ systems are recognized: integumentary, skeletal, muscular, nervous, endocrine, circulatory, respiratory, digestive, urinary and reproductive.

Size, Shape, and the External Environment

Every cell must be in an aqueous medium to allow for oxygen, nutrient, and waste exchange across its plasma membranes. Cellular size is limited by the need for a sufficient surface area-to-volume ratio. Single-celled organisms, or animals with two-layered saclike bodies or thin flat bodies, can maintain sufficient cellular contact with the environment.

Animals with compact bodies must provide extensive moist internal membranes for exchanging materials with the environment. The surface area of the air chambers of the lung is about 100 m². The millions of tubules forming the internal surface of the kidney are specialized to filter wastes from blood, and the extensive, convoluted lining of the digestive tract absorbs nutrients. The circulatory system connects these exchange surfaces with the aqueous environment bathing the body's cells.

The Internal Environment

The internal environment of vertebrates is the *interstitial fluid* surrounding the cells through which oxygen, nutrients, and wastes are exchanged. The ability of an organism to control the "steady state" or constancy of this environment is called *homeostasis*. Most of the mechanisms by which animals maintain homeostasis involve *negative feedback*. When some variable moves above or below a *set point*, a control mechanism is turned on or off to return the condition to normal. One of the negative feedback mechanisms controlling body temperature in humans involves the sweat glands, which are signaled to increase their activity when the hypothalamus senses a rise in body temperature. When the temperature falls below the set point, the hypothalamus stops sending "sweat" signals.

Positive feedback in a physiological function is a mechanism in which a change in a variable serves to amplify rather than reverse the change. The stimulation of uterine contractions during childbirth is an example.

STRUCTURE YOUR KNOWLEDGE

1. Fill in the following table on the structure and function of animal tissues.

2. Organs usually are composed of layers of several different tissues. Which of the four major animal tissues do you think would be included in all organs? In what types of organs would you predict the remaining tissues would be included?

3. How do compact animals deal with the need for gas, nutrient, and waste exchange between each of their cells and the environment?

TEST YOUR KNOWLEDGE

MULTIPLE CHOICE: *Choose the one best answer.*

1. Which of the following is *not* an organ system?
   a. skeletal
   b. connective
   c. digestive
   d. nervous

2. A stratified squamous epithelium would be composed of

Tissue	Structural Characteristics	General Functions	Structure and Function of Specific Types

a. several layers of flat cells attached to a basement membrane.

b. a layer of ciliated, mucus-secreting flattened cells.

c. a hierarchical arrangement of boxlike cells.

d. an irregularly arranged layer of pillarlike cells.

3. Which of the following is *not* true of connective tissue?

a. It consists of few cells surrounded by fibers and a ground substance.

b. It includes such diverse tissues as bone, cartilage, tendons, adipose, and loose connective tissue.

c. It connects and supports other tissues.

d. It forms the internal and external lining of many organs.

4. Which of the following are incorrectly paired?

a. fibrous connective tissue—chondrocytes embedded in chondrin

b. bone—osteocytes connected by canaliculi in Haversian systems

c. loose connective tissue—collagenous, elastic, reticular fibers

d. adipose tissue—loose connective tissue with fat-storing cells

5. The best description of visceral muscle is

a. striated, branching cells, under involuntary control.

b. spindle-shaped cells, under involuntary control.

c. spindle-shaped cells connected by intercalated discs.

d. striated cells containing overlapping filaments, under involuntary control.

6. The diaphragm

a. is a mesentery.

b. increases the surface area of the lungs.

c. is part of the mammalian reproductive system.

d. separates the thoracic and abdominal cavities in mammals.

7. The interstitial fluid of animals

a. is the internal environment within cells.

b. provides for the exchange of nutrients and wastes between cells and blood.

c. is composed of blood.

d. is found at the major exchange areas in lungs, kidneys, and intestine.

8. Negative feedback circuits are

a. mechanisms that maintain homeostasis.

b. activated when a physiological variable deviates from a set point.

c. analogous to a thermostat that controls room temperature.

d. all of the above.

CHAPTER 37

ANIMAL NUTRITION

FRAMEWORK

Animals eat other organisms in order to obtain fuel for respiration, organic raw materials, and essential nutrients such as essential amino acids, vitamins, and minerals. Digestion is the enzymatic hydrolysis of macromolecules into monomers that can be absorbed across cell membranes.

Gastrovascular cavities, found in cnidarians and flatworms, are digestive sacs in which some extracellular digestion takes place before food particles are phagocytized by cells lining the cavity. All the more complex animal groups have alimentary canals—one-way tracts with specialized regions for mechanical breakdown of food, storage, digestion, absorption of nutrients, and elimination of wastes.

This chapter details the structures, functions, enzymes, and hormones of the human digestive tract.

CHAPTER SUMMARY

Animals, as heterotrophs, rely on the organic compounds in their food for energy and raw materials. In the long span of evolutionary history, animals have developed diverse mechanisms for obtaining and processing their food.

Feeding Mechanisms

Most animals are holotrophs, ingesting plants or animals either whole or in large pieces. *Herbivores* eat plants; *carnivores* eat animals; and *omnivores* consume both plants and animals.

Many aquatic animals are *filter-feeders*, sifting small food particles from the water. *Substrate-feeders* are found in or on their food, eating their way through it.

Deposit-feeders salvage pieces of decaying organic matter in detritus. *Fluid-feeders* suck fluids from a living plant or animal host. Most animals eat relatively large pieces of food.

Digestion: A Comparative Introduction

Animals must digest their food. The proteins, fats, and carbohydrates that make up food are too large to pass through cell membranes, and these macromolecules must be reorganized to form the macromolecules that make up each heterotroph. *Digestion* breaks apart food particles and splits macromolecules into monomers: polysaccharides into simple sugars, fats into glycerol and fatty acids, proteins into amino acids, and nucleic acids into nucleotides.

Enzymatic Hydrolysis Macromolecules are broken down by *hydrolysis*, the enzymatic addition of a water molecule when the bond between monomers is broken. Hydrolysis must occur in a separate compartment, to protect the animal's own cells from its hydrolytic enzymes. Following digestion, monomers and small molecules are absorbed across the membrane of the digestive compartment and enter the animal's cells. The undigested remainder of the food is then eliminated.

Intracellular Digestion in Food Vacuoles Large food molecules may be taken into cells by endocytosis and digested in the resulting food vacuoles. This *intracellular digestion* is typical of protozoa and sponges. Most animals, however, at least initially break down their food by *extracellular digestion* within a separate compartment of the body.

Digestion in Gastrovascular Cavities The simplest animals have single-opening sacs called *gastrovascular cavities*, which function in both digestion and transport of nutrients throughout the body. Using tentacles equipped with stinging nematocysts, *Hydra*, a cnidarian, captures prey and stuffs it through its expandable mouth. Cells of the gastrodermis secrete digestive enzymes, which initiate food breakdown, and beating flagella circulate the smaller food particles throughout the cavity. Gastrodermal cells take in food particles by phagocytosis, and hydrolysis of macromolecules occurs by intracellular digestion within food vacuoles. Undigested materials are expelled through the mouth. Flatworms, such as planaria, also have gastrovascular cavities in which digestion is begun extracellularly and continues within cells that take up food particles by phagocytosis.

Digestion in Alimentary Canals   Nematodes, annelids, mollusks, arthropods, echinoderms, and chordates all have *complete digestive tracts* or *alimentary canals*, which have two openings: a mouth and an anus. Specialized regions allow the sequential digestion and absorption of nutrients. Food, ingested through the mouth and pharynx, passes through an esophagus that leads to a crop, stomach, or gizzard, organs that are specialized for storing or grinding food. In the lengthy intestine, digestive enzymes hydrolyze macromolecules, and nutrients are absorbed across the tube lining. Undigested material exits through the *anus*.

The Mammalian Digestive System

The mammalian digestive tract has a four-layered wall: an inner mucous membrane (the mucosa), a connective-tissue layer, smooth muscle, and a sheath of connective tissue attached to the body-cavity membrane. The rhythmic waves of contraction called *peristalsis* push food through the tract. Ringlike valves called *sphincters* regulate the passage of material between some specialized segments.

Accessory glands, including *salivary glands*, the *pancreas*, and the *liver* with its *gall bladder*, deliver digestive enzymes to the alimentary canal through ducts.

Oral Cavity   Physical and chemical digestion begins in the mouth, where teeth chew food to expose a greater surface area to enzyme action. The presence of food in the oral cavity, or learned associations with other stimuli, triggers a nervous reflex that causes the salivary glands to deliver saliva through ducts to the mouth.

Saliva contains several components: mucin, a glycoprotein that protects the mouth lining from abrasion and lubricates the food for swallowing; buffers, which help neutralize the cavity-causing acidic environment; antibacterial agents; and *salivary amylase*, which hydrolyzes starch into smaller polysaccharides

and prevents a build up of starch between the teeth. The tongue is used to taste, to manipulate food, and to push the food ball or *bolus* into the pharynx for swallowing.

Pharynx   The *pharynx* is the intersection leading to both the esophagus and trachea. During swallowing, the windpipe moves so that its opening is blocked by the cartilaginous *epiglottis*.

Esophagus   Food moves down through the narrow esophagus to the stomach, squeezed along by a wave of smooth muscle contraction called peristalsis.

Stomach   The J-shaped stomach, with its elastic wall and folds called rugae, can expand to hold abour 2 liters of food and fluid. The epithelium lining the lumen (cavity) secretes *gastric juice*, a digestive fluid containing a high concentration of hydrochloric acid and *pepsin*, an enzyme that hydrolyzes specific peptide bonds in proteins. The low pH value of the gastric juice breaks down food tissues, kills bacteria, and denatures proteins, which increases the exposure of peptide bonds to the action of pepsin.

Pepsin is synthesized and secreted in an inactive form called *pepsinogen*. The HCl secreted by other stomach epithelia cells activates pepsinogen, and pepsin itself activates pepsinogen—an example of positive feedback. Protein-digesting enzymes are often secreted in inactive forms, called *zymogens*. The mucous coating secreted by the epithelium protects the stomach lining from digestion. Lesions called gastric ulcers develop where the lining is eroded faster than it can be regenerated by mitosis.

The sight, smell, or taste of food sends a nervous message from the brain to the stomach that initiates secretion of gastric juice. Substances in the food stimulate the stomach wall to release the hormone *gastrin* into the circulatory system. When gastrin recirculates back to the stomach, it stimulates further secretion of gastric juice. If the pH of the stomach contents becomes too low, the release of gastrin is inhibited, which decreases secretion of gastric juice—an example of negative feedback.

Smooth muscles mix the stomach contents about every 20 seconds. The stomach is usually closed off by two sphincters, one that prevents backflow into the esophagus, and one that regulates passage into the small intestine of the *acid chyme* produced by the action of the stomach and its secretions on the ingested food.

Small Intestine   Most enzymatic hydrolysis of macromolecules and absorption of nutrients into the blood takes place in the small intestine, the longest section of the alimentary canal.

The pancreas produces hydrolytic enzymes and a bicarbonate-rich alkaline solution that offsets the acidity of the chyme. The liver produces *bile*, which is stored in the gall bladder. Bile aids in the digestion and absorption of fats and also contains pigments that are byproducts of the breakdown of red blood cells in the liver.

Bile, digestive juices from the pancreas, and secretions from gland cells of the intestinal wall are mixed with the chyme in the first section of the small intestine called the *duodenum*. Various regulatory hormones coordinate the release of digestive secretions. *Secretin*, which is released by cells in the intestinal wall in response to the acidic pH of the chyme, stimulates the secretion of bicarbonate ions from the pancreas. *Cholecystokinin* (CCK), also produced by lining cells of the duodenum, stimulates both the contraction of the gall bladder and the release of pancreatic enzymes. A fat-rich chyme causes the duodenum to release *enterogastrone*, a hormone that slows down entry of chyme into the duodenum by inhibiting peristalsis in the stomach.

The digestion of starch, begun by salivary amylase, is continued by a pancreatic amylase, which hydrolyzes starch into maltose. The enzyme maltase splits maltose into two glucose molecules. *Disaccharidases*, enzymes specific for hydrolysis of different disaccharides, are built into the membranes and glycocalyx of epithelial cells, facilitating sugar absorption through the intestinal wall.

Pepsin in the stomach breaks proteins into smaller pieces. Protein digestion is completed in the small intestine by *trypsin* and *chymotrypsin*, enzymes specific for peptide bonds adjacent to certain amino acids; *carboxypeptidase*, which splits amino acids off the free carboxyl end of the polypeptide; and *aminopeptidase*, which works on the peptide bond at the amino end of the chain. Protein-digesting enzymes from the pancreas are secreted as zymogens, which are activated by the intestinal enzyme *enterokinase*.

Nucleases are enzymes that hydrolyze DNA and RNA into their nucleotide monomers.

The digestion of fats is aided by bile salts. In a process called *emulsification*, bile salts coat tiny fat droplets so the latter do not coalesce. *Lipase* is an enzyme that hydrolyzes fat molecules.

Most digestion is completed while the chyme is still in the duodenum. The *jejunum* and *ileum* are regions of the small intestine specialized for the absorption of nutrients. The huge surface area of the small intestine is created by large folds covered with fingerlike projections called *villi*, on which the epithelial cells have microscopic extensions called *microvilli*. The core of each villus has a net of capillaries and a lymph vessel called a *lacteal*. Nutrients are absorbed across the epithelium of the villus and then across the single-celled walls of the capillaries or lacteal. Transport may be passive (the nutrient moving down its concentration gradient) or active (the nutrient pumped against a gradient). The active transport of sodium into the lumen of the intestine and its passive reentry into epithelial cells seem to drive the uptake of certain nutrients.

Amino acids and sugars enter capillaries and are carried to the liver by the bloodstream. Glycerol and fatty acids are absorbed by epithelial cells and then recombined to form fats, which are coated with proteins to make tiny globules called *chylomicrons*. These packages are transported by exocytosis out of the epithelial cells and enter a lacteal. Some fat molecules, bound to specialized proteins, are transported as *lipoproteins* into capillaries.

The nutrient-laden blood from the small intestine is carried directly to the liver by the large *hepatic portal vein*. The liver interconverts and stores various molecules and regulates the nutrient content of the blood.

Large Intestine   The small intestine leads into the large intestine, or colon, at a junction with a sphincter. A blind pouch, called the *cecum*, attaches at the juncture and has a fingerlike extension, the *appendix*. One of the colon's functions is to finish the reabsorption of the large quantity of water secreted into the digestive tract along with the digestive enzymes. An irritation or infection of the colon lining may result in less absorption of water and lead to diarrhea; whereas an excess of reabsorption, occuring when peristalsis moves too slowly, may result in constipation. The wastes of the digestive tract are called *feces*.

A rich flora of *Escherichia coli* and other mostly harmless bacteria live on organic material remaining in the feces. Some of these bacteria produce vitamin K, which is absorbed by the host.

The feces contain cellulose, other undigested ingredients of food, bile pigments, salts excreted by the colon, and a large proportion of intestinal bacteria. Feces are stored in the *rectum*. A voluntary and involuntary sphincter between the rectum and anus control the elimination of feces, which is initiated by strong contractions of the colon.

Some Adaptations of Vertebrate Digestive Systems

Dentition, the types and arrangement of teeth in the mouth, is correlated with diet. Carnivores are characterized by sharp incisors and fanglike canines; herbivores have broad molars for grinding plant material; omnivores, such as humans, have a relatively unspecialized dentition.

Herbivores have longer alimentary canals because plant material, with its cell walls, is more difficult to digest than meat is. The extra length also provides more area for absorption of the less concentrated nutrients in vegetation. Specialized structures, such as the spiral valve in a shark's intestine, functionally increase intestinal length by providing additional surface area. Many herbivorous mammals also have special fermentation chambers filled with symbiotic bacteria and protozoa. These microorganisms, often housed in the cecum, both digest cellulose to simple sugars and produce a variety of essential nutrients for the animal. Since many of the symbiotic bacteria of rabbits and some rodents live in the large intestine, these animals may ingest their feces so that the nutrients produced by these symbionts may be absorbed as their food goes through the small intestine a second time.

Ruminants have the most elaborate adaptations for a herbivorous diet. The stomach is divided into four chambers, two of which contain symbiotic bacteria that digest cellulose. The cud is regurgitated, rechewed and swallowed into the other chambers, where both the microbially digested cellulose and the microorganisms themselves are digested and their nutrients absorbed.

Nutritional Requirements

Food provides fuel for cellular respiration, organic raw materials for the construction of the animal's molecules, and *essential nutrients*, which the animal cannot synthesize and must obtain in prefabricated form.

Food as Fuel   The monomers of carbohydrates, fats, and proteins can be used as fuel for cellular respiration, although the first two are used preferentially. Energy content of food is measured in *calories*. The calorie, as used by nutritionists, is actually a *kilocalorie*. Fat supplies about two times as many kcal/g as carbohydrate or protein.

Metabolism must supply energy continuously to maintain breathing, heart beat, and, in some animals, stable body temperature. The *basal metabolic rate (BMR)* is the number of kilocalories a resting organism requires for these processes for a given time. Birds and mammals, which are endothermic and use metabolic energy to maintain a constant body temperature, have higher basal metabolic rates than do the ectothermic fish, reptiles and amphibians, which absorb their body heat from the environment. In endotherms, body size is inversely related to the number of calories required to maintain each gram of body weight. The smaller the animal, the greater its surface area-to-volume ratio, the greater the loss of heat to the surroundings, and the greater the energy cost of maintaining a stable body temperature. This inverse relationship between body size and metabolic rate, however, also applies to ectothermic vertebrates and even protozoa.

The BMR for humans averages 1600 to 1800 kcal per day for males and about 1300 to 1500 kcal for females. BMR is determined by multiplying the O_2 consumption of a resting subject by the 4.83 kcal of energy produced by respiration for each liter of O_2 consumed. BMRs are standardized as kcal/h per kg body weight. Any activity increases the caloric requirement above this level.

When an animal consumes more calories than are needed to meet its energy requirements, the excess calories are stored. The liver and muscles store energy as glycogen, a polymer of glucose. When the gylcogen stores are full, additional calories are stored in adipose tissue as fat.

An *undernourished* person or other animal has a diet deficient in calories. If a calorie deficiency continues, the body breaks down its own proteins for energy, eventually causing irreversible damage. Incidents of undernourishment are usually associated with drought or war, although the condition anorexia nervosa can result in undernourishment.

Overnourishment, or obesity, increases the risk of heart attack, diabetes, and other disorders. "Dieting" is a billion dollar industry in the United States. Successful weight watching requires balancing caloric intake with caloric demand. To lose weight one must eat less, exercise more, or do both.

Food for Fabrication   Animals can fabricate most of their organic molecules using enzymes to rearrange the carbon skeletons and organic nitrogen acquired from food. In vertebrates, the conversion of nutrients to organic molecules occurs mainly in the liver.

Essential Nutrients   Molecules that an animal requires but cannot make are called essential nutrients. These requirements vary from species to species, depending on biosynthetic capabilities. When the diet is missing one or more essential nutrients, the animal is said to be *malnourished*. Malnutrition is more common than is undernutrition in human populations.

Eight of the 20 amino acids required to make proteins are essential amino acids in the human diet. Protein deficiency develops from a diet that lacks one or more essential amino acid. In Africa, the resulting syndrome of retarded mental and physical development in children is called *kwashiorkor*, which may arise when a child is weaned from mother's milk to a starchy diet. The trend away from breast-feeding in some less-developed countries has increased the incidence of protein deficiency.

Meat, eggs, and cheese contain complete proteins

with all essential amino acids in roughly balanced proportions. Most plant proteins are incomplete, and diets built on a single staple, such as corn, beans or rice, can result in protein deficiency. The body cannot store amino acids, and a deficiency of a single essential amino acid may prevent protein synthesis and limit the use of other amino acids. A combination of plant foods, complementary in amino acids and consumed at the same meal, can prevent protein deficiencies.

Animals are able to make most of the fatty acids they need. Linoleic acid, an unsaturated fatty acid used to make some phospholipids found in membranes, is required in the human diet. Deficiencies of essential fatty acids are rare.

Vitamins are organic molecules required in small amounts in the diet. Most vitamins are used as coenzymes or parts of coenzymes, and deficiencies can cause severe syndromes. The first vitamin isolated was thiamine; a deficiency of thiamine results in the disease beriberi.

Thirteen vitamins essential to humans have been identified. Water-soluble vitamins include the B-complex, most of which function as coenzymes in key metabolic processes, and vitamin C, required for production of connective tissue. Excesses of water-soluble vitamins are excreted. The fat-soluble vitamins are A, incorporated into visual pigments; D, aiding in calcium absorption and bone formation; E, seeming to protect phospholipids in membranes from oxidation; and K, required for blood clotting. Excesses of fat-soluble vitamins are deposited in body fat, and overdoses may cause toxic accumulations. There is debate between those scientists who believe that the recommended

daily allowances (RDAs) are sufficient and those who believe that optimal intakes of certain vitamins are much higher.

A compound that is a vitamin for one species may not be essential for a second species that can synthesize it. Symbiotic intestinal microorganisms may produce vitamins that are used by the host.

Minerals are inorganic nutrients and usually are needed in very small amounts. Requirements vary with species. Vertebrates require relatively large quantities of calcium and phosphorus for bone construction. Calcium is necessary for normal nerve and muscle functioning, and phosphorus is needed in ATP and nucleic acids. Iron is a component of the cytochromes and hemoglobin. Other minerals function as cofactors. Iodine is needed by vertebrates to make thyroxin, a metabolic regulatory hormone. Sodium, potassium, and chlorine are important in nerve function and osmotic balance.

Animal nutrition provides calories for energy needs, organic raw materials for biosynthesis, and essential amino acids, vitamins, and minerals.

STRUCTURE YOUR KNOWLEDGE

1. Food provides fuel, organic raw materials, and essential nutrients. Develop a concept map to organize your understanding of these nutritional needs of animals.

2. Fill in the following chart on the processes that occur in the sections or organs of the human diges-

Organ or Section	Processes Occurring	Associated Glands, Organs, and Digestive Fluids
Oral cavity		
Pharynx		
Esophagus		
Stomach		
Small intestine		
Large intestine		

tive tract, and list the glands and digestive fluids associated with these processes.

TEST YOUR KNOWLEDGE

MATCHING: *Match the description with the correct enzyme or hormone.*

1. ___ enzyme that hydrolyzes peptide bonds, works in the stomach

2. ___ hormone that stimulates secretion of gastric juice

3. ___ hormone that causes the gall bladder to contract and the pancreas to release enzymes

4. ___ enzyme that begins digestion of starch in mouth

5. ___ enzymes specific for hydrolyzing disaccharides

6. ___ enzyme that hydrolyzes peptide bonds at amino end of polypeptide

7. ___ enzyme that hydrolyzes fats

8. ___ intestinal enzyme that activates zymogens

9. ___ enzyme specific for peptide bonds adjacent tocertain amino acids, works in duodenum

10. ___ hormone that stimulates secretion of bicarbonate ions from pancreas

A. aminopeptidase

B. bile salts

C. cholecystokinin

D. chymotrypsin

E. disaccharidases

F. enterokinase

G. enterogastrone

H. gastrin

I. hydrochloric acid

J. lipase

K. maltase

L. pepsin

M. salivary amylase

N. secretin

MULTIPLE CHOICE: *Choose the one best answer.*

1. Deposit-feeders
   a. feed mostly on mineral substrates.
   b. filter small organisms from water.
   c. eat plants.
   d. feed on detritus.

2. The energy content of fats
   a. is released by bile salts.
   b. is inversely related to body size.
   c. is approximately two times that of carbohydrate or protein.
   d. can reverse the effects of malnutrition.

3. BMR
   a. stands for bottom metabolic rate.
   b. is higher for ectotherms than for endotherms.
   c. is higher per gram for smaller animals than for larger ones.
   d. All of the above are correct.

4. Which of the following statements is *not* true?
   a. The average human has enough stored glycogen to supply calories for several weeks.
   b. Eating less and/or exercising more can result in weight loss.
   c. Conversion of glucose and glycogen takes place in the liver.
   d. Excessive calories are stored as fat, regardless of their food source.

5. Kwashiorkor
   a. results from protein deficiency.
   b. is an example of malnutrition.
   c. is a syndrome involving retarded mental and physical development.
   d. All of the above are correct.

6. Incomplete proteins are
   a. lacking in essential vitamins.
   b. a cause of undernourishment.
   c. found in proper proportion in meat, eggs, and cheese.
   d. lacking in one or more essential amino acids.

7. Vitamins
   a. may be produced by intestinal microorganisms.
   b. are constant from one species to the next.
   c. are toxic in excess because they are deposited in body fat.
   d. are inorganic nutrients, needed in small amounts, that usually function as cofactors.

8. Which of the following is most prevalent in animals with gastrovascular cavities?

a. absorption of predigested nutrients
b. extracellular digestion
c. intracellular digestion
d. filter feeding

9. Ruminants
   a. have teeth adapted for an herbivorous diet.
   b. must use microorganisms to digest plant starch.
   c. eat their feces to obtain nutrients digested from cellulose by microorganisms.
   d. house symbiotic bacteria and microorganisms in a cecum.

10. The stomach can expand during eating due to its
    a. villi.
    b. rugae.
    c. cecae.
    d. sphincters.

11. The acid pH of the stomach
    a. hydrolyzes proteins.
    b. is regulated by the release of gastrin.
    c. is neutralized by gastric juice.
    d. is produced by pepsin.

12. Acid chyme is the
    a. nutrient broth that moves from the stomach to the duodenum through a sphincter.
    b. nutrient broth that passes through the jejunum and ileum.
    c. gastric secretions of epithelial cells in the stomach.
    d. mixture of macromolecules passing into the stomach.

13. Chylomicrons are
    a. lipoproteins transported by the circulatory system.
    b. small branches of the lymphatic system.
    c. protein-coated fat globules excreted out of epithelial cells.
    d. small peptides acted upon by chymotrypsin.

14. Which of the following is *not* a common component of feces?
    a. intestinal bacteria
    b. cellulose
    c. bile salts
    d. undigested ingredients of food

15. The hepatic portal vein
    a. supplies the capillaries of the intestines.
    b. carries absorbed nutrients to the liver for processing.
    c. carries blood from the liver to the heart.
    d. drains the lacteals of the villi.

CHAPTER 38

CIRCULATION AND GAS EXCHANGE

FRAMEWORK

This chapter surveys the basic approaches to circulation and gas exchange found in the animal kingdom, with special attention to the human systems and the problems of cardiovascular disease. The following concept map organizes some of the chapter's key ideas.

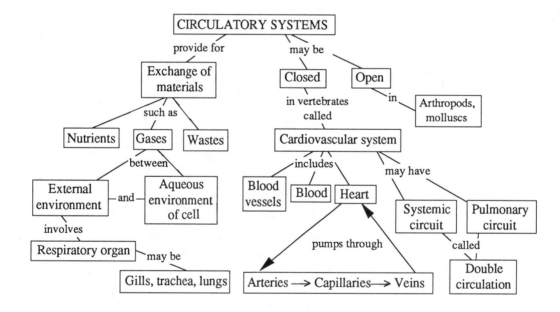

CHAPTER SUMMARY

The exchange of materials is essential for the cells of an organism. Every living cell must reside in an aqueous environment, which provides oxygen and nutrients and permits disposal of carbon dioxide and metabolic wastes.

The diffusion time of a substance is proportional to the square of the distance covered. Diffusion cannot possibly transport substances over the macroscopic distances in animals. All but the simplest animals have internal transport systems for body fluids. The body fluid, usually blood, exchanges materials with the environment across the thin and extensive epithelia of organs specialized for gas exchange, nutrient absorption, and waste removal. The blood then exchanges chemicals with the interstitial fluid that bathes the cells.

Internal Transport in Invertebrates

Gastrovascular Cavities An internal transport system is unnecessary in *Hydra* and other cnidarians. The central gastrovascular cavity inside the two-cell-thick body wall makes exchange between cells and their environment direct. Planarians and other flatworms also have gastrovascular cavities that ramify throughout the body.

Open and Closed Circulatory Systems In the *open circulatory system* found in insects and other arthropods, and in most mollusks, there is no real distinction between blood and interstitial fluid because the *hemolymph*, as it is called, bathes the internal tissues directly. Hemolymph is circulated as body movements squeeze the *sinuses*, or spaces between organs, and by the beating of a heart that is usually part of a dorsal vessel. Hemolymph, drawn into the vessel through ostia when the heart relaxes, is pumped into the interconnected system of sinuses during contraction.

Annelids, some mollusks, and vertebrates have *closed circulatory systems*, in which the blood remains in vessels and exchanges materials with the interstitial fluid bathing the cells. In an earthworm, the dorsal vessel functions as a heart by pumping blood forward. Five pairs of vessels, which loop around the digestive tract and connect the dorsal and ventral main vessels, pulsate and function as auxiliary hearts. Blood circulates more rapidly and efficiently through a closed circulatory system.

Circulation in Vertebrates

The closed circulatory system of vertebrates, also called the *cardiovascular system*, consists of the heart, blood vessels, and blood. The heart has one or more *atria*, which receive blood, and one or more *ventricles*, which pump blood out of the heart. *Arteries*, carrying blood away from the heart, branch into tiny *arterioles* within organs, which then divide into the microscopic *capillaries*. Exchange of substances between blood and interstitial fluid occurs across the thin capillary walls. Capillaries rejoin to form *venules*, which meet to form the *veins* that return blood to the heart.

Vertebrate Circulatory Schemes The ventricle of a fish's two-chambered heart pumps blood first to the capillary beds of the gills, from which the oxygenated blood flows through a vessel to the capillary beds in the rest of the body. Veins return the blood to the atrium.

In the three-chambered heart of amphibians, the single ventricle pumps blood through a forked artery into the *pulmonary circuit*, which leads to lungs and skin and then back to the left atrium, and the *systemic circuit*, which carries blood to the rest of the body and back to the right atrium. This *double circulation* repumps blood after it returns from the capillary beds of the lungs, ensuring a strong flow of oxygenated blood to the brain, muscles, and body organs. A ridge in the ventricle diverts most of the oxygenated blood from the left atrium into the systemic circuit and the deoxygenated blood from the right atrium into the pulmonary circuit. The three-chambered reptilian heart has a septum that partially divides the single ventricle. In crocodiles, the septum completely divides the ventricle into two chambers.

Delivery of oxygen for cellular respiration is most efficient in birds and mammals, which, as endotherms, have high oxygen demands. The left side of the four-chambered heart handles only oxygenated blood, whereas the right side receives and pumps deoxygenated blood. In the human circulatory system, oxygenated blood returning from the lungs in the pulmonary veins flows to the left atrium and then into the left ventricle, from which it is pumped through the aorta in the systemic circulation. The aorta gives rise to branch arteries to the heart, brain, limbs, and organs of the body, where arterioles break into thin-walled capillaries, and the blood gives up its oxygen. Deoxygenated blood from the capillary beds returns through venules into veins and is collected by the anterior (or superior) and posterior (or inferior) vena cavae. These large vessels empty into the right atrium; blood flows into the right ventricle and it is pumped in the pulmonary arteries to the capillary beds of the lungs. The blood completes the pulmonary circuit by returning to the heart in the pulmonary veins.

The Heart The human heart is enclosed in a two-layered, fluid-filled sac just beneath the sternum. The atria have relatively thin muscular walls, whereas the ventricles have thicker muscular walls. The *heart cycle* lasts about 0.8 sec and consists of the *systole*, during which the heart muscle contracts and the chambers pump blood, and the *diastole*, when the heart is relaxed. The atria are filling with blood during the heart cycle except for the first 0.1 sec of the systole when they contract. The slow and powerful contraction of the ventricles lasts 0.3 sec. Blood flows passively into the ventricles during diastole, and atrial contraction finishes filling them.

Atrioventricular valves between each atrium and ventricle are snapped shut when the ventricles contract. Strong fibers prevent these connective tissue flaps from turning inside out. The *semilunar valves* at the exit of the aorta and pulmonary artery are forced open by ventricular contraction and close when the ventricles relax and the elastic walls of the arteries recoil. The heart-beat sounds are caused by the closing of these valves. A heart murmur is the detectable hissing sound of blood leaking back through a defective valve.

The average human heart rate, or *pulse*, is 65 to 75 beats per minute. The inverse relationship between size and pulse is attributable to the higher metabolic rate per gram of tissue of smaller animals.

The *cardiac output*, or volume of blood pumped per minute into the systemic circuit, depends on the heart rate and the *stroke volume*, or quantity of blood pumped by each contraction of the left ventricle.

Cells of cardiac muscle are self-excitable or myogenic; they have an intrinsic ability to contract. The rhythm of contractions is coordinated by the *SA (sinoatrial) node*, or *pacemaker*. When nodal tissue contracts, it generates electrical impulses. The contraction of the SA node, located in the wall of the right atrium near the anterior vena cava, initiates a wave of excitation that signals the cells of the two atria to contract. (Remember that cardiac muscle cells are electrically connected by the intercalated discs.) The *AV (atrioventricular) node*, located at the base of the two atria, relays the impulse (after a slow transit through transition fibers that results in a 0.1 sec delay) to the ventricles. The electrical currents produced during the heart cycle can be detected by electrodes placed on the skin and recorded in an electrocardiogram (EKG or ECG). The SA node is controlled by two sets of nerves with antagonistic signals and is influenced by hormones, temperature, and stimulation from increased blood flow during exercise.

Blood Flow The wall of an artery or vein consists of three layers: an outer connective tissue zone with elastic fibers; a middle layer of smooth muscle and more elastic fibers; and an inner lining of *endothelium* made of simple squamous epithelium, its basement membrane, and a thin connective-tissue ring. The middle layer is especially thick in arteries, which must be stronger and more elastic. Capillaries have only the endothelial layer.

The flow of blood decelerates after leaving the heart as a result of the increasing cross-sectional area of the branching vessel system. The enormous number of capillaries creates a total diameter much greater than other parts of the system, and resistance is also greater in the narrow capillaries. Blood flow is very slow in capillaries, improving opportunity for exchange with interstitial fluid. Blood flow speeds up within veins, due to the decrease in their total cross-sectional area.

Blood pressure, the hydrostatic force exerted against the wall of a blood vessel, is much greater in arteries than in veins and is greatest in systole. *Peripheral resistance*, caused by the narrow openings of the arterioles impeding the exit of blood from arteries, causes the swelling of the arteries during systole. The snapping back of the elastic arteries during diastole maintains a continuous blood flow into arterioles and capillaries. Blood pressure may be measured with a sphygmomanometer. The higher number is the pressure during systole (when blood first spurts through the artery that was closed off by the cuff), and the second is the pressure during diastole (when blood flows smoothly through the artery).

Both cardiac output and peripheral resistance determine blood pressure. Contraction of smooth muscles in arteriole walls increases resistance and thus increases blood pressure, whereas dilation of arterioles lowers blood pressure. Neural and hormonal signals control these muscles.

Blood pressure drops to almost zero after the capillary beds. The one-way valves in veins and the contraction of skeletal muscles between which veins are embedded force blood to flow back to the heart. Pressure changes during breathing also draw blood into the large veins in the thoracic cavity.

Capillaries branch off *thoroughfare channels*, the direct connections between arterioles and venules. A sphincter regulates the passage of blood into a capillary. Only about 5% to 10% of the body's capillaries have blood flowing through them at any one time. The brain, kidneys, liver, and heart are usually heavily supplied with blood; the distribution to the other areas of the body varies with need.

Capillary Exchange The exchange of substances between blood and interstitial fluid may involve bulk transport by endocytosis and exocytosis of the endothelial cells making up the capillary wall, passive diffusion of small substances (water, sugars, salts, oxygen, and urea) through both the cells themselves and the clefts between adjoining cells, and the forcing of fluid out of the capillary by hydrostatic pressure. Blood cells and proteins are too large to pass easily through the endothelium. Blood pressure at the upstream end of a capillary forces fluid out, whereas osmotic pressure at the downstream end tends to draw about 99% of that fluid back in.

The Lymphatic System The fluid that does not return to the capillary, and any proteins that may have leaked through the capillary wall, are returned to the blood through the *lymphatic system*. Fluid diffuses into lymph capillaries intermingled in the capillary net. The fluid, called *lymph*, moves through lymph vessels with one-way valves as a result of the movement of skeletal muscles and the rhythmic contractions of the vessel walls. In *lymph nodes*, the lymph is filtered, and white blood cells attack viruses and bacteria. The accumulation of interstitial fluid in tissues causes a condition known as edema.

Blood

Vertebrate blood is a type of connective tissue with cells in a liquid matrix called *plasma*. When *whole blood* is centrifuged, the *formed elements* or cells form a dense red pellet that makes up about 45% of the volume of blood.

Plasma   The plasma consists of a large variety of solutes dissolved in water. The collective and individual concentrations of *electrolytes*, or inorganic salts in the form of dissolved ions, are important to osmotic balance between blood and interstitial fluid and to the functioning of muscles and nerves. The kidney is responsible for the homeostatic regulation of ion concentrations. Plasma proteins function as buffers, osmotic components of the blood, antibodies, escorts for lipids, and clotting factors. Blood plasma from which the fibrinogens or clotting factors have been removed is called serum. Nutrients, metabolic wastes, gases, and hormones are also transported in the plasma. Except for its higher protein concentration, plasma is very similar in composition to the interstitial fluid.

Blood Cells   The numerous (about 25 trillion in the body's 5 liters of blood) *erythrocytes* transport oxygen. Mammalian erythrocytes lack nuclei, and all red blood cells lack mitochondria and generate their ATP by anaerobic metabolism. The small size and biconcave shape of red blood cells creates a large surface area of plasma membrane across which oxygen can diffuse. Erythrocytes are packed with *hemoglobin*, an iron-containing protein that binds oxygen.

Erythrocytes are formed from stem cells in the red marrow of bones, and their production is controlled by a negative feedback mechanism involving the hormone *erythropoietin* secreted by the kidney in response to low oxygen supply in tissues. Red blood cells circulate for about 3 to 4 months before they are phagocytized by cells in the liver and their components recycled.

Leukocytes, or *white blood cells*, fight infections. Some white cells are phagocytes; some are lymphocytes that give rise to cells that produce antibodies. Most leukocytes are found in the interstitial fluid and lymph nodes. Leukocytes also arise from stem cells in the bone marrow. Lymphocytes mature in lymphoid organs (spleen, thymus, tonsils, adenoids, and lymph nodes).

Platelets, pinched-off fragments of large cells in the bone marrow, are involved in the blood-clotting mechanism.

Blood Clotting   The blood clotting process usually begins when platelets, clumped together along a damaged endothelium, release clotting factors. By a series of steps, the blood protein *fibrinogen* is converted to *fibrin*. The threads of fibrin weave into a patch. An inherited defect in any step of the complex clotting process causes *hemophilia*, a disease characterized by prolonged bleeding from even minor injuries. Anti-clotting factors normally prevent clotting of blood in the absence of injury. A *thrombus* is a clot that occurs within a blood vessel and blocks the flow of blood.

Cardiovascular Disease

More than half of all deaths in the United States are caused by *cardiovascular disease*, usually a result of heart attack or stroke. A heart attack (an arrhythmia or cessation of beating) may occur when a thrombus or an *embolus*, which is a clot formed elsewhere that moves through the circulatory system, blocks a coronary artery. The cardiac muscle that was served by the artery dies, and the conduction of electrical impulses through the cardiac muscle may be interrupted. Strokes result from a thrombus or embolus blocking an artery in the brain.

Most heart-attack and stroke victims had been suffering from a chronic disease known as *atherosclerosis*, in which growths called *plaques* develop within arteries and narrow the vessels. Plaques that become hardened by calcium deposits result in *arteriosclerosis*, or "hardening of the arteries." An embolus is more likely to be trapped in narrowed vessels, and plaques are common sites of thrombus formation. Occasional chest pains, known as angina pectoris, may warn that a coronary artery is partially blocked.

Hypertension, or high blood pressure, is thought to damage the endothelium and initiate plaque formation, promoting atherosclerosis and increasing the risk of heart attack and stroke. This condition can be easily diagnosed and controlled by drugs, diet, and exercise. Hypertension and atherosclerosis tend to be inherited. Smoking, not exercising, and eating a fat-rich diet have been correlated with an increased risk of cardiovascular disease.

The cholesterol level in blood plasma is an indicator of potential atherosclerosis. Most cholesterol is carried as low-density lipoproteins (LDLs), which are removed from the blood when they bind to receptors on cells of the liver and other organs. Genetic defects may result in a reduction in the number of receptors and a dangerously high level of blood cholesterol. High-density lipoproteins (HDLs) are another form of cholesterol carriers that appear to reduce cholesterol deposition in plaques. The ratio of LDLs to HDLs is an indication of potential cardiovascular disease.

The death rate from cardiovascular disease has been significantly declining in the United States. Diagnosis and treatment of hypertension, improved intensive care for cardiovascular patients, and healthier life styles may be contributing to this trend.

Gas Exchange

General Problems of Gas Exchange Gas exchange provides the continuous supply of oxygen needed for cellular respiration and removes the waste product carbon dioxide. The *respiratory medium*, which supplies oxygen, is air for a terrestrial animal and water for an aquatic one. The *respiratory surface*, the portion of an animal's body where gas exchange with the respiratory medium occurs, must be moist and large enough to supply the whole body.

General Structure and Function of Respiratory Organs A localized region of the body surface is usually specialized as a respiratory surface with a thin, moist epithelium separating a rich blood supply and the respiratory medium. When the entire outer skin serves as a respiratory organ, as in an earthworm, the animal must live in damp places and have a relatively small, long and thin body to provide a high ratio of surface area to volume. Most animals use an extensively branched or folded localized region of the body surface for gas exchange. Aquatic animals have *gills*; terrestrial vertebrates use *lungs*; and insects have *tracheae*.

Gills: Respiratory Adaptations of Aquatic Animals Gills are evaginations of the body surface, ranging from the simple bumps on echinoderms, to the complex, finely divided gills of mollusks, crustaceans, fishes, and some amphibians. Delicate gills are usually sheltered by a protective cover. Due to the low oxygen concentration in a water environment, gills usually require *ventilation*, or movement of the respiratory medium across the respiratory surface—often an energy-intensive process.

Blood flows through the capillaries in a direction opposite to the flow of water over the gills of a fish. This arrangement sets up a *countercurrent exchange*, in which the diffusion gradient favors the movement of oxygen into the blood throughout the length of the capillary.

Tracheae: Respiratory Adaptations of Insects Air's advantages as a respiratory medium are its higher concentration of oxygen, faster diffusion rate of O_2 and CO_2, and lower density. To prevent the disadvantageous water loss from large, moist surfaces, the respiratory surfaces of most terrestrial organisms are invaginated.

The trachea of insects are tiny air tubes that ramify throughout the body to come into contact with nearly every cell. Openings to the tracheal system are *spiracles*. Some large insects use rhythmic body movements to ventilate their tracheal systems. The efficiency of gas exchange within the tracheal system counteracts the rather slow moving hemolymph of an insect's open circulatory system.

Lungs: Respiratory Adaptations of Terrestrial Vertebrates Lungs are invaginated respiratory surfaces restricted to one location from which oxygen is transported to the rest of the body by the circulatory system. Frog lungs are balloonlike, but mammalian lungs have a spongy texture with a much greater surface area of epithelium for gas exchange. The lungs of mammals are located in the thoracic cavity and enclosed in a double-walled sac. The two layers adhere tightly to each other, the lungs, and the chest-cavity wall.

Air, entering through the nostrils, is filtered, warmed, humidified, and smelled in the nasal cavity. Air passes through the pharynx and enters the windpipe through the glottis. When food is being swallowed, the glottis is pushed against the epiglottis. The *larynx* functions as a voicebox in humans and many other mammals; exhaled air vibrates a pair of *vocal cords*. The *trachea*, or windpipe, branches into two *bronchi*, which then branch repeatedly into *bronchioles* when they reach the lungs. Ciliated, mucus-coated epithelium lines much of the respiratory tree and removes dust and other particles from the respiratory system.

Multilobed air sacs encased in a web of capillaries are at the tips of the tiniest bronchioles. Gas exchange takes place across the thin moist epithelium of these *alveoli*.

Vertebrate lungs are ventilated by breathing, the alternate inhalation and exhalation of air. A frog breathes by expanding its mouth cavity, which draws air into the mouth, and then raising the jaw, which forces air into the lungs.

Mammals ventilate their lungs by *negative pressure breathing*. The volume of the lungs and the thoracic cavity is increased by expansion of the rib cage and contraction of the *diaphragm*. Air pressure is reduced within this increased volume, and air flows through the nostrils down to the lungs. Relaxation of the rib muscles and diaphragm compresses the lungs, increases the pressure, and forces air out.

Tidal volume is the normal volume of air inhaled and exhaled by an animal. The maximum volume that can be inhaled and exhaled by forced breathing is called *vital capacity*. The *residual volume* is the air that remains in the alveoli and lungs after forceful exhaling.

Birds have air sacs that penetrate their abdomen,

neck, and wings. The air sacs and lungs are ventilated when the bird breathes, through a circuit that includes a one-way passage through tiny channels, called *parabronchi*, in the lungs. Air sacs act as bellows to maintain air flow through the lungs, lower the density of the bird, and help to dissipate heat formed by the metabolism of flight muscles.

Breathing is under automatic control. The *breathing center* in the medulla at the stem of the brain sends nerve impulses to the rib muscles and diaphragm to contract. In a negative feedback loop, stretch sensors in the lungs respond to the expansion of the lungs and send nervous impulses that inhibit the breathing center. When CO_2 concentration increases in the blood, more carbonic acid is formed. The corresponding drop in the pH of the blood, sensed by the breathing center, increases the messages to breathe. Hyperventilation can cause breathing to stop temporarily when excessive deep breathing has rid the blood of most CO_2. Oxygen sensors in key arteries react to severe deficiencies of O_2.

The concentration of gases in air or dissolved in water is measured as *partial pressure*. The partial pressure of oxygen, which makes up 21% of the atmosphere, is 160 mm Hg (0.21 x 760 mm—atmospheric pressure at sea level). The partial presure of CO_2 is 0.23 mm Hg. Partial pressure is proportional to concentration; a gas will diffuse from a region of higher partial pressure to lower partial pressure.

Blood entering the lungs has a lower P_{O_2} and a higher P_{CO_2} than does the air in the alveoli, and oxygen diffuses into the capillaries and carbon dioxide diffuses out. In the systemic capillaries, pressure differences favor the diffusion of O_2 out of the blood into the interstitial fluid and of CO_2 into the blood.

In most animals, O_2 is carried by *respiratory pigments* in the blood. Hemoglobin, usually located in red blood cells, is the respiratory pigment of almost all vertebrates. Copper is the oxygen-binding component in the blue respiratory protein *hemocyanin*, common in arthropods and many mollusks.

Hemoglobin is composed of four subunits, each of which has a cofactor heme group with iron at its center. The binding of O_2 to the iron atom of one subunit induces a shape change in the other subunits, and their affinity for oxygen increases. Likewise, the unloading of the first O_2 effects a conformational change that lowers the other subunits' affinity for oxygen.

The *dissociation curve* for hemoglobin shows the relative amounts of oxygen bound to hemoglobin under varying oxygen concentrations. At low partial pressures of oxygen, little O_2 is bound, but the percent saturation of hemoglobin rises sharply after the concentration rises high enough for the first oxygen to be bound to a hemoglobin molecule. Thus, in the steep part of this S-shaped curve, a slight change in partial pressure will cause hemoglobin to load or unload a substantial amount of oxygen. A drop in pH lowers the affinity of hemoglobin for O_2. Since a rapidly metabolizing tissue produces more CO_2, which lowers pH, hemoglobin will unload more of its O_2 to that tissue.

Over 20% of CO_2 is transported in blood bound to amino groups of hemoglobin. Seventy percent is transported as bicarbonate ions. Carbon dioxide enters into the red blood cells where it first reacts to form carbonic acid and then dissociates into H^+ and a bicarbonate ion. The hydrogen ions are bound to hemoglobin and other proteins, and the pH value of the blood is not greatly lowered during the transport of carbon dioxide. In the lungs, the diffusion of CO_2 out of the blood shifts the equilibrium in favor of the conversion of bicarbonate back to CO_2, and CO_2 is unloaded from the blood.

Special physiological adaptations have enabled some air-breathing mammals to make long underwater dives. The Weddell seal stores twice the amount of O_2/kg body weight as do humans, mostly by having a larger volume of blood, a huge spleen that stores blood, and a higher concentration of *myoglobin*, an oxygen-storing muscle protein. A diving reflex slows the pulse of diving mammals, reroutes blood to the brain and essential organs, and restricts blood supply to the muscles.

STRUCTURE YOUR KNOWLEDGE

1. Briefly outline or sketch the circulation of blood through the human heart to the systemic and pulmonary circuits.

2. Develop a concept map or table that shows the components of blood and their functions.

3. Create a concept map that shows your understanding of blood pressure, what it does, what causes it, and where and when it is highest.

4. Briefly outline or sketch the exchange of gases in the human body, indicating the major structures and processes involved.

TEST YOUR KNOWLEDGE

MULTIPLE CHOICE: *Choose the one best answer.*

1. A gastrovascular cavity
   a. is found in cnidarians and annelids.
   b. functions to pump fluids throughout the body.

c. functions in both digestion and distribution of nutrients.

d. involves all of the above.

2. Which of the following is *not* a *similarity* between open and closed circulatory systems?
   a. Some sort of pumping device helps to move blood through the body.
   b. Some of the circulation of blood is a result of movements of the body.
   c. The blood and interstitial fluid are indistinguishable from each other.
   d. All tissues come into close contact with the circulating body fluid so that the exchange of nutrients and wastes can take place.

3. In a system with double circulation,
   a. blood is pumped at two locations as it circulates through the body.
   b. there is a countercurrent exchange within the gills.
   c. there is no mixing of oxygenated and deoxygenated blood in the heart.
   d. blood is repumped after it returns from the capillary beds of the gas-exchange organ.

4. During diastole,
   a. the atria fill with blood.
   b. blood flows passively into the ventricles.
   c. the elastic recoil of the arteries maintains hydrostatic pressure on the blood.
   d. all of the above are occurring.

5. An atrioventricular valve prevents the backflow or leakage of blood
   a. from a ventricle into an atrium.
   b. between ventricles.
   c. from the aorta into the left ventricle.
   d. from the pulmonary vein into the right atrium.

6. During heavy exercise,
   a. stroke volume increases.
   b. heart rate increases.
   c. cardiac output increases.
   d. all of the above occur.

7. Heart beat is initiated by contraction of the
   a. SA node.
   b. AV node.
   c. right atrium.
   d. myogenic cardiac muscle cells.

8. Blood flows more slowly in the arterioles than in the arteries because the arterioles
   a. have thoroughfare channels to venules that are often closed off.

b. collectively have a larger cross-sectional area than do the arteries.
   c. must provide opportunity for exchange with the interstitial fluid.
   d. All of the above are correct.

9. If all the body's capillaries were open at the same time,
   a. blood pressure would fall dramatically.
   b. peripheral resistance would increase.
   c. blood would move too rapidly through the capillary beds.
   d. the amount of blood returning to the heart would increase.

10. Which of the following is *not* a factor in the exchange of substances in capillary beds?
    a. endocytosis and exocytosis
    b. passive diffusion through clefts between endothelial cells
    c. hydrostatic pressure
    d. bulk flow through the apoplast

11. Edema is a result of
    a. swollen lymph glands.
    b. an accumulation of interstitial fluid.
    c. swollen feet.
    d. too high a concentration of blood proteins.

12. A function of the kidney is
    a. control of the lymphatic system.
    b. maintenance of electrolyte balance.
    c. production of lymphocytes.
    d. destruction and recycling of red blood cells.

13. Fibrinogen is
    a. a blood protein that escorts lipids through the circulatory system.
    b. a cell fragment involved in the blood-clotting mechanism.
    c. a blood protein that is converted to fibrin to form a blood clot.
    d. one of the formed elements of blood.

14. Angina pectoris
    a. causes atherosclerosis.
    b. is a mild form of stroke.
    c. promotes hypertension.
    d. may be a warning sign that a coronary artery is partially blocked by plaque.

15. A thrombus
    a. may form at a region of plaque in an artery.
    b. is a traveling embolism.
    c. may cause hypertension.
    d. may calcify and result in arteriosclerosis.

16. The trachea of insects

a. are stiffened with rings and lead into the lungs.
b. are filled by positive pressure breathing.
c. ramify along the circulatory system for gas exchange.
d. are highly branched, coming into contact with almost every cell for gas exchange.

17. Which of the following is *not* involved in speeding up breathing?
    a. a drop in the pH of the blood
    b. stretch receptors in the lungs
    c. impulses from the breathing center in the medulla
    d. severe deficiencies of oxygen

18. The binding of an O_2 to the first iron atom of a hemoglobin molecule
    a. occurs at a very low partial pressure of oxygen.
    b. produces a conformational change that lowers the other subunits' affinity for oxygen.
    c. is an example of cooperativity that then results from a conformational change.
    d. occurs more readily at a lower pH value.

19. Most carbon dioxide is transported in the blood
    a. attached to the iron of hemoglobin.
    b. as bicarbonate ions.
    c. at a higher partial pressure than oxygen is.
    d. as carbonic acid.

20. According to the dissociation curve for hemoglobin,
    a. the percent saturation of hemoglobin changes with the amount of oxygen to which it is bound.
    b. there is a range in which a slight change in P_{O2} will cause hemoglobin to load or unload a large amount of oxygen.
    c. the P_{O2} in the tissues is lower than that in the alveoli.
    d. a drop in pH value raises the affinity of hemoglobin for O_2.

21. In which animal does blood flow through vessels from the respiratory organs to the rest of the body without returning to the heart?
    a. fish
    b. insect
    c. frog
    d. both a and b

22. Blood leaving the right ventricle of a mammal's heart will pass through how many capillary beds before it returns to the right ventricle?
    a. one
    b. two
    c. one or two, depending on the circuit it takes
    d. at least two, but frequently three

23. Fluid is moved back into the capillaries at the downstream end of a capillary bed by
    a. active transport.
    b. hydrostatic pressure.
    c. osmotic pressure.
    d. bulk transport.

24. In countercurrent exchange,
    a. the flow of fluids or gases in opposite directions maintains a favorable diffusion gradient along the length of an exchange surface.
    b. oxygen is exchanged for carbon dioxide.
    c. a double circulation keeps oxygenated and deoxygenated blood separate.
    d. oxygen moves from a region of high partial pressure to one of low partial pressure, but carbon dioxide moves in the opposite direction.

25. Negative pressure is created in the lungs of mammals by
    a. exhalation of air.
    b. inhalation of air.
    c. relaxation of the diaphragm and rib muscles.
    d. contraction of the diaphragm and rib muscles.

CHAPTER 39

THE IMMUNE SYSTEM

FRAMEWORK

The *immune system* protects the body from external and internal threats such as bacteria, viruses, and early-stage cancer cells. *Nonspecific defense mechanisms* include the physical and chemical barriers of the skin and mucous membranes, phagocytes and natural killer cells, the inflammatory response, and antimicrobial proteins such as interferon and complement. The immune system also reacts to foreign antigens with *specific defenses* including *humoral immunity* produced by antibodies from B cells, and the *cell-mediated immunity* resulting from the production of cytotoxic T cells, helper T cells and suppressor T cells.

The immune system is characterized by the ability to distinguish self from nonself, high specificity to a huge number of antigenic determinants, coordinated action of various cells and chemical messages (histamines, interferon, interleukins I and II), and immunological memory. Disorders of the immune system include autoimmune diseases, cancers, allergies, immunodeficiencies, and AIDS.

CHAPTER SUMMARY

Nonspecific Defense Mechanisms

The nonspecific defense mechanisms of the body, which include the skin and mucous membranes, phagocytes, the inflammatory response, interferon, and complement proteins, are effective against a variety of microbial assaults.

The Skin and Mucous Membranes   The physical barrier of the skin is reinforced by oil and sweat secretions that create a low pH value and contain *lysozyme*, an enzyme that attacks bacterial cell walls. Lysozyme is also present in tears and saliva. Gastric juice kills most bacteria that reach the stomach. The tiny hairs of the nostrils filter out particles that may carry microorganisms, and the ciliated mucus-coated epithelial lining of the respiratory tract traps and removes particles.

Phagocytes and Natural Killer Cells   *Macrophages* are amoeboid cells that engulf bacteria, viruses, and cellular debris. Developing from monocytes in the blood, most macrophages are found in the interstitial fluid, although some permanently reside in organs. *Neutrophils*, another class of white blood cells, become phagocytic in infected tissue. *Natural killer cells* destroy the body's own infected or aberrant cells, causing them to lyse by attacking their membranes.

The Inflammatory Response   *Inflammation*, in response to damage caused by physical injury or microorganisms, includes the dilation of small blood vessels to increase blood flow, the leakiness of capillaries, and the congregation of phagocytic white blood cells in the interstitial fluid. The pus that may accumulate is a combination of dead cells and body fluids. Clotting proteins help to seal off the infected region.

Histamine, released by injured cells, causes blood vessels to dilate and initiates the inflammatory response. Other chemical signals attract phagocytes, and leukocytosis inducing factor stimulates neutrophil release from bone marrow.

An inflammatory response may be localized or systemic, which would include an increase in the number of circulating white blood cells and a fever. Fever, which may be triggered by toxins produced by pathogens or by *pyrogens* released by certain white blood cells, may stimulate phagocytosis and inhibit growth of microorganisms.

Antimicrobial Proteins   *Interferons* are proteins produced by virus-infected cells that help other cells to resist virus infection. Interferon, diffusing from an infected cell to neighboring cells, stimulates the production of proteins that inhibit cells from making proteins for an infecting virus. In an inflammatory response, interferons activate macrophages.

Recombinant DNA technology has made it possible to produce large quantities of interferons, which are being tested for their effectiveness in treating viral infections and cancer. The action of interferon in cancer treatment is not completely understood: it may slow protein production if a tumor is of viral origin; one type of interferon stimulates natural killer cells that may attack tumor cells; and it changes cell membranes such that tumor cells are less likely to metastasize.

Complement is a group of at least 20 proteins that cooperate with other defense mechanisms. Complement proteins are activated by the onset of a specific immune response or by contact with microorganisms. Activated proteins may attract phagocytes and stimulate histamine release, thus amplifying the inflammatory response. Some proteins coat microorganisms in a process called *opsonization*, which facilitates the invader's destruction by macrophages. In a *membrane attack complex*, other complement proteins cause a foreign cell to lyse by inserting into its membrane.

Specific Defense Mechanisms: The Immune Response

The immune system distinguishes "self" from "nonself" by recognizing differences in the cell-surface markers of invading bacteria and viruses as well as cells of tissue grafts. The immune system responds to foreign molecules, called *antigens*, by proliferating cells that either attack the invader or produce *antibodies*, which combat the antigen. The specific immune response is primed by a particular antigen and produces defenses against only that antigen.

The immune system "remembers" antigens and can react against them more promptly upon re-exposure, a property known as *immunity*. In *vaccination*, the immune system is exposed to an inactive or attenuated pathogen, which initiates the system's long-term capability to respond quickly to the infective agent. Whether by vaccination or exposure to viral or bacterial pathogens, the body is stimulated to produce antibodies and develops *active immunity*. In *passive immunity*, antibodies are supplied by an antibody injection or through the placenta to a fetus, and these antibodies provide temporary immunity.

Duality of the Immune System Humoral immunity involves the production of antibodies that circulate in the blood and lymph and defend against free bacteria and viruses. *Cell-mediated immunity*, the second component of the immune system, involves specialized cells that react against body cells infected by bacteria or viruses and against fungi and protozoa. The cell-mediated system responds to tissue transplants and may protect the body from its own cancerous cells.

Cells of the Immune System Lymphocytes mediate both humoral and cell-mediated immune responses. *B cells* produce antibodies; *T cells* attack infected or defective cells in cell-mediated immunity. All blood cells, including these lymphocytes, develop from stem cells located in the bone marrow. These cells either migrate to the thymus and differentiate into T cells or continue to develop in the bone marrow as B cells. As they mature, T and B cells develop *immunocompetence*, the ability to recognize one specific antigen. Specific receptor proteins are attached to the surface of the lymphocyte—a bound antibody on B cells or a receptor complementary to a specific antigen on T cells. These specific receptor proteins are synthesized before the body has ever encountered the corresponding antigen—the immune system is prepared for an incredible variety of potential infections.

Final maturation of lymphocytes occurs in the lymph nodes or spleen, where the immunocompetent cells lie waiting to encounter antigens. Macrophages also congregate in lymph nodes, ingesting pathogens and triggering the activation of lymphocytes. B and T cells, once they encounter antigens, multiply into *effector cells* that move through circulatory and lymphatic vessels to fight specific infections.

Antigens In general, antigens are proteins or polysaccharides that are part of the capsules, cell walls, or coats of bacteria or viruses. Antigens also may be parts of membranes of other cells. Antibodies recognize localized regions, called *antigenic determinants*, of an antigen.

Clonal Selection Remember that during differentiation, each B and T cell develops a single type of receptor that can respond to a single antigenic determinant. The ability of the immune system to respond specifically to an enormous variety of different antigens is explained by the presence of a great diversity of immunocompetent lymphocytes. When an antigen binds to a lymphocyte with a complementary receptor on its surface, the lymphocyte is activated to proliferate, creating a clone of effector cells. *Clonal selection* is the mechanism by which the body mounts an immune response against a specific antigen. Since a foreign molecule may have several antigenic determinants, several different clones of B and T cells may be mobilized.

Immunological Memory Several days after the first encounter with an antigen, enough B and T effector cells have proliferated to produce a *primary immune response*. If the body reencounters the same antigen, the *secondary immune response* is more rapid, effective, and longer in duration. In the initial encounter with an antigen, *memory cells* are produced along with effector cells. Memory cells are long-lived cells, able to produce effector cells and more memory cells rapidly when stimulated by the same antigen. In some cases, memory cells confer a lifetime immunity to a disease.

The Humoral Immune Response

Every B cell acquires a capacity to produce a specific antibody during its development, but exposure to that antigen is necessary to activate antibody production. An antigen may activate a B cell by *capping*, in which multiple copies of the antigenic determinant bind to a cluster of antibody receptors, and the complex is taken into the cell by endocytosis. More commonly, B cells are activated with the help of special T cells. Activated cells differentiate into clones of some memory cells and many *plasma cells*. These cells produce and secrete huge numbers of antibodies, which circulate in the blood and lymph, and bind to antigens.

Antibodies are a class of proteins called *immunoglobulins*, abbreviated *Ig*. The *variable (V) region* of an antibody permits it to recognize and bind to a specific antigen; the *constant (C) region* interacts with an effector mechanism that destroys and eliminates the antigen. The Y-shaped antibody molecule consists of two pairs of polypeptide chains: two identical short *light (L) chains* and two identical longer *heavy (H) chains*. Both H and L chains have variable sections at the ends of the two arms of the Y, forming two *antigen-binding sites*. The amino acid compositions of the variable regions create the unique contours and binding potential of the sites. Weak chemical bonds that form between an antigenic determinant and the antigen-binding site of an antibody account for the specificity of antibody–antigen binding. The five types of the constant regions of the H and L chains create five classes of antibodies in mammals: IgM, IgG, IgA, IgD, and IgE. Each class plays a different role in the immune response.

Antibodies label foreign molecules and cells for destruction by one of several effector mechanisms. The formation of *antigen-antibody complexes* may neutralize the harmful chemical groups of toxins, agglutinate bacteria, or precipitate antigen molecules. The resulting complexes or clumps are then engulfed by phagocytes. Antigen-antibody complexes also activate the complement system, a set of blood proteins that attack a foreign cell's membrane, opsonize antigens, and activate the inflammatory response. Note that most pathogens are destroyed by nonspecific defenses, such as phagocytes and complement, after being labeled as foreign by the specific binding of antibodies.

Antibodies are used in biological research and clinical testing to detect specific antigens. Techniques for making *monoclonal antibodies*, developed in the late 1970s, supply quantities of identical antibodies specific for an antigenic determinant. Such antibodies allow widescale clinical testing for sexually transmitted diseases and pregnancy and can be coupled with toxins, which may soon be used to specifically target and destroy cancer- or disease-causing cells. Cells derived from a tumor line are fused with antibody-producing lymphocytes, obtained from the spleen of an animal exposed to the antigenic determinant of interest. These hybrid cells, called *hybridomas*, are cultured to produce a mass of cells, which produce the same antibody.

Cell-Mediated Immunity

Whereas the humoral immune response reacts to free pathogens, cell-mediated immunity acts against pathogens that have entered cells of the body. Immunocompetent T cells have membrane-bound *T-cell receptors*, specific proteins that recognize a complex of antigenic determinant and "self" markers displayed by infected cells. The *major histocompatibility complex (MHC)* is a group of cell-surface proteins unique to each individual. *Antigen-presenting cells (APCs)*, such as macrophages, attach portions of antigens to MHC proteins on their cell surface. T-cell receptors recognize and respond to specific "self–nonself" complexes; this *histocompatibility restriction* results from the development of the T-cell's immunocompetence. Upon recognizing its specific MHC-antigen complex, a T cell proliferates clones of cells, including memory cells and *cytotoxic T cells*, which kill infected cells.

When *helper T cells* recognize and bind to an antigen-presenting macrophage, the macrophage releases *interleukin I*, which stimulates the division of the T cell. This signal is an example of a *cytokine*, a chemical secreted by one cell that regulates neighboring cells. The activated T cells release *interleukin II*, a cytokine that stimulates both the helper T and cytotoxic T cells to grow more rapidly. Helper T cells also contribute to the mounting of the humoral immunity response. Whereas some *T-independent antigens* can activate B cells directly and cause capping, most common are *T-dependent antigens*, which only activate B cells that have been stimulated by interleukin II and other cytokines.

Cytotoxic T cells, whose proliferation has been increased by the interleukin II released by helper T cells, attach to cells displaying specific antigen–MHC complexes and release *perforin*, a protein that forms a hole in the target cell's membrane. Cytotoxic T cells kill infected cells displaying specific antigenic determinants and probably also attack cancer cells and foreign tissue transplant cells.

Late in the immune response, *suppressor T cells* release cytokines that suppress the now unnecessary action of other T and B cells.

Self versus Nonself

The ability of the immune system to distinguish self from nonself develops during fetal growth, when lymphocytes capable of reacting against molecules that tag "self" cells are apparently destroyed, leaving only immunocompetent cells specific for foreign molecules.

Blood Groups The chemical markers that determine blood groups are important in the immune response against foreign cells and must be considered in blood transfusions. Individuals with type A blood have the A antigen of the surface of their red blood cells and produce antibodies against the B antigen. Likewise, individuals with type B blood produce anti-A antibodies and have B antigens on their blood cells; group AB individuals have both A and B antigens and no antibodies. Type O individuals, who produce A and B antibodies, are called universal donors because their blood cells carry no antigens and will not agglutinate when exposed to the antibodies of a blood recipient.

An Rh-negative mother may develop antibodies against the *Rh factor*, another blood-group antigen, if fetal blood from an Rh-positive child leaks across the placenta. Should she carry a second Rh-positive fetus, her immunological memory may result in the production of antibodies that cross the placenta and agglutinate fetal red blood cells. Treatment of the mother with anti-Rh antibodies just after delivery destroys any fetal cells that may have leaked into her circulation and prevents the mother's immunological response to the antigen.

The Major Histocompatibility Complex This set of several cell-surface antigens is involved in the immune system's ability to distinguish self from nonself. This biochemical fingerprint, coded for by several MHC genes, each of which has 50 or more alleles, is unique to each individual. Transplanted tissues and organs may be rejected because the foreign antigens of the MHC trigger T cell responses. Various drugs are used with transplant operations to suppress the immune response. Cyclosporine is a drug that suppresses only cell-mediated immunity.

Disorders of the Immune System

Autoimmunity Sometimes the immune system turns against self, leading to *autoimmune diseases*, such as lupus erythematosis and probably rheumatoid arthritis, rheumatic fever, and juvenile diabetes. Autoimmune disorders may develop when antibodies or effector T cells that are responding to a foreign invader cross-react with the individual's own tissues. In other cases, the surface components of some cells, affected by viruses, drugs, or mutations, may change enough that they become recognized as foreign.

Immunological Aspects of Cancer Cancer encompasses a variety of diseases, all of which involve changes in normal body cells. Some changes to the outer membrane surface of cells may result in the immune system identifying these cancer cells as foreign and destroying them. Individuals with deficient immune systems are more susceptible to cancer. The *immune surveillance theory* suggests that the immune system constantly finds and destroys cancer cells in early stages of development. The failure of this surveillance system when tumors develop is being researched.

Allergy *Allergies* are hypersensitivities to certain environmental antigens, or *allergens*. Allergic reactions, which are usually quite rapid, can occur in the respiratory system, gastrointestinal tract, and skin. Antibodies of the IgE family that are bound to mast cells (stationary cells found in connective tissue) trigger allergic reactions. When antigens bind to cell-surface antibodies, the mast cells undergo *degranulation*, releasing histamines, which create an inflammatory response. Antihistamines are drugs that counteract histamines. *Anaphylaxis* is a severe allergic response in which the abrupt dilation of peripheral blood vessels (caused by a rapid release of histamines) leads to a life-threatening drop in blood pressure and thus to shock. Severe allergic reactions can be counteracted with injections of the hormone epinephrine.

Immunodeficiency A defect in any of the components of the immune system leads to increased susceptibility to viruses, bacterial infections, and cancer. In the rare congenital disease known as severe combined immunodeficiency (SCID), both T and B cells are absent or inactive. Afflicted individuals must receive successful bone marrow transplants or live in completely sterilized, isolated environments. Certain cancers, such as Hodgkin's disease, and various drug and radiation therapies can depress the immune system.

AIDS Acquired immune deficiency syndrome (AIDS) is an immune disorder involving a great reduction in the number of T helper cells, which, as you have just learned, activate the other T and B lymphocytes. The immune system is severely weakened, and most afflicted individuals die within three years, usually from opportunistic infections or cancers. A rare form of pneumonia, severe diarrhea, various viral and fungal infections, and some rare cancers are among the diseases associated with AIDS. *AIDS-related complex (ARC)* is a transitional stage of the disease that produces swollen lymph nodes, fever, night sweats, and weight loss.

AIDS was first recognized in 1981, when the increasing incidence of Kaposi's sarcoma in homosexual men was investigated by the Centers for Disease Control (CDC) in Atlanta. In America, high-risk groups include sexually active homosexual and bisexual males, intravenous drug users, recipients of blood products, and sexual partners of members of the high-risk groups. Transmission of the disease may occur by intimate sexual contact, contaminated needles, and blood-to-blood contact.

The infectious agent responsible for AIDS, now known as *HIV (human immunodeficiency virus)*, was identified independently in the United States and France in 1984. A glycoprotein on the surface of this virus binds to a receptor called *CD4* on the surface of helper T cells, entering and eventually killing these crucial cells. Most individuals exposed to HIV have circulating antibodies to the virus. A test has been developed that identifies low levels of AIDS antibody and is being used to screen blood donations. A small fraction of the individuals who have antibodies against the virus display symptoms of ARC or AIDS, perhaps partly due to the long incubation time of the disease. AIDS is an incurable, lethal disease. Newly developed drugs may slow the development of the virus or treat the opportunistic infections, but no cure has been found. The rapid genetic changes of HIV have complicated the development of a vaccine against it. With the increasing spread of the disease, education about unsafe sex and sharing needles is critical.

Stress and Immunity There is growing evidence that general emotional health and immunity are related. Hormones secreted during stress affect the numbers of white blood cells; nerve fibers penetrate deep into lymphoid tissue, and receptors for chemical signals from nerve cells have been found on lymphocytes. The new field of psychoneuroimmunology looks at these connections between state-of-mind and immunity.

Defense in Invertebrates

The ability to distinguish self from nonself has been found to be well developed in invertebrates. In many invertebrates, amoeboid cells called coelomocytes can identify and destroy foreign substances. Tissue graft experiments in earthworms have established that their defense systems both reject and develop a memory response to these grafts.

STRUCTURE YOUR KNOWLEDGE

This chapter contains a wealth of information that is probably fairly new to you. If you take a little time and pull out the key players of the immune system and organize them first into very basic concept clusters and then develop more interrelated concept maps, you will find that this information is both understandable and fascinating.

1. Fill in the chart on the following page on some of the molecules involved in the immune system.

2. Create a concept map that outlines nonspecific and specific defense mechanisms and shows how these two types of defenses interact and cooperate in protecting the body.

3. Now create a concept map on humoral and cell-mediated immunities, showing the cells involved and their functions.

4. Describe the structure of an antibody molecule and how this structure relates to its function. Briefly explain the clonal selection theory.

TEST YOUR KNOWLEDGE

MULTIPLE CHOICE: *Choose the one best answer.*

1. Which of the following is incorrectly paired with its effect?
   a. gastric juice—kills bacteria in the stomach
   b. fever—stimulates phagocytosis and inhibits microbial growth
   c. histamine—causes blood vessels to dilate
   d. vaccination—creates passive immunity

2. Monoclonal antibodies
   a. are used to treat AIDS.
   b. explain the ability of the immune system to create a large number of identical T or B cells.

Molecule	Where Produced or Found	Action
Lysozyme		
Histamine		
Interferon		
Complement		
Antibody		
Interleukin I		
Interleukin II		
Perforin		

c. are produced in tissue culture by hybridomas.
d. initiate the secondary immune response.

3. Antigens are
   a. proteins that consist of two light and two heavy polypeptide chains.
   b. proteins or polysaccharides usually found on the cell surfaces of invading bacteria or viruses.
   c. proteins found in the blood that cause foreign blood cells to clump.
   d. proteins embedded in T-cell membranes.

4. A secondary immune response is more rapid and greater in effect than a primary immune response because
   a. memory cells respond to the pathogen and rapidly clone more effector cells.
   b. the second response is an active immunity, whereas the primary one was a passive immunity.
   c. helper T cells are available to activate other blood cells.
   d. interleukins cause the rapid accumulation of phagocytic cells.

5. Lymphocytes capable of reacting against "self" molecules

a. are usually not a problem until a woman's second pregnancy.
b. are probably destroyed before birth.
c. are usually kept separate from the immune system.
d. contribute to immunodeficiency diseases.

6. The major histocompatibility complex
   a. is involved in the ability to distinguish self from nonself.
   b. is a set of several cell-surface antigens.
   c. may trigger T-cell responses after transplant operations.
   d. All of the above are correct.

7. In opsonization,
   a. proteins coat microorganisms and help macrophages bind to and engulf the invading cell.
   b. a set of proteins lyses a hole in the foreign cell's membrane.
   c. antibodies cause cells to agglutinate, and the resulting clumps are engulfed by phagocytes.
   d. a flood of histamines is released that may result in anaphylactic shock.

8. Severe combined immunodeficiency

a. is an autoimmune disease.

b. is a form of cancer in which the membrane surface of the cell has changed.

c. is a disease in which both T and B cells are absent or inactive.

d. is an immune disorder in which the number of T helper cells is greatly reduced.

9. A transfusion of B-type blood given to a person who has A-type blood would result in

a. the recipient's anti-B antibodies clumping the donated red blood cells.

b. the recipient's B antigens reacting with the donated anti-B antibodies.

c. the recipient forming both anti-A and anti-B antibodies.

d. no reaction, because B is a universal donor type of blood.

10. Which of the following are incorrectly paired?

a. variable region—antibody specificity for an antigenic determinant

b. helper T cells—production of plasma cells

c. cytotoxic T cells—destruction of foreign cells

d. immunoglobulins—antibodies

CHAPTER 40

CONTROLLING THE INTERNAL ENVIRONMENT

FRAMEWORK

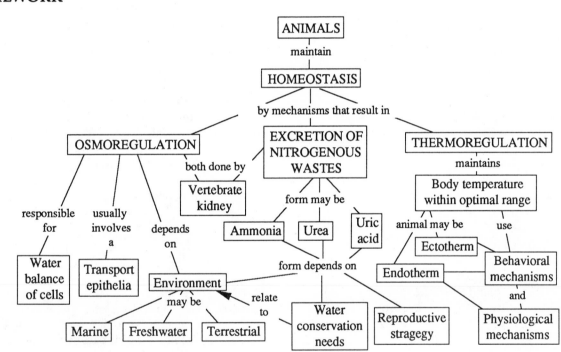

CHAPTER SUMMARY

Animals are able to survive large fluctuations in their external environment by maintaining a relatively constant internal environment, a condition known as homeostasis. Changes in the internal body fluid—hemolymph or interstitial fluid serviced by blood—that bathes the cells are tempered by regulatory systems, usually involving feedback mechanisms. The physiological adjustments that maintain homeostasis,

such as *osmoregulation, excretion,* and *thermoregulation,* enable organisms to cope with short-term environmental changes and have developed as a result of natural selection as populations have evolved in specific environments.

Osmoregulation

The environment largely determines the osmoregulation problems an animal faces. But whether an animal lives in saltwater, freshwater, or on land, the water gain must balance the water loss in the cells of its body.

Osmoconformers and Osmoregulators

Osmosis is the diffusion of water across a selectively permeable membrane separating two solutions that differ in *osmolarity* (moles of solute particles per liter). Osmolarity is expressed in units of milliosmoles per liter (mosm/l). Isosmotic solutions are equal in osmolarity, and there will be no net osmosis between them. Water flows across a membrane from a hypoosmotic (more dilute) to a hyperosmotic (more concentrated) solution.

Osmoconformers are isosmotic with their aqueous surroundings and do not regulate their osmolarity. *Osmoregulators* must get rid of excess water if they live in a hypoosmotic medium or take in water to offset osmotic loss if they inhabit a hyperosmotic environment.

Problems of Osmoregulation in Different Environments

Most marine invertebrates are osmoconformers, whereas most marine vertebrates are osmoregulators. Sharks maintain an osmolarity slightly higher than that of seawater by retaining urea—a nitrogenous waste product—and trimethylamine oxide (TMAO)—a protection from the damaging effects of urea—within their bodies. They produce a large quantity of urine to balance the osmotic uptake of water. Their internal salt concentrations are lower than those of seawater because they use rectal glands to pump salt out of the body.

Many marine bony fishes, having evolved from freshwater ancestors, are hypoosmotic to seawater. They must drink large quantities of seawater to replace the water they lose by osmosis. Excess salt is pumped out through salt glands located in the gills. Many marine birds get rid of excess salt through nasal salt glands, and marine reptiles also have osmoregulating salt glands.

Freshwater animals constantly take in water by osmosis because their internal fluids are hyperosmotic to their medium. Protozoa use contractile vacuoles to pump out excess water. Freshwater animals excrete large quantities of dilute urine. Salt supplies are replaced from their food, or, in some fish, by active uptake of ions across the gills.

Most animals are *stenohaline* and are able to tolerate only small changes in external osmolarity. Animals that move between fresh- and saltwater environments, or live in brackish water, can survive substantial osmotic changes in their environments. These *euryhaline* animals often can maintain a constant internal osmolarity by changing osmoregulatory mechanisms in the different environments.

Some animals are capable of *anhydrobiosis*, or cryptobiosis—surviving dehydration in a dormant state. One mechanism of this adaptation is the production of multiple dissacharides, whose many hydroxyl groups form hydrogen bonds, replacing water, and protecting cellular structures during dehydration.

The most important problem confronting terrestrial animals is the threat of desiccation. Only the arthropods and vertebrates have successfully colonized land. Adaptations to prevent dehydration include impervious coverings, drinking and eating food with high water content, nervous and hormonal control of thirst, behavioral adaptations, and water-conserving excretory organs.

Transport Epithelia and Osmoregulation

The osmolarity of the internal environment is usually regulated by the transport of salt, followed by the osmotic movement of water, across a *transport epithelium*. Located at the tissue-environment boundary, this single sheet of cells, linked by impermeable tight junctions, regulates the passage of solutes between the extracellular fluid and the environment. Variations in the molecular composition of the plasma membrane of the transport epithelium determine its passive permeability to water and salts and its active transport by various membrane proteins.

The transport epithelia of salt glands function exclusively in osmoregulation, whereas the transport epithelia of excretory organs, often arranged in tubular networks, function additionally in metabolic waste excretion.

Excretory Systems of Invertebrates

Protonephridia: Flame-Cell System of Flatworms

The *flame-cell system* of flatworms consists of a branched system of tubules embedded in the body tissues. Extracellular fluids pass into the bulbous flame cells at the ends of the smallest tubules. Cilia projecting into the tubule propel the fluid along into excretory ducts that empty by way of nephridiopores. The flame-cell system of freshwater flatworms is primarily osmoregulatory; metabolic wastes are excreted through the gastrovascular cavity.

Protonephridia, simple excretory systems of closed tubules lacking internal openings, are found in rotifers, some annelids, molluscan larvae, and lancelets, as well as in flatworms.

Metanephridia of Earthworms

Excretory tubules with internal openings that collect body fluids, called *metanephridia*, are found in most annelids. The metanephridia of earthworms occur in pairs in each segment of the worm. An open ciliated funnel called a nephrostome collects coelomic fluid, which then moves through a folded tubule encased in capillaries. The transport epithelium of the tubule pumps salts out of the tubule, and the salts are reabsorbed into the blood. The copious dilute urine, which helps to offset the osmotic uptake of water from the moist soil, exits by way of nephridiopores.

Malpighian Tubules of Insects In insects and other terrestrial arthropods, *Malpighian tubules* remove nitrogenous wastes from the body fluid (hemolymph) and function in osmoregulation. Transport epithelia lining these blind sacs, which open into the digestive tract at the end of the midgut, pump salts and wastes from the hemolymph into the tubule. The fluid passes through the hindgut and into the rectum, where a transport epithelium pumps most of the salt back into the hemolymph, and water follows by osmosis. In this water-conserving system, nitrogenous wastes are eliminated as dry matter along with the feces.

The Vertebrate Kidney

Nephrons, the excretory tubules of vertebrates, are arranged into compact organs, the *kidneys*. The vertebrate *excretory system* consists of the kidneys, associated blood vessels, and excretory ducts.

Anatomy of the Excretory System Blood enters the pair of bean-shaped kidneys through the *renal arteries* and leaves by way of the *renal veins*. Urine exits through the *ureter* and is temporarily stored in the *urinary bladder*. The body periodically releases urine by micturition (urination) through the *urethra*.

Structure of the Nephron Each human kidney contains about a million nephrons, each of which consists of a *renal tubule* and associated blood vessels. Water, urea, salts, and small molecules flow from capillaries into the renal tubule, forming the *filtrate*, which is processed by the transport epithelium of the tubule to form urine.

The blind end of the renal tubule is formed into a cuplike *Bowman's capsule* that encloses a ball of capillaries called the *glomerulus*. Filtrate enters the tubule and passes through its three specialized regions: a *proximal convoluted tubule*; the *loop of Henle* with a *descending limb* and an *ascending limb*, and a *distal convoluted tubule*. *Collecting ducts* receive filtrate from many renal tubules, and pass the urine into the renal pelvis, the chamber that drains into the ureter.

Bowman's capsules and the proximal and distal tubules are restricted to the outer *cortex* of the kidney. *Cortical nephrons* have reduced loops of Henle and are located entirely in the cortex. *Juxtamedullary nephrons*, found only in mammals and birds, have long loops of Henle that extend into the *medulla*.

An *afferent arteriole* supplies each nephron, subdividing to form the capillary ball and converging as it leaves the capsule to form an *efferent arteriole*. This vessel then forms a second capillary network—the *peritubular capillaries*—which surrounds the proximal and distal convoluted tubules. Other capillaries form the *vasa recta*, the descending and ascending capillaries that parallel the loop of Henle. Exchange between the tubules and capillaries takes place through the interstitial fluid.

General Physiology of the Nephron *Filtration* occurs when blood pressure forces water and small solutes out of the porous capillaries, through clefts between *podocytes*, the capsule lining cells, and into the nephron tubule. The filtrate contains salts, nitrogenous wastes, glucose, and other small molecules. Blood cells and plasma proteins remain in the capillary.

Secretion of substances from the interstitial fluid across the tubule epithelium occurs as the filtrate moves through the proximal and distal convoluted tubules. This selective secretion involves both passive and active transport.

Filtration is nonselective, and many neccesary molecules pass into the filtrate. Most of the water, sugar, vitamins, and other organic nutrients in the filtrate are *reabsorbed* across the tubule epithelium into the interstitial fluid, from which they are returned to the blood. Selective secretion and reabsorption control the balance of water and various salts in body fluids.

Transport Properties of the Renal Tubule (1) The transport epithelium of the proximal convoluted tubule selectively secretes ammonia, drugs or poisons processed by the liver, and hydrogen ions into the filtrate and actively transports glucose and amino acids out of the filtrate in the tubule. Potassium is also reabsorbed. About 3/4 of the NaCl and water forced into the filtrate is reabsorbed in this area. The *brushborder* of the epithelia cells provides a huge surface area across which salt and water diffuse into the epithelial cells. The cells actively transport Na^+ across the membrane into the interstitial fluid outside the tubule; Cl^- is transported passively out of the cell to balance the increasing positive charge; and water follows the salt by osmosis. The salt and water then diffuse into the peritubular capillaries.

(2) The descending limb of the loop of Henle is freely permeable to water but not to salt or other small solutes. Water moves into the interstitial fluid, which becomes increasingly hyperosmotic toward the inner medulla, leaving a filtrate with high salt concentrations.

(3) The ascending limb of the loop of Henle is not very permeable to water but is permeable to salt, which diffuses out of the lower thin segment of the loop and adds to the high osmolarity of the medulla. Cl⁻ is actively transported out of the thick upper portion of the ascending limb, and Na⁺ follows the negative charge passively. As salt leaves, the filtrate becomes less concentrated.

(4) The distal convoluted tubule is specialized for selective secretion and reabsorption, contributing to homeostasis of the body fluids. It regulates K⁺ concentration by controlling its secretion, and helps to regulate pH by secreting H⁺ and reabsorbing the buffering compound bicarbonate.

(5) As the filtrate moves in the collecting duct back through the increasing osmotic gradient of the medulla, more and more water exits by osmosis. The duct is not permeable to salt, but its lower region is permeable to urea. Some urea diffuses out of the duct and eventually diffuses back into the ascending limb of the loop. Urea and salts form the osmotic gradient that enables the kidney to produce urine that is hyperosmotic to the blood.

How the Mammalian Kidney Conserves Water: A Closer Look

The osmolarity gradient of NaCl and urea in the interstitial fluid, which is produced by the juxtamedullary nephrons, enables the kidney to produce urine up to four times as concentrated as blood and normal interstitial fluid. Filtrate leaving Bowman's capsule has an osmolarity of about 300 mosm/l, the same as blood. Both water and salt are reabsorbed in the proximal convoluted tubule; the volume of filtrate is reduced but the osmolarity remains about the same. In the trip down the descending limb of the loop of Henle, water exits by osmosis, and the filtrate becomes more concentrated. Salt, which is now in high concentration in the filtrate, diffuses out as the filtrate moves up the salt-permeable but water-impermeable ascending limb—helping to create the osmolarity gradient.

The vasa recta has a countercurrent flow to the movement of filtrate in the loop of Henle. As the blood in these capillaries moves down into the inner medulla, water leaves by osmosis and salt enters; as the blood moves back up toward the cortex, water moves back in and salt diffuses out. Thus the capillaries can carry oxygen and other supplies to the medulla without disrupting the osmolarity gradient by carrying away NaCl or watering down the interstitial fluid.

The filtrate makes one final pass through the medulla, this time in the collecting duct, which is permeable to water and urea but not to salt. Water flows out by osmosis, and, as the filtrate becomes more concentrated, urea leaks out into the interstitial fluid, signifi-

cantly adding to the osmolarity of the inner medulla. The resulting urine is isosmotic to the interstitial fluid of the inner medulla, which can be as high as 1200 mosm/l.

Regulation of the Kidneys

The osmolarity of the urine can vary from 70 to 1200 mosm/l, depending on the hydration needs of the body. Osmoregulation in vertebrates depends on the nervous and hormonal control of water and salt reabsorption in the kidneys.

Antidiuretic hormone, or *ADH*, is produced in the hypothalamus and stored in the pituitary gland. *Osmoreceptor cells* in the hypothalamus send signals to the pituitary gland when the osmolarity of the blood becomes too high, triggering the release of ADH. This hormone increases the water permeability of the distal convoluted tubules and collecting ducts, resulting in the reabsorption of more water, lowering the osmolarity of the blood, and increasing the osmolarity of the urine. Negative feedback decreases the release of ADH. When blood osmolarity is low, the absense of ADH results in the production of large volumes of dilute urine, a condition called diuresis. Alcohol inhibits the release of ADH and can cause dehydration.

The *juxtaglomerular apparatus (JGA)*, located near the afferent arteriole, responds to a drop in blood pressure or decrease in Na⁺ concentration in the blood by the release of *renin*, an enzyme that activates the plasma protein *angiotensin*. The active angiotensin II functions as a hormone to constrict arterioles and thus raise blood pressure and to stimulate the adrenal glands to release *aldosterone*. This hormone stimulates the reabsorption of Na⁺ in the distal convoluted tubules, which leads to the osmotic flow of water from the filtrate and an increase in blood volume and blood pressure. This mechanism prevents a loss of total volume of water in the body should the salt concentration and/or blood volume drop. ADH and aldosterone are both hormones that increase water reabsorption, but they respond to different osmoregulatory problems: ADH to an increase in blood osmolarity and aldosterone to a drop in blood pressure or deficiency of Na⁺.

Atrial natriuretic protein (ANP) counters the renin–angiotensin–aldosterone mechanism. ANP, released by the atrium of the heart in response to increased blood volume and pressure, inhibits the release of renin from the juxtaglomerular apparatus.

Comparative Physiology of the Kidney

Modifications of nephron structure and function in various classes of vertebrates relate to the requirements for osmoregulation and excretion of nitrogenous wastes in various habitats. Mammals who live in dry habitats and need to excrete hyperosmotic urine have long loops of Henle. Birds also have juxtamedullary nephrons that can produce osmotic gradients and hyperosmotic urine. The kidneys of reptiles have only cortical nephrons and excrete a urine isosmotic to body fluids. Water is conserved, however, by the production of an insoluble nitrogenous waste, uric acid.

Freshwater fishes are hyperosmotic to their environment and must excrete large quantities of very dilute urine. Salts are conserved by the efficient reabsorption of ions from the filtrate. The kidneys of marine bony fishes excrete very little urine and function mainly to get rid of divalent cations taken in by the drinking of sea water. Most nitrogenous wastes and monovalent ions are excreted by the epithelia of the gills.

Nitrogenous Wastes

Ammonia, a small and toxic molecule, is produced when proteins and nucleic acids are broken down for energy or conversion to carbohydrates or fats. Some animals excrete ammonia; others convert it to less toxic wastes such as *urea* or *uric acid*.

Ammonia Aquatic animals can excrete nitrogenous wastes as ammonia because it is very soluble and easily permeates membranes. Soft-bodied invertebrates lose ammonia across the whole body surface. In freshwater fishes, most of the ammonia is passed as NH_4^+ ions across the epithelia of the gills in exchange for Na^+ taken up from the water.

Urea Mammals and most adult amphibians produce urea, a much less toxic compound that can be tolerated in more concentrated form and excreted with less loss of water. Ammonia and carbon dioxide are combined in the liver to produce urea, which is then carried by the circulatory system to the kidneys.

Uric Acid Land snails, insects, birds, and some reptiles produce uric acid, a compound of low solubility in water that can be excreted as a precipitate after nearly all the water has been reabsorbed from the urine.

The mode of reproduction of terrestrial animals may have determined the evolution of their form of nitrogenous waste product. Vertebrates that produce shelled eggs excrete uric acid, which can be stored safely within the egg while the embryo develops. Mammals and amphibians produce urea, which can be removed by the placental blood or diffuse out of a shell-less egg.

The form of nitrogenous wastes that is excreted also relates to habitat. Terrestrial reptiles excrete uric acid, whereas crocodiles mainly excrete ammonia, and aquatic turtles excrete both urea and ammonia. Some animals actually shift their nitrogenous waste product depending on enrivonmental conditions.

Regulation of Body Temperature

Metabolism is very sensitive to internal changes in temperature. Many animals can maintain their internal temperature within an optimal range even as the environmental temperature fluctuates.

Heat Production and Transfer between Organisms and their Environments Heat is exchanged between an organism and the environment by the physical processes of *conduction*, the direct transfer of thermal motion between surfaces in contact; *convection*, the mass flow of air or water past a body; *radiation*, the emission of electromagnetic waves by all objects warmer than absolute zero; and *evaporation*, the loss of heat due to the conversion of surface molecules of a liquid to a gas. Convection and radiation account for the bulk of heat loss under relatively mild conditions. Convection and evaporation are the most variable causes of heat loss; wind and the evaporation of sweat can greatly increase the loss of heat.

Ectotherms and Endotherms Thermoregulation can be classified on the basis of heat source. Animals such as invertebrates, fishes, reptiles, and amphibians absorb their body heat from their surroundings and are called *ectotherms*; whereas mammals and birds, who derive their body heat from their own metabolism, are called *endotherms*. Some ectotherms in very stable environments have a more constant temperature than endotherms; other ectotherms may have body temperatures higher than endotherms; and some endotherms may absorb heat from the environment.

Endothermy allows terrestrial animals to maintain a constant body temperature in an environment where severe temperature fluctuations may be common. A warm body temperature requires active metabolism, but high levels of aerobic metabolism facilitate the energetically expensive movement of animals on land.

Thermoregulation in Terrestrial Mammals Metabolism generates heat; fat and fur insulation helps retain that heat. Heat production can be increased by contraction of muscles (by moving or shivering) and by a rise in metabolic rate. Certain hormones can increase the metabolic rate in a process called *nonshivering thermogenesis*. Some mammals have *brown fat* that is specialized for this form of rapid heat production.

The physiological and behavioral processes that enable a land mammal to maintain a constant body temperature include: (1) adjustment of the rate of metabolic heat production; (2) adjustment of the amount of blood flowing to the skin—*vasodilation* increases the rate of heat exchange, whereas *vasoconstriction* decreases it; (3) control of evaporative heat loss through panting or sweating; and (4) changes in behavioral responses, such as moving to warmer or cooler areas of the environment.

In the hypothalamus of the brain are two thermoregulatory areas: a *heating center* that controls vasoconstriction of superficial vessels, shivering, and nonshivering thermogenesis; and a *cooling center* that controls vasodilation and sweating or panting. Temperature-sensing nerve cells are located in the skin, hypothalamus, and some other areas of the nervous system and stimulate one of the thermoregulatory areas while inhibiting the other. Control of body temperature involves feedback mechanisms, with the hypothalamus functioning as the thermostat.

Some Thermoregulatory Adaptations in Other Animals

Birds usually have very high body temperatures. A bird uses panting to promote evaporative heat loss. Feathers serve as insulation against heat loss. The close contact between arteries and veins going to and from the limbs creates a *countercurrent heat exchanger* that reduces heat loss. In some birds, blood can detour the exchanger should the animal not need to conserve body heat.

Loss of heat to water occurs 50 to 100 times more rapidly than loss to air. Marine mammals maintain their high body temperatures by efficient heat-conserving mechanisms, including a thick layer of insulating blubber and countercurrent heat exchange between arterial and venous blood. When in warmer waters, these animals can dissipate metabolic heat by dilating superficial blood vessels.

The metabolic rate of the ectothermic reptiles is very low and contributes little to body temperature. Behavioral adaptations, such as orientating the body to the sun or finding suitable microclimates, allow these animals to regulate their temperature within an optimal range. In diving reptiles, more blood is routed to the body core to conserve heat. Reptiles can also increase thermogenesis.

Amphibians produce little heat and easily lose heat by evaporation from their moist skin. Behavioral adaptations enable them to maintain their internal temperature within optimal ranges.

The body temperature of fishes is controlled mainly by the water temperature. Some large, active fish are able to retain the heat produced by their swimming muscles and thus maintain an elevated body-core temperature. A countercurrent heat exchanger, called the *rete mirabile*, keeps the deep swimming muscles warmer than the surface tissues. These animals are considered to be partial endotherms.

Some invertebrates may adjust their temperatures by behavioral or physiological mechanisms. Large flying insects may "warm up" before taking off by contracting their flight muscles. Winter moths have a countercurrent heat exchanger that elevates the temperature of their flight muscles. The social organization of honeybees enables them to maintain body heat in cold temperatures by huddling together in the hive. They also can cool their hive by bringing in water and fanning it with their wings to promote evaporation and convection.

Temperature Acclimation

Many animals are capable of *acclimation*, a physiological adjustment to new temperatures, such as occur in seasonal changes. The increased production of enzymes may offset their lowered activity at nonoptimal temperatures. Variants of enzymes, which have different optimal temperatures, may also be produced. Changes in the proportions of saturated and unsaturated lipids keep membranes fluid at different temperatures.

Torpor

Torpor is a physiological state characterized by decreases in metabolic, heart, and respiratory rates. In *hibernation*, the body temperature is maintained at a lower level. A hibernating animal is able to withstand long periods of cold temperatures and decreased food supplies. *Aestivation*, another type of torpor, allows animals to survive long stretches of elevated temperatures and diminished water supplies. *Diurnation* is a torpor that lasts for much shorter periods of time than hibernation or aestivation.

Hibernation and aestivation may be triggered by changes in the length of daylight. Many ectotherms enter a state of slowed metabolism and inactivity when their food supply decreases, and environmental changes linked to an increase in food supply cause them to come out of torpor. Some hummingbirds and shrews, which have very high metabolic rates due to their small size, enter a daily torpor during the periods in which they are not feeding. This daily cycle is controlled by the biological clock. Sleep may be a remnant of a more pronounced daily torpor in our early mammalian ancestors.

Interaction of Regulatory Systems

A major challenge of animal physiology is determining

	Osmoregulation	Excretory System	Thermoregulation
Marine invertebrate			
Marine bony fish			
Freshwater fish			
Flatworm			
Earthworm			
Insect			
Reptile			
Bird			
Mammal			

how the various regulatory systems are controlled and how they interact to maintain homeostasis in the internal environment. The feedback circuits of homeostasis involve nervous communication and hormones.

STRUCTURE YOUR KNOWLEDGE

1. Fill in the table above with a brief description of the osmoregulatory, excretory, and thermoregulatory mechanisms for the animals listed.

2. Look at the table you have completed and draw some generalizations relating the control mechanisms of these animals to their environments and their evolutionary history. In particular, think about the problems of osmoregulation presented by marine, freshwater, and terrestrial habitats, the

types of nitrogenous waste products suited to these environments, and the thermoregulatory considerations in an aquatic or terrestrial habitat. Now fill in the following table with your conclusions.

3. In the sketch on the following page of a nephron, label the parts on the indicated lines. Label the arrows to indicate the movement of salt, water, and urea out of the tubule. What enters the tubule through Bowman's capsule? List the substances that are filtered, secreted, and reabsorbed as the nephrons of the vertebrate kidney produce urine.

4. The osmolarity of urine can vary from 70 to 1200 mosm/l. Develop a concept map to help organize your understanding of the physiological control of

Environment	Osmoregulation	Nitrogenous Waste	Thermoregulation
Seawater			
Freshwater			
Terrestrial			

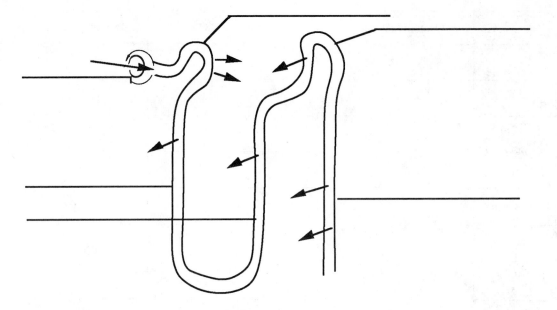

Substances Filtered	Substances Secreted	Substances Reabsorbed

water and salt reabsorption in the kidneys. Include the organs or structures involved, the hormones or other proteins released, and their effects.

TEST YOUR KNOWLEDGE

MULTIPLE CHOICE: *Choose the one best answer.*

1. There is a net water flow by osmosis through a membrane from
   a. a solution that is hyperosmotic to one that is hypoosmotic.
   b. a solution with a lower osmolarity to one with a higher osmolarity.
   c. one isosmotic solution to another.
   d. cells in a freshwater environment to the surrounding medium.

2. Contractile vacuoles most likely would be found in protists
   a. in a freshwater environment.
   b. in a marine environment.
   c. that are internal parasites.
   d. that are hypoosmotic to their environment.

3. Transport epithelia are responsible for
   a. pumping water across a membrane.
   b. forming an impermeable boundary at an interface with the environment.
   c. the exchange of solutes between the extracellular fluid and the environment, followed by the osmotic movement of water.
   d. pumping salt into a marine bony fish.

4. Which of the following is incorrectly paired with its excretory system?
   a. earthworm—protonephridia
   b. flatworm—flame-cell system
   c. insect—Malpighian tubules
   d. amphibian—kidneys

5. A freshwater fish would be expected to
   a. pump salt in through salt glands in the gills.
   b. produce copious quantities of dilute urine.
   c. diffuse ammonia out across the epithelium of the gills.
   d. do all of the above.

6. Which of the following is *not* part of the filtrate entering Bowman's capsule?
   a. water, salt, and electrolytes
   b. glucose
   c. plasma proteins
   d. urea

7. Which is the correct pathway for the passage of urine in vertebrates?
   a. collecting tubule → ureter → bladder →urethra
   b. renal vein →renal ureter →bladder →urethra
   c. nephron →urethra →bladder →ureter
   d. cortex →medulla →bladder →ureter

8. Aldosterone
   a. is a hormone that stimulates thirst.
   b. is secreted by the adrenal glands in response to a high osmolarity of the blood.
   c. stimulates the active reabsorption of Na⁺ in the nephrons.
   d. causes diuresis.

9. Which of the following statements is *incorrect*?
   a. Long loops of Henle are associated with steep osmotic gradients and the production of hyperosmotic urine.
   b. Ammonia is a toxic nitrogenous waste molecule that invertebrates must transport out of their bodies.
   c. The form of nitrogenous waste that requires the least amount of water to excrete is uric acid.
   d. In the mammalian kidney, urea diffuses out of the collecting duct and contributes to the osmotic gradient within the medulla.

10. The process of secretion in the formation of urine insures that
    a. a constant pH is maintained in body fluids.
    b. drugs and other poisons are removed from the blood.
    c. potassium balance is maintained in body fluids.
    d. all of the above

11. The peritubular capillaries
    a. form the ball of capillaries inside the glomerulus from which filtrate is forced by blood pressure into the renal tubule.
    b. intertwine with the proximal and distal convoluted tubules and function in secretion and reabsorption.
    c. form a countercurrent flow of blood through the medulla that supplies oxygen and other materials without interfering with the osmolarity gradient.
    d. surround the collecting ducts and reabsorb water, helping to create a hyperosmotic urine.

12. Consumption of alcohol may result in diuresis because
    a. the JGA (juxtaglomerular apparatus) releases renin, which activates angiotensin, resulting in an increase in blood pressure.
    b. alcohol inhibits the release of ADH, causing excessive water loss in the urine.
    c. the atrium of the heart is stimulated to release ANP (atrial natriuretic protein), which inhibits the release of renin and aldosterone and thus lowers blood volume and pressure.
    d. the kidneys must work harder to remove alcohol from the system.

13. The diffusion of urea from the collecting tubules
    a. helps to maintain the osmotic gradient in the medulla.
    b. indirectly contributes to the production of a hyperosmotic urine.
    c. must be offset by its active secretion into the distal convoluted tubules.
    d. Both a and b are correct.

14. Which of the following is used by terrestrial animals as a mechanism to dissipate heat?
    a. hibernation
    b. countercurrent exchange
    c. evaporation
    d. vasoconstriction

15. The heating center of the hypothalamus
    a. responds to messages from temperature-sensing nerve cells in the skin.
    b. controls shivering and vasoconstriction of superficial vessels.
    c. functions along with a cooling center to thermoregulate by feedback mechanisms.
    d. does all of the above.

16. A countercurrent heat exchange between arterial and venous blood
    a. warms the blood going to the extremities.
    b. warms the blood returning to the body core.
    c. maintains a constant body temperature.
    d. is used by marine mammals when they journey to warmer waters.

17. Physiological adjustments to cooler seasonal temperatures may include
    a. production of enzymes with lower optimal temperatures.
    b. changes in the lipid components of cell membranes.
    c. increased production of enzymes to offset their reduced efficiency.
    d. all of the above.

18. Which of the following sections of the mammalian nephron is *incorrectly* paired with its function?

a. Bowman's capsule and glomerulus—filtration of blood

b. proximal convoluted tubule—secretion of ammonia and H^+ into filtrate and transport of glucose and amino acids out of tubule

c. descending limb of loop of Henle—diffusion of urea out of filtrate

d. ascending limb of loop of Henle—diffusion and pumping of Na^+ and Cl^- out of filtrate

CHAPTER 41

CHEMICAL COORDINATION

FRAMEWORK

This chapter introduces the intricate system of chemical control and communication within animals. The endocrine system produces hormones, often in response to nervous messages, which regulate homeostasis, growth, and development. Steroid hormones bind with receptors within the nuclei of their target cells and influence gene expression. Peptide hormones bind with cell surface receptors, and second messengers mediate their effects in target cells.

The hypothalamus and pituitary gland play coordinating roles by integrating the nervous and endocrine systems and producing many tropic hormones that control the synthesis and secretion of hormones in other endocrine glands and organs.

CHAPTER SUMMARY

Coordination and communication among the specialized parts of complex animals are achieved by the *nervous system* and the *endocrine system*. The nervous system conveys high-speed messages along neurons, whereas the endocrine system produces chemical messengers called hormones, which travel more slowly and influence development, metabolism, behavior, and homeostasis.

Chemical Messengers of the Body

Hormones are chemical messages that travel between organs of the body; pheromones communicate between different individuals; and some chemical messengers, such as neurotransmitters, carry information between cells.

Hormones   An animal *hormone* can be defined as a molecule that is synthesized by a group of specialized cells, secreted into the circulatory system, and elicits a response from *target cells* in another area of the body. Target cells have molecular *receptors* that determine the specificity and action of hormones. *Endocrine glands* are ductless secretory organs that produce hormones and release them into the circulatory system. The endocrine system interacts and cooperates closely with the nervous system.

Endocrinology, the study of hormones, is a rapidly growing area of research. Hormones can be grouped into three general classes based on chemical structure: fat-soluble *steroid hormones*, derived from cholesterol; smaller, water-soluble hormones derived from amino acids; and *peptide* hormones, which are chains of amino acids of varying lengths.

Pheromones   *Pheromones* are chemical signals that communicate between animals. Serving as mate attractants, territory markers, or alarm substances, these usually small and volatile molecules are active in minute amounts.

Local regulators   *Local regulators* are chemical signals that affect target cells near their points of secretion. Neurotransmitters, histamine, and interleukins, as well as growth factors and prostaglandins, are examples of local regulators.

Growth Factors   Experiments with culturing mammalian cells on artificial media led to the discovery of *growth factors*, extracellular proteins necessary for certain cells to grow and develop. Growth factors bind to surface receptors on target cells, but it is not known how this binding triggers the cell's response. Some oncogenes code for growth-factor-like receptors, but these receptors stimulate growth and division without requiring the binding of a growth factor.

Prostaglandins Prostaglandins (PGs) are modified fatty acids that are released from most cells and have a wide range of effects on nearby cells. Small differences in the molecular structures of some of these identified protaglandins can result in profound differences in their effects. Some prostaglandins have opposite effects, and the balance of these antagonistic signals are an important regulatory mechanism. In mammals, PGs stimulate uterine contractions and help to induce labor. Aspirin inhibits the secretion of prostaglandins and thus reduces their fever-inducing and pain-intensifying actions.

Mechanisms of Hormone Action

Hormones act at very low concentrations, can have varying effects on different target cells, and can differ greatly in effect in different species. Two general mechanisms of hormone action have been identified, both involving receptor molecules to which the hormones bind: steroid hormones enter the nucleus and influence gene expression, and most nonsteroid hormones attach to the cell surface and stimulate release of intermediaries called second messengers.

Steroid Hormones and Gene Expression Steroids are thought to diffuse through the target cell membrane and into the nucleus, where they may bind to receptor proteins. The steroid-receptor complex then attaches to a specific acceptor protein located on the chromatin, initiating transcription of mRNA, which is used for synthesis of new proteins.

Two types of cells can respond differently to the same hormone; for example, estrogen induces cells to synthesize ovalbumin in a bird's reproductive system but induces liver cells in the same animal to synthsize different proteins. The specific placement of the chromosomal acceptor proteins may account for the differing cellular action of the same hormone.

Peptide Hormones and Second Messengers Peptide hormones and most hormones derived from amino acids do not enter their target cells but bind outside the cell to specific receptor proteins embedded in the plasma membrane. This binding results in the production by the cytoplasmic surface of the plasma membrane of a *second messenger*, which triggers a cascade of biochemical reactions in the cell.

Through his work with epinephrine and glycogen hydrolysis in liver and muscle cells, Sutherland developed the second messenger model of hormone action, for which he received a Nobel prize in 1971. He found that the binding of epinephrine to a liver cell's plasma membrane signaled the membrane protein, *adenylate*

cyclase, (an enzyme with its active site facing the cytoplasm) to convert ATP to cyclic adenosine monophosphate (also known as *cyclic AMP* or *cAMP*). Cyclic AMP activates glycogen hydrolysis in the cell. The hormone acts as the first messenger when it binds to a receptor molecule in the membrane. Cyclic AMP, as the second messenger, relays the message to cytoplasmic enzymes. Enzymatic degradation of cAMP terminates the response to the hormone. Many peptide hormones and hormones derived from amino acids use cAMP as their second messenger.

A third membrane protein is an intermediary between the hormone receptor and adenylate cyclase. Binding of a hormone to its receptor activates *G protein*, which hydrolyzes GTP to GDP, providing the energy to activate adenylate cyclase. The same G protein may stimulate cAMP production in response to several different hormone signals. Another G protein inhibits adenylate cyclase in response to the binding of an inhibitory hormone. The balance of these antagonistic hormones allows the cell to precisely regulate its metabolism.

The effect of a hormone is amplified when the second messenger sets off an *enzyme cascade*, in which a sequence of enzyme molecules is activated. First, cAMP activates *cAMP-dependent protein kinase*, an enzyme that catalyzes the phosphorylation of a protein, which, depending on the protein, either increases or decreases its activity. In the liver cell's response to epinephrine, cAMP-dependent protein kinase activates phosphorylase kinase, an enzyme that then adds a phosphate group to glycogen phosphorylase—the enzyme that hydrolyzes glycogen. (Or should we say that cAMP-dependent protein kinase phosphorylates phosphorylase kinase, which phosphorylates glycogen phosphorylase!) The end result of this tongue-twister is that each step in the cascade greatly increases the number of activated products. A few molecules of epinephrine can result in the release of millions of sugar molecules from glycogen.

This single basic mechanism is used to mediate responses of many different cell types to different hormones. The specificity of hormone action is maintained because different hormone receptors are found on specific target cells, cAMP-dependent protein kinases may vary from tissue to tissue, and different cell types have different complements of proteins that can be regulated by cAMP-dependent protein kinases.

Inositol triphosphate (IP$_3$) is another second messenger that relays and amplifies a hormonal signal. The binding of a hormone to its receptor activates a G protein that stimulates phospholipase C, a membrane enzyme that cleaves a plasma membrane phospholipid into IP$_3$ and diacylglycerol. Diacylglycerol activates protein kinase C, which phosphorylates specific proteins. IP$_3$ causes the release of Ca^{++} from the endoplasmic reticulum, and the calcium either directly alters the activities of certain enzymes or binds to a protein called *calmodulin*, which then binds to and changes the activities of other enzymes and proteins.

Steroid hormones, which enter the nucleus of the cell and alter gene expression, result in the synthesis of proteins, causing a slower but longer-lasting hormone response. Peptide hormones, acting through second messengers, alter the activity of already present proteins and enzymes and produce a faster hormone response.

Invertebrate Hormones

Invertebrate hormones regulate homeostasis, development, and reproduction.

The hormones of insects, which have been extensively studied, control molting and development of adult characteristics. The steroid hormone *ecdysone*, secreted from a pair of prothoracic glands, stimulates the transcription of specific genes and functions to trigger molts and to promote development of adult characteristics. *Brain hormone* stimulates the prothoracic glands to secrete ecdysone. *Juvenile hormone (JH)*, secreted by a pair of small glands, the corpora allata, counters the action of ecdysone and promotes retention of larval characteristics.

The Vertebrate Endocrine System

The Hypothalamus and the Pituitary Gland The *hypothalamus*, situated in the lower brain, plays a key role in integrating the endocrine and nervous systems. *Neurosecretory cells* of the hypothalamus receive nerve signals and release hormones. One set of neurosecretory cells produces the hormones of the posterior pituitary; another produces *releasing factors* that regulate the anterior pituitary.

The *pituitary gland*, a small appendage at the base of the hypothalamus, consists of two lobes. A *posterior lobe*, or *neurohypophysis*, stores and secretes two hormones produced by the hypothalamus; the *anterior lobe*, or *adenohypophysis*, produces its own hormones, several of which are tropic hormones, whose targets are other endocrine glands.

Oxytocin and *antidiuretic hormone (ADH)* are the small peptide hormones synthesized by neurosecretory cells of the hypothalamus that move within extensions of these cells to the posterior lobe. Oxytocin induces uterine contractions during birth and milk ejection during nursing. ADH functions in osmoregulation. Osmoreceptor cells in the hypothalamus respond to an increase in blood osmolarity by sending impulses to neurosecretory cells of the hypothalamus, which release ADH from their tips located in the posterior pituitary. ADH binds to receptors on cells lining the collecting ducts in the kidney and sets off a cAMP second-messenger system that results in an increased permeability of the tubule epithelia to water. The reabsorption of water from the urine lowers the osmolarity of the blood and causes the hypothalamus to slow the release of ADH, completing a negative feedback loop.

The anterior pituitary produces a variety of protein and peptide hormones. *Growth hormone (GH)* affects a variety of target tissues, promoting growth directly and stimulating other growth factors, such as *somatomedins*, which are produced by the liver and stimulate bone and cartilage growth. Several human growth disorders are related to abnormal GH production, including gigantism and acromegaly, both related to excessive GH, and hypopituitary dwarfism, caused by GH deficiency. GH deficiency can be treated with GH produced by bacteria genetically engineered to contain the human gene for this hormone.

The hormone *prolactin (PRL)*, a protein hormone very similar to GH, exhibits effects in different vertebrate species that range from control of mammary gland growth and milk synthesis in mammals to delay of metamorphosis in amphibians to regulation of salt and water balance in fish.

Three of the tropic hormones produced by the anterior pituitary are glycoproteins. *Thyroid stimulating hormone (TSH)* regulates the release of thyroid hormones. *Follicle stimulating hormone (FSH)* and *luteinizing hormone (LH)*, also called *gonadotropins*, stimulate the activities of the gonads.

Several hormones come from *pro-opiomelanocortin*, a single large protein that is cleaved into several short fragments inside the pituitary cells. *Adrenocorticotropin (ACTH)* is a tropic hormone that stimulates the adrenal cortex to produce and secrete its steroid hormones. *Melanocyte-stimulating hormone (MSH)* regulates the activity of pigment-containing cells in the skin, as seen in the color changes of amphibians. *Endorphins* and *enkephalins*, two other classes of pro-opiomelanocortin derivatives, are also produced by certain neurons in the brain. These recently discovered hormones have effects similar to those of morphine and appear to inhibit pain reception.

Releasing factors, which stimulate or inhibit the secretion of hormones by the anterior lobe of the pituitary, are produced at the base of the hypothalamus and released into capillaries in an area called the *median eminence*. Veins draining the median eminence lead to capillary beds in the anterior pituitary, creating a portal system that delivers releasing factors directly to the pituitary.

The Thyroid Gland   The thyroid gland produces two hormones, triiodothyronine (T_3) and thyroxine (T_4), which contain respectively three and four iodine atoms. Thyroid hormones, critical to the development and maturation of vertebrates, are required for normal functioning of bone-forming cells and for branching of nerve cells during embryonic development. An inherited deficiency of thyroid hormone in humans results in retarded skeletal growth and mental development, a condition known as cretinism.

The thyroid gland also helps regulate metabolism. Excess thyroid hormone results in a condition known as hyperthyroidism, the symptoms of which include weight loss, irritability, and high body temperature and blood pressure. Hypothyroidism can cause cretinism or result in weight gain and lethargy in adults. A goiter is an enlarged thyroid gland, associated with a lack of iodine in the diet.

A negative feedback loop controls secretion of thyroid hormones. Secretion of TSH-releasing hormone, or TRH, by the hypothalamus stimulates the anterior pituitary to secrete TSH, which binds to receptors on the thyroid gland, generating cAMP and triggering the synthesis and release of thyroid hormones. High thyroid hormone level inhibits the secretion of TSH.

The thyroid gland also secretes *calcitonin*, a polypeptide hormone that lowers calcium levels in the blood.

The Parathyroid Glands   The four parathyroid glands, located on the thyroid, secrete *parathyroid hormone (PTH)*, which stimulates Ca^{++} reabsorption in the kidney, absorption of Ca^{++} in the intestine, and release of Ca^{++} from bone to the blood stream, all of which raise blood calcium levels. A balance between the antagonistic PTH and calcitonin maintains calcium homeostasis. Vitamin D is needed for proper PTH function.

The Pancreas   *Insulin* is produced in response to increased blood sugar levels by *ß cells* within clusters of pancreatic cells called *islet cells*. The *a cells* of the islets secrete the hormone *glucagon*. These antagonistic hormones regulate carbohydrate metabolism.

Insulin lowers blood sugar levels by promoting the movement of glucose into fat, muscle, and most other cells; the formation and storage of glycogen in the liver; the synthesis of protein, and the storage of fat. Glucagon raises glucose concentrations by stimulating the breakdown of fats and proteins and the conversion of glycogen to glucose in the liver. The secretion of both these hormones is controlled by the blood sugar level.

Insulin is the only hormone that reduces blood sugar levels. In diabetes, the absence of insulin in the bloodstream restricts glucose uptake by cells, and the body must use fats and body proteins for fuel. Glucose concentration in the blood rises, and its excretion in the urine can result in dehydration as water is excreted with the concentrated urine. Diabetes can be controlled by diet or regular injections of insulin. Insulin, previously obtained from animal pancreases, is now produced from human genes inserted in genetically engineered bacteria.

The Adrenal Glands   In mammals, the *adrenal glands*, located on top of the kidneys, consist of two different glands: the outer *cortex* and the central *medulla*. The adrenal medulla synthesizes *epinephrine* (adrenalin) and *norepinephrine* (noradrenalin)—both compounds known as *catecholamines* that are synthesized from the amino acid tyrosine. Epinephrine, released in response to positive or negative stress, mobilizes glucose from skeletal muscle and the liver and stimulates the release of fatty acids from fat cells. Epinephrine and norepinephrine increase the rate and volume of the heart beat and influence the contraction of smooth muscles to increase the blood supply to the heart, brain, and skeletal muscles, while reducing the supply to the skin, gut, and kidneys. The sympathetic division of the autonomic nervous system stimulates the release of epinephrine. Epinephrine and norepinephrine also function as neurotransmitters in the nervous system.

The adrenal cortex responds to endocrine signals of stress. The hypothalamus secretes a releasing factor that causes the anterior pituitary to release ACTH. This tropic hormone stimulates the adrenal cortex to synthesize and secrete *corticosteroids*, a group of steroid hormones.

The two main human corticosteroids are the *glucocorticoids*, such as cortisol, and the *mineralocorticoids*, such as aldosterone. Glucocorticoids promote synthesis of glucose from noncarbohydrates, such as protein, and thus increase energy supplies during stress. High doses of glucocorticoids suppress parts of the body's immune system, including the inflammatory reaction. Cortisone has been used to treat serious inflammatory conditions such as arthritis, but its immunosuppressive effects can be dangerous.

The mineralocorticoids affect salt and water balance. Aldosterone stimulates kidney cells to reasorb sodium ions from the filtrate. Aldosterone secretion is regulated by hormones produced in the liver and kidneys in response to plasma ion concentrations.

The Gonads   The testes of males and ovaries of females produce steroids that affect growth, development, and reproductive cycles and behaviors. The three major categories of gonadal steroids—androgens, estrogens, and progestins—are found in different proportions in males and females.

The testes primarily synthesize *androgens*, such as

testosterone, which stimulate development of the male reproductive system and the secondary sex characteristics. *Estrogens*, such as *estradiol*, are responsible for the development and maintenance of the female reproductive system and secondary sex characteristics. In mammals, *progestins* help prepare and maintain the uterus for the growth of an embryo.

The gonadotropins from the anterior pituitary gland (follicle-stimulating hormone and luteinizing hormone) control the synthesis of both estrogens and androgens. A hypothalamic regulatory hormone, GnRh (gonadotropic releasing hormone) controls secretion of FSH and LH.

Other Endocrine Organs Many other organs whose main functions are nonendocrine, such as the digestive tract, kidney, and heart, secrete important hormones.

The *pineal gland*, located near the center of the mammalian brain, secretes a modified amino-acid hormone, *melatonin*, which regulates functions related to light and changes in daylength. Melatonin is secreted at night, and its production may function as a biological clock for daily or seasonal activities such as reproduction.

The *thymus* plays a role in the immune system in young animals. It secretes *thymosin* and other messengers that stimulate development and differentiation of T lymphocytes.

Endocrine Glands and the Nervous System

The endocrine system and nervous system function together to control chemical communication and coordination in animals. They are *structurally* related in that many endocrine glands, such as the hypothalamus, posterior pituitary, and the insect brain, are made of nervous tissue. Other endocrine glands, such as the adrenal medulla, have evolved from the nervous system.

The endocrine and nervous systems are *chemically* related in that several vertebrate hormones, such as epinephrine, are used by both systems. And the two systems are *functionally* related. Many physiological processes are coordinated by both nervous and hormonal communications, often in a type of neuroendocrine reflex. The nervous system controls endocrine glands, and hormones affect the development and functioning of the nervous system.

STRUCTURE YOUR KNOWLEDGE

1. Create a concept map that illustrates your understanding of the mechanisms of steroid and peptide hormones.

2. The hypothalamus and pituitary gland produce a number of hormones, tropic hormones, and releasing factors. Fill in the table on the following page with the major products of each of the regions of these glands.

3. Develop a concept map or diagram of the hormones and mechanisms involved in the body's control of blood sugar levels.

TEST YOUR KNOWLEDGE

MATCHING: *Match the hormone and gland or organ that produces it to the descriptions of hormone action. Not all choices are used.*

Hormones	**Gland or Organ**
A. adrenocorticotropin (ACTH)	a. adrenal cortex
B. androgens	b. adrenal medulla
C. antidiuretic hormone (ADH)	c. hypothalamus
D. calcitonin	d. pancreas
E. epinephrine	e. parathyroid
F. glucagon	f. pineal
G. glucocorticoids	g. pituitary
H. insulin	h. testis
I. melatonin	i. thymus
J. oxytocin	j. thyroid
K. thyroxine	

Hormone	Gland	Hormone action
1. _____	_____	involved in biological clock and seasonal activities
2. _____	_____	involved in synthesis of glucose, suppress inflammatory reaction
3. _____	_____	involved in glycogen breakdown in liver, increase blood sugar

GLAND	HORMONES	MAIN ACTION
HYPOTHALAMUS Neurosecretory cells (drain to posterior pituitary)		
(drain to median eminence)		
PITUITARY Posterior lobe (neurohypophysis)		
Anterior lobe (adenohypophysis)		

4. ____ ____ stimulate development of male reproductive system

5. ____ ____ stimulate adrenal cortex to synthesize corticosteroids

6. ____ ____ increase available energy supplies, increase heart rate and blood supply to skeletal muscles, heart and brain

7. ____ ____ regulate metabolism, development of bone-forming and nerve cells

8. ____ ____ lower blood calcium levels by inhibiting calcium release from bone

9. ____ ____ influence reabsorption of water from collecting ducts of kidney

10. ____ ____ stimulate contraction of uterine muscles and mammary gland cells

MULTIPLE CHOICE: *Choose the one best answer.*

1. Which of the following is *not* an accurate statement about hormones?
   a. All hormones are secreted by endocrine glands.
   b. Hormones move through the circulatory system to their destination.
   c. Target cells have specific molecular receptors for hormones.
   d. Hormones are essential to homeostasis.

2. The difference between pheromones and hormones is that
   a. pheromones are small, volatile molecules, whereas hormones are steroids.
   b. pheromones are involved in reproduction, whereas hormones are not.
   c. pheromones are a form of olfactory communication; hormones are a form of chemical communication.
   d. pheromones are signals that function between animals, whereas hormones communicate among the parts within an animal.

3. Which one of the following examples is *incorrectly* paired with its class?
   a. local regulators—neurotransmitters
   b. steroid hormone—estrogen
   c. prostaglandin—modified fatty acid
   d. peptide hormone—calmodulin

4. In the second-messenger model of hormone action,
   a. a tropic hormone signals an endocrine gland to release a hormone.
   b. another molecule relays the message from the membrane-bound hormone to the cytoplasmic enzymes.
   c. negative feedback turns off the production of the hormone.
   d. genes are turned on and new proteins are synthesized.

5. Ecdysone
   a. is a steroid hormone found in insects that promotes retention of larval characteristics.
   b. is responsible for the color changes in amphibians.
   c. is produced by the corpora allata.
   d. is secreted by prothoracic glands in insects and triggers molts and development of adult characteristics.

6. The role of cAMP in hormone action is to
   a. act as the second messenger and activate other enzymes.
   b. activate adenylate cyclase.
   c. bind a specific hormone to the plasma membrane.
   d. bind to the protein calmodulin and initiate an enzyme cascade.

7. Neurosecretory cells
   a. are found in the hypothalamus.
   b. receive signals from nerve cells and release hormones.
   c. produce releasing factors that drain into the median eminence.
   d. All of the above are correct.

8. Antidiuretic hormone (ADH)
   a. is produced by cells in the kidney and liver.
   b. stimulates the reabsorption of Na^+ from the urine.
   c. acts through a second messenger system to increase the permeability of kidney collecting tubules to water.
   d. is a steroid hormone produced by the adrenal cortex.

9. The adenohypophysis or anterior lobe of the pituitary
   a. stores oxytocin and ADH produced by the hypothalamus.
   b. is connected by a portal system to the median eminence.
   c. produces several releasing factors.
   d. does all of the above.

10. Acromegaly is caused by
    a. an excess of thyroxine.
    b. an excess of growth hormone.
    c. an excess of glucocorticoids.
    d. an abnormally high androgen-to-estrogen ratio.

11. Which of the following hormone sequences results in the secretion of estrogens in mammals?
    a. GNRH → FSH → estrogen
    b. LH → FSH → estrogen
    c. CRH → ACTH → FSH → estrogen
    d. gonadotropin → LH → estrogen

12. Pro-opiomelanocortin
    a. is produced by the hypothalamus.
    b. is cleaved into several hormones, including ACTH, melanocyte-stimulating hormone, endorphins, and enkephalins.
    c. produces several releasing factors.
    d. All of the above are correct.

13. Hyperthyroidism
    a. is responsible for cretinism.
    b. results in the formation of a goiter.
    c. produces weight loss and a high metabolic rate.
    d. is associated with a deficiency of iodine.

14. Calcium homeostasis is maintained by a balance between
    a. calcitonin, secreted by the thyroid, and PTH, secreted by the parathyroid, which respectively lower and raise blood calcium levels.
    b. PTH and vitamin D.
    c. calmodulin, calcitonin, PTH, and vitamin D.
    d. aldosterone, which stimulates the kidney to resorb calcium, and thymosin, which lowers blood calcium levels.

15. Which of the following is *not* true of norepinephrine?
    a. It is secreted by the adrenal medulla.
    b. Its action is to increase the rate and volume of heart beat and constrict selected blood vessels.
    c. Its release is stimulated by ACTH, which is in turn released in response to the release of CRH.
    d. It serves as a neurotransmitter in the nervous system.

CHAPTER 42

ANIMAL REPRODUCTION

FRAMEWORK

This chapter covers the patterns and mechanisms of animal reproduction. In sexual reproduction, fertilization may occur externally or internally, both of which may require behavioral and physical adaptations. Development of the zygote may occur internally within the female, externally in a moist environment, or protected in a resistant egg. Placental mammals provide nourishment as well as shelter for the embryo and nurse their young.

In the discussion of mammalian reproduction, the human reproductive system is described, including its organs, glands, hormones, gamete formation, and sexual physiology. The chapter also covers pregnancy and birth, contraception, and recent reproductive technologies.

CHAPTER SUMMARY

Modes of Reproduction

In *asexual reproduction*, a single individual produces offspring that are genetically identical to itself. In *sexual reproduction*, two individuals contribute genes to the offspring.

Asexual Reproduction   Many invertebrates can reproduce asexually by *budding*, in which a new individual grows out from the parent's body, or by *fragmentation*, in which the body is broken into several pieces, each of which develops into a complete animal. Specialized groups of cells may be released that grow into new individuals, as in the *gemmules* produced by sponges. Regeneration is a means of asexual reproduction when a piece of an animal develops into a new organism.

The advantages of asexual reproduction are that it may be more rapid, easier for sessile or widely separated animals, and most effective for perpetuating successful genotypes.

Sexual Reproduction   In sexual reproduction, the two haploid *gametes*, which are usually a relatively large, nonmotile female *ovum* and a small, flagellated male *spermatozoon*, fuse to form a diploid *zygote*.

Sexual reproduction produces offspring with varying genotypes and phenotypes, which may enhance reproductive success of parents in fluctuating environments. Debate continues over the evolution and relative advantages of sexual reproduction.

Reproductive Cycles and Patterns   Most animals show periodic cycles of reproductive activity. A combination of seasonal and hormonal cues control the timing of these cycles, which may be linked to favorable environmental conditions or energy supplies.

The freshwater crustacean *Daphnia*, like aphids and rotifers, produces eggs that develop by *parthenogenesis*, without being fertilized, as well as eggs that are fertilized. In bees, wasps, and ants, sterile females are produced parthogenetically, whereas males and reproductive females are produced sexually. In a few parthenogenic fishes, amphibians, and lizards, doubling of chromosomes creates diploid "zygotes." In some species of whiptail lizards in which there are no males, the females of the "mating" pair take turns impersonating a male in courtship and mating behavior.

In *hermaphroditism*, found in sessile barnacles, burrowing earthworms, and parasitic tapeworms, each individual has functioning male and female reproductive systems. Mating results in fertilization of both individuals. Some species are capable of self-fertilization. In some fishes and oysters, individuals reverse their sex during their lifetimes, a pattern known as *sequential hermaphroditism*. Sex reversal may be related to age or to the relative advantage conferred by size. In a *protandrous* species, an organism is first a male; in a *protogynous* species, it is first a female.

Mechanisms of Sexual Reproduction

Patterns of Fertilization and Development In *external fertilization*, eggs and sperm are shed from the body, and fertilization occurs in the environment. This type of reproduction occurs almost exclusively in moist habitats, where the gametes and developing zygote are not in danger of desiccation. Environmental or pheromone signals may ensure that gamete release is synchronized. Fishes and amphibians that use external fertilization show mating behaviors, which triggers the release of gametes and provides a means for mate selection.

Internal fertilization involves the placement of sperm in or near the female reproductive tract, and egg and sperm unite internally. Behavioral cooperation, as well as copulatory organs and sperm receptacles, is required for internal fertilization.

Types of protection of the developing embryo and young include resistant eggs, development inside the female reproductive tract, and parental care. The amniote eggs of birds and reptiles protect the embryo in a terrestrial environment. The embryos of placental mammals develop within the uterus, nourished from the mother's blood supply through the placenta.

Diversity in Reproductive Systems Some invertebrates, such as polychaete annelids, do not have distinct *gonads*; eggs and sperm develop from cells lining the coelom. Gametes are stored in the coelom and shed through excretory organs or released by the splitting of the parent.

Insects have separate sexes and complex reproductive systems. Sperm develop in the testes, are stored in the seminal vesicles, and exit through an ejaculatory duct in the penis. Accessory glands may add fluid to the semen. Eggs, produced in the ovaries, pass through the oviducts to the vagina. A *spermatheca*, or sperm-storing sac, may be present.

Flatworms are hermaphrodites with complex reproductive systems. In addition to ovaries, oviducts, and vagina, the female reproductive system includes yolk and shell glands and a uterus, where the eggs are fertilized and may begin development. The male system includes a complex copulatory apparatus that may be inserted in the vagina or may inject sperm into the body through hypodermic impregnation.

With the exception of most mammals, vertebrates have a common opening, called the *cloaca*, for the digestive, excretory, and reproductive systems. The uterus of most vertebrates is bicornate—having two separate branches. Nonmammalian vertebrates do not have penes and may evert the cloaca to ejaculate.

Mammalian Reproduction

Human anatomy is used as the example of a mammalian reproductive system.

Anatomy of Reproduction The external male *genitalia* include the *scrotum*, a fold of skin enclosing the testes, and the *penis*. Sperm are produced in the highly coiled *seminiferous tubules* of the *testes*. *Interstitial cells* produce testosterone and other androgens. In most mammals, a cooler temperature than that of the internal body is needed for sperm production, so the scrotum suspends the testes below the abdominal cavity.

Sperm pass from the testes into the coiled *epididymis*, in which they mature and are stored. During *ejaculation*, sperm are propelled through the *vas deferens*, into a short *ejaculatory duct*, and out through the *urethra*.

Three glands add secretions to the *semen*. The *seminal vesicles* contribute a fluid containing mucus, amino acids, fructose (an energy source for the sperm), and prostaglandins, which stimulate uterine contractions. The *prostate gland* produces an alkaline secretion that balances the acidity of the urethra and the vagina. The *bulbourethral glands* produce a small amount of viscous fluid of uncertain function.

The penis is composed of spongy tissue that engorges with blood during sexual arousal, producing an erection that facilitates insertion of the penis into the vagina. Some mammals have a *baculum*, a bone that stiffens the penis. The head of the penis, called the *glans penis*, is covered by a fold of skin called the *prepuce*, or foreskin.

The female gonads, the *ovaries*, consist of many *follicles*, which are sacs of cells that nourish and protect the egg cell contained within each of them. The follicle produces the primary female sex hormones, the estrogens. Following *ovulation*, the follicle forms a solid mass called the *corpus luteum*, which secretes progesterone and a small amount of estrogen.

The egg cell is expelled into the abdominal cavity and swept by cilia into the *oviduct*, or fallopian tube, through which it is transported to the *uterus*. The *endometrium*, or lining of the uterus, is highly vascularized. The narrow neck of the uterus, the cervix, opens into the *vagina*, the thin-walled birth canal and repository for sperm during copulation.

The separate openings of the vagina and urethra are enclosed by two pairs of skin folds, which form a *vestibule*. The *glans clitoris* is at the top of the vestibule. Both the labia minora and the clitoris are composed of erectile tissue. The vaginal orifice is initially covered by a thin membrane called the *hymen*.

The *mammary gland*, or breast, is composed of deposits of fatty tissue and a series of *alveoli*, small milk-secreting sacs that drain into ducts that join together and open at the nipple.

The external genitalia of both sexes arise from common *primordia*, or undifferentiated embryonic tissue, whose development into male or female structures is dependent on the presence or absence of androgens.

Hormonal Control of Mammalian Reproduction An-drogens, the most important of which is *testosterone*, are responsible for the primary (associated with reproduction) and secondary (associated with "male" traits of voice, hair growth, muscle growth) sex characteristics. Androgens are also determinants of sexual and other behaviors. Androgen secretion and sperm production are controlled by hormones from the anterior pituitary and hypothalamus.

Female humans and many primates have *menstrual cycles*, during which the endometrium lining thickens to prepare for the implantation of the embryo and then is shed if fertilization does not occur. This bleeding, called *menstruation*, occurs on a cycle of approximately 28 days in humans. Other mammals have *estrous cycles*, during which the endometrium thickens, but if fertilization does not occur, it is reabsorbed and there is no bleeding. Estrous cycles are often coordinated with season or climate, and females are receptive to sexual activity only during *estrus*, or heat, the period surrounding ovulation.

The human menstrual cycle refers to changes in the uterus. The cycle consists of the *flow phase*, the few days of menstrual bleeding; the *proliferative phase*, the week or two during which the endometrium begins to thicken; and the *secretory phase*—about two weeks during which the endometrium becomes more vascularized and secretes a glycogen-rich fluid.

The *ovarian cycle* begins with the *follicular phase*, during which an egg cell enlarges, its follicle cells become multilayered and enclose a fluid-filled cavity, and the large, mature follicle forms a bulge near the ovary surface. The *luteal phase* begins when ovulation occurs with the rupture of the follicle and adjacent ovary wall, and the remaining follicular tissue develops into the hormone-secreting corpus luteum.

Five hormones coordinate the menstrual and ovarian cycles in an elaborate scheme involving positive and negative feedback. During the follicular phase, the hypothalamus secretes GnRH (gonadotropin-releasing hormone), which stimulates the anterior pituitary to secrete FSH (follicle-stimulating hormone) and LF (luteinizing hormone). FSH stimulates follicular growth, and the cells of the follicle secrete estrogen. When the level of estrogen secretion rises sharply, the hypothalamus is stimulated to increase GnRH output, which results in a rise in LH and FSH release.

By positive feedback, the increase in LH, caused by increased estrogen secretion from the follicle, induces maturation of the follicle, and ovulation occurs. In the luteal phase, LH stimulates the transformation of the ruptured follicle and maintains the corpus luteum, which secretes estrogen and progesterone. The rising level of these two hormones exerts negative feedback on the hypothalamus and pituitary, inhibiting secretion of LH and FSH. The corpus luteum, deprived of its LH stimulation, degenerates, thereby dropping the levels of estrogen and progesterone. This drop releases the inhibition of the hypothalamus and pituitary, and FSH secretion begins again, stimulating growth of new follicles and the start of the next follicular phase.

The hormones of the ovarian cycle synchronize the menstrual cycle and the preparation of the uterus for possible implantation of an embryo. Estrogen causes the endometrium to begin to thicken in the proliferative phase. After ovulation, estrogen and progesterone stimulate increased vascularization and gland development of the endometrium. Thus, the luteal phase corresponds with the secretory phase of the menstrual cycle. The rapid drop of ovarian hormones caused by the degeneration of the corpus luteum reduces blood supply to the endometrium and begins its degeneration.

Estrogens are also responsible for female secondary sex characteristics. They induce fat deposition in the breasts and hips, affect water retention and calcium metabolism, and mediate sexual behavior.

Gamete Formation (Gametogenesis) Spermatogenesis, the production of mature sperm cells, occurs in the seminiferous tubules of the testes. The haploid nucleus of a sperm is contained in a thick head, tipped with an *acrosome*, which contains enzymes that help the sperm penetrate the egg. Mitochondria in the neck of the spermatazoon provide ATP for movement of the flagellum, or tail.

In *oogenesis*, the meiotic cytokinesis is unequal, and one large egg is produced along with three small haploid polar bodies, which disintegrate. In spermatogenesis, all four meiotic products develop into mature sperm. Also, at birth, an ovary already contains all its presumptive egg cells, whereas the precursor cells of sperm continue to mitotically divide throughout a male's reproductive years. Oogenesis happens in stages: the first meiotic division occurs within maturing follicles, and the second division occurs after ovulation. In humans, this second division is triggered by penetration of the egg cell by the sperm.

Sexual Maturation In humans, the onset of reproductive ability, termed puberty, is a gradual process that is completed around the age of 12 to 14 and includes the development of secondary sex characteristics.

The human sexual response cycle includes two types of physiological reactions: *vasocongestion*, or increased blood flow to a tissue, and *myotonia*, or increased muscle tension. During the *excitement* phase, vasocongestion of the penis, testes, labia, vagina, and breasts occurs; myotonia results in nipple erection and tension of the arms and legs. The *plateau* phase continues vasocongestion and myotonia, and breathing rate and heart rate increase. In both sexes, *orgasm* is characterized by rhythmic, involuntary contractions of the reproductive system. In males, emission deposits semen in the urethra, and ejaculation occurs when the urethra contracts and semen is expelled. In the *resolution* phase, vasocongested organs return to normal size, muscles relax, and nonreproductive reactions (breathing and heart rates) return to normal.

Conception, Pregnancy, and Birth

The development of an embryo in the uterus, beginning with *conception* and ending with birth, is called *pregnancy* or *gestation*. The gestation period correlates with body size and the degree of development of the young at birth. Human pregnancy averages 266 days, or 38 weeks.

Human gestation can be divided into three *trimesters*. Following fertilization in the oviduct, the *zygote* travels, in about 3 to 5 days, to the uterus. After about a week of *cleavage*, or cell division, the zygote develops into a hollow ball of cells called a *blastocyst* and implants in the endometrium. Tissues grow out of the developing embryo to meet with the endometrium and form the *placenta*, a disk-shaped organ in which gas and nutrient exchange and waste removal take place between the maternal and fetal circulations. Differentiation leads to *organogenesis*, and by the eighth week, the embryo has all the rudimentary structures of the adult and is called a *fetus*.

The embryo secretes *human chorionic gonadotropin (HCG)*, which maintains the corpus luteum's secretion of progesterone and estrogen through the first trimester. High progesterone levels initiate growth of the maternal part of the placenta, enlargement of the uterus, and cessation of ovulation and menstrual cycling.

During the second trimester HCG declines, the corpus luteum degenerates, and the placenta secretes its own progesterone, which maintains the pregnancy. The fetus grows rapidly and is quite active. The third trimester is a period of rapid fetal growth.

Labor consists of a series of strong contractions of the uterus and results in birth, or *parturition*. During the first stage of labor, the cervix dilates. The second stage consists of the contractions that force the fetus out of the uterus and through the vagina. The placenta is expelled in the final stage of labor.

Lactation is unique to mammals. Prolactin secretion stimulates milk production, and oxytocin controls the release of milk during nursing.

Several hypotheses attempt to explain why a mother does not reject an embryo, which has paternal as well as maternal chemical markers. There is evidence that the trophoblast, a protective layer that develops from the blastocyst and surrounds the embryo, induces the development of white blood cells that suppress other white cells from mounting an immune attack. One hypothesis suggests that this suppression can occur only after an initial immune response to the trophoblast. Some researchers suggest that if the initial response is too weak, then suppression may not occur, and the continued immunological attack results in spontaneous abortion of the embryo.

Contraception, the deliberate prevention of pregnancy, can be accomplished by one of several methods: preventing fertilization, preventing implantation, or preventing release of egg or sperm. The *rhythm method* is based on refraining from intercourse during the period that conception is most likely, the few days before and after ovulation. The failure rate of the rhythm method is 10% to 20%; 10 to 20 women become pregnant during a year out of every hundred using the method. Coitus interruptus, or withdrawal of the penis from the vagina prior to ejaculation, is an undependable method of contraception.

Several *barrier methods* of contraception that prevent fertilization have failure rates of less than 10%. Condoms and diaphragms, used in conjunction with spermicidal foam or jelly, present a physical and chemical barrier to fertilization.

The intrauterine device (IUD) probably prevents implantation by irritating the endometrium. IUDs have a low failure rate but have been associated with serious side effects.

The release of gametes may be prevented by chemical contraception, as in birth control pills, and sterilization, as in *tubal ligation* in women or *vasectomy* in men. Birth control pills are combinations of synthetic estrogen and progestin, which act by negative feedback to stop the release of GnRH by the hypothalamus and FSH and LH by the pituitary. The absence of these hormones results in a cessation of ovulation and follicle development. Cardiovascular problems are a potential side effect of birth control pills. The development of male chemical contraception is currently under research.

Abortion is the termination of a pregnancy. Spontaneous abortion, or miscarriage, occurs in as many as one-third of all pregnancies. Induced abortions and contraception are the focus of political and religious controversy.

Reproductive Technology   Some genetic diseases and congenital defects can be detected while the fetus is in the uterus. The use of *ultrasound* produces an image of the fetus from the echoes of high-frequency sound waves. In *amniocentesis*, a sample of amniotic fluid is withdrawn with a long needle, and fetal cells are cultured and analyzed for chromosomal defects. *Chorionic villi sampling* removes a small sample of tissue from the fetal part of the placenta and examines the cells for genetic defects.

A new technique of *in vitro fertilization* involves the removal of ova from a woman whose oviducts are blocked, fertilization within a petri dish, and implantation of the developing embryo in the uterus.

STRUCTURE YOUR KNOWLEDGE

1. Trace the path of a human sperm from the point of production to the point of fertilization, briefly commenting on the functions of both the structures it passes through and the associated glands.

2. Describe the path of a human ovum that does not become fertilized.

3. List the three general classes, along with examples, of birth control methods. Which examples are most likely to prevent pregnancy? Which are least likely to do so?

TEST YOUR KNOWLEDGE

FILL IN THE BLANKS

1. _____ type of asexual reproduction in which a new individual grows while attached to the parent's body

2. _____ group of genetically identical offspring

3. _____ development of egg without fertilization

4. _____ individual with functioning male and female reproductive systems

5. _____ common opening of digestive, excretory, and reproductive systems in nonmammalian vertebrates

6. _____ cycle in which thickened endometrium is reabsorbed

7. _____ hormone that prepares the uterus for pregnancy

8. _____ common duct for urine and semen

9. _____ filling of a tissue with blood due to increased blood flow

10. _____ period when reproductive ability begins in humans

MULTIPLE CHOICE: *Choose the one best answer.*

1. Which of the following is an explanation for the periodicity of reproductive cycles in animals?
   a. Reproduction may correspond with periods of increased food supply, during which energy can be invested in gamete formation.
   b. Seasonal cycles may allow offspring to be produced during favorable environmental conditions when chances of survival are highest.
   c. Hormonal control of reproduction may be tied to biological clocks and seasonal cues.
   d. All of these may contribute to periodic reproductive activity.

2. Which of the following is *not* required for internal fertilization?
   a. internal development of the embryo
   b. copulatory organ
   c. sperm receptacle
   d. behavioral interaction

3. Insect reproductive systems may include all of the following except
   a. seminal vesicles.
   b. an ejaculatory duct.
   c. a spermatheca.
   d. an endometrium.

4. Which of the following is incorrectly paired with its function?
   a. semiferous tubules—add fluid containing mucus, fructose, and prostaglandins to semen
   b. scrotum—encase testes, hold below abdominal cavity
   c. epididymis—store sperm
   d. prostrate gland—add alkaline secretion to semen

5. The function of the corpus luteum is to
   a. nourish and protect the egg cell.
   b. produce prolactin in the alveoli.
   c. produce progesterone and estrogen.
   d. convert into a hormone-producing follicle following ovulation.

6. In an estrous cycle,

a. the endometrium does not thicken unless fertilization occurs.

b. females are not receptive to sexual activity during estrus.

c. the endometrial lining is reabsorbed if fertilization does not occur.

d. the thickened endometrium lining is shed and the cycle is coordinated with season or day length.

7. Which of the following hormones is incorrectly paired with its function?
   a. GnRH—control release of FSH and LH
   b. prolactin—stimulate production of milk in alveoli of mammary gland
   c. estrogen—responsible for primary and secondary female sex characteristics
   d. human chorionic gonadotropin—stimulates growth of placenta

8. Myotonia is
   a. a congenital birth defect.
   b. the hormone responsible for breast development.
   c. muscle tension.
   d. rhythmic contractions of reproductive organs during orgasm.

9. The secretory phase of the menstrual cycle
   a. is associated with dropping levels of estrogen and progesterone.

b. is when the endometrium begins to degenerate and menstrual flow occurs.

c. corresponds with the luteal phase of the ovarian cycle.

d. corresponds with the follicular phase of the ovarian cycle.

10. Examples of birth-control methods that prevent release of gametes from the gonads are
    a. sterilization and chemical contraception.
    b. birth control pills and IUDs.
    c. condoms and diaphragms.
    d. abstinence and chemical contraception.

11. The ability of a pregnant woman not to reject her "foreign" fetus may be due to
    a. the fact that fetal and maternal blood never mix.
    b. the suppression of her immune response to the paternal chemical markers on fetal tissue.
    c. the protection of the fetus in the placenta.
    d. the production of human chorionic gonadotropin that maintains the pregnancy.

12. Which of the following is the least invasive technique of obtaining fetal cells for culturing to check for birth defects?
    a. ultrasound
    b. amniocentesis
    c. chorionic villi sampling
    d. in vitro fertilization

CHAPTER 43

ANIMAL DEVELOPMENT

FRAMEWORK

Fertilization initiates physical and molecular changes, such as the formation of fertilization membranes and increased metabolism and protein synthesis. The early processes of embryonic development include cleavage, gastrulation, and organogenesis. The forms of these cell divisions and morphogenetic movements depend on the amount of yolk in the egg and on the taxonomic position of the animal.

Development may be mosaic, in which the fate of blastomeres is fixed by the first cleavage divisions, or regulative, in which blastomeres remain totipotent for a longer time. Determination and differentiation of cells are the result of the control of gene expression by both cytoplasmic determinants and the positional information a cell receives within a developing structure. Developmental biologists, using transplant experiments and other techniques, are gradually unraveling some of the mechanisms underlying the complex processes of animal development.

CHAPTER SUMMARY

An animal's form and function develop throughout its lifetime and include such events as embryonic development, postembryonic growth and maturation, metamorphosis, regeneration, wound-healing, and aging. The three key processes of embryonic development are cell division, the production of large numbers of cells; *differentiation*, the formation of specialized cells that are then arranged in tissues and organs; and *morphogenesis*, the development of body shape and organization.

Fertilization

Fertilization, the union of egg and sperm, *activates* the egg by initiating metabolic reactions that trigger embryonic development.

The Acrosomal Reaction In sea urchin fertilization, the acrosome at the tip of the sperm discharges hydrolytic enzymes when in contact with the jelly coat of an egg, allowing the *acrosomal process* to penetrate this layer. Fertilization of egg and sperm of the same species is assured when a protein called *bindin* on the surface of the acrosomal process attaches to specific receptor molecules on the egg's *vitelline layer*, external to the egg's plasma membrane.

Enzymes digest a hole through the vitelline layer, and the sperm's plasma membrane fuses with that of the egg, allowing the sperm nucleus to enter the egg. Fusion of the membranes opens membrane ion channels, and Na^+ ions flow into the egg. The resulting depolarizing of the membrane prevents other sperm cells from fusing with the egg and provides a *fast block to polyspermy*.

The Cortical Reaction The depolarization initiates the release of calcium into the egg cytoplasm, causing *cortical granules*, located in the *cortex* (the outer zone of cytoplasm), to fuse with the plasma membrane. These vesicles release their contents by exocytosis and cause the vitelline layer to loosen from the plasma membrane and the resulting space to swell by the osmotic uptake of water. The elevated vitelline layer, known as the *fertilization membrane*, functions as a *slow block to polyspermy*.

Activation of the Egg The release of calcium ions into the cytoplasm results in a change in pH that *activates* the egg by increasing the rates of cellular respiration and protein synthesis. Injection of calcium into an egg, or temperature shock, can turn on these metabolic responses and initiate parthenogenetic development. Even an enucleated egg can be activated to begin protein synthesis, showing that inactive mRNA has been stockpiled in the egg.

After the sperm nucleus fuses with the egg nucleus, DNA replication begins in preparation for the cleavage division that begins the development of the embryo.

Early Stages of Embryonic Development

The early development of a sea urchin and amphioxus, a cephalochordate, illustrates the steps of cleavage, gastrulation, and organogenesis.

Cleavage *Cleavage* is a succession of rapid cell divisions during which the embryo becomes partitioned into many small cells, called *blastomeres*.

The axis of the amphioxus egg is defined by the *animal pole*, which is the point at which the polar body budded from the egg during meiosis. The opposite end is called the *vegetal pole*. The first two cleavage divisions are vertical, or polar; the third division is equatorial. The deuterostomes—the echinoderms and chordates—exhibit radial cleavage, in which each tier of cells aligns directly with the cells of the lower tier. Most protostomes—annelids, arthropods, and mollusks— follow a pattern known as spiral cleavage.

Further cleavage produces a *morula*, a solid ball of cells, followed by the arrangement of cells into a hollow ball, the *blastula*, surrounding a fluid-filled cavity called the *blastocoel*.

Gastrulation In amphioxus, *gastrulation* occurs when one side of the blastula buckles inward by a process known as *invagination*. The *gastrula* is a cup-shaped embryo having two layers of cells that surround the *archenteron*, the cavity that will become the digestive tract. The archenteron opens by way of the *blastopore*, which develops into the anus in deuterostomes and the mouth in protostomes.

A third layer of cells develops from pouches that bud from the archenteron and expand into the space between the outer and inner layers. The triploblastic body plan, characteristic of chordates and most other animals, now consists of the *primary germ layers*—the outer *ectoderm*, the middle *mesoderm*, and the inner *endoderm*.

Organogenesis Rudiments of organs develop from the primary germ layers in the process known as *organogenesis*. The *notochord*, characteristic of chordate embryos, is formed from dorsal mesoderm along the

roof of the archenteron. Ectoderm above the developing notochord thickens to form a *neural plate*, which sinks and rolls into a tube, forming the *neural tube*. This tube develops into the central nervous system. The ectoderm also gives rise to the epidermis, inner ear, and eye lens.

The mesoderm forms the notochord, lining of the coelom, muscles, skeleton, gonads, kidneys, and most of the circulatory system. The linings of the digestive tract and organs that form as outpocketings of the archenteron (lungs, liver, pancreas) arise from endoderm.

Comparative Embryology of Vertebrates

Amphibian Development Yolk *platelets*, which store lipids and protein, are concentrated in the vegetal hemisphere of a frog egg, and *pigment granules* are located in the cortex of the animal hemisphere. The first two cleavage divisions are polar. The horizontal third division is displaced toward the top of the embryo. Because the yolk impedes divisions in the vegetal hemisphere, the frog blastula has its blastocoel, enclosed in a wall that is several cells thick, restricted to the animal hemisphere.

A change in the shape of "bottle cells" begins gastrulation, forming a tuck that will become the *dorsal lip* of the blastopore. In a process called *involution*, surface cells roll over the lip and migrate along the roof of the blastocoel. The blastopore is first crescent shaped and then it forms a circle surrounding a *yolk plug* of large, yolk-filled cells. Gastrulation produces an archenteron, surrounded by the three primary germ layers.

In organogenesis the notochord, derived from dorsal mesoderm, forms first, followed by the neural tube, derived from a plate of ectoderm above the notochord. Mesoderm along the side of the notochord separates into blocks, called *somites*, which give rise to the vertebrae and muscles associated with the axial skeleton. Lateral to the somites, the mesoderm forms the lining of the coelom.

A band of cells called the *neural crest* forms above the neural tube. These cells later migrate to form pigment cells in the skin, skull bones, teeth, the adrenal medulla, and components of the peripheral nervous system.

Avian Development The yolk of a bird egg is so dense that cleavage occurs in only a small disc of cytoplasm at the animal pole, producing a cap of cells called the *blastodisc*. *Meroblastic cleavage* is this incomplete division of a yolk-filled egg, whereas *holoblastic cleavage* is the complete division through an egg with little to modest amounts of yolk.

The blastodisc separates into an upper and lower layer, forming the blastocoel in between. A straight invagination, called the *primitive streak*, marks the anterior–posterior axis of the embryo and is the site of gastrulation. Rolling in of the top layer of cells forms the mesoderm and contributes to the endoderm. The borders of this flat sheet fold down and join, forming a triple-layered tube from which organogenesis proceeds.

The primary germ layers also give rise to four *extraembryonic membranes*, essential to development within the avian egg shell. The *yolk sac* grows to enclose the yolk and develops blood vessels to carry nutrients to the embryo. The *amnion* encloses a fluid-filled sac, which provides an aqueous environment for development and acts as a shock absorber. The *allantois*, growing out of the hindgut, serves as a receptacle for the nitrogenous waste, uric acid, and expands to press the *chorion* against the lining of the eggshell. The vascularized allantois, in conjunction with the chorion, provides gas exchange for the developing embryo.

Mammalian Development   The egg of placental mammals has little yolk, and cleavage of a mammalian zygote is holoblastic. Gastrulation and early organogenesis, however, are similar in pattern to those of birds and reptiles.

The blastocyst consists of an outer epithelium, the *trophoblast*, surrounding a cavity into which protrudes a cluster of cells, called the *inner cell mass*, which will develop into the embryo and some extraembryonic membranes. The trophoblast secretes enzymes that enable the blastocyst to embed in the endometrium and extends projections that develop into the fetal portion of the placenta.

The four extraembryonic membranes are homologous to those of reptiles and birds. The chorion surrounds the embryo and the other membranes. The amnion encloses the embryo in a fluid-filled amniotic cavity. The yolk-sac membrane, enclosing a small fluid-filled cavity, is the site of early formation of blood cells. The allantois forms blood vessels to connect the embryo with the placenta through the umbilical cord.

The inner cell mass forms a flat *embryonic disc*, which resembles the two-layered blastodisc of birds and reptiles. In gastrulation, cells from the upper layer roll through a primitive streak to form the mesoderm and endoderm. Organogenesis begins with the formation of the notochord, neural tube, and somites. In the human embryo, all major organs have begun development by the end of the first trimester.

Mechanisms of Development

Polarity of the Embryo   A bilaterally symmetrical animal has a left and right side, an anterior–posterior

axis, and a dorsal–ventral axis. These three polarities may be established at the time of fertilization.

In the amphibian egg, an animal–vegetal axis exists due to the yolk platelets concentrated in the vegetal pole and cortical pigment granules concentrated near the animal pole. During fertilization, the cortex of the egg is moved toward the point of sperm entry, and the edge of the pigmented layer opposite that point is pulled toward the animal pole, exposing a lighter-colored cytoplasm. This half-moon shaped mark is the *gray crescent*, which will be bisected by the first cleavage division and marks the future location of the dorsal lip of the blastopore. The gray crescent determines the anterior–posterior axis, the left–right axis of the embryo, and the animal–vegetal axis of the egg becomes the dorsal–ventral axis of the embryo.

Mammalian eggs, in contrast, do not have an apparent polarity, and early cleavage divisions are randomly oriented.

Cytoplasmic Determinants   *Cytoplasmic determinants* are unevenly distributed cytoplasmic components that fix the developmental fates of cells. In mollusks, a polar lobe that forms on one of the first two blastomeres is necessary for the subsequent formation of mesodermal structures in the embryo. If separated, the blastomere without the polar lobe does not develop normally.

Early blastomeres that can be separated and still produce complete embryos are said to be *totipotent*. Even though the cytoplasmic determinants are asymmetrically distributed in the frog egg, the first cleavage division equally separates these determinants. If an experimentally induced cleavage plane does not bisect the gray crescent, only the blastomere that receives the gray crescent will develop into a normal tadpole.

Mosaic and Regulative Development   Development in which the early blastomeres are not totipotent is said to be *mosaic*. In *regulative* development, cells are totipotent longer and their developmental fates can be experimentally altered. Mammalian embryos remain regulative longer than do those of most other animals.

Fate Maps   In the 1920s, Vogt developed a *fate map* for the amphibian embryo by labeling regions of the blastula with vital dyes and determining where these dyes showed up in later developmental stages. The developmental history of every cell has been determined for the nematode, *C. elegans*.

Morphogenetic Movements   Extension, contraction, and adhesion are involved in various morphogenetic movements of cells. Reorganization of the cytoskeleton produces the wedge shape of cells that leads to the formation of invaginations and evaginations. Amoeboid movement is also involved in morphogenesis. Cells at the leading edge of migrating tissue may extend pseudopodia and then drag along neighboring cells, attached to them by intercellular junctions. Amoeboid cells wander individually from the neural crest to various parts of the embryo.

An extracellular matrix of adhesive substances and fibers may help to guide the movement of cells. The orientation of *fibronectin* fibrils corresponds to the arrangement of contractile microfilaments of the cytoskeleton of migrating cells. Substances on the surfaces of cells, called *cell adhesion molecules (CAMS)*, function to hold the cells of specific tissues and organs together during morphogenesis.

Induction

In *induction,* one group of cells influences the development of another. If the chordamesoderm that forms the notochord is transplanted, a neural plate and tube will form in an abnormal region of ectoderm. Using transplant experiments, Mangold and Spemann established that the dorsal lip is a primary *organizer* of the embryo because of its influence in early organogenesis.

Determination and Differentiation

Cells differentiate, or develop characteristic structures and functions, because they express different portions of a common genome. Once a cell is committed to a particular path of development, it is said to be *determined*—its developmental potential has become restricted. Transplant experiments show whether tissues at various stages of development have been determined. *Imaginal disks* are groups of undifferentiated yet determined cells in insect larvae that are destined to form specific organs during metamorphosis. Determination appears to develop progressively; the developmental options of cells become more and more narrow. Differentiation is a product of the developmental history of the cell. Transplanted ectoderm from an early gastrula will form a neural plate above the developing notochord; whereas ectoderm from a late-stage gastrula transplanted into an early gastrula will not be induced and no neural plate will form.

The cytoplasmic environment of the cell apparently controls the expression of the genome. Cytoplasmic determinants may preset which genes will be expressed when the cell differentiates later. Morphogenetic movements also contribute to determination and differentiation by exposing cells to differing chemical and physical environments.

Pattern Formation

Pattern formation is the ordering of cells and tissues into three-dimensional structures that make up the various parts of an animal. The importance of genes in regulating body pattern is illustrated by the misplaced structures caused by homeotic mutations in *Drosophila*. But more than genetic information guides the formation of body parts. The cells of a rudimentary organ are initially similar, and they differentiate based on their position within the field of cells. Gradients of chemical signals moving along three planes would provide the *positional information* needed for a cell to determine its location in the three-dimensional developing organ.

Wolpert has been studying positional information in chick limb development by doing transplant experiments. The "zone of polarizing activity" (ZPA) appears to communicate position along the anterior–posterior axis in a developing limb bud. Another region of mesoderm at the tip of the wing bud may assign position along the proximal–distal axis. The substances producing the chemical gradients to which developing cells respond are called *morphogens*. Experiments have isolated *retinoic acid* as the possible morphogen produced by the ZPA.

Other experiments indicate that positional information contributes to overall body form, not just the pattern of individual parts. Undifferentiated mesoderm from the thigh region developed into a toe when transplanted to the tip of a wing bud. The mesoderm had already been determined to form a part of the leg, but it still responded to its location at the tip of a developing limb. Installments of positional information are provided to cells throughout their development, and determination becomes more and more specific.

STRUCTURE YOUR KNOWLEDGE

1. Create a flow chart that shows the sequence of events in the fertilization of a sea urchin egg. Include the functions of key events.

2. Fill in the table on the following page, briefly describing the early stages of development for amphioxus, frog, bird, and mammalian embryos.

3. Use the following terms to create a concept map that illustrates the relationships among these concepts. Add additional concepts as needed.

   development   cytoplasmic determinants

   differentiation   pattern formation

   determination   positional information

Stage	Amphioxus	Amphibian	Bird	Mammal
Cleavage				
Blastula				
Gastrulation				

TEST YOUR KNOWLEDGE

MULTIPLE CHOICE: *Choose the one best answer.*

1. The vitelline layer
   a. is inside the fertilization membrane.
   b. releases calcium, which initiates the cortical reaction.
   c. has receptor molecules that are specific for bindin proteins on the acrosomal process of sperm.
   d. All of the above are correct.

2. The fast block to polyspermy
   a. prevents sperm from other species from fertilizing the egg.
   b. is produced by the depolarization of the membrane.
   c. is a result of the formation of the fertilization membrane.
   d. is caused by the exocytosis of cortical granules.

3. The activation of an enucleated egg by pricking shows that
   a. calcium ions initiate activation.
   b. inactive mRNA had been stockpiled in the egg cell.
   c. parthenogenic development may replace fertilization in sea urchins.
   d. the nucleus is not needed for the development of a sea urchin embryo.

4. Which of the following groups does not have eggs with definite animal and vegetal poles?
   a. amphioxus
   b. frog
   c. bird
   d. mammal

5. The archenteron develops into the
   a. anus in deuterostomes.
   b. blastocoel.
   c. neural tube.
   d. digestive tract.

6. Organogenesis
   a. produces the gastrula by invagination.
   b. is the movements of cells that produce body structures.
   c. is the rudimentary development of organs from the primary germ layers.
   d. begins with the formation of the neural plate.

7. Which of the following is incorrectly paired with its primary germ layer?
   a. muscles—mesoderm
   b. central nervous system—ectoderm
   c. lens of the eye—mesoderm
   d. liver—endoderm

8. In a frog embryo, gastrulation
   a. is impossible because of the large amount of yolk.
   b. proceeds by involution as cells roll over the dorsal lip of the blastopore.
   c. produces a blastocoel displaced into the animal hemisphere.
   d. occurs along the primitive streak.

9. The embryonic disc of mammalian embryos
   a. develops from the inner cell mass.
   b. resembles the blastodisc of birds.
   c. develops into the embryo and some extraembryonic membranes.
   d. All of the above are correct.

10. The function of the allantois in birds is to
    a. provide for nutrient exchange between the embryo and yolk sac.
    b. store nitrogenous wastes in the form of urea.
    c. form a respiratory organ in conjunction with the chorion.
    d. provide an aqueous environment for the developing embryo.

11. Which of the following is a *correct* statement of a difference between avian and mammalian early embryology?
    a. The bird egg has meroblastic cleavage whereas the mammalian egg has holoblastic cleavage.
    b. The bird embryo is surrounded by an amnion membrane, whereas the mammalian embryo is enclosed by the placenta.
    c. Gastrulation occurs through the primitive streak in birds but by way of the blastopore in mammals.
    d. The bird egg has four extraembryonic membranes, whereas the mammalian embryo develops only two.

12. Somites are
    a. blocks of mesoderm circling the archenteron.
    b. cells from which the notochord arises.
    c. serially arranged mesoderm blocks that develop into vertebrae and skeletal muscles.
    d. mesodermal derivatives of the notochord that lines the coelom.

13. Which of the following is an example of induction?
    a. formation of the inner cell mass in a mammalian embryo
    b. interactions between optic vesicle and ectoderm to form optic cup and lens
    c. movement of cells over the dorsal lip of the gray crescent and into the interior of frog embryo
    d. budding off of mesoderm from the endoderm of the archenteron

14. The imaginal disks of an insect larva
    a. are islands of cells that are determined but not differentiated and will form various organs of the adult following metamorphosis.
    b. establish the polarity of the embryo.
    c. contain all three primary germ layers as a result of invagination along the primitive streak.
    d. are like somites that develop into the segmented structure of the adult insect.

15. The gray crescent
    a. is the future location of the dorsal lip of the blastopore.
    b. is created when the pigmented layer is pulled toward the animal pole at fertilization.
    c. is bisected by the first cleavage division, creating an equal division of cytoplasmic determinants.
    d. is all of the above.

16. In regulative development,
    a. cells remain totipotent until the gastrula stage.
    b. early blastomeres that are separated produce complete embryos.
    c. blastomeres without the polar lobe do not develop normally.
    d. one group of cells regulates or influences the development of another.

17. Cytoplasmic determinants
    a. are unevenly distributed cytoplasmic components that fix the developmental fates of cells.
    b. are involved in the regulation of gene expression.
    c. are involved in the determination of cells and tissues.
    d. are all of the above.

18. Cells of specific tissues and organs are held together during morphogenesis by
    a. fibronectin fibrils.
    b. cell adhesion molecules.
    c. induction.
    d. morphogens.

19. Pattern formation appears to be determined by
    a. positional information a cell receives from gradients of chemical signals called morphogens.
    b. differentiation of cells, which then migrate into developing organs.
    c. the transplant experiment performed during development
    d. the induction of cells from chordamesoderm cells found in the center of organs.

20. The zone of polarizing activity (ZPA)
    a. communicates positional information along the anterior–posterior axis in a developing limb bud.
    b. appears to produce retinoic acid as a morphogen.
    c. has been studied by Wolpert and his colleagues.
    d. All of the above are correct.

CHAPTER 44

NERVOUS SYSTEMS

FRAMEWORK

This chapter describes the structural components of nervous systems and their functional transmission of information as they integrate and coordinate information from and responses to the internal and external environment.

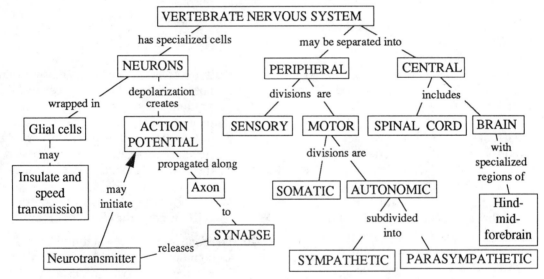

CHAPTER SUMMARY

The nervous system allows an animal to respond rapidly and appropriately to environmental stimuli. The coordination of behavior is a product of both the nervous and endocrine systems. Three characteristics that distinguish the nervous system from the endocrine system are more rapid communication, specific control with messages traveling directly to individual target cells, and a structural complexity that provides for the integration of more kinds of information and responses.

The three main functions of the nervous system are sensory input, integration, and motor output. Information from sensory receptors is carried to integration centers, where it is interpreted and connected with appropriate responses through the motor output that is conducted to *effector cells* such as muscle or gland cells. Information is transmitted by electrical and chemical signals along the specialized nerve cells known as *neurons*.

Cells of the Nervous System

The cells of the nervous system include the neurons, which transmit the signals, and *supporting cells*, which support, insulate, and protect the neurons.

Neurons   A neuron consists of a cell body, which contains the nucleus and most of the organelles and cytoplasm, and long fiberlike processes, *axons* and *dendrites*, which conduct impulses. A neuron usually has one axon, which transmits signals away from the cell body. The more numerous, shorter, and branched dendrites carry signals toward the cell body.

Axons originate from a cone-shaped region of the cell body, the *axon hillock*, and are often wrapped in *Schwann cells*, which collectively form the insulating *myelin sheath*. The ends of axons terminate in branchlets called teledendria, whose bulbous tips, called *synaptic knobs*, release the *neurotransmitters* that relay a nervous signal across the *synapse* between the axon and another neuron or effector.

The *central nervous system* consists of the brain and spinal cord. The cell bodies of most neurons are located in the central nervous system. Ganglia are clusters of cell bodies of some neurons that are located outside the brain and spinal cord. The *peripheral nervous system* contains the nerves, or bundles of neurons, that communicate between the central nervous system and the rest of the body. *Sensory neurons* transmit information from the internal and external environments to the brain or spinal cord. *Motor neurons* carry information from the brain or spinal cord to effector organs. *Interneurons*, or *association neurons*, transmit signals from sensory to motor neurons within the central nervous system.

Supporting Cells   Supporting cells in the central nervous system are known as *glial cells*. Among the types of glial cells are astrocytes, lining the capillaries in the brain and contributing to the *blood-brain barrier*, which restricts the passage of most substances into the brain, and oligodendrocytes, which insulate axons in a myelin sheath. Oligodendrocytes and the Schwann cells in the peripheral nervous systems wrap to form concentric membrane layers around axons. The resulting myelin sheaths provide electrical insulation between neurons.

Transmission Along Neurons

Electric currents, created by ion fluxes across the plasma membrane, transmit signals along the length of neurons.

The Resting Potential   A polarized membrane, which is a result of the charge difference between the cytoplasm and extracellular fluid, is a fundamental feature of all cells. Electrophysiologists can measure the magnitude of the charge separation across the membrane by using microelectrodes placed inside and outside a cell and connected to a voltmeter. The voltage across the membrane is a measure of its membrane potential. The membrane potential or *resting potential* of a nontransmitting neuron is about –70 mV (millivolts). *Excitable cells*, such as neurons and muscle cells, use changes in their membrane potentials to conduct signals.

The resting potential depends on the different ionic concentrations on either side of the membrane and the abilities of ions to cross the membrane. The sodium–potassium pump moves sodium ions out of the cell and potassium ions into the cell, creating steep concentration gradients for these ions across the membrane. The concentration gradients will drive the ions back across the membrane, but their movement is influenced by electrical interactions with other positively and negatively charged ions and the relative permeabilities of the membrane to each ion.

Stimulation and Graded Potentials   A stimulus, or change in an environmental factor that affects the permeability of a membrane to ions, may reduce a membrane's potential and thus *depolarize* the membrane, or increase the membrane's voltage to greater than the resting potential and *hyperpolarize* the membrane. The magnitude of a local voltage change of a *graded potential* is proportional to the strength of the stimulus. Graded potentials last only about a millisecond and may either depolarize or hyperpolarize the membrane.

The Action Potential   An *action potential* is a rapid depolarization of the neuron's membrane, that produces the nerve impulse. The resting potential of –70 mV can change to about +35 mV and return to the resting potential within a few milliseconds of stimulation. The depolarization of a neuron's plasma membrane in response to a stimulus causes a local increase in the permeability of Na^+ ions. Sodium ions enter the cell, driven by their concentration gradient and attraction to the negatively charged interior of the cell. The voltage change caused by the influx of Na^+ ions opens *voltage-sensitive gates* that create a large change in the membrane's permeability to Na^+.

The short duration of the action potential is due to the rapid closing of the sodium gates, which will not reopen until the membrane has returned to its original resting potential. The *refractory period* is the short time after an action potential when the neuron cannot respond to another stimulus. The slightly delayed opening of voltage-sensitive potassium channels helps to restore the resting potential, and the increased permeability to K^+ may temporarily hyperpolarize the membrane and produce a voltage more negative than the resting potential. The sodium–potassium pump quickly restores the ion gradients so that future action potentials are possible.

The action potential operates on the *all or none* principle; an action potential always creates the same voltage spike, regardless of the intensity of the stimulus. The initial depolarization, however, must reach a *threshold potential* before it generates an action potential. The stimulus must be intense enough to depolarize the membrane from the resting potential of –70 mV to the threshold of –50 mV (for many neurons); the sodium gates then open and the action potential is generated.

Though the amplitudes of action potentials are always the same, their frequencies vary with the intensity of a stimulus. A neuron will "fire" repeatedly after each refractory period in response to a strong stimulus. The dendrites or cell body usually receive the stimulus, but the action potential is generated at the axon hillock.

Propagation of the Nerve Impulse

As the sodium ions moving into the cell during the action potential spread out, they depolarize adjacent sections of the membrane to the threshold potential, causing voltage-sensitive channels to open and the generation of an action potential farther along the membrane, which in turn incites another action potential. Thus local depolarizations and action potentials across the membrane result in the propagation of serial action potentials along the length of the neuron. Because of the brief refractory period, the action potential is usually propagated in one direction, from the axon hillock to the tips of the axon.

Rate of Transmission

Resistance to current flow is inversely proportional to the cross-sectional area of the conducting "wire." The greater the axon diameter, the lower the resistance and the faster the rate of conduction of the action potential. Some invertebrates have giant axons, which conduct impulses rapidly.

In vertebrates, supporting cells form a thick, myelin sheath around many axons, with small gaps, called *nodes of Ranvier*, where these cells abut. Voltage-sensitive ion channels are concentrated in the node regions where the neuron membrane has contact with the extracellular fluid. Action potentials can only be generated at these nodes, and a nerve impulse "jumps" from node to node, resulting in a faster mode of transmission known as *saltatory conduction*.

The Synapse: Transmission between Cells

Synapses between neurons conduct impulses from the axon of a *presynaptic cell* to a dendrite or cell body of the *postsynaptic cell*.

Electrical Synapses

Passage of the impulse from presynaptic to postsynaptic cell may occur across an electrical synapse, in which cytoplasmic connections called gap junctions allow ions to flow between two cells and transmit the action potential across the synapse. Electrical synapses are found in the giant neurons of some crustaceans.

Chemical Synapses

In a chemical synapse, a chemical message travels from the presynaptic cell across a cleft to the postsynaptic cell. The synaptic knob, at the tip one of the axon's teledendria, contains numerous *synaptic vesicles*, in which thousands of molecules of the chemical messenger, called a *neurotransmitter*, are stored. The depolarization of the *presynaptic membrane* of an axon opens voltage-sensitive calcium channels in the membrane. The influx of Ca^{2+} causes the synaptic vesicles to fuse with the presynaptic membrane and release their neurotransmitter into the cleft.

The *postsynaptic membrane* of the dendrite or cell body of the postsynaptic cell contains receptor proteins for different neurotransmitter molecules, which are associated with particular ion channels. These chemically sensitive gates allow ions, such as Na^+, K^+, or Cl^-, to cross the membrane, altering the membrane potential by either bringing it closer to the threshold potential or hyperpolarizing it. Enzymes in the synaptic cleft or on the postsynaptic membrane rapidly break down the neurotransmitter.

Summation: Nervous Integration at the Cellular Level

A neuron must integrate the information it receives at thousands of excitatory and inhibitory synapses from numerous neighboring neurons. At an excitatory synapse, binding of the neurotransmitter to receptors opens a gated channel that allows Na^+ to flow into and K^+ to flow out of the cell. A net flow of positive charge into the cell depolarizes the membrane, creating an *excitatory postsynaptic potential (EPSP)* and bringing the membrane potential closer to an action potential threshold. At an inhibitory synapse, binding of the neurotransmitter opens ion gates to allow K^+ to flow out and/or Cl^- to move into the cell, hyperpolarizing the membrane and producing an *inhibitory postsynaptic potential (IPSP)*, making the triggering of an action potential more difficult. The response to a neurotransmitter depends on the type of receptor and the gated ion channels associated with it on the postsynaptic membrane.

EPSPs and IPSPs are graded potentials, whose magnitudes depend on the number of neurotransmitter molecules that bind to receptors. *Summation* of several postsynaptic potentials usually determines whether the postsynaptic cell fires. *Temporal summation* occurs when repeated release of neurotransmitters from one or more synaptic knobs creates postsynaptic potentials that affect the membrane before it can return to the resting potential. *Spatial summation* occurs when several different presynaptic knobs, usually from different neurons, release neurotransmitter simultaneously. Several EPSPs arriving simultaneously may depolarize the axon hillock to the threshold potential and generate an action potential. Summated IPSPs will hyperpolarize the membrane. Concurrent EPSPs and IPSPs also summate, countering their electrical effects. The membrane potential of the axon hillock at any given time is determined by the average of all EPSPs and IPSPs.

Neurotransmitters and Receptors About ten molecules have been identified as neurotransmitters, and many more may be added to the list. The criteria for neurotransmitters are as follows: (1) the compound must be contained in synaptic vesicles and discharged when the presynaptic cell is stimulated, and it must alter the membrane potential of the postsynaptic membrane; (2) when experimentally injected into the synapse, it must cause an EPSP or IPSP; and (3) the compound must be rapidly degraded or removed from the synapse.

Acetylcholine is a common neurotransmitter in invertebrates and vertebrates. In vertebrate neuromuscular junctions, acetylcholine triggers contraction in skeletal muscle cells. It can also be inhibitory, slowing down the heart rate of vertebrates and mollusks.

Biogenic amines, neurotransmitters derived from amino acids, usually function within the central nervous system. *Epinephrine, norepinephrine*, and *dopamine* are biogenic amines derived from the amino acid tyrosine. Dopamine and *serotonin* affect sleep, mood, attention, and learning, and imbalances of these transmitters have been associated with mental illness.

Amino acids may also be neurotransmitters. *Glycine, glutamate*, and *gamma aminobutyric acid (GABA)* function in the central nervous system. GABA is the most common inhibitory transmitter in the brain.

The *endorphins* and *enkephalins* are *neuropeptides* that are produced in the brain during physical or emotional stress and function as natural pain killers. An endorphin that is released from the anterior pituitary also functions as a hormone.

Acetylcholine and the amino acid transmitters bind to receptors on the postsynaptic membrane, affect ion permeability, and create either EPSPs or IPSPs. Biogenic amines and neuropeptides, when they bind to their receptors, create a second messenger response that affects the postsynaptic cell's metabolism.

The action of a neurotransmitter depends on the response of its specific receptor on the postsynaptic membrane to its binding. The structure of the acetylcholine receptor at excitatory synapses has been identified as a donut-shaped protein consisting of four polypeptide subunits. The mechanism of its function as an ion channel has not yet been determined.

Neural Circuits and Clusters Groups of neurons that interact and carry information along specific pathways are called circuits. In *convergent circuits*, several neurons come together along a pathway and feed information into a few cells. *Divergent circuits* spread out information from one pathway in several directions; and *reverberating circuits* are circular paths in which signals return to their source.

Nerve cell bodies are arranged in functional groups called *ganglia*. A ganglion in the brain is often called a *nucleus*.

Invertebrate Nervous Systems

The cnidarian *nerve net* is a loosely organized system of nerves, connected by electrical synapses, in which impulses are conducted in both directions. Some centralization is seen in jellyfish in which clusters of nerve cells around the margin of the bell coordinate swimming movements. Modified nerve nets are also found in echinoderms.

Bilateral animals with more active lifestyles show the evolutionary trend of *cephalization*—the concentration of sense organs and feeding structures in the head, along with the enlargement of anterior ganglia leading to the formation of a brain that processes sensory information and controls motor responses.

Flatworms have a simple brain and two or more nerve trunks, with ladderlike transverse nerves. Annelids and arthropods have a prominent brain and a ventral nerve cord, which may have ganglia within each segment of the body. Sessile mollusks have little cephalization and only simple sense organs. Cephalopods, such as the octopus, have large brains, image-forming eyes, and giant axons, all of which contribute to their active predatory life and their ability to learn and remember.

The Vertebrate Nervous System

Peripheral Nervous System The peripheral nervous system consists of the *sensory* or *afferent nervous system*, which brings information to the central nervous system (CNS), and the *motor* or *efferent nervous system*, which carries signals away from the CNS to effector organs. The human peripheral nervous system contains 12 pairs of cranial nerves and 31 pairs of spinal nerves.

The sensory nervous system brings in messages from both the external and internal environments. The motor nervous system has two divisions, one of which governs responses to the external environment, while the other coordinates functions of the the internal organs. The *somatic nervous system* carries signals to skeletal muscles, the movement of which is largely under conscious control. Many skeletal muscle movements, however, are determined by *reflexes*, which are automatic, subconscious reactions to stimuli mediated by the spinal cord or lower brain. The *autonomic nervous system* controls smooth and cardiac muscles and various organs.

The autonomic nervous system is subdivided into the *sympathetic* and the *parasympathetic nervous systems*. In general, the parasympathetic division slows the heart, stimulates digestion, and enhances activities that gain and conserve energy. The sympathetic division accelerates the heart rate and increases the metabolic rate, preparing an organism for action.

Central Nervous System   The central nervous system, which bridges the sensory and motor components of the peripheral nervous system, consists of the *spinal cord* and the *brain*. Protective layers of connective tissue called the *meninges* cover these bilaterally symmetrical organs. Axons and dendrites are located in bundles or tracts; the *white matter* is named for the white color of their myelin sheaths. The cell bodies of the neurons make up the *gray matter*, located on the outer layer in the brain but on the inside of the white matter in the spinal cord.

Spaces called *ventricles* in the brain, which are continuous with the narrow *central canal* of the spinal cord, are filled with *cerebrospinal fluid*. This fluid, which is formed by filtration of the blood, cushions the brain and carries out circulatory functions.

The spinal cord integrates simple responses to some stimuli (in the form of reflexes) and carries information to and from the brain. The knee-jerk reflex involves a stretch receptor, a sensory neuron, and a motor neuron. Most reflexes have interneurons between sensory and motor neurons.

The vertebrate brain evolved as bulges in the anterior end of the spinal cord. Three regions—the *rhombencephalon* or *hindbrain*, the *mesencephalon* or *midbrain*, and the *prosencephalon* or *forebrain*—are present in all vertebrates, although these regions may be subdivided to provide a greater capacity for integration of complex activities.

Three evolutionary trends in the vertebrate brain include increases in the relative size of the brain, in the compartmentalization of function, and in the complexity of the forebrain. More sophisticated behaviors are correlated with increased size of one region of the forebrain, the *cerebrum*. The folding or convolution of the cerebral cortex increases the surface area of this layer.

The human brain develops from the three primary regions—the hindbrain, midbrain, and forebrain—which differentiate into specialized structures. The *brainstem*, which includes the hind- and midbrain, extends from the spinal cord to the middle of the brain.

The hindbrain includes the *medulla oblongata*, which contains control centers for homeostatic functions including respiration, heart and blood-vessel actions, and digestion; the *pons*, which functions with the medulla in some of these activities and in conducting information between the rest of the brain and the spinal cord; and the *cerebellum*, which functions in coordination of movement. The tracts of motor neurons from the mid- and forebrain cross in the medulla, so that the right side of the brain controls much of the movement of the left side of the body and vice versa. The cerebellum integrates information from the auditory and visual systems with sensory input from the joints and muscles as well as motor pathways from the cerebrum to provide unconscious coordination of movements and balance.

The upper portion of the brainstem, or *midbrain*, receives and integrates sensory information and sends this information to specific regions of the forebrain. Fibers involved in hearing pass through or terminate in the *inferior colliculi*. The *superior colliculi* are important visual centers. The *reticular formation*, which includes a major group of nuclei in the midbrain, regulates states of arousal.

The *forebrain* is the region of the most intricate neural processing. The lower *diencephalon* contains the thalamus and hypothalamus, and the *telencephalon* contains the cerebrum.

The *thalamus* is a major projection area; most of the input to the cerebral cortex comes from neurons with cell bodies in the thalamus. The thalamus also contains nuclei of the reticular formation, which filters the information sent to the cerebral cortex.

The *hypothalamus* is the major site for regulation of homeostasis. It produces the posterior pituitary hormones and the releasing factors that control much of the hormone secretion of the anterior pituitary. The hypothalamus contains the regulating centers for many autonomic functions and also plays a role in sexual response, mating behaviors, the alarm response, and pleasure.

The *basal ganglia*, a cluster of nuclei below the cortex, are important in relaying motor impulses and coordinating motor responses. Parkinson's disease is associated with degeneration of cells entering the basal ganglia.

The *cerebral cortex* has changed most during vertebrate evolution. The human cortex is divided into five lobes. The two hemispheres of the cortex are connected by the *corpus callosum*, a thick band of fibers. The cortex contains both primary sensory and motor areas, which directly process information, and association areas, which integrate information from several sources.

Bilateral regions of the cerebral cortex are specialized as the primary sensory area, which receives impulses from touch, pressure, and pain receptors throughout the body, and the primary motor area, which sends impulses to the skeletal muscles. The proportion of the primary sensory and motor areas devoted to controlling each part of the body is correlated with the importance of that area. The special senses of vision, hearing, smell, and taste are controlled by other regions of the cortex.

Integration and Higher Brain Functions Nerve impulses are integrated on all levels of the nervous system, from the spinal reflex to the intellectual creations of the cerebral cortex.

Arousal is a state in which an individual is aware of the external world, whereas sleep is a state in which the individual is not conscious of external stimuli. Different patterns in the electrical activity of the brain may be recorded in an *electroencephalograph*, or *EEG*. Slow, synchronous alpha waves are produced by a person lying quietly with eyes closed. Faster beta waves are associated with opened eyes or thinking about a complex problem. Quite slow and highly synchronized delta waves occur during sleep. Periods of delta waves, however, alternate with periods of a desynchronized EEG and rapid eye movements. Most dreaming occurs during this REM sleep.

The *reticular formation* filters the sensory information reaching the cortex. Arousal is related to the amount of input the cortex receives. Sleep-producing centers are located in the pons and medulla, and serotonin may be the neurotransmitter involved in these areas. A center that causes arousal is found in the midbrain.

Human emotions have been tied to interactions between the cerebral cortex and a group of nuclei in the lower forebrain called the *limbic system*. Destruction of this portion of the brain may result in docility.

The association areas of the cerebral cortex are not bilaterally symmetrical. The left hemisphere controls speech, language, and calculation, whereas the right hemisphere controls artistic ability and spatial perception. Much of the information on this *lateralization* of the brain comes from Sperry's work with "split-brain" patients—people whose corpus callosums have been severed to control some forms of epilepsy.

Language and speech are controlled by areas on the left cerebral hemisphere. *Wernicke's area* stores information required for speech content and construction, whereas *Broca's area* involves information required for speech production by the lips, tongue, and other muscles. Different types of aphasia, the inability to speak coherently, occur if one or the other of these regions is damaged.

Human memory consists of *short-term memory*, the immediate sensory perception of an object or idea, and *long-term memory*, the recall of these perceptions after the passage of time. Transferring information from short-term to long-term memory is facilitated by rehearsal, a favorable emotional state, and associations with previously learned and stored information. Fact memories can be consciously recalled from your memory bank. Skill memories are formed by repetition of motor activities and do not require conscious recall of specific information.

Neuroscientists have determined that, in what appears to be a memory circuit, sensory information is transmitted from sensory regions of the cerebral cortex to the *hippocampus* and *amygdala*, components of the limbic system, from which impulses are relayed to an integrating area called the basal forebrain and then back to the sensory areas of the cortex. Chemical or structural changes in the sensory cortex may store the information as a memory. Some neuroscientists suggest that changes in dendrite structure enhance future transmission across a synapse, thus facilitating learning.

Simpler organisms are often used to study the mechanisms involved in memory and learning. The habituation of the sea hare *Aplysia* to mild touches, as well as its sensitization to mild touches associated with harmful stimuli, seem to involve changes in the ion channels at the synapses of interneurons.

STRUCTURE YOUR KNOWLEDGE

1. Develop a flow chart, diagram, or description of the sequence of events in the creation and propagation of an action potential, and the transmission of this potential across a chemical synapse.

2. The vertebrate nervous system can be separated into a sequence of divisions and subdivisions. Create a concept map that shows these subsets and their functions.

3. Fill in the table on the following page by briefly describing the functions of the parts of the human brain and identifying the section (hind-, mid-, or forebrain) in which they are found.

Part of brain	Section	Function
Medulla		
Pons		
Cerebellum		
Inferior, superior colliculi		
Reticular formation		
Thalamus		
Hypothalamus		
Cerebral cortex		

TEST YOUR KNOWLEDGE

MULTIPLE CHOICE: *Choose the one best answer.*

1. Which of the following is *not* a difference between the nervous system and the endocrine system?
   a. Nervous communication is more rapid.
   b. Endocrine system uses chemical communication; nervous system uses electrical communication.
   c. Nervous system messages are delivered directly to target cells or organs.
   d. The structural complexity of the nervous system allows for integration of more information and responses.

2. Interneurons
   a. may connect sensory and motor neurons.
   b. are confined to the peripheral nervous system.
   c. do not have cell bodies.
   d. are electrical synapses between neurons.

3. Nodes of Ranvier are
   a. gaps where Schwann cells abut at which action potentials are generated.
   b. neurotransmitter-containing vesicles located in the synaptic knobs.
   c. the major components of the blood-brain barrier that restrict the passage of substances into the brain.
   d. clusters of receptor proteins located on the postsynaptic membrane.

4. Which of the following is *not* true of the resting potential of a neuron?
   a. The inside of the cell is more negative than the outside is.
   b. There are concentration gradients with more sodium outside the cell and a higher potassium concentration inside the cell.
   c. It is about –70 mV and can be measured by using microelectrodes placed inside and outside the cell.
   d. It is formed by the sodium–potassium pump and the opening of voltage-sensitive channels.

5. After an action potential, the resting potential is restored by
   a. the opening of voltage-sensitive potassium channels.
   b. the opening of sodium gates.
   c. the refractory period in which the membrane is hyperpolarized.
   d. the delay in the action of the sodium–potassium pump.

6. The threshold potential of a membrane
   a. is a positive value, often equal to about 35mV.
   b. opens voltage-sensitive channels and permits the rapid outflow of sodium ions.
   c. is the depolarization that is needed to generate an action potential.
   d. is a graded potential that is proportional to the strength of a stimulus.

7. Which of the following is *not* true of chemical synapses?
   a. Synaptic knobs at the ends of branching dendrites contain synaptic vesicles, which enclose the neurotransmitter.
   b. The influx of calcium when an action potential reaches the presynaptic membrane causes synaptic vesicles to release their neurotransmitter into the cleft.
   c. The binding of neurotransmitter to receptors on the postsynaptic membrane changes the membrane's permeability to certain ions.
   d. An excitatory postsynaptic potential forms when sodium channels open and the membrane potential moves closer to an action potential threshold.

8. In spatial summation,
   a. the sum of simultaneously arriving neurotransmitters from different presynaptic nerve cells determines whether the postsynaptic cell fires.
   b. several action potentials arrive in fast succession from the same presynaptic cell.
   c. several IPSPs arrive concurrently, bringing the presynaptic cell closer to its threshold.
   d. the voltage spike of the action potential that is initiated is higher than normal.

9. An inhibitory postsynaptic potential occurs when
   a. sodium flows into the postsnyaptic cell.
   b. binding of the neurotransmitter opens ion gates that result in the membrane becoming hyperpolarized.
   c. enzymes do not break down the neurotransmitter in the synaptic cleft.
   d. acetylcholine is the neurotransmitter.

10. Which of the following would *not* function as a neurotransmitter?
    a. acetylcholine
    b. neuropeptides
    c. biogenic amines
    d. steroids

11. Groups of neurons that interact and carry information along pathways are called
    a. ganglia.
    b. nuclei, if they occur in the brain.
    c. circuits.
    d. nerve nets.

12. Which of the following animals is mismatched with its nervous system?
    a. sea star—modified nerve net, central nerve ring with radial nerves
    b. hydra (cnidarian)—ring of ganglia, paired ventral nerve cords
    c. annelid worm—brain, ventral nerve cord with segmental ganglia
    d. vertebrates—dorsal central nervous system of brain and spinal cord

13. Which of the following is *not* true of the autonomic nervous system?
    a. It is a subdivision of the somatic nervous system.
    b. It consists of the sympathetic and parasympathetic divisions.
    c. It is part of the peripheral nervous system.
    d. It controls smooth and cardiac muscles.

14. If you needed to obtain nuclei from neurons for an experiment, you would want to make preparations of
    a. astrocytes of the brain.
    b. the gray matter of the brain.
    c. the inner portion of the spinal cord.
    d. both b and c.

15. Cerebrospinal fluid
    a. cushions the brain.
    b. supplies the brain with nutrients and oxygen and removes wastes.
    c. is a filtrate of blood.
    d. is or does all of the above.

16. Which of the following structures is incorrectly paired with its function?
    a. pons—conducts information between spinal cord and brain
    b. cerebellum—contains pyramidal tracts that cross motor neurons from one side of the brain to the other side of the body
    c. thalamus—major projection area, screens and relays incoming impulses
    d. corpus callosum—band of fibers connecting left and right hemispheres

17. During REM sleep,
    a. a desynchronized EEG is produced.
    b. dreaming occurs.
    c. rapid eye movement occurs.
    d. all of the above occur.

18. The reticular formation
    a. filters sensory information and produces arousal.
    b. relays motor impulses and coordinates motor responses.
    c. is responsible for sleep and uses serotonin as its neurotransmitter.
    d. is severed in "split brain" patients.

19. The limbic system
    a. controls speech patterns and prevents aphasia.
    b. is responsible for lateralization of the brain.

c. is a group of nuclei in the lower forebrain associated with emotions.

d. regulates sexual response, mating behaviors, and pleasure.

20. A simple reflex, such as that of rapidly pulling a finger from a scalding tea kettle, includes

a. the transfer of information to and from the white matter of the spinal cord.

b. an exchange of information between interneurons.

c. a sensory receptor, a sensory neuron, and a motor neuron.

d. an interneuron and a stretch receptor in the skin.

CHAPTER 45

SENSORY AND MOTOR MECHANISMS

FRAMEWORK

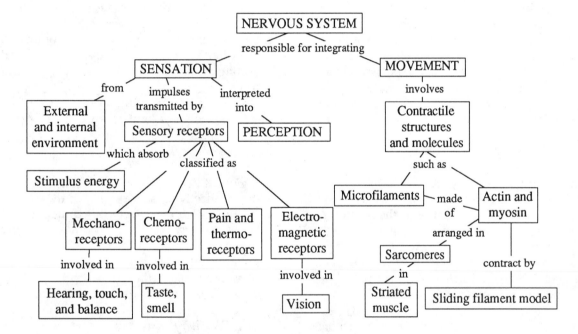

CHAPTER SUMMARY

Sensory Receptors

Sensation and Perception Information is transmitted as action potentials in the nervous system. *Sensations*, or impulses traveling along sensory neurons, are routed to different parts of the brain that interpret them into *perceptions*. The destination of the impulses determines what is perceived.

General Function of Sensory Receptors *Sensory receptors* are usually modified neurons that collect and transmit information from environmental stimuli. *Reception* is the absorption of the energy of a particular stimulus. The conversion of that energy into the electrochemical energy of changes in membrane potentials and action potentials is called *transduction*. The stimulus energy may need to undergo *amplification*, either by accessory structures of sense organs or as part of the transduction process, in order to enter the nervous system.

Transmission of the changes in the sensory cell membrane's electrical potential, called a *receptor potential*, to the nervous system may occur either as action potentials, if the receptor is a sensory neuron, or by the release of neurotransmitter into a synapse with a sensory neuron, which then translates into action potentials in the neuron. The intensity of the receptor potential correlates with the frequency of action potentials or the quantity of neurotransmitter released. Changes in the spontaneous firing rate of sensory cells provides information on both the presence or absense of the stimulus, as well as its intensity.

The *integration* of information from a stimulus begins at the receptor level, by the summation of signals, through *sensory adaptation* of the receptor cell to continued stimulation, and by changes in the sensitivity of receptors to different conditions.

Types of Receptors   Sensory receptors may be *exteroreceptors*, which receive information from the outside environment, or *interoreceptors*, which provide information from the inside of the body. Receptors may also be categorized on the basis of the type of energy stimulus to which they respond.

Mechanoreceptors respond to mechanical energy such as pressure, touch, motion, and sound. Bending or stretching of the mechanoreceptor cell membrane increases its permeability to sodium and potassium ions, creating a receptor potential. In humans, *Pacinian corpuscles*, modified dendrites of sensory neurons found in deep skin layers, respond to strong pressure, whereas the *Meissner's corpuscles* and *Merkel's discs* are closer to the surface and detect light touch. The position of body parts is monitored by *muscle spindles*, or stretch receptors, which are stimulated by stretching of the muscles. *Hair cells*, mechanoreceptors that detect motion, are found in the vertebrate ear, the lateral line organs of fishes and amphibians, and the balance organs of arthropods. When motion produces bending in the cilia or microvilli projecting upward from a hair cell, the membrane stretches, ion permeabilities change, and the rate of action potential firing is affected.

Chemoreceptors may be general receptors that monitor the total solute concentration in a solution or specific receptors that respond to individual kinds of molecules. Stimulus molecules bind to membrane sites on the receptor cell and initiate changes in membrane permeability. *Gustatory* (taste) and *olfactory* (smell) *receptors* respond to groups of related chemicals.

Thermoreceptors respond to heat or cold and help to regulate body temperature. *Ruffini's end organs* and *Krause's end bulbs*, receptors consisting of encapsulated, branched dendrites in mammalian skin, have been proposed as heat and cold receptors, respectively.

Naked dendrites called *nociceptors* detect pain. Different groups of receptors respond to excess heat, pressure, or chemicals released from damaged tissues. Histamines and acids trigger pain receptors, and prostaglandins increase pain by sensitizing receptors. *Photoreceptors* detect the electromagnetic radiation of visible light and are often organized into eyes. Some animals have receptors that detect infrared rays and electric currents.

Vision

Light Receptors and Vision of Invertebrates Receptors containing light-absorbing pigments are used by most invertebrates to detect light. The *eye cup* of planaria, which detects light intensity and direction, consists of receptor cells within a darkly pigmented cup opening to one side. Impulses from the left and right eye cups are compared to help the animal navigate a direct path away from a light source.

The compound eye and single-lens eye are two types of image-forming eyes found in invertebrates. The *compound eye* of insects and crustaceans contains up to thousands of light detectors called *ommatidia*, each of which has a cornea and lens. The different intensities of light entering the many ommatidia produce a mosaic image. Compound eyes are adept at detecting movement and may detect color and ultraviolet radiation.

In a *single-lens eye*, found in some jellyfish, spiders, and many mollusks, light is focused through the single lens onto the retina, consisting of a bilayer of photosensitive receptor cells.

Vertebrate Vision   The eyeball consists of the tough, outer, connective tissue layer called the *sclera*, and the thin, pigmented inner layer called the *choroid*. At the front of the eye, the sclera becomes the transparent *cornea*, and the choroid forms the colored *iris*, which regulates light entering the *pupil*. The *retina*, layered on the choroid, contains the photoreceptor cells. The optic nerve attaches to the eye at the optic disc.

The transparent *lens* focuses an image onto the retina. The *ciliary body* produces the *aqueous humor* that fills the anterior eye cavity. Glaucoma is a condition in which pressure in the eye is increased by an accumulation of aqueous humor. The jellylike *vitreous humor* fills the posterior cavity of the eye. Many fishes focus by moving the lens backward or forward. Most vertebrates focus by *accommodation*, in which muscles of the ciliary body change the shape of the lens.

Rod cells and *cone cells* are the photoreceptors found in the retina. The proportion of these two types of receptors correlates with the activity pattern of the animal. Rods are more light sensitive and are important in night vision, whereas cones distinguish colors. In the human eye, rods are concentrated toward the edge of the retina, and the center of the visual field, the *fovea*, is filled with cones. Birds of prey, such as hawks, have a high density of cones and keen eyesight.

Both rods and cones have visual pigments embedded in an outer stack of folded membranes. *Retinal* is the light-absorbing molecule, and it is bonded to a membrane protein called an *opsin*. Rods contain the molecule *rhodopsin*, which dissociates when struck by light. Enzymes recombine the retinal and opsin in the dark; in bright light, the rhodopsin remains bleached and cones are responsible for vision. The three subclasses of red, green, or blue cones, each with its own opsin that binds with retinal to form *photopsin*, are named for the colors of light they are best at absorbing.

When retinal in a rod or cone is stimulated by light, it initiates metabolic changes that decrease the permeability of the cell membrane to Na$^+$ ions, resulting in the creation of a receptor potential that hyperpolarizes the cells and reduces the release of neurotransmitter. Rods and cones synapse with *bipolar cells* in the retina, which in turn synapse with *ganglion cells*, whose axons transmit action potentials to the brain. *Horizontal cells* and *amacrine cells* are neurons in the retina that help to integrate visual information.

Signals from rods and cones may follow the vertical pathway directly from the receptor cells to bipolar cells to ganglion cells. (Several receptor cells synapse with each bipolar cell, and several bipolar cells pass information to one ganglion cell.) Lateral integration of visual signals involves both horizontal cells, which carry signals from one receptor cell to others and to several bipolar cells, and amacrine cells, which relay information from one bipolar cell to several ganglion cells. The *lateral inhibition* of nonilluminated receptors and bipolar cells by the horizontal cells and of ganglion cells by amacrine cells enhances the contrast between a spot of light and its surroundings.

The receptive field of a ganglion cell (the photoreceptors supplying information to the cell—information that is integrated by bipolar, horizontal, and amacrine cells) may be of two types, one of which responds to light spots surrounded by darkness (on-center fields), and the other to spots of darkness surrounded by light (off-center fields). Both of these patterns enhance the patterns of bright and dark spots.

The axons of the ganglion cells form the optic nerve. The left and right optic nerves meet at the *optic chiasma* at the base of the cerebral cortex, and information from the left sides of both eyes travels to the right side of the brain, whereas what is sensed in the right fields of view go to the left side of the brain. Most axons of the ganglion cells go to the *lateral geniculate nuclei* of the thalamus, where neurons lead to the *primary visual cortex* in the occipital lobe. Other interneurons carry information to other visual centers in the cortex.

Somehow the coded spots, lines, and movements that are projected onto the visual cortex are integrated into our perception and recognition of objects.

Hearing and Balance

The Mammalian Ear In mammals and most terrestrial vertebrates, the mechanoreceptors for hearing and balance are located within the ear. The ear consists of three regions. The external *pinna* and the *auditory canal* make up the outer ear. The *tympanic membrane* (eardrum) of the *middle ear* transmits sound waves to three small bones—the *malleus* (hammer), *incus* (anvil), and *stapes* (stirrup)—which conduct the waves to the *inner ear* by way of a membrane beneath the stapes, called the *oval window*. The *auditory (Eustachian) tube*, connecting the pharynx and the middle ear, equalizes pressure within the middle ear. The inner ear consists of a fluid-filled labyrinth of channels within the temporal bone.

The coiled *cochlea* of the inner ear has two large chambers—an upper vestibular canal and a lower tympanic canal—separated by a smaller cochlear canal. The *organ of Corti*, located on the floor of the cochlear canal, contains the receptor hair cells. The tips of the hair cells extend into the cochlear canal, and some attach to the tectorial membrane, which overhangs the organ of Corti.

Sound waves—which were transmitted and amplified by the tympanic membrane, the three bones of the middle ear, the oval window, and pressure waves in the cochlear fluid—are transduced into action potentials in the cochlea. Pressure waves, traveling from the vestibular canal through the tympanic canal and dissipating when they strike the *round window*, vibrate the basilar membrane on which the organ of Corti lies. The bending of the hairs against the tectorial membrane stretches the cell membranes, which makes them more permeable to sodium. The resulting depolarization increases neurotransmitter release from the hair cell and increases the frequency of action potentials generated in the sensory neuron and carried through the auditory nerve to the brain.

Volume is a result of the *amplitude*, or height, of the sound wave; a stronger wave bends the hair cells more and results in an increase in action potentials. *Pitch* is related to the *frequency* of sound waves, usually expressed in Hertz (Hz), or vibrations per second. Different regions of fibers on the basilar membrane vibrate more vigorously in response to different frequencies and transmit their impulses to specific regions of the cerebral cortex, where the sensation is perceived as a particular pitch.

Within the inner ear, two chambers, the *utricle* and *saccule*, and three *semicircular canals* make up the vestibular apparatus responsible for balance and equilibrium in humans and most mammals. The hairs of the hair cells in the utricle and saccule project into a gelatinous material containing calcium carbonate particles called otoliths. Changes in head position change the pull on the hair cells, increasing or decreasing their output of action potentials and thus signalling the position of the head. The inertia of the endolymph—the fluid in the semicircular canals, which extend in three planes in space—causes hair cells to bend and increase impulse frequency in sensory neurons in response to the rotation of the head.

Hearing and Equilibrium in Other Vertebrates

Mechanoreceptor units called *neuromasts*—clusters of hair cells with sensory hairs embedded in a gelatinous cap or cupula—are contained in the *lateral line system* of fish and aquatic amphibians. Water moving through a tube past these mechanoreceptors bends the cupula and stimulates the hair cells, enabling the fish to perceive its movement, water currents, and pressure waves generated by other moving objects.

Fish also have inner ears, consisting of a saccule, utricle, and semicircular canals, within which sensory hairs are stimulated by the movement of otoliths or granules. The vibrations of sound waves in the water are passed through the skeleton of the head to the inner ears. Some fishes have a *Weberian apparatus*, a series of bones that transmits vibrations from the swim bladder to the inner ear.

In amphibians and reptiles, sound vibrations are conducted by a tympanic membrane and single bone to the inner ear. A cochlea has evolved in birds, but sound is conducted from the tympanic membrane to the inner ear by a single bone.

Sensory Organs for Hearing and Balance in Invertebrates

Many arthropods use body hairs that vibrate in response to sound waves of specific frequencies to sense sounds. Localized "ears" are often found on the legs of insects and consist of a tympanic membrane with attached receptor cells that is stretched over an internal air chamber.

Invertebrates have mechanoreceptors, called *statocysts*, which often consist of a layer of hair cells around a chamber containing statoliths, dense granules or grains of sand. The stimulation of the hair cells under the statocysts provides positional information to the animal.

Taste and Smell

The sense of taste detects chemicals that are present in a solution, whereas smell detects airborne chemicals. Chemical senses are used by animals to find mates, navigate, communicate, and feed. *Setae*, or sensory hairs containing chemoreceptor taste cells, are found on the feet and mouthparts of insects. Insects also have olfactory setae, usually located on their antennae.

The mammalian chemical senses of gustation and olfaction are produced when a molecule binds to a receptor protein in a receptor cell membrane and triggers a membrane depolarization and the release of neurotransmitter.

Taste buds, which contain groups of receptor cells, are scattered on the tongue and mouth. The four primary taste sensations—sweet, sour, salty, and bitter—are detected in distinct regions of the tongue. The brain integrates the input from the taste buds as they differentially respond to different chemicals and creates the perception of a complex flavor.

Olfactory receptor cells line the upper part of the nasal cavity and send their axons to the olfactory bulb of the brain. Chemicals bind to specific receptor molecules on the cilia from the receptor cells that extend into the mucous layer of the nasal cavity.

Introduction to Animal Movement

Animals move to obtain food, escape from danger, and find mates. Modes of locomotion vary; animals may swim, crawl, walk, run, hop, or fly. Skeletal and muscle design and body shape reflect evolutionary adaptations to the challenges of overcoming friction and gravity in a water, land, or air environment.

All animal movement depends on protein strands moving past each other, either in microtubules, which are responsible for beating of cilia and flagella, or in microfilaments, which are involved in amoeboid movement and muscle contraction.

Skeletons and Their Roles in Movement

Skeletons function in support of the body, protection of soft tissues, and movement.

Hydrostatic Skeletons

Fluid under pressure in a closed body compartment creates a *hydrostatic skeleton*. As muscles contract and change the shape of these fluid-filled compartments, the animal moves. Hydrostatic skeletons are found in most cnidarians, flatworms, nematodes, and annelids.

Exoskeletons

Exoskeletons, typical of mollusks and arthropods, are hard coverings deposited on the surface of animals. The *cuticle* of an arthropod contains fibrils of *chitin*, a polysaccharide similar to cellulose, embedded in a protein matrix. This flexible support is hardened where protection is needed by the cross-linking of proteins and the addition of calcium salts but remains pliable at the joints of the skeleton.

Endoskeletons

The supporting elements of an *endoskeleton* are embedded in the soft tissues of the animal.

Sponges have endoskeletons composed of spicules or fibers, and echinoderms have ossicles or hard plates beneath the skin. The endoskeletons of chordates are composed of cartilage and/or bone. The vertebrate axial skeleton consists of skull, vertebral column, and rib cage; the appendicular skeleton contains the pectoral and pelvic girdles and the limb bones.

Muscles

Animal movement involves the contraction of muscles working against some type of skeleton. Movement in oppositive directions requires antagonistic pairs of muscles.

Structure and Physiology of Vertebrate Skeletal Muscle   Vertebrate *skeletal muscle* consists of a bundle of multinucleated muscle cells, called fibers, running the length of the muscle. Each fiber is a bundle of *myofibrils*, each composed of two kinds of *myofilaments*: *thin filaments*, consisting of two strands of actin coiled with a strand of regulatory protein, and *thick filaments*, made of myosin molecules.

The regular arrangement of myofilaments produces repeating light and dark bands; thus skeletal muscle is also called striated muscle. The repeating units are called *sarcomeres* and are delineated by *Z lines*. Thin filaments attach to the Z line and project toward the center of the sarcomere. Thick filaments lie free in the center, overlapping the thin filaments in an *A band*, which is broken by an *H zone* in the center where the thin filaments do not reach. At the edges of the sarcomere, where the thick filaments do not extend when the muscle is at rest, are *I bands* of only thin filaments.

According to the *sliding filament model* of muscle contraction, the thick filaments ratchet the thin filaments toward the center of the sarcomere, pulling the Z lines together, shortening the I bands, and eliminating the H zone. The heads of myosin molecules spontaneously bind to specific sites on the actin filament, forming *cross-bridges* between the thick and thin filaments, and then bend, pulling the thin filaments toward the center of the sarcomere. The myosin molecule hydrolyzes an ATP molecule, releasing energy to break the cross-bridge, and straightens out. The "attach, bend, detach, straighten" cycle pulls the thin filaments past the thick filaments. The synchronous shortening of sarcomeres in the myofibrils of the fibers results in the contraction of the muscle.

When the muscle is at rest, the myosin-binding sites of the actin molecules are blocked by the *tropomyosin* strand in the thin filament and by a set of regulatory proteins, called the *troponin complex*, at each binding site. When troponin binds calcium ions, the tro-pomyosin-troponin complex changes shape, and the actin filament is exposed.

Calcium ions are actively transported into the *sarcoplasmic reticulum* of a muscle cell. When an action potential of a motor neuron causes the release of acetylcholine into the neuromuscular junction (and the resulting graded depolarization is large enough), an action potential spreads across the cell membrane and into *transverse (T) tubules*, extensions of the plasma membrane that project into the muscle cell. The action potential carried by the T tubules depolarizes the sarcoplasmic reticulum membrane, and calcium ions are released into the cell. Calcium binding with troponin exposes the actin sites and contraction begins. The sarcoplasmic reticulum pumps calcium back out of the cytoplasm when the action potential passes, and the tropomyosin–troponin complex again blocks the actin sites.

Most of the energy for muscle contraction comes from *phosphagens*, such as *creatine phosphate* in vertebrates, which transfer high-energy phosphate bonds to ADP to replenish ATP supplies.

A contraction of a single muscle fiber is an all-or-none action. Graded muscular contractions result from the involvement of additional fibers. A branched motor neuron and the muscle fibers it innervates make up a *motor unit*. The strength of a muscle contraction can be increased by the stimulation of additional motor units, called *multiple motor unit summation*, or *recruitment*.

A *muscle twitch* results from a single stimulus. More graded contraction of muscles occurs by *wave summation*, in which stimuli are given so rapidly that the muscle does not relax between them, and each contraction is stronger than the last. Motor neurons usually deliver a volley of action potentials, producing a state of smooth and sustained contraction called *tetanus*.

The muscle fatigue of extended tetanus is caused by the depletion of ATP, loss of ion gradients required for depolarization, and accumulation of lactic acid resulting from fermentation in the muscle. Some support muscles are always in a state of partial contraction, known as *tonus*, in which different motor units take turns contracting.

Slow muscle fibers have less sarcoplasmic reticulum, and thus calcium remains in the cytoplasm longer and twitches last longer. Slow fibers are common in muscles that must sustain long contractions. They have many mitochondria, a good blood supply, and *myoglobin*, which extracts and stores oxygen from the blood. Fast muscle fibers are used for rapid and powerful contractions.

Other Types of Muscles   Vertebrate *cardiac muscle* cells are branched and connected by *intercalated discs*, across which action potentials spread to all the cells of the heart. Cardiac muscle cells have high sodium permeability, and they can generate action potentials without nervous input. Both their action potentials and their refractory periods last a long time.

The spiral arrangement of actin and myosin filaments in *smooth muscle* accounts for its nonstriated appearance. Smooth muscle lacks a T-tubule system and a well-developed sarcoplasmic reticulum. Its relatively slow contractions can extend over a greater length.

Invertebrates have muscles similar to the skeletal and smooth muscles of vertebrates. The flight muscles of insects are capable of independent and rapid contraction.

An animal's behavior is a function of the integration and interpretation of inputs from sensory receptors by a nervous system, which then signals muscles to move and respond.

STRUCTURE YOUR KNOWLEDGE

1. Label the parts of the eye.

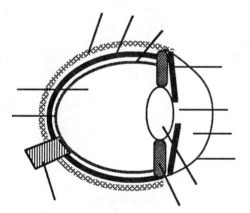

2. Trace the path of a light stimulus from where it enters the eye to its transmission as an impulse through the optic nerve.

3. Arrange the following structures in an order that enables you to trace the passage of sound waves from the external environment to the brain.

round window	cochlea	vestibular and tympanic canal
oval window	auditory	tectorial membrane
organ of Corti	basilar membrane	tympanic membrane
auditory nerve	cerebral cortex	malleus, incus, stapes

4. Trace the sequence of events in muscle contraction from an action potential in a motor neuron to the relaxation of the muscle.

TEST YOUR KNOWLEDGE

MATCHING: *Match the term with its description.*

1. _____ oxygen-storing compound in muscles

2. _____ receptor on muscle, monitors position

3. _____ conversion of stimulus energy into electrochemical energy

4. _____ deep pressure receptors in humans

5. _____ naked dendrites that detect pain

6. _____ sensory hairs with taste receptors on feet and mouthparts of insects

7. _____ invertebrate mechanoreceptors for position; chamber of hair cells

8. _____ energy-storing compound in muscles

9. _____ focusing by changing shape of lens

10. _____ numerous light detectors making up compound eye

11. _____ decrease in sensitivity of receptor during constant stimulation

12. _____ fibers in muscle cells

A. accommodation

B. adaptation

C. ampulla

D. creatine phosphate

E. cupula

F. ganglion cells

G. muscle spindle

H. myofibrils

I. myoglobin

J. nociceptors

K. ommatidia

L. Pacinian corpuscles

M. setae

N. statocyst

O. transduction

P. transmission

Q. tropomyosin

MULTIPLE CHOICE: *Choose the one best answer.*

1. The absorption of energy from a particular stimulus is called?
   a. reception
   b. transduction
   c. integration
   d. transmission

2. Prostaglandins
   a. increase perception of pain by sensitizing pain receptors.
   b. act like histamines to trigger pain receptors.
   c. are released by nociceptors in response to a pain stimulus.
   d. are stimulated by aspirin and decrease pain.

3. The compound eye found in insects and crustaceans
   a. is composed of thousands of prisms that focus light onto a few photoreceptor cells.
   b. cannot sense colors.
   c. contains many ommatidia, each of which consists of a cornea and lens that focus light from a tiny portion of the field of view.
   d. all of the above are correct.

4. Which of the following statements is *not* true?
   a. The iris regulates the amount of light entering the pupil.
   b. The choroid is the thin pigmented inner layer of the eye that contains the photoreceptor cells.
   c. The ciliary body changes the shape of the lens.
   d. The thin aqueous humor fills the anterior eye cavity.

5. Which of the following is incorrectly paired with its function?
   a. cones—respond to different light wavelengths producing color vision
   b. horizontal cells—carry signals between receptor cells and bipolar cells
   c. bipolar cells—stimulate ganglion cells
   d. ganglion cells—produce lateral inhibition of receptor and bipolar cells

6. The fovea is
   a. the blind spot.
   b. the center of the visual field, containing only cones.
   c. the layer that contains photoreceptor cells.
   d. the optic disc, where the optic nerve attaches to the eye.

7. The molecule rhodopsin
   a. is involved with color vision.
   b. dissociates when struck by light and is only functional in complete dark.
   c. is recombined by enzymes from retinal and opsin in the light.
   d. is more sensitive to light than are photopsins and is involved in black and white vision in dim light.

8. Rotation of the head is sensed by
   a. the Weberian apparatus located in the utricle and saccule.
   b. changes in the action potentials of hair cells in response to the pull on them by calcium carbonate particles.
   c. bending of hair cells in the semicircular canals in response to the inertia of the endolymph.
   d. hair cells in response to movement of fluid in the coiled cochlea.

9. The tympanic membrane
   a. conducts sound waves to the fluid in the vestibular canal.
   b. sets up vibrations in the basilar membrane.
   c. transmits sound waves to the malleus, incus, and stapes.
   d. equalizes pressure between the middle ear and the outside.

10. Fish can perceive pressure waves from moving objects with their
    a. lateral line system containing mechanoreceptor units called neuromasts.
    b. tectorial membrane as it is stimulated by the bending of sensory hairs embedded in a gelatinous cap.
    c. inner ears, which consist of a saccule, utricle, and semicircular canals, within which sensory hairs are stimulated by the movement of otoliths.
    d. Weberian apparatus, a series of bones that transmits vibrations from the swim bladder to the inner ear.

11. The binding of a molecule to a receptor protein on a receptor cell membrane
    a. triggers a membrane depolarization and release of neurotransmitter.
    b. is the mechanism of reception and transduction in chemoreceptors.
    c. is involved in both gustatory and olfactory sensation.
    d. involves all of the above.

12. Hydrostatic skeletons are used for movement by all of the following except
    a. cnidarians.
    b. echinoderms.
    c. nematodes.
    d. annelids.

13. When striated muscle fibers contract,
    a. the Z lines are pulled closer together.
    b. the I band is eliminated.
    c. the A and I bands shorten.
    d. all of the above occur.

14. Tetanus
    a. is the result of puncture wounds.
    b. is the all-or-none contraction of a single muscle fiber.
    c. is the result of stimulation of additional motor neurons, called multiple motor unit summation.
    d. is the result of waveform summation, which produces increased and sustained contraction of a muscle.

15. What is the role of ATP in muscle contraction?
    a. to form cross-bridges between thick and thin filaments
    b. to break the cross-bridge between the head of the myosin molecule and a binding site on the actin filament
    c. to remove the tropomyosin–troponin complex from blocking the binding site on the actin filament
    d. to bend the cross-bridge and pull the thin filaments toward the center of the sarcomere

16. When an action potential spreads into the transverse tubules,
    a. calcium is released from the tubules into the sarcoplasmic reticulum.
    b. acetlycholine is released into the neuromuscular junction.
    c. the sarcoplasmic reticulum membrane is depolarized and calcium ions are released into the cell.
    d. calcium binds with troponin and the actin sites are exposed.

17. Most of the readily available energy for muscle contraction in vertebrates comes from
    a. phosphagens.
    b. myoglobin.
    c. glycogen.
    d. both a and b.

18. Muscle fatigue may be caused by
    a. a prolonged state of tonus.
    b. an accumulation of lactic acid resulting from fermentation in the muscle.
    c. the depletion of calcium in a muscle.
    d. wave summation in which motor neurons deliver a volley of action potentials.

19. Which of the following is *not* a characteristic of cardiac muscle?
    a. spiral arrangement of actin and myosin filaments
    b. intercalated discs that spread action potentials between cells
    c. action potentials and refractory periods that last a long time
    d. high sodium permeability and the ability to generate action potentials without nervous input

20. Smooth muscle contracts relatively slowly because
    a. the only ATP available is supplied by fermentation.
    b. its contraction is stimulated by hormones, not motor neurons
    c. it does not have a well-developed sarcoplasmic reticulum, and calcium enters the cell through the plasma membrane during an action potential.
    d. it is not striated.

FILL IN THE BLANKS: *Use this diagram of a section of a skeletal muscle fiber to answer the following.*

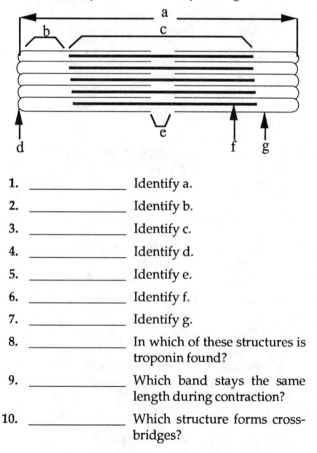

1. _____ Identify a.
2. _____ Identify b.
3. _____ Identify c.
4. _____ Identify d.
5. _____ Identify e.
6. _____ Identify f.
7. _____ Identify g.
8. _____ In which of these structures is troponin found?
9. _____ Which band stays the same length during contraction?
10. _____ Which structure forms cross-bridges?

CHAPTER 36
INTRODUCTION TO ANIMAL STRUCTURE AND FUNCTION

Suggested Answers to Structure Your Knowledge

1.

TISSUE	STRUCTURAL CHARACTERISTICS	GENERAL FUNCTIONS	STRUCTURE AND FUNCTION OF SPECIFIC TYPES
Epithelial	Packed cells, basement membrane, cuboidal, columnar, squamous, simple or stratified	Protection, absorption, secretion, lines body surfaces	Mucous membrane: goblet cells secrete mucus, creating lubricated surface, may have cilia. Glands: exocrine excretes into ducts, endocrine secrete hormones into blood.
Connective	Few cells that secrete extracellular matrix with fibers in liquid, gel or solid ground substance	Connect and support other tissues	Loose connective: loose weave of collagenous, elastic fibers, hold organs in place. Adipose: adipose cells store fat. Fibrous connective: collagenous fibers, form tendons, ligaments. Cartilage: chondrocytes secrete chondrin, rubbery matrix, flexible support. Bone: osteocytes in Haversian system, mineralized matrix, strong support. Blood: erythrocytes, leukocytes, platelets, plasma
Muscle	Long cells with large numbers of micro-filaments of actin and myosin	Contraction, movement	Skeletal: striated, voluntary movement. Visceral: smooth, spindle-shaped cells, in organs, involuntary. Cardiac: striated, branched cells, intercalated discs, in heart.
Nervous	Neurons with cell body, axons, dendrites	Sense stimuli, conduct impulses	

2. All organs are covered with epithelial tissue, and internal lumens, ducts, and blood vessels will be lined with epithelium. Some types of connective tissue would be found in most organs, usually serving to connect other tissues and supply blood to the nonvascularized epithelium. Muscle tissue would be present if the organ must contract. Nervous tissue innervates most organs. The small intestine is an example of an organ composed of many tissues. An epithelial mucosa lines the lumen; the submucosa contains blood vessels and loose connective tissue; a layer of circular and longitudinal smooth muscle comes next; and the outer layer is an epithelium.

3. A compact animal must provide gas exchange, nutrients, and waste removal for each cell. Large internal surface areas are specialized for obtaining oxygen, for digestion and absorption, and for removal of waste from the blood. The circulatory system usually provides these resources to all the cells of the body through the interstitial fluid.

Answers to Test Your Knowledge

Multiple Choice:

1. b   3. d   5. b   7. b
2. a   4. a   6. d   8. d

CHAPTER 37
ANIMAL NUTRITION

Suggested Answers to Structure Your Knowledge

1.

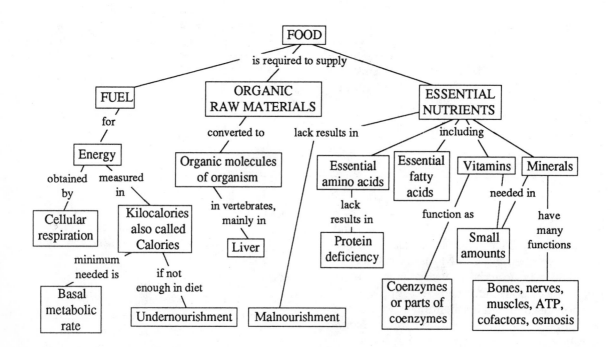

2.

Organ or Section	Processes Occurring	Associated Glands, Organs, and Digestive Fluids
Oral cavity	Mechanical breakup of food, mix with saliva, begin breakdown of starch, mucin to lubricate	Salivary glands secrete salivary amylase, mucin, buffers
Pharynx	Swallowing of bolus, epiglottis closes windpipe.	
Esophagus	Bolus through esophageal sphincter, peristalsis moves bolus to stomach	
Stomach	Food tissues broken down by low pH, beginning of protein hydrolysis, churning creates acid chyme, exit by sphincter	Epithelial glands secrete gastric juice with pepsin, HCl; hormone gastrin regulates gastric-juice release
Small intestine	Complete digestion of starch and proteins, fats emulsified and hydrolyzed, absorption of nutrients across epithelial lining into capillaries and lacteals, large surface area due to villi and microvilli	Pancreatic enzymes: amylase, trypsin, and other protein-hydrolyzing enzymes; bicarbonate neutralizes; liver produces bile salts, stored in gall bladder, emulsify fats; lipase hydrolyzes fats Hormones: secretin, CCK, enterogastrone control secretions and peristalsis
Large intestine	Reabsorption of water, feces stored in rectum; some microorganisms may produce vitamin K	

Answers to Test Your Knowledge

Matching:

1. L	**5.** E	**9.** D	
2. H	**6.** A	**10.** N	
3. C	**7.** J		
4. M	**8.** F		

Multiple Choice:

1. d	**5.** d	**9.** a	**13.** c
2. c	**6.** d	**10.** b	**14.** c
3. c	**7.** a	**11.** b	**15.** b
4. a	**8.** c	**12.** a	

CHAPTER 38
CIRCULATION AND GAS EXCHANGE

Suggested Answers to Structure Your Knowledge

1.

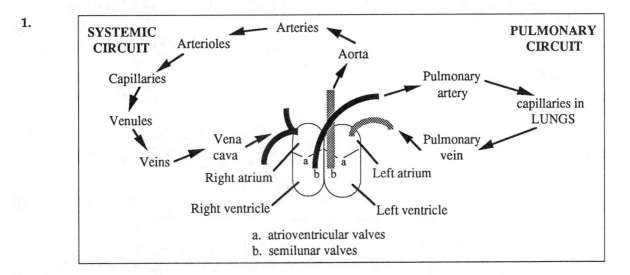

a. atrioventricular valves
b. semilunar valves

2.

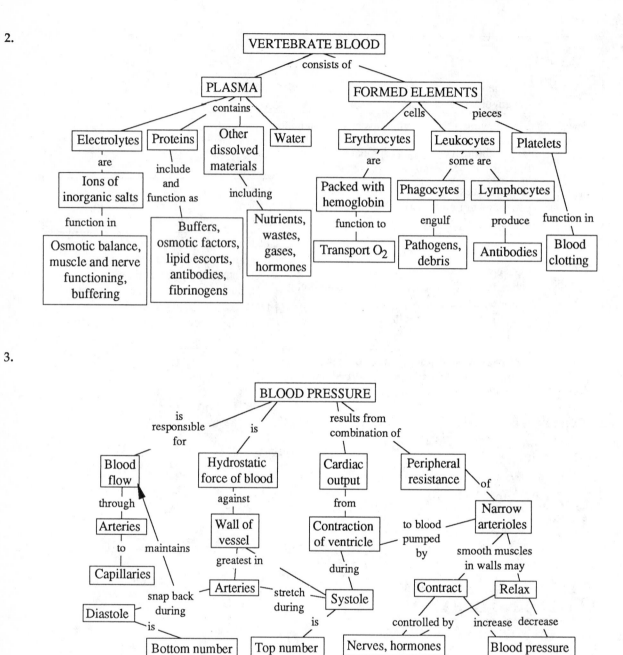

3.

4.

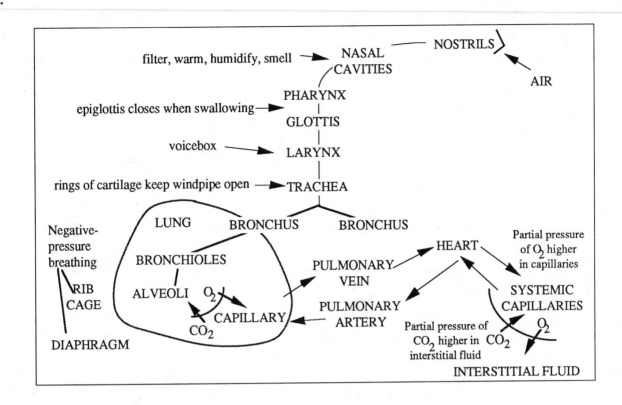

Answers to Test Your Knowledge

Multiple Choice:

1. c	6. d	11. b	16. d	21. a
2. c	7. a	12. b	17. b	22. b
3. d	8. b	13. c	18. c	23. c
4. d	9. a	14. d	19. b	24. a
5. a	10. d	15. a	20. b	25. d

CHAPTER 39
THE IMMUNE SYSTEM

Suggested Answers to Structure Your Knowledge

1.

Molecule	Where Produced or Found	Action
Lysozyme	Enzyme in perspiration, tears, and saliva	Attacks cell walls of many bacteria
Histamine	Released by injured cells and by degranulation of mast cells	Causes dilation and leakiness of small blood vessels, initiates inflammatory response
Interferon	Production initiated by virus infection, messenger that diffuses to other cells	Stimulates production of proteins that inhibit cells from making viral proteins, may recruit natural killer cells and macrophages
Complement	Set of blood proteins, activated by antigen–antibody complexes in immune response	Attract phagocytes, stimulate histamine release, coat microorganisms in opsonization, form membrane attack complex & lyse cells
Antibody	Protein (Ig) made by plasma cells derived from B cells, has V and C regions, circulate in blood and lymph	Bind to specific antigen, mark foreign cells and molecules for destruction, antigen-antibody complexes activate complement
Interleukin I	Messenger molecule (cytokine) secreted by antigen-presenting macrophage (APC) after binding with helper T cell	Stimulates division of T cell
Interleukin II	Cytokine released by activated T cells	Stimulates helper T and cytotoxic T cells to reproduce, helps activate B cells
Perforin	Released by cytoxic T cells when attached to cells displaying antigen-MHC complexes	Forms hole in target cell's membrane

2.

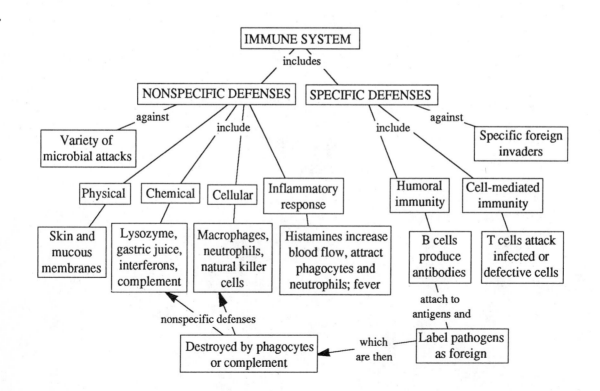

3.

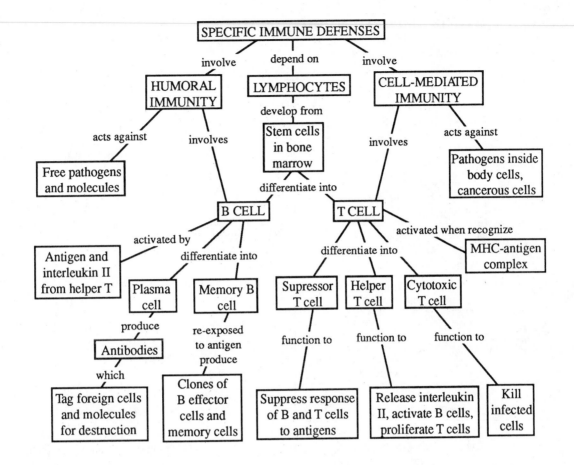

SPECIFIC IMMUNE DEFENSES

involve — HUMORAL IMMUNITY

depend on — LYMPHOCYTES

involve — CELL-MEDIATED IMMUNITY

HUMORAL IMMUNITY:
- acts against — Free pathogens and molecules
- involves

LYMPHOCYTES:
- develop from — Stem cells in bone marrow
- differentiate into — B CELL / T CELL

CELL-MEDIATED IMMUNITY:
- involves
- acts against — Pathogens inside body cells, cancerous cells

B CELL:
- activated by — Antigen and interleukin II from helper T
- differentiate into — Plasma cell / Memory B cell

Plasma cell — produce — Antibodies — which — Tag foreign cells and molecules for destruction

Memory B cell — re-exposed to antigen produce — Clones of B effector cells and memory cells

T CELL:
- activated when recognize — MHC-antigen complex
- differentiate into — Supressor T cell / Helper T cell / Cytotoxic T cell

Supressor T cell — function to — Suppress response of B and T cells to antigens

Helper T cell — function to — Release interleukin II, activate B cells, proliferate T cells

Cytotoxic T cell — function to — Kill infected cells

4. An antibody is a protein that typically consists of two identical light polypeptide chains and two identical heavy chains, held together by disulfide bonds in a Y-shaped molecule. The amino acid sequences in the variable sections of the L and H chains in the arm portions of the Y account for the specificity in binding between antibody and antigenic determinant. The constant region of the antibody determines its effector function: five types of constant regions correspond to five classes of antibodies. As an example, IgE immunoglobulins are attached to mast cells, which create allergic responses, and IgA is found in body secretions and is the major antibody in colostrum, the first breast milk produced.

According to the clonal selection theory, T and B cells become fixed early in their development to produce a specific antigen receptor or antibody. The body's ability to respond to a great variety of antigens depends of a lymphocyte population with a huge diversity of receptor specificities. When a T or B cell encounters its antigen, it is selectively activated to proliferate and produce a clone of lymphocytes, all having the same antigenic specificity.

Answers to Test Your Knowledge

Multiple Choice:

1.	d	5.	b	9.	a
2.	c	6.	d	10.	b
3.	b	7.	a		
4.	a	8.	c		

CHAPTER 40
CONTROLLING THE INTERNAL ENVIRONMENT

Suggested Answers to Structure Your Knowledge

1.

	Osmoregulation	Excretory System	Thermoregulation
Marine invertebrate	Isosmotic to seawater, osmoconformer	Diffusion of ammonia through body wall	Ectothermic
Marine bony fish	Drink water to compensate for loss to hyperosmotic seawater, salt glands in gills pump out salt	Little urine excreted, ammonia diffuses out	Ectothermic, may have heat exchanger
Freshwater fish	Dilute urine to compensate for osmotic gain, may pump salts in through gills	Kidneys produce copius, dilute urine, ammonia lost across epithelium of gills	Ectothermic
Flatworm	Flame-cell system, dilute fluid excreted; protonephridia	Most wastes excreted into gastrovascular cavity	Ectothermic
Earthworm	Metanephridia, excrete dilute urine to offset osmosis from habitat	Metanephridia	Ectothermic, behavior responses
Insect	Malpighian tubules, salt pumped back and most water reabsorbed	Malpighian tubules, uric acid conserves water	Ectothermic, behavior responses
Reptile	Nephron in kidney, uric acid conserves water	Nephron in kidney, uric acid in most, related to shelled egg	Ectothermic, behavior responses
Bird	Nephron in kidney, nasal salt glands in sea birds, uric acid conserves water	Nephron in kidney, juxtamedullary nephron, uric acid related to shelled egg	Endothermic, high metabolism, pant, heat exchanger
Mammal	Nephron in kidney, concentration of urine related to habitat, hormonal, feedback control	Nephron in kidney, juxtamedullary nephron, urea	Endothermic, vasodilation and constriction, heat production, behavior responses

2.

Environment	Osmoregulation	Nitrogenous Waste	Thermoregulation
Seawater	Most are osmoconformers, hypoosmotic fish lose water, drink sea water, secrete salt	Ammonia diffuses out easily into water, excreted by gills in fish	Stable environment mostly ectotherms, mammals have blubber, countercurrent
Freshwater	Animals are hyperosmotic, gain water, excrete large quantities of urine, ion uptake across gills in fish	Ammonia diffuses easily into water, exchanged across gills for Na^+ in fish	Fairly stable environment, mostly ectotherms
Terrestrial	Lose water by evaporation, drink, regulate water lost through excretion	Urea less toxic, needs less water to excrete, uric acid most concentrated, found in animals with shelled egg	Variable environment, behavioral adaptations if ectotherms, endotherms can maintain temperature and move rapidly

3.

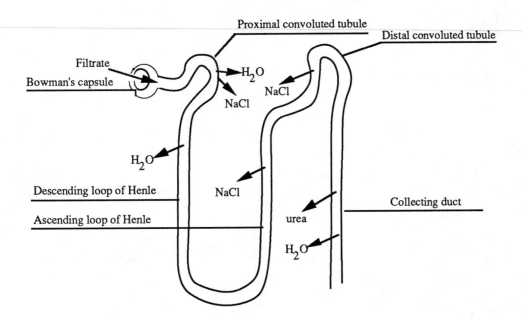

Substances filtered	Substances secreted	Substances Reabsorbed
water	ammonia	water
salts	hydrogen ions	glucose
nitrogenous wastes	potassium (maintains balance)	amino acids
glucose	drugs and poisons	vitamins
vitamins		potassium
urea		NaCl
other small molecules		bicarbonate

4.

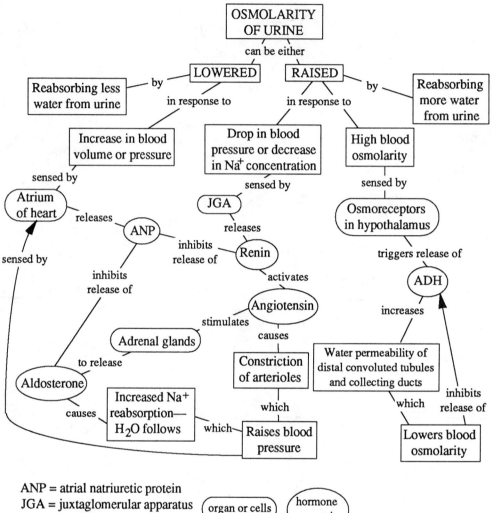

ANP = atrial natriuretic protein
JGA = juxtaglomerular apparatus
ADH = antidiuretic hormone

Answers to Test Your Knowledge

Multiple Choice:

1. b	**5.** d	**9.** b	**13.** d	**17.** d
2. a	**6.** c	**10.** d	**14.** c	**18.** c
3. c	**7.** a	**11.** b	**15.** d	
4. a	**8.** c	**12.** b	**16.** b	

CHAPTER 41
CHEMICAL COORDINATION

Suggested Answers to Structure Your Knowledge

1.

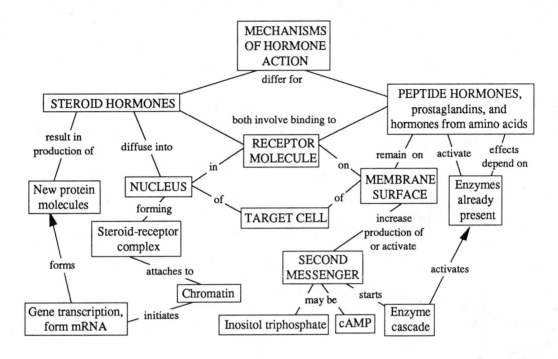

2.

GLAND	HORMONES	MAIN ACTION
HYPOTHALAMUS Neurosecretory cells (drain to posterior pituitary)	Oxytocin	Stimulate uterine muscle contraction, mammary glands
	Antidiuretic hormone (ADH)	Increase water permeability of collecting ducts, promote reabsorption of water from urine
(drain to median eminence)	Releasing factors	Carried by portal system to anterior pituitary, control release of hormones from pituitary
	TSH-releasing hormone (TRH)	Stimulate anterior pituitary to secrete TSH
	Gonadotropic-releasing hormone (GnRH)	Stimulate anterior pituitary to secrete FSH and LH
PITUITARY Posterior lobe (neurohypophysis)	Stores and secretes oxytocin and ADH	See actions listed for oxytocin and ADH above.
Anterior lobe (adenohypophysis)	Growth hormone (GH)	Promote growth, stimulate other growth factors
	Prolactin (PRL)	Various effects, depending on species—mammary gland growth, osmoregulation, metamorphosis
	Thyroid-stimulating hormone (TSH)	Stimulate thyroid gland to secrete hormones
	Follicle-stimulating, lutenizing hormones (FSH and LH)	Stimulate activities of gonads, ovulation, and sperm production
	Adrenocorticotropin (ACTH)	Stimulate adrenal cortex to produce and secrete glucocorticoids
	Melanocyte-stimulating hormone (MSH)	Regulate activity of pigment-containing cells in skin
	Endorphins and enkephalins	Effects similar to morphine, reduce pain

3.

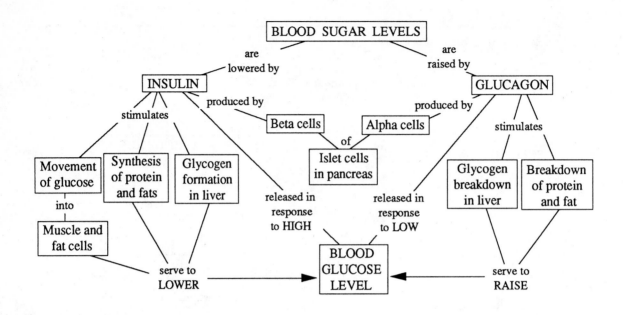

Answers to Test Your Knowledge

Matching:

1. I f	**5.** A g	**9.** C	c, stored in g
2. G a	**6.** E b	**10.** J	c, stored in g
3. F d	**7.** K j		
4. B h	**8.** D j		

Multiple Choice:

1. a	**4.** b	**7.** d	**10.** b	**13.** c
2. d	**5.** d	**8.** c	**11.** a	**14.** a
3. d	**6.** a	**9.** b	**12.** b	**15.** c

CHAPTER 42
ANIMAL REPRODUCTION

Suggested Answers to Structure Your Knowledge

1.

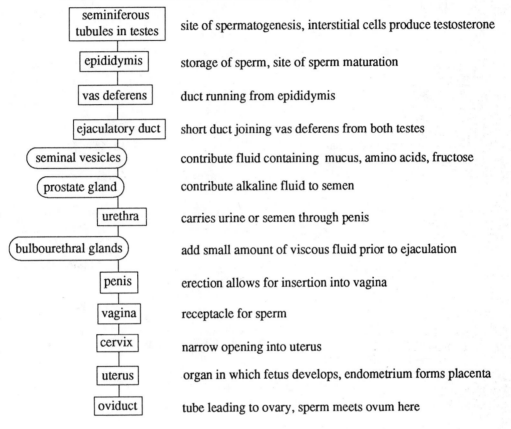

PATH OF SPERM FROM FORMATION TO FERTILIZATION

seminiferous tubules in testes	site of spermatogenesis, interstitial cells produce testosterone
epididymis	storage of sperm, site of sperm maturation
vas deferens	duct running from epididymis
ejaculatory duct	short duct joining vas deferens from both testes
seminal vesicles	contribute fluid containing mucus, amino acids, fructose
prostate gland	contribute alkaline fluid to semen
urethra	carries urine or semen through penis
bulbourethral glands	add small amount of viscous fluid prior to ejaculation
penis	erection allows for insertion into vagina
vagina	receptacle for sperm
cervix	narrow opening into uterus
uterus	organ in which fetus develops, endometrium forms placenta
oviduct	tube leading to ovary, sperm meets ovum here

2. A mature egg cell within a follicle is released during ovulation and swept into the fallopian tube or oviduct. It travels to the uterus, where it flows out through the vagina with the bleeding associated with menstruation.

3. Birth control methods (a) prevent fertilization—abstinence, rhythm method, condom, diaphragm, coitus interruptus; (b) prevent implantation of embryo—IUD; (c) prevent release of gametes from gonads—sterilization, chemical contraception such as birth control pills. The most effective methods are abstinence, sterilization, chemical contraception. Least effective are the rhythm method and coitus interruptus.

Answers to Test Your Knowledge

Fill in the Blanks:

1.	budding	6.	estrous cycle
2.	clone	7.	progesterone
3.	parthenogenesis	8.	urethra
4.	hermaphrodite	9.	vasocongestion
5.	cloaca	10.	puberty

Multiple Choice:

1.	d	5.	c	9.	c
2.	a	6.	c	10.	a
3.	d	7.	d	11.	b
4.	a	8.	c	12.	c

CHAPTER 43
ANIMAL DEVELOPMENT

Suggested Answers to Structure Your Knowledge

1.

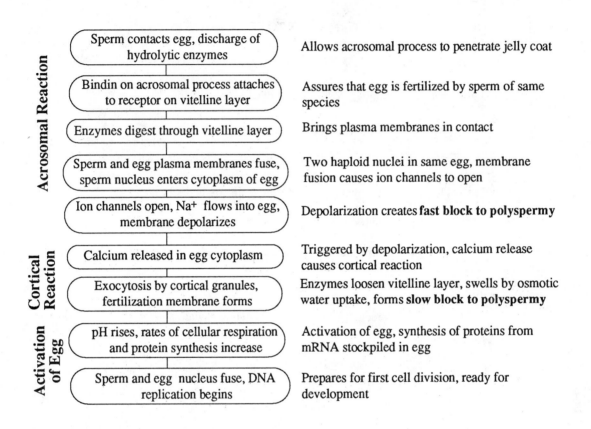

2.

	Amphioxus	Amphibian	Bird	Mammal
Cleavage	Holoblastic, animal pole at polar body buds	Holoblastic, but yolk slows divisions, creates uneven cells	Meroblastic, only in cytoplasmic disc on top of yolk	Holoblastic, no polarity to egg, blastomeres equal
Blastula	Hollow ball of cells, one cell thick	Blastocoel displaced to animal hemisphere, wall several cells thick	Cavity between two layers of blastodisc	Embryonic disc, resembles blastodisc of birds and reptiles
Gastrulation	Invagination to form cup-shaped gastrula, mesoderm forms from pouches off archenteron	Invagination by bottle cells, involution of cells at dorsal lip of blastopore, located at gray crescent, 3 germ layers formed	Cells roll in at primitive streak, forming mesoderm, flat, 3-cell layered embryonic disc folds under at edges	Flat embryonic disc, gastrulation through primitive streak, similar to birds

3.

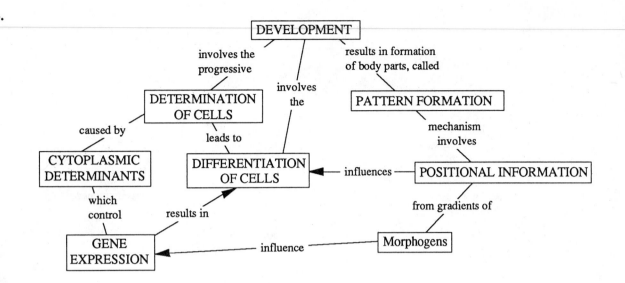

Answers to Test Your Knowledge

Multiple Choice:

1.	c	**5.**	d	**9.**	d	**13.**	b	**17.**	d
2.	b	**6.**	c	**10.**	c	**14.**	a	**18.**	b
3.	b	**7.**	c	**11.**	a	**15.**	d	**19.**	a
4.	d	**8.**	b	**12.**	c	**16.**	b	**20.**	d

CHAPTER 44
NERVOUS SYSTEMS

Suggested Answers to Structure Your Knowledge

1.

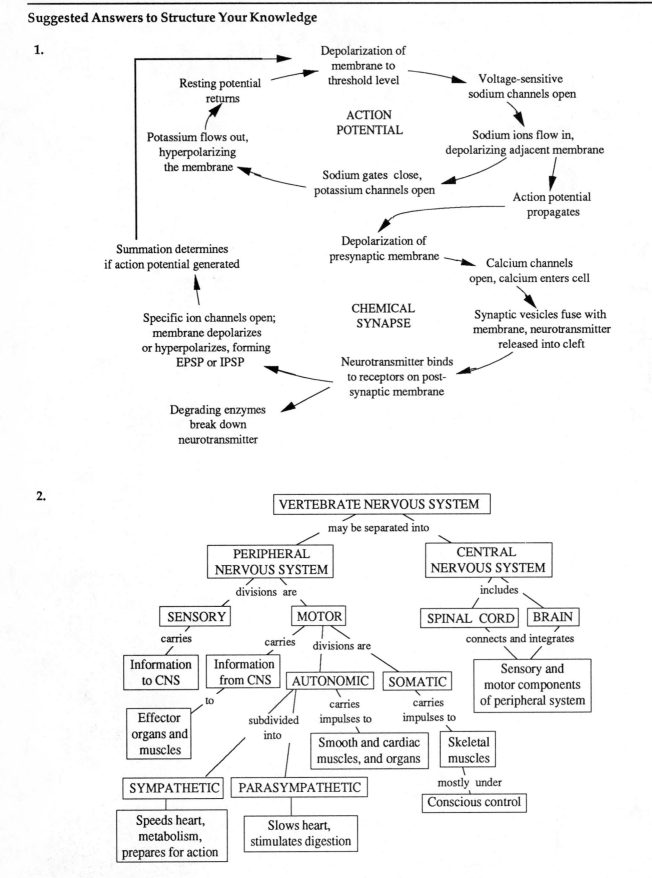

2.

3.

Part of brain	Section	Function
Medulla	Hindbrain	Control centers for homeostatic functions–respiration, heart and blood vessel actions, digestion; region where pyramidal tracts cross
Pons	Hindbrain	Aids in some of functions of medulla; conducts information between brain and spinal cord
Cerebellum	Hindbrain	Controls unconscious coordination of movement and balance
Inferior, superior colliculi	Midbrain	Inferior–involved with hearing; superior–visual center
Reticular formation	Midbrain	Regulates state of arousal; filters all sensory information going to cerebral cortex
Thalamus	Forebrain	Major projection area; relays input to cerebral cortex; contains nuclei of reticular formation
Hypothalamus	Forebrain	Produces hormones of posterior pituitary; produces releasing factors that control anterior pituitary; regulates pleasure, alarm response, sexual response and mating behaviors, autonomic functions
Cerebral cortex	Forebrain	Processes sensory and motor information; integrates information; responsible for thinking, memory, emotions, speech

Answers to Test Your Knowledge

Multiple Choice:

1. b	5. a	9. b	13. a	17. d
2. a	6. c	10. d	14. d	18. a
3. a	7. a	11. c	15. d	19. c
4. d	8. a	12. b	16. b	20. c

CHAPTER 45
SENSORY AND MOTOR MECHANISMS

Suggested Answers to Structure Your Knowledge

1.

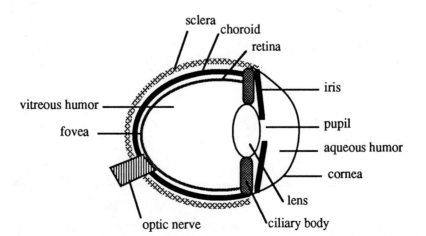

2. Light enters the eye through the pupil and is focused by the lens onto the retina. When rhodopsin, the photopigment in rods and cones, absorbs light energy, it isomerizes and changes the receptor cell's permeability to sodium. The reduction in the release of neurotransmitter by rods and cones serves to depolarize or hyperpolarize connected bipolar cells, which in turn stimulate action potentials in ganglion cells—the sensory neurons that form the optic nerve.

3. Sound waves travel down the *auditory canal* to the *tympanic membrane*, from which they are amplified and transmitted to the *malleus, incus, and stapes.* Vibration of the stapes against the *oval window* sets up pressure waves in the fluid in the *vestibular* and *tympanic canal* within the *cochlea.* These pressure waves are dissipated when they strike the *round window.* The pressure waves vibrate the *basilar membrane*, on which the *organ of Corti* is located; and the tips of the hair cells, embedded in the *tectorial membrane*, are bent, triggering a depolarization, release of neurotransmitter, and initiation of action potentials in the sensory neurons of the *auditory nerve*, which leads to the *cerebral cortex.*

4. A motor neuron releases acteylcholine into the neuromuscular junction, initiating an action potential within the muscle fiber, which spreads into the cell through the T tubules. The action potential depolarizes the sarcoplasmic reticulum membrane and releases calcium ions. The calcium binds with troponin, changes the shape of the tropomyosin–troponin complex, and exposes the myosin-binding sites of the actin molecules. Heads of myosin molecules bind to these sites, bend, and pull the thin filament toward the center of the sarcomere. The myosin molecule hydrolyzes an ATP to break the cross-bridge, straightens out, and forms another cross-bridge, continuing this sequence as long as there is ATP and until calcium is pumped back into the sarcoplasmic reticulum, and the tropomyosin–troponin complex again blocks the actin sites.

Answers to Test Your Knowledge

Matching:

1.	I	4.	L	7.	N	10.	K
2.	G	5.	J	8.	D	11.	B
3.	O	6.	M	9.	A	12.	H

Multiple Choice:

1.	a	6.	b	11.	d	16.	c
2.	a	7.	d	12.	b	17.	a
3.	c	8.	c	13.	a	18.	b
4.	b	9.	c	14.	d	19.	a
5.	d	10.	a	15.	b	20.	c

Fill in the Blanks:

1.	sarcomere	6.	thick filament, myosin
2.	I band	7.	thin filament, actin and troponin
3.	A band	8.	in g, thin filament
4.	Z line	9.	part c, A band
5.	H zone	10.	part f, myosin of thick filament, attaches to g

CHAPTER 46

DIVERSE ENVIRONMENTS OF THE BIOSPHERE: AN INTRODUCTION TO ECOLOGY

FRAMEWORK

Ecology deals with the distribution and abundance of organisms and the interrelationships between organisms and both the abiotic and biotic factors in their environments. This chapter describes the organizational levels at which ecological questions are asked, the abiotic factors to which organisms have adapted in both an ecological and an evolutionary time frame, and the major world communities, or biomes, in which adaptations to climate and abiotic factors have produced similar and characteristic life forms.

CHAPTER SUMMARY

Ecology is the scientific study of the interactions of organisms with their environments. Ecology focuses on interactions because organisms are affected by, and, in turn, affect both the *abiotic* and *biotic* factors of their environments. Ecology is a multidisciplinary science, incorporating aspects of genetics, evolution, physiology, and behavior. Applying the usual methods of scientific experimentation is often challenging because of the scope and complexity of the questions that are asked.

The science of ecology is not synonomous with the growing "ecological" consciousness of our current environmental problems, but it helps us to understand these problems and their possible solutions. Most ecologists feel a commitment to educate legislators and the public about environmental issues and decisions.

The Scope and Development of Ecology

The Questions of Ecology Ecology encompasses the study of the distribution and abundance of organisms and considers the factors that determine these phenomena. Questions of distribution and abundance can be considered at the various levels of ecology. *Physiological ecology* considers an organism's responses to its physiochemical environment. Population ecology is concerned with the factors that control the size of *populations*, which are groups of individuals of the same species in an area. The *community* includes all the populations of organisms in an area, and ecology on this level considers interactions such as predation and competition. *Biomes* are typical communities that are found in major climatic regions. The *ecosystem* includes the abiotic factors as well as the communities that exist in an area, and ecological questions include the flow of energy and chemical cycling.

Ecology as an Experimental Science Historically ecology was a descriptive science, but currently an experimental approach, in both the laboratory and the field, is being used to investigate ecological questions. Creative approaches are often required to control variables and manipulate communities in field experiments. Ecologists may develop mathematical models and computer simulations that help to study the interactions of variables and answer ecological questions.

The theoretical framework of ecology is still being constructed and debated as ecologists add to basic knowledge with descriptive research, field and lab experiments, and mathematical modeling.

Ecology and Evolution Interactions between organisms and their environments occur within ecological time, whereas the cumulative effects of these interactions are realized on the scale of evolutionary time. The distribution and abundance of organisms are the result of both past history and present interactions with the environment.

Environmental Diversity of the Biosphere

The relatively thin layer of Earth that is inhabited by life is called the *biosphere* and extends from a few meters below the soil surface to a few kilometers into the atmosphere. The patchiness of the biosphere is a result of abiotic factors, such as temperature, rainfall, and light, which create a variety of habitats. Natural selection has resulted in diverse adaptations to the wide spectrum of abiotic factors. The ranges of environmental variables that organisms can tolerate is a major factor determining the distribution of species.

Some Important Abiotic Factors   *Temperature* is an important environmental factor because of its effects on metabolism and enzyme activity. Most organisms cannot maintain body temperatures that vary appreciably from the environmental temperature. Some organisms have extraordinary adaptations that enable them to live outside the temperature range of 0° to 50°C.

The availability of *water* is reflected in the adaptations organisms have to regulate their osmolarity in aquatic environments and to obtain water and reduce water loss in terrestrial habitats.

Light energy drives almost all ecosystems. Within forests, the availability of light influences the distribution of species. The intensity and quality of light are important abiotic factors in aquatic environments. Most photosynthesis occurs relatively near the surface of the water; deeper-dwelling photosynthetic organisms have accessory pigments that can absorb the wavelengths of light that selectively penetrate water. Many plants and animals are sensitive to photoperiod, which serves as an indicator of seasonal changes.

Variations in the physical structure, pH, and mineral composition of *soil* affect the distribution of plants, which, in turn, affects the distribution of animals. *Wind* increases the rates of heat and water loss in organisms. Wind may also affect the morphology of plants.

Fire and other disturbances can drastically affect biological communities. Many plants have evolved adaptations to periodic fires.

An Integrated Approach to the Physical Environment   Organisms must cope with the entire set of environmental factors within their habitats. The *principle of allocation* is used to analyze the amount of energy that an organism uses for its various processes,

such as reproduction, eating, growing, escaping from predators, and dealing with environmental fluctuations. The distribution of organisms is related to different systems of energy allocation. In stable environments, organisms can allot more energy for growth and reproduction, but their resulting intolerance of environmental changes restricts them to such environments. Organisms that put more of their energy into dealing with environmental fluctuations can successfully live in a wider range of habitats.

Responses of Organisms to Environmental Change

Individuals may use behavioral, physiological, or morphological mechanisms to respond to changes in their environments.

Behavioral Responses   Behavioral responses may include a relocation to a more favorable environment—either local or more distant—as in seasonal migrations. Cooperative social behavior, such as huddling, can help organisms respond to unfavorable conditions.

Physiological Responses   Physiological responses to environmental change are usually slower than are behavioral responses. *Regulators* are organisms whose physiological responses allow them to maintain constant internal conditions, whereas *conformers* cannot regulate their internal environments, which vary along with external environmental changes. Organisms living in stable environments are more likely to be conformers.

The optimal environmental conditions and tolerance limits for an organism can be determined experimentally by varying abiotic factors. Tolerance limits help to determine the spatial distribution of organisms. Physiological adjustment to environmental changes, which can extend tolerance limits, is called *acclimation*.

Morphological Responses   Morphological responses to environmental changes, such as changes in growth that result in altered body form, are most common in plants and provide these rooted organisms with a means of adapting to their environments.

Adaptation over Evolutionary Time   Behavioral, physiological, and morphological responses occur within an ecological time scale but are based on adaptations that have developed by natural selection over an evolutionary time span. The adaptation of organisms to localized environments restricts their geographic distribution and often their ability to respond to changes within their environments. The presence or absence of a species within a location may result from its ability to tolerate local abiotic factors or from its history of dispersal.

Terrestrial Biomes

Ecologists have recognized the major large-scale communities, or biomes, by various schemes; the textbook shows a map of twelve terrestrial biomes. The general appearance of a biome may be similar over large areas, but the actual species composition varies locally. Biomes may also have several communities present. This patchiness may result from the natural succession of communities resulting from disturbances. Terrestrial biomes can also be characterized by specific types of animals that are adapted to that particular environment.

The geographic distribution of the world's major biomes is related to the prevailing climate—to the temperature, rainfall, and seasonal fluctuations of the various altitudes and latitudes. Climatic patterns on Earth result from the interaction of a variety of factors, including the angle of the sun's rays; the tilting of the earth's axis; movement of large air masses, which vary in moisture content, major warm or cold ocean currents, and topographical variations such as mountains. Biomes can be mapped on a climograph according to annual means of temperature and rainfall, and a strong correlation with these variables is evident. Overlaps of biomes on a climograph indicate that seasonal patterns of variations, such as the distribution of rainfall or the range of temperatures, may affect the type of biome that develops.

Tropical Forest   *Tropical forests*, found near the equator, have little variation in temperature (averaging around 25°C) or day length. Variations in the amount of rainfall result in the existence of tropical thorn forests, where rainfall is scarce; tropical deciduous forests, in which trees and shrubs drop their leaves during the long dry season that alternates with the monsoon; or tropical rain forests, where rainfall is abundant.

The *tropical rain forest*, with its dense canopy formed by tall trees, lianas (woody vines), and epiphytes, has the largest diversity of species and is the most complex of all communities. The soil is generally poor and thin because the high temperatures and rainfall lead to rapid decomposition and recycling of nutrients back into plant material. The animals, typically tree-dwellers, have mutualistic interactions with plants as they disperse pollen, fruits, and seeds while foraging. Human destruction of the tropical rain forest by mining, lumbering, and farming may greatly reduce species diversity and cause large-scale climate changes.

Savanna   *Savannas*, tropical and subtropical grasslands with only scattered trees, typically have three distinct seasons: cool and dry, hot and dry, and warm and wet. Growth forms are restricted by frequent fires to grasses and *forbs*, or small broad-leaf plants. Large herbivores and burrowing animals are most common. Areas where forest and grasslands intergrade are also referred to as savannas.

Desert   *Deserts* are characterized by low precipitation (less than 30 cm/yr). Temperatures may be hot or cold, depending on location. Where perennial vegetation occurs, it consists of widely scattered shrubs, cacti, or succulents. Rainy periods are marked by rapid blooms of annual plants. Seed eaters, and their reptilian predators, are common animals. Desert animals have physiological and behavioral adaptations to dry conditions and extreme temperatures.

Chaparral   The *chaparral*, or shrubland, is common along coast lines in mid-latitudes that have mild, rainy winters and hot, dry summers. The dense, spiny evergreen shrubs and annual vegetation is adapted to periodic fires; species may have deep root systems that permit quick regeneration or seeds that germinate only after a fire. Animals are typically browsers, fruit-eating birds, seed-eating rodents, and reptiles.

Temperate Grassland   *Temperate grasslands* are found in relatively cold regions. Occasional fires and drought prevent the invasion of woody shrubs and trees. Large grazing mammals, along with their large carnivore predators, are typical animals.

Temperate Forest   *Temperate forests*, characterized by broad-leaved deciduous trees, grow in mid-latitude regions that have adequate moisture to support the growth of large trees. Cold winters alternate with hot summers. Herbs, shrubs, and one or two strata of trees are typical, and a rich diversity of animal life is associated with the variety of food and habitat. Human activity has greatly reduced these worldwide forests.

Taiga   The *taiga*, also known as the coniferous or boreal forest, is characterized by harsh winters and short summers and is found in northern areas and higher elevations. The soil is thin and acidic. Low temperatures and the waxy covering of needles lead to slow decomposition rates. The growth of coniferous trees is usually so thick that little undergrowth is present. The heavy snowfall insulates the soil and protects small mammals. Animals include seed-eaters, insect herbivores, larger browsers, and predators.

Tundra   The *tundra* is characterized by low shrubby or matlike vegetation. The arctic tundra, circling the

North Pole, has long periods of very little light, interspersed with a brief warm summer marked by long days and rapid plant growth. The *permafrost* of the Arctic tundra prevents roots from penetrating very far into the continually saturated soil. In the alpine tundra, found above the tree line on high mountains, day length is more even and plant growth is slow but steady.

Animals of the tundra adapt to the cold by living in burrows or having large, well-insulated bodies. Some large herbivores are found, and many animals are migratory.

Aquatic Communities

As in terrestrial biomes, similar aquatic environments have similar communities with organisms that show analogous adaptations and body forms.

Freshwater Communities   Lakes and ponds may be stratified in temperature, community structure, and availability of light. The *photic zone* supports a variety of phytoplankton, which are fed upon by zooplankton. Nitrogen and phosphorous are often limiting nutrients, except when supplied in excess by agricultural run-off. Changes in temperature stratification may cause mixing of water layers and the recycling of nutrients from decomposing *detritus* that had settled into the *aphotic zone*. Phytoplankton and zooplankton form the base of a food web that includes fishes, aquatic insects, snakes, amphibians, and fish-eating birds. *Benthic* communities in shallow areas include aquatic plants, algae, clams, snails, crustaceans, and numerous insects and their larvae.

Rivers and streams are freshwater, flowing habitats, whose physical and chemical characteristics vary from their source to point of entry into ocean or lake. Clear, fast-flowing, and rocky-bottomed headwaters gradually change to more turbid, slower-flowing, and nutrient-rich downstream areas. The biotic communities reflect these abiotic factors. Plankton communities are absent, and food chains are built on the photosynthesis of attached algae or the addition of organic matter and nutrients from the land.

Marine Communities   Marine communities populate the oceans, where algae produce a large portion of the world's oxygen. The marine environment can be classified into a photic zone, in which phytoplankton, zooplankton, and may fishes are found, and an aphotic zone, where light does not penetrate.

Marine communities also may be divided on the basis of depth. The *intertidal zone* is the shallow area where the water meets the land. A rocky intertidal zone is inhabited by organisms that compete for attachment sites to avoid being washed away by the tides. A sandy or mud intertidal zone is home to burrowing worms and clams. The *neritic zone* is the shallow area over the continental shelves. The *oceanic zone*, past the continental shelf includes the open water of the *pelagic zone*. The bottom surface is the *benthic zone*. The *abyssal zone* is the benthic region where light does not penetrate. Organisms in these areas, adapted to darkness, cold, and high pressures, feed on the detritus that sifts down from above.

Estuaries, where a river or stream meets the ocean, usually are highly productive areas. Estuaries serve as spawning grounds for many marine organisms and feeding and breeding areas for crustaceans, mollusks, fish, and waterfowl.

STRUCTURE YOUR KNOWLEDGE

1. Describe ecology. What types of questions are considered? What methods are used to answer those questions? What theory guides the interpretation of data.

2. What are biomes? What are the key abiotic factors that determine them? What accounts for the similarities in life forms throughout a biome?

TEST YOUR KNOWLEDGE

MULTIPLE CHOICE: *Choose the one best answer.*

1. Which level of ecology considers energy flow and chemical cycling?
   a. community
   b. ecosystem
   c. physiological
   d. population

2. Evolutionary history
   a. influences the distribution of organisms.
   b. involves the cumulative effects of the interactions between organisms and their environments.
   c. has produced the adaptations of organisms to different environments.
   d. involves all of the above.

3. Which abiotic factor has the least direct impact on the distribution of animals?
   a. temperature
   b. water
   c. soil
   d. light

4. An animal that expends most of its energy in growth and reproduction probably lives in
   a. a stable environment.
   b. an unstable environment.
   c. a terrestrial environment.
   d. an aquatic environment.

5. Acclimation
   a. is a morphological response to an environmental change.
   b. can extend the tolerance limits of an organism.
   c. involves behavioral responses to environmental change.
   d. occurs more rapidly in animals than in plants.

6. Which of the following is incorrectly paired with its description?
   a. neritic zone—shallow area over continental shelves
   b. abyssal zone—benthic region where light does not penetrate
   c. littoral zone—area of open water
   d. intertidal zone—shallow area at edge of water

7. A conformer is more likely to be successful in
   a. a desert.
   b. a tropical rain forest.
   c. the taiga.
   d. savanna.

8. Two communities have the same mean temperature and rainfall but very different compositions and characteristics. The best explanation for this phenomenon is that the two
   a. have a different range of temperatures and pattern of rainfall throughout the year.
   b. are composed of species that have very low dispersal rates.
   c. are found on different continents.
   d. receive different amounts of sunlight.

9. Ecologists use mathematical models and computer simulations because
   a. ecological experiments are too difficult to perform.
   b. most of them are mathematicians.
   c. they are trying to change the image of ecology as simply a descriptive science.
   d. these approaches allow them to study the interactions of variables and simulate large-scale experiments.

10. A biome that is characterized by three seasons — cool and dry, hot and dry, and warm and wet—is a
    a. tropical forest.
    b. savanna.
    c. chaparral.
    d. temperate forest.
    e. taiga.

11. The permafrost of the Artic tundra
    a. prevents plants from getting established and growing.
    b. helps to keep the soil from getting wet since water cannot soak in.
    c. prevents plant roots from penetrating very far into the soil, probably restricting plants to low growing forms.
    d. both b and c.

12. Many plant species have adaptations for dealing with the periodic fires typical of a
    a. savanna.
    b. chaparral.
    c. temperate grassland.
    d. All of the above may be correct.

MATCHING: Match the biotic and abiotic descriptions with the biomes.

Biome

1. ____ chaparral

2. ____ desert

3. ____ savanna

4. ____ taiga

5. ____ temperate forest

6. ____ temperate grassland

7. ____ tropical rain forest

8. ____ tundra

Biotic Description

A. broad-leaved deciduous trees

B. lush growth, trees, lianas, epiphytes

C. dense shrubs, fire-adapted vegetation

D. tropical grasslands, grasses and forbs

E. coniferous forests

F. low shrubby or matlike vegetation

G. grasslands in relatively cold regions

H. widely scattered shrubs, cacti, succulents

CHAPTER 47

POPULATION ECOLOGY

FRAMEWORK

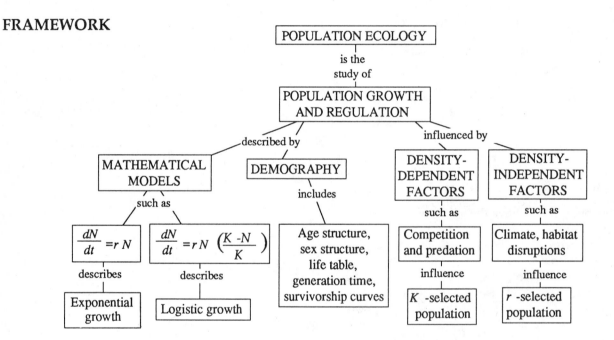

POPULATION ECOLOGY

is the
study of

POPULATION GROWTH
AND REGULATION

described by — influenced by

MATHEMATICAL
MODELS — DEMOGRAPHY — DENSITY-
DEPENDENT
FACTORS — DENSITY-
INDEPENDENT
FACTORS

such as — includes — such as — such as

$$\frac{dN}{dt} = r N$$ $$\frac{dN}{dt} = r N \left(\frac{K - N}{K} \right)$$ — Age structure, sex structure, life table, generation time, survivorship curves — Competition and predation — Climate, habitat disruptions

describes — describes — influence — influence

Exponential growth — Logistic growth — K-selected population — r-selected population

CHAPTER SUMMARY

The continuing growth of the human population in the face of limited and even dwindling resources is a dangerous biological phenomenon. Population ecology is the study of the fluctuations in and regulation of population size. This branch of ecology overlaps physiological and community ecology.

Density and Dispersion

Ecologists determine the geographic boundaries of the population they are studying in relationship to the type of question they ask. The number of individuals per unit area or volume is the population *density*; the spacing of those individuals within the boundaries of the population is referred to as *dispersion*.

Measuring Density Measuring density is done most often by estimates based on a number of sample areas, or *quadrats*. Indirect indicators, such as burrows or nests, also may be used. In a *mark–recapture* technique, animals are trapped, marked, and released. The proportion of marked and unmarked animals captured in a second trapping is used to estimate population density.

Patterns of Dispersion   Individuals may be dispersed in the population's *range*, or geographic area, in several patterns. *Clumping* may indicate a heterogeneous environment, with organisms congregating in suitable microenvironments. Clumping also may be related to social interactions between individuals. *Uniform* distribution is usually related to competition for resources, resulting in antagonistic interactions between organisms or the establishment of territories. *Random* spacing, indicating the absence of strong attractions or repulsions between individuals, is not very common.

Demography

Population size changes in response to the relative rates of births and immigrations versus deaths and emigrations. *Demography* is the study of vital statistics that affect population growth, such as births and deaths, the ratio of males to females, and the age structure.

Age and Sex Structure   The *age structure* of a population is the relative number of individuals of each age. Each age group in a population has a characteristic fertility and death rate. The young and old members may be most likely to die, and intermediate-age individuals may have the highest birth rate. Populations with age structures with the greatest proportion of members of reproductive age or slightly younger will grow the fastest. A shorter *generation time* will also result in faster population growth.

The proportion of females in the population may affect growth, depending on the mating patterns of the species.

Life Tables and Survivorship Curves   *Life tables* summarizing mortality rates for each age group can be constructed by following a *cohort*, or group, of newborn organisms throughout their lives.

Age-specific mortality can also be represented on a *survivorship curve*, which shows the number of members of a cohort that are still alive at each age. The three types of survivorship curves are shown below.

Type I, typical of modern human populations, shows that mortality is low during early and middle age and increases rapidly with old age. Populations that produce relatively few offspring and provide parental care usually have Type I curves. In a Type II curve, death rates are relatively constant throughout the life span. A Type III curve is typical of populations that produce many offspring, most of which die off rapidly. The few that survive are likely to reach adulthood.

The Logistic Model of Population Growth

Mathematical modeling is a common tool that ecologists use to describe rates of population growth, to study variables affecting growth, and to predict population sizes.

Growth of a small population in a very favorable environment will only be restricted by the physiological ability of that species to reproduce. The change in a population's size during a specific time period may be signified as $\Delta N/\Delta t$. The formula describing instantaneous change in the population number, $dN/dt = rN$ includes the *intrinsic rate of increase* (r) multiplied by the starting number (N) in the population. Each species has a characteristic *r*, the difference between births and deaths estimated for ideal growing conditions. According to this exponential model of growth, the larger the population (N) becomes, the faster the population grows.

A population may grow exponentially for only a short time. Growth slows as the population approaches its *carrying capacity* (K) within that particular environment. The *logistic equation* includes a term that reflects the impact of the increasing population size as it approaches the carrying capacity. The formulas and graphs for exponential and logistic growth are:

$$\frac{dN}{dt} = rN$$

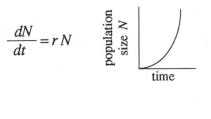

$$\frac{dN}{dt} = rN\left(\frac{K-N}{K}\right)$$

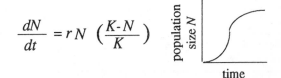

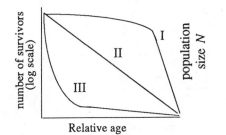

In the logistic equation when N is small, the $(K-N)/K$ term is close to 1, and growth is approximately exponential (rN). As N approaches K, the carrying capacity, the $(K-N/K)$ term becomes a small fraction, and exponential growth (rN) is reduced by that value. When N reaches K, the term $(K-N)/K$ is 0, and population growth (dN/dt) is 0. At this point, birth rate equals death rate, and the size of the population does not increase.

Maximum increase in population numbers occurs when N is intermediate—when the population is at about one-half the carrying capacity of the environment.

The logistic model makes the assumption that any increase in population numbers will have a negative effect on population growth. For some populations, however, too low a population size may be detrimental. According to the *Allee effect*, individuals may actually benefit as the population grows, from physical support, as in plants, or from social interactions important to reproduction or protection from predators.

Many populations overshoot their carrying capacity because of a lag time before the negative effects of increasing numbers are realized. Thus populations may not smoothly approach a maximum population size and level off but may oscillate above and below a general carrying capacity.

There are also populations to which the concept of a carrying capacity does not apply. These populations may be reduced by environmental conditions before resources have a chance to become limiting.

Regulation of Populations

Density-Dependent Factors   As a population grows, *density-dependent factors*, such as dwindling resources or increasing predation, may affect a larger proportion of individuals and act to slow population growth. In laboratory studies of flour beetles, population growth ceased as the food supply became limiting. Studies of mice have shown that even when food or shelter are not limiting, population size stabilizes as high densities of mice cause physiological changes that inhibit reproduction.

Predation may be a density-dependent factor when a predator shows switching behavior and feeds preferentially on a prey population that has reached a high density.

Density-Independent Factors   Density-independent factors, such as weather or habitat disruption, reduce the population by the same proportion whether the population is large or small. Density-independent factors may reduce population size before resources become limiting or other density-dependent factors take effect.

Interaction of Regulating Factors   Many populations remain fairly stable in size, apparently influenced by density-dependent factors that maintain the populations close to their carrying capacity. There may be short-term or seasonal fluctuations in population size, however, that are due to density-independent factors.

Population Cycles   Some populations of birds, mammals, and insects show regular density fluctuations called *population cycles*. One hypothesis is that crowding may affect the endocrine systems of organisms and cause these cycles. The time lag in a population's response to density-dependent factors is another explanation for the regular oscillation about the carrying capacity. Cycles may be caused by the effects of predators on the increasing density of prey. Alternately, the effect an increasing herbivore population has on its food supply may regulate population size more than increasing predator pressure.

Evolution of Life Histories

The *life histories* of organisms—their pattern of birth, reproduction, and death—affect population growth in ecological time, but evolved as a result of natural selection operating in evolutionary time.

Three factors of life histories influence the intrinsic rate of increase (r): (1) the number of reproductive episodes; (2) clutch size; and (3) age at first reproduction. *Semelparous* organisms reproduce only once, whereas *iteroparous* individuals reproduce more than once in a lifetime. Clutch size usually is inversely related to the size of offspring. The shorter the generation time, the faster the population growth: r is inversely related to generation time. Because organisms have a finite energy budget, they cannot maximize all three of these factors simultaneously.

An earlier age of reproduction may be selected for in a population in which lifespans are short due to predation.

Darwinian fitness depends on the number of offspring that survive to produce their own offspring. Ecologists developed the concepts of *r-selection* and *K-selection* to explain life histories in relationship to evolutionary selective pressures. Populations that have a high r value are said to be r-selected: individuals reproduce at an early age, are usually semelparous, and have large clutch sizes of relatively small offspring. This type of life history is most common in variable environments, where density-independent factors regulate population size. Reproductive success is assured by producing large numbers of offspring.

K-selected populations are found in stable environments with limited resources, where density-dependent factors tend to maintain population sizes around the carrying capacity. Offspring are fewer, and reproductive success depends on the competitive capabilities of the offspring.

The concepts of *r*-selection and *K*-selection, while now seen as oversimplifications, have led to many field and laboratory studies that show a relationship between environmental stability and life histories. Some populations, such as those of dandelions, show different life history characteristics in response to the severity of habitat disruptions.

The concept of *stress*, the general difficulty of environmental conditions, has been suggested as an additional environmental characteristic that influences life history. Most populations have life histories with a mix of both *r*- and *K*-selected characteristics that have resulted from the interactions of several factors.

Human Population Growth

The human population has been growing exponentially for centuries. In the 80 years between 1850 and 1930, it doubled to 2 billion; it doubled again in the next 45 years and is projected to reach 8 billion by the year 2017 if the present growth rate is maintained. In 1989, the world population stood at about 5.2 billion. The growth since the Industrial Revolution has resulted mainly from a drop in death rates, especially among infants.

Human population growth is unique in that it can be consciously controlled by voluntary contraception or government sanctions. In most developed countries, women are delaying marriage and reproduction, thus slowing population growth. High birth rates are more common in underdeveloped and developing countries. The age structure of a population influences growth rate; a large proportion of individuals of reproductive age or younger results in more rapid growth.

Technology has increased the carrying capacity of the earth for humans, but it is hard to predict what this carrying capacity is, and how and when we will approach it.

STRUCTURE YOUR KNOWLEDGE

1. Create a concept map to organize your understanding of the exponential and logistic equations—the mathematical models of population growth.

2. Describe the demographic factors that influence population growth rates.

TEST YOUR KNOWLEDGE

MATCHING: *Match the term with its description.*

1. ____	area from which sample is taken	A.	age structure
2. ____	table with mortality rates for each age group	B.	clutch size
3. ____	number of offspring produced at one time	C.	cohort
4. ____	group followed from birth to death	D.	demography
5. ____	relative number in each age group of a population	E.	generation time
6. ____	graph of age-specific	F.	iteroparous mortality
7. ____	having one reproductive event per lifetime	G.	life table
8. ____	regular fluctuations in population size	H.	quadrat
9. ____	span between an individual's birth and birth of offspring	I.	semelparous
10. ____	study of vital statistics that affect population growth cycle	J.	population
		K.	survivorship curve

MULTIPLE CHOICE: *Choose the one best answer.*

1. In a range with a heterogeneous distribution of suitable habitats, the dispersion pattern of a population probably would be
   a. clumped.
   b. uniform.
   c. random.
   d. unpredictable.

2. In a mark–recapture study,
   a. all the organism within several quadrats are counted.
   b. only the organisms that are recaptured are counted.

c. the proportion of marked individuals that are recaptured is used to estimate the population size.

d. 10% of all the organisms in an area are captured and marked.

3. The age structure of a population influences population growth because
   a. younger females have more offspring than do older females.
   b. populations with shorter generation times grow more rapidly.
   c. different age groups have different reproductive capabilities.
   d. life tables show that mortality rates change with age.

4. A survivorship curve of type I—level at first with a rapid increase in mortality in old age—is
   a. typical of many invertebrates that produce large numbers of offspring.
   b. typical of humans and other large mammals.
   c. found most often in *r*-selected species.
   d. almost never found in nature.

5. The middle of the S growth curve in the logistic growth model
   a. is a period when the population is increasing the fastest.
   b. is best described by the term rN.
   c. shows that reproduction ceases when the population size reaches K and dN/dt. becomes 0.
   d. is the period when competition for resources is highest.

6. The shorter the generation time,
   a. the smaller the carrying capacity.
   b. the greater the value of r.
   c. the more that density-independent factors affect growth.
   d. the larger the K value possible.

7. A few members of a population have reached a very favorable habitat (few predators, unlimited resources), but their population growth rate is slower than that of the parent population. What is a possible explanation for this situation?
   a. The genetic makeup of these founders may be less favorable than that of the parent population.
   b. The parent population may still be in an exponential part of its growth curve and not yet limited by density-dependent factors.
   c. The Allee effect may be operating; there are not enough population members present for successful reproduction.
   d. All of the above may apply.

8. The term $(K-N)/K$
   a. is the carrying capacity for a population.
   b. is greatest when K is very large.
   c. decreases in value as N approaches K.
   d. represents the Allee effect.

9. Density-independent factors
   a. tend to maintain a population around the carrying capacity.
   b. may be involved in the population cycles seen in some mammals.
   c. are involved in the regulation of K-selected populations.
   d. include climatic events and habitat disruptions.

10. Which of the following is *not* true? A population with a large r value
    a. probably has a short generation time.
    b. probably has large clutch sizes.
    c. is most likely found in variable environments.
    d. is most likely to be regulated by density-dependent factors.

11. The age structure of a population
    a. can be inferred from a survivorship curve.
    b. is a result of age-specific fertility.
    c. greatly influences the population growth rate.
    d. is most uniform in populations with high growth rates.

12. The clutch size
    a. is inversely related to the number of offspring produced.
    b. influences the intrinsic rate of increase of a population.
    c. is directly proportional to the size of offspring.
    d. is greatest for iteroparous individuals.

13. Population cycles
    a. may result from the physiological effects of crowding on reproductive behavior and success.
    b. are most common in predator populations.
    c. are linked with seasonal changes.
    d. all of the above are correct.

14. A population in which the few young that are produced are well equipped to compete in that environment would most likely be
    a. found in a highly variable environment.
    b. described as *r*-selected.
    c. described as *K*-selected.
    d. subject to intense predation that keeps the average life span short.

15. The human population is growing at such an alarmingly fast rate because

a. technology has increased our carrying capacity and, thus, density-dependent factors have not slowed reproduction.
b. the death rate has greatly decreased since the Industrial Revolution.
c. the age structure of underdeveloped countries is highly skewed toward younger ages.
d. all of the above are correct.

CHAPTER 48

COMMUNITIES

FRAMEWORK

Communities are composed of populations of various species that may interact through competition, predation, and symbiosis. The structure of a community—its species diversity, prevalent form of vegetation, trophic relations, and stability—is determined by these interactions, the past history of the community, and its successional stage. Disturbances and chance events contribute to the dynamic equilibrium (or nonequilibrium) of a community.

Biogeography studies the distribution of species throughout the world. Island biogeography permits the study of community structure and equilibrium, especially in relationship to size of habitat and immigration rate, within a more simple system.

CHAPTER SUMMARY

A collection of populations living close enough to allow for interaction is called a *community*. Community ecology studies the factors that are involved in structuring a community—in determining the diversity and relative abundance of species in a community.

Two Views of Communities

Ecologists in the 1920s and 1930s developed two divergent views on community composition. Gleason described communities as chance groupings of species found in the same area because of their similar requirements for environmental factors. Clements claimed that a community was a kind of superorganism that functioned as an integrated unit and was organized by prescribed interactions among species.

The plant species of most communities appear to vary on a continuum, with each species independently distributed along environmental gradients according to its tolerance range. The distribution of animal species may be more dependent on the presence of other species. Community composition is most likely determined by both abiotic gradients and biological interactions.

Properties of Communities

The community level of organization has several unique properties: (1) species *richness*, the number of species found in a community; species *equitability*, the relative numbers of individuals in each species; and species *diversity*, which considers both richness and equitability; (2) *prevalent form of vegetation*, the vertical profile or the most common vegetative growth forms; (3) *trophic structure*, the feeding relationships including plant–herbivore and predator–prey interactions that may determine community structure; and (4) *stability*, the ability of a community to return to its original makeup following a disturbance.

Community Interactions

Adaptations to Biotic Factors: An Overview Both the abiotic and biotic components of the environment act as selective factors for evolution. *Coevolution* refers to reciprocal adaptations, in which a change in one species acts as a selective agent on another species, whose adaptations in turn act as selective forces on the first species. The adaptations of flowers and their exclusive pollinators are examples of coevolution.

A complex example of coevolution may be found in the passion-flower vines, with their toxic compounds and nectaries that attract ants, and the butterfly

Heliconius, whose larvae are specialized feeders on these vines. The female butterflies tend to avoid depositing their bright yellow eggs on leaves that already have eggs. Some species of *Passiflora* have leaves with yellow dots that mimic these eggs and protect the vine from a heavy infestation of *Helioconius* larvae. Ants and wasps, which are attracted to these yellow nectaries and which prey on *Heliconius* eggs and larvae, may discourage butterflies from laying eggs.

Coevolution is marked by reciprocity and specificity. Whereas it is often difficult to determine whether intertwined adaptations among several species have served as selective factors for each other, it is generally accepted that interactions in ecological time are linked to adaptations over evolutionary time.

Competition Between Species Intraspecific competition (among individuals of the same species) for limited resources limits population growth. If two species use the same resource, *interspecific competition* should affect the density of both species. Competition for shared resources is believed to be important in determining species diversity.

As a result of his laboratory experiments with *Paramecium*, Gause formulated the *competitive exclusion principle*: two species that compete for the same limiting resource cannot coexist in the same habitat. Whether competitive exclusion occurs in natural communities and whether extinction, emigration, and/or further adaptations result from these competitions are areas of research by community ecologists.

An organism's *niche* is described as its "role" in an ecosystem—its habitat and its usage of the biotic and abiotic resources. The *fundamental niche* includes the resources an organism theoretically could use; the *realized niche* is the resources that it actually can use as determined by the biological constraints of competition or predation.

The competitive exclusion principle holds that two species with the same niche cannot coexist in a community. Slight variations in niche may allow closely related species to coexist. MacArthur's study of the feeding domains of various warbler species illustrates the concept of *resource partitioning*.

Many ecologists do not accept the generalization that competition is the major factor that structures communities. Niches are so complex that it is difficult to determine whether two species actually are competing and even more difficult to know what has occurred in evolutionary history.

Predation *Predators* are species that eat other organisms; *prey*, whether plants or animals, are the species that are eaten. Adaptations to increase success in predation on animals include acute senses, speed and agility, and physical structures such as claws, fangs, and stingers.

Plants may defend against animal predators by camouflage, mechanical devices such as thorns or microscopic spines, or chemical compounds. Distasteful or toxic chemicals, called *secondary compounds*, include such well-known compounds as strychnine, morphine, nicotine, mescaline, and tanins. Counteradaptations may enable certain herbivores to overcome these plant defenses.

Animals can defend against predation by hiding, escaping, or defending. A prey may blend in with its background by shape, behavior, or *cryptic coloration*. Deceptive markings, such as fake eyes or false heads, may discourage or confuse predators. The porcupine's mechanical defenses and the skunk's chemical defenses discourage predation. Some animals passively accumulate toxins from their plant diets and then use *aposematic coloration*, or bright warning colors, to help predators learn not to eat them.

Mimicry, in which one species resembles another, may be used by prey or predators. In *Batesian mimicry*, a harmless species resembles a poisonous or distasteful species. *Mullerian mimicry* is mutual imitation by two distasteful species. One predatory firefly species mimics another species' signal flashes, thereby attracting males as prey.

The role of predation in structuring communities may be to stabilize species diversity. As one prey species dwindles in numbers, most predators will feed on another species. This *switching behavior* may serve to maintain a variety of species within a community. Paine's study of the predatory sea star in the intertidal community helped to develop the concept of a *keystone predator* and its role in maintaining species diversity by reducing the density of a highly competitive prey species.

Other Interactions *Symbiotic* interactions influence community structure. In *parasitism*, the host is harmed; in *commensalism*, one individual is benefitted and the other unharmed; and in *mutualism*, both symbionts benefit from the relationship.

Both the parasites best adapted to find and feed on their hosts and the hosts best able to defend against parasites would be favored by natural selection. Some secondary plant products and the immune system of vertebrates are examples of host defenses. Many parasites have evolved adaptations that allow them to feed without killing their specific host. Coevolution may stabilize host–parasite relationships.

It is difficult to establish that one member of a commensal relationship is completely unaffected by the other. Cattle egrets and the cattle that flush the insects on which the egrets feed illustrate commensalism. Coevolution cannot be involved in commensal relationships, because the fitness of the one species is not affected by the species that benefits from the association.

Coevolution may be involved in mutualistic relationships, as changes in one species affect the fitness of the other. Examples of mutualism include the nitrogen-fixing bacteria within legumes, cellulose-digesting microorganisms within the guts of termites and ruminants, mychorrhizae, flowering plants and their pollinators, and the ants on acacia trees. Some mutualistic interactions may have evolved when organisms became able to derive some benefit from their predator or parasite.

Complex Effects of Species Interactions on Community Structure　A variety of community interactions affect species diversity within a community. It may be difficult to determine whether predator pressure or competition has shaped the resource partitioning observed in some communities. Multiple interactions between organisms and their biotic and abiotic environment help determine community structure.

Succession

A transition in species composition in a community, usually following some disturbance, is known as *succession*. If no previous organisms were present due to the absence of soil, the process is called *primary succession*. *Secondary succession* occurs when an existing community has been disrupted by fire, logging, or farming. Both the primary glacial-till succession and the secondary old field succession take about 200 years to reach a stable community. Many ecologists used to claim that succession follows a definite sequence, leading to a predictable climax community that was the final, stable stage of succession that would occur within each particular abiotic environment.

Causes of Succession　Early successional stages are dominated by colonizing plants that disperse readily and grow rapidly. Species whose tolerances best fit the specific abiotic environment of an area are the most successful colonizers.

Successional changes may be *autogenic* and attributed to changes induced by the biotic community. Early colonizers may influence the growth of other plants. An example of *inhibition* is a rapidly growing plant that shades the growth of most other plants. Transitions occur when latecomers are able to outcompete the pioneers.

Autogenic succession may occur because of *facilitation*; one stage of vegetation alters the environment to make it more suitable for species in the next stage.

Facilitation is observed in the progression from barren ground, left by retreat of a glacier, to a stable spruce-hemlock forest with the associated accumulation of humus in the soil.

Allogenic factors that disturb some communities may prevent the succession to a typical climax community. Grasslands may be maintained because periodic fires prevent the invasion of trees.

Human Disturbance　Human activities have altered the structure and succession of communities all over the world. Logging and agriculture disrupt mature communities and restart successional growth. The disappearance of the tropical rain forests and the creation of vast barren areas in Africa have resulted from human disturbances.

Equilibrium and Species Diversity　Species diversity generally increases during succession, partly as a result of more extensive and complex community interactions. According to the equilibrium model, succession reaches a climax community when interactions are so intricate that no new niches are available for additional species.

The nonequilibrium model views communities as being in continual nonequilibrium, with the number of species changing even in the so-called climax stage. Chance events such as dispersal and disturbance are given major roles in the process of succession. Succession depends both on the species that happen to colonize the area and on environmental disturbances that prevent species diversity from becoming constant. Severe and frequent disturbances may restrict the community to good colonizers, whereas infrequent disturbances may allow late-successional, highly competitive species to become dominant. According to the *intermediate disturbance hypothesis*, species diversity may be greatest when disturbances are intermediate in severity and frequency, allowing organisms from different successional stages to be present.

Diversity and Community Stability　Ecological theory once held that increased diversity led to increased stability, to communities either more resistant to change or more resilient following disturbance. Some mathematical models developed in the 1970s predict the opposite—that increased diversity leads to decreased stability. Ecologists continue to debate the relationship between species diversity and community stability.

Biogeographical Aspects of Diversity

Biogeography, the study of past and present distributions of species, is concerned with both the number and the identity of species in an area. Biogeographic realms, roughly correlated with the patterns of continental drift after the breakup of Pangaea and separated in places by deserts or mountain ranges, may have unique taxonomic groups. Species distribution reflects both past history and present interactions.

Limits of Species Ranges   A species may be limited to a particular range because it has never dispersed beyond that range or because pioneers to areas outside the range have failed to tolerate the physical environment or to compete with the resident species. Transplant experiments can be used to determine which of these two reasons has limited the species' range. The rabbits in Australia, gypsy moths in the northeastern United States, and Africanized "killer" bees of South America are examples of successful transplants that have had negative ecological effects.

The northern expansion of the ranges of cardinals and some mammals in the past 100 years has not been satisfactorily explained.

Global Clines in Diversity   Two major global clines show increases in diversity in many groups as one moves from northern and southern latitudes toward the equator and from shallow coastal waters to the deep ocean. The high diversity of the tropics may be correlated with increased photosynthetic productivity, but the shallow-to-deep water cline is not correlated with increased productivity.

Island Biogeography   Islands, or any habitat surrounded by an inhospitable habitat, allow ecologists to study the role of dispersal and habitat size in determining species diversity. In the 1960s, MacArthur and Wilson developed a general theory of island biogeography, stating that the size of the island and its closeness to the mainland (or source of dispersing species) are important factors that directly correlate with species diversity. Following a period of initial immigration and colonization, fewer new species can become established, and the extinction rate of present species increases. When immigration and extinction rates are equal, an equilibrium in species diversity develops, although species composition may continue to change. Species diversity will be greater both for larger islands in which more niches are available and for closer islands in which a higher immigration rate accounts for higher diversity.

Observations and experiments, including the work of Simberloff and Wilson with arthropod diversity, provide support for the island biogeography theory. These experiments have shown that, whereas diversity relates to size and closeness of islands, chance events such as dispersal affect the actual species composition of communities. Ecologists now find that community structure and dynamics are much less predictable and deterministic than once believed.

STRUCTURE YOUR KNOWLEDGE

1. Develop a concept map organizing your understanding of the important factors that structure a community.

2. Describe succession and the factors that contribute to this process.

TEST YOUR KNOWLEDGE

MULTIPLE CHOICE: *Choose the one best answer.*

1. Which of the following is not part of Gleason's concept of communities?
   a. Communities are chance collections of species that are in the same area because of similar environmental requirements.
   b. There should be no distinct boundaries between communities.
   c. The consistent composition of a community is based on interactions that cause it to function as an integrated unit.
   d. Species are distributed independently along environmental gradients.

2. Boundaries of communities
   a. occur along abrupt changes in abiotic factors.
   b. are wherever community ecologists choose to make them.
   c. are not discrete or distinct.
   d. may be all of the above.

3. The species equitability of a community refers to
   a. the relative numbers of individuals in each species.
   b. the number of different species found in a community.
   c. the feeding relationships or trophic structure within the community.
   d. the species diversity that is characteristic of that community.

4. Which of the following is not an example of coevolution?
   a. adaptations of flowers and their exclusive pollinators
   b. passion-flower vines and the butterfly *Heliconius*

c. development of a mutualistic symbiosis between former host and parasite

d. aposematic coloration of monarch butterfly and predators that learn not to eat them

5. Through resource partitioning,
   a. two species can compete for the same prey item.
   b. slight variations in niche allow closely related species to coexist in the same habitat.
   c. two species can share the same realized niche in a habitat.
   d. competitive exclusion results in the success of the superior species.

6. Which of the following is an example of a predator using mimicry?
   a. a firefly that flashes another species' signal flashes to attract males
   b. the porcupine's quills
   c. Mullerian mimicry of two distasteful species
   d. wing markings of io moth that look like large eyes

7. The ability of some herbivores to eat plants that have toxic secondary compounds may be an example of
   a. coevolution.
   b. mutualism.
   c. commensalism.
   d. parasitism.

8. A good-tasting prey species may defend against predation by
   a. Mullerian mimicry.
   b. Batesian mimicry.
   c. secondary compounds.
   d. aposematic coloration.

9. The role of predation within a community may be to
   a. maintain species diversity by switching between prey species as each becomes more numerous.
   b. increase the relative abundance of the dominant species.
   c. stabilize host–parasite relationships.
   d. encourage resource partitioning and coevolution.

10. A highly successful parasite
    a. will not harm its host.
    b. may benefit its host.
    c. will be able to feed without killing its host.
    d. will kill its host fairly rapidly.

11. The most important factor in determining community structure

a. may change from one community to another.
b. is predation.
c. is competition.
d. is history.

12. Primary succession
    a. involves the first colonists that arrive to an area.
    b. occurs when no previous organisms were present in an area.
    c. is a result of facilitation.
    d. occurs following a severe disruption of a community.

13. Which of the following would *not* be an autogenic factor in succession?
    a. early colonizers that shade other species
    b. accumulation of humus from decomposition
    c. periodic fires that restart secondary succession
    d. changes in soil pH that make soil unsuitable for present inhabitants

14. Inhibition
    a. may prevent the achievement of a climax community.
    b. is evidence for the equilibrium theory of succession.
    c. is one of the factors that determines the most tolerant species in an area.
    d. eventually may be overcome by competitively successful species.

15. According to the nonequilibrium model of succession,
    a. chance events such as dispersal and disturbance play major roles in succession.
    b. species diversity is greatest in the climax community.
    c. when succession reaches a climax community, only extinctions make room for new colonists.
    d. the communities with the greatest diversity have the greatest resistance to change and resiliency.

16. The taxonomic similarity of species within biogeographic realms
    a. is due to the similarity in environmental characteristics.
    b. is mostly a reflection of past history of species' distributions.
    c. is greatest on larger islands.
    d. reflects the inability of species to live outside their ranges.

17. A species may be restricted to a particular range because
    a. it cannot tolerate environmental conditions outside that range.
    b. it has never dispersed beyond that range.

c. it would outcompete native species if it were transplanted to their habitat.

d. of both a and b.

18. The increasing diversity of the benthic community in the deep seas compared to that in shallow coastal waters is an example of

a. the diversity of an area correlating with its productivity.

b. a global cline.

c. the equilibrium model of community structure.

d. a biogeographical realm formed by the breakup of Pangaea.

19. An island that is small and far from the mainland, as compared to a large island close to the mainland,

a. would be expected to have a low species diversity.

b. would be expected to be in an early successional stage.

c. would have a small species diversity but a large abundance of organisms.

d. would have a low rate of colonization and a low rate of extinction.

20. The island recolonization experiment of Simberloff and Wilson showed that

a. species diversity returns very slowly to an island after a disturbance.

b. the species diversity was highest when disturbances were intermediate in frequency and severity.

c. whereas the same numbers of species of arthropods returned to each island, the species composition was different, indicating the importance of chance events in community composition.

d. the islands that were closest to the mainland had the greatest numbers of arthropods recolonize, and their community composition and diversity were the same as they were prior to fumigation.

CHAPTER 49

ECOSYSTEMS

FRAMEWORK

This chapter describes the flow of energy and chemical cycling that occurs through the trophic structures found at the organizational level of ecosystems. Producers convert light energy into chemical energy, which is then passed, with a loss of energy at each level, from consumer to consumer, and to detritivores. Energy makes a one-way trip through ecosystems. Chemical elements, such as carbon, oxygen, nitrogen, and phosphorus, are cycled in the ecosystem from reservoirs in the atmosphere or soil through producers, consumers, and detritivores, and back to the atmosphere or soil.

The burning of fossil fuels, deforestation, and the dumping of toxic chemicals into the environment and excess nutrients into lakes are human activities that have altered normal ecosystem dynamics and pose threats of future ecological disruption.

CHAPTER SUMMARY

An *ecosystem* is a community and its physical environment; it includes all the biotic and abiotic components within an area. The two processes of energy flow and chemical cycling are associated with the ecosystem level of organization. Most ecosystems are powered by the input of sunlight energy, which is transformed to chemical energy by autotrophs, passed to a series of heterotrophs in the organic compounds of food, and continually dissipated in the form of heat. Energy flows through the system; it is not recycled. Chemical elements, such as carbon and nitrogen, are cycled between the abiotic and biotic portions of the ecosystem as autotrophs incorporate them into organic compounds, and the processes of metabolism return them to the soil and air.

Trophic Levels and Food Webs

The *trophic structure* of an ecosystem determines the flow of energy and chemicals through different feeding (trophic) levels. The *primary producers* are autotrophs, which usually use light energy to photosynthesize sugars for use as fuel in respiration and as building materials for other organic compounds. The *primary consumers* are herbivores, which eat plants and algae. *Secondary consumers* are carnivores that eat herbivores; *tertiary consumers* eat other carnivores. The *detritivores* consume organic wastes and dead organisms.

Plants are the main producers in terrestrial ecosystems; photosynthetic protoctists and cyanobacteria form the basis for most aquatic ecosystems. Chemosynthetic bacteria found around thermal vents deep in the seas depend on geothermal rather than solar energy.

Fungi and bacteria are the most important decomposers in most ecosystems. Earthworms and such scavengers as crayfish, cockroaches, and bald eagles are also detritivores.

A *food chain* shows the transfer of food between trophic levels. Omnivores eat producers as well as consumers of various levels. Most ecosystems have complex, branching food chains called *food webs*.

Energy Flow

Primary Productivity Of the visible light that reaches plants and algae, less than 1% is converted to chemical energy by photosynthesis. Nevertheless, the world-wide photosynthetic production is about 170 billion tons/year of organic material.

Different ecosystems vary in productivity. The rate of conversion of light to chemical energy in an ecosystem is called *primary productivity*. The *net primary productivity* (NPP) is the *gross primary productivity* (GPP) minus the amount used by plants in their own cellular respiration. For most plants, 50% to 90% of gross primary productivity remains as net primary productivity.

Gross primary productivity can be measured in aquatic habitats by the comparison of oxygen concentrations in incubated dark and transparent bottles in which, respectively, just respiration or both respiration and photosynthesis have been occurring. In more unproductive waters, the level of radioactive carbon that has been incorporated into plankton when incubated with labeled bicarbonate can be used to estimate net primary production.

Primary productivity can be expressed as energy per unit area per unit time (kcal/m^2/yr), or as biomass (g/m^2/yr). *Biomass* is measured in terms of the dry weight of organic material. The *standing crop* is the total biomass of plants. Primary productivity is the rate at which new biomass is synthesized. Tropical rain forests are the most productive ecosystems; estuaries and coral reefs also have high rates of productivity.

Productivity in terrestrial ecosystems is influenced by precipitation, heat, light intensity, length of growing season, mineral content of the soil, and even CO_2 supply. A *limiting nutrient* is one that is not present in adequate amounts and limits productivity.

Productivity in the seas is greatest in shallow waters and along coral reefs. The open ocean has generally low productivity because mineral nutrients are limited near the surface where light is available for photosynthesis. Phytoplankton communities are most productive where upwellings bring nitrogen, phosphorus, and other organic nutrients to the surface. In freshwater ecosystems, light intensity, temperature, and availablility of minerals may affect productivity.

Pyramids of Energy and Biomass *Secondary productivity* is the rate at which the consumers of an ecosystem—herbivores, carnivores, and detritivores—produce new biomass from organic material. Each trophic level sees a decline in productivity, partly because of a loss of energy to heat as each consumer converts organic fuel into its own molecules or into energy in cellular respiration.

Herbivores cannot digest much of the cellulose that they eat, and two-thirds of the energy they do absorb is used for cellular respiration. In a grass field, insects may convert about 4% of the net primary productivity to secondary productivity. Carnivores can digest and absorb more of the organic compounds in their food, but often they use more of it for cellular respiration.

Ecological efficiency is a measure of the percent of the potential energy on one trophic level that makes it to the next level. Approximately 10% of the chemical energy is transferred to the next trophic level. This progressive loss of energy can be represented as an *energy pyramid*. A pyramid of biomass illustrates the total dry weight of organisms at each trophic level, an indication of the amount of chemical energy stored in organic compounds, and thus, this pyramid also steps steeply from the producers to the top trophic level. Some aquatic ecosystems may have inverted biomass pyramids. The zooplankton primary-consumer level may outlive and outweigh the highly productive but heavily consumed phytoplankton primary-producer level. Although more biomass is present at any one time in the zooplankton, the energy pyramid is normal in shape; more energy is converted to organic material by phytoplankton than reaches the zooplankton level.

Because of the loss of energy at each trophic level, most food chains are limited to three to five links. Only about one-thousandth of the chemical energy produced in photosynthesis makes it to a tertiary consumer. Eating grain-fed beef as a main source of calories is an inefficient means of obtaining the energy trapped by photosynthesis.

Chemical Cycling

Chemical elements are passed between the abiotic and biotic components of the ecosystems through *biogeochemical cycles*. Plants and other autotrophs use these inorganic nutrients to build new organic matter, which then is passed through the food chain. Atoms of organisms are returned to the atmosphere, water, or soil through respiration and the action of decomposers.

The route of a chemical cycle depends on the element involved. Global cycles, which involve atmospheric reservoirs, occur for carbon, oxygen, sulfur, and nitrogen. Less mobile elements, such as phosphorus, potassium, calcium, and the trace elements, have a more localized cycle in which the soil is the main abiotic reservoir.

The actual movement of elements through biogeochemical cycles is quite complex, with influx and loss of nutrients from an ecosystem occurring in many ways. Ecologists have worked out general schemes in several ecosystems by adding radioactive tracers to chemical elements.

The Carbon Cycle Carbon moves between the abiotic and biotic portions of ecosystems through the processes of photosynthesis and cellular respiration. Carbon generally cycles at a fast rate as plants take up CO_2 for photosynthesis and organisms release it in respiration. Wood, coal, and petroleum can store carbon for long periods of time. The relatively low concentration of CO_2 in the atmosphere fluctuates slightly with seasonal changes in photosynthetic activity in the northern hemisphere. The combustion of fossil fuels is increasing the amount of atmospheric CO_2.

In aquatic environments, CO_2 interacts with water

and limestone to reversibly form bicarbonate, which acts as a carbon reservoir. The ocean may absorb some of the excess CO_2 added by the combustion of fossil fuels.

The Nitrogen Cycle   Plants cannot assimilate atmospheric nitrogen. Only certain prokaryotes can fix or reduce N_2 into ammonia (NH_3), which can then be incorporated into amino acids and other nitrogenous organic compounds. Soil bacteria and symbiotic bacteria in root nodules fix nitrogen in terrestrial ecosystems; cyanobacteria do so in aquatic ecosystems. Fertilizers add a significant amount of nitrogen to agricultural soils and associated waters. Some soil bacteria, in a process called *nitrification*, oxidize ammonia to NO_2^- (nitrite) and NO_3^- (nitrate). Plants can absorb either ammonia or nitrate. Animals obtain their nitrogen in organic form from plants or other animals.

Detritivores, in a process called *ammonification*, decompose nitrogenous compounds from organic wastes and dead organisms and return ammonia to the soil. Some bacteria convert nitrate to N_2 to obtain their oxygen. This *denitrification* returns N_2 to the atmosphere.

Nitrogen fixation and denitrification are minor processes in the nitrogen cycle. Most assimilated nitrogen comes from nitrates in the soil and water recycled from organic compounds by ammonification and nitrification.

The Phosphorus Cycle   Weathering of rock adds phosphorus to the soil, usually in the form of PO_4^{-3}, which can be absorbed by plants. Organic phosphate is transferred from plants to consumers and returned to the soil through the action of decomposers or by excretion by animals. Humus and soil particles usually bind phosphorus, keeping it available locally for recycling. The phosphorus that leaches into the water table eventually travels to the sea, where sediments become incorporated into rocks. These rocks may eventually return to terrestrial ecosystems as a result of geological processes. The sedimentary cycle works within geological time, whereas the local cycle between soil, plants, and consumers operates in ecological time.

Variations in Nutrient-Cycling Time   Nutrients cycle rapidly in a tropical rain forest; forest-floor litter contains only about 1% to 2% of the total biomass. More nutrients are stored in the detritus found in temperate forests. Nutrient-cycling times in ecosystems relate to decomposition rate, size of standing crop, soil chemistry, and frequency of fires.

Human Intrusions in Ecosystem Dynamics

The trophic structure, energy flow, and chemical cycling of most ecosystems have been influenced by human activities and technology. This widespread ecological impact is often harmful.

Agricultural Effects on Nutrient Cycling   The harvesting of crops removes nutrients that would normally recycle into the soil. After depleting the organic and inorganic reserves of nutrients in a newly cleared area, crops require the addition of expensive synthetic fertilizers. Many nutrients are shunted into aquatic systems through sewage and runoff from fields, where they are lost to terrestrial systems and may stimulate excessive algal growth. Advances in sewage treatment may allow such nutrients to be safely returned to the land, thus reducing the need for fertilizer.

Effects of Deforestation on Chemical Cycling: The Hubbard Brook Forest Study   A team of scientists, coordinated by Bormann and Likens, has looked at nutrient cycling in this forest ecosystem for 30 years. The mineral budget for each of six valleys was determined by measuring the influx of key nutrients in rainfall and their outflux through the creek that drained each watershed. About 60% of the precipitation exits through the stream; the rest is lost by transpiration and evaporation. Mineral influx and outflow were nearly balanced, and most minerals were recycled within the forest ecosystem.

The effect of deforestation on nutrient cycling was measured for three years in a valley that was completely logged. Compared with a control valley, water runoff from the deforested valley increased by 30% to 40%; net losses of minerals, such as Ca^{2+}, K^+, and nitrate, were huge. The sixtyfold increase in nitrate in the creek created unsafe drinking water.

Accelerated Eutrophication of Lakes   Lakes undergo a natural succession in which their primary productivity increases as mineral nutrients are added from runoff and recycled through the lake's food chain. Lakes eventually become *eutrophic*, or very productive and nutrient-rich. Sewage, factory wastes, and runoff of animal wastes or fertilizers from agricultural lands often accelerate this process. The rapid increase in nutrients can cause an explosive increase in producers. Oxygen shortages, due to plant and algal respiration at night, and the metabolism of decomposers that work on the accumulating organic material, kill off many fish and other lake organisms.

Poisons in Food Chains Toxic chemicals have been dumped into many ecosystems. Many of these are synthetic and nonbiodegradable; some may become more harmful as they react with abiotic and biotic environmental factors. One of the most serious environmental threats is radioactive fallout from nuclear accidents.

Toxic substances may be retained within the tissues of organisms. In a process known as *biological magnification*, the concentration of such compounds increases in each succesive link in the food chain. DDT, a classic example of the problem of biological magnification, has been found in nearly every organism tested, and the high concentrations found in birds of prey have interfered with successful egg shell production and reproduction.

Intrusions in the Carbon Cycle The concentration of CO_2 in the atmosphere has been increasing since the Industrial Revolution as a result of the combustion of fossil fuels and wood, much of which has been removed by deforestation. The level has increased 7% in the past 30 years.

The increase in CO_2 may influence agricultural crop choices and, as C_3 plants become able to outcompete C_4 plants, affect species composition in natural communities.

Through a phenomenon known as the *greenhouse effect*, CO_2 in the atmosphere absorbs infrared radiation and slows its escape from Earth, causing an increase in temperature. The continued increase in atmospheric CO_2 concentration may have far-reaching consequences in weather changes, in the amount and distribution of precipitation, and in rising sea levels due to the melting of the polar ice. Scientists continue to debate the extent, consequences, and solutions to increasing CO_2 levels.

Biotic Effects at the Biosphere Level The Gaia hypothesis, as first proposed by Lovelock, maintained that the biosphere, as a sort of superorganism, shapes the climate and atmosphere of Earth and is capable of regulating the environment. An example of this regulation is the fairly constant concentration of O_2 over the past 200 million years, resulting partly from balances of photosynthesis and respiration but also from the production of methane by certain bacteria and termites, which reacts with O_2 to form CO_2 and H_2O.

More recent proponents of this hypothesis stress the natural feedback and regulation inherent in the dynamics of ecosystems. The Gaia metaphor points out the interconnectedness of the trophic levels, energy cycling, and chemical cycles within and between ecosystems, and the unknown dangers of our technological disruptions of these ecological balances.

STRUCTURE YOUR KNOWLEDGE

1. Two processes that emerge at the ecosystem level of organization are energy flow and chemical cycling. Develop a concept map that explains, compares, and contrasts these two processes.

2. Sketch out a generic food web, and then include examples of each link that could be found in an aquatic ecosystem.

3. Describe three or four human intrusions in ecosystem dynamics that have detrimental effects.

TEST YOUR KNOWLEDGE

MULTIPLE CHOICE: *Choose the one best answer.*

1. Which of the following organisms and trophic levels are mismatched?
   a. algae—producer
   b. phytoplankton—primary consumer
   c. earthworm—detritivore
   d. bobcat—secondary consumer

2. Chemosynthetic bacteria found around deep sea vents are examples of
   a. producers.
   b. detritivores.
   c. chemical cycling.
   d. secondary productivity.

3. In an ecosystem,
   a. energy is recycled through the trophic structure.
   b. energy is usually captured from sunlight by primary producers, passed to secondary producers in the form of organic compounds, and lost to detritivores in the form of heat.
   c. chemicals are recycled between the biotic and abiotic sectors, whereas energy makes a one-way trip through the food web.
   d. there is a continuous process by which energy is lost as heat, and chemical elements leave the ecosystem through runoff.

4. Primary productivity
   a. is equal to the standing crop of an ecosystem.
   b. is greatest in freshwater ecosystems.
   c. is the rate of conversion of light to chemical energy in an ecosystem.
   d. is all of the above.

5. Net primary productivity

a. is less than gross primary productivity.
b. takes into account the energy used by plants in respiration.
c. can be measured in an aquatic habitat by the incorporation of radioactive carbon from labeled bicarbonate into phytoplankton.
d. All of the above are correct.

6. The productivity in freshwater ecosystems is affected by
   a. temperature.
   b. light intensity.
   c. availability of nutrients.
   d. all of the above.

7. Secondary productivity
   a. is measured by the standing crop.
   b. is the rate of production of biomass in consumers.
   c. is greater than primary productivity.
   d. is only slightly less than primary productivity.

8. Which of the following is *not* true of an energy pyramid?
   a. Only about 10% of the energy in one trophic level is passed into the next level.
   b. Because of the loss of energy at each trophic level, most energy pyramids are limited to three to five steps.
   c. The energy pyramid of some aquatic ecosystems is inverted because of the large zooplankton primary-consumer level.
   d. Eating grain-fed beef is an inefficient means of obtaining the energy trapped by photosynthesis.

9. Biogeochemical cycles are global for
   a. elements that are found in the atmosphere.
   b. elements that are found mainly in the soil.
   c. carbon, nitrogen, and phosphorus.
   d. elements that have been marked with radioactive tracers.

10. Which of these processes is incorrectly paired with its description?
    a. nitrification—oxidation of ammonia in the soil to nitrite and nitrate
    b. nitrogen fixation—reduction of atmospheric nitrogen into ammonia
    c. denitrification—removal of nitrogen from organic compounds
    d. ammonification—decomposition of organic compounds into ammonia

11. Carbon cycles relatively rapidly except when it is
    a. dissolved in aquatic ecosystems.
    b. released by respiration.
    c. converted into sugars.
    d. stored in petroleum, coal, or wood.

12. The geological time scale of phosphorus cycling involves
    a. its incorporation into shells.
    b. the weathering of rock to add PO_4^{-2} to the soil.
    c. sedimentation to form rocks in the sea bed.
    d. the incorporation of phosphorus into fossils.

13. Which of the following was *not* shown by the Hubbard Brook Forest Study?
    a. Most minerals recycle within a forest ecosystem.
    b. Deforestation results in a large increase in water runoff.
    c. Mineral losses from a valley were great following deforestation.
    d. Following deforestation, mineral influx and outflow were nearly balanced; minerals left the ecosystem as fast as they were added.

14. The high concentrations of DDT found in birds of prey is an example of
    a. biological magnification.
    b. eutrophication.
    c. the Gaia hypothesis.
    d. chemical cycling.

15. The greenhouse effect
    a. could change global weather and lead to the flooding of coastal areas.
    b. could result in more C_4 plants in plant communities.
    c. causes an increase in temperature when CO_2 absorbs more sunlight entering the atmosphere.
    d. can do all of the above.

16. The Gaia hypothesis
    a. shows that atmospheric O_2 concentration is regulated by methane.
    b. was developed by Bormann and Likens.
    c. has shown that the biosphere is a superorganism, able to regulate the climate and atmospheric CO_2 concentration.
    d. may serve as a useful metaphor for the interactions between the biosphere and the physical parameters of Earth.

FRAMEWORK

This chapter introduces the complex and fascinating subject of animal behavior. The study of behavior involves the integration of the study of biochemistry, genetics, physiology, evolutionary theory, and ecology.

Behaviors can range from simple fixed action patterns in response to specific stimuli to insight learning in novel situations. Behaviors are the result of interactions between environmental stimuli, experience, and the genetic makeup of the individual. The parameters of behavior are controlled by genetics and thus are acted upon by natural selection. The ultimate cause of reproductive fitness can be used to interpret foraging behavior, agonistic interactions, mating patterns, and altruistic behavior. Sociobiology extends evolutionary interpretations to human social interactions.

CHAPTER SUMMARY

Behavior can be defined as any observable movement or action of an animal in response to a stimulus. Animal behavior includes such activities as feeding, courtship, mating, and communication. *Ethology* is the study of how animals behave in their natural environments. The interactions of animals with their environments is related to all levels of ecological organization: physiological, population, and community.

Studying Behavior

Scientists have begun to investigate and debate the complex and variable mechanisms that explain the *how* and *why* of animal behavior.

The Problem of Anthropomorphism   The graylag goose example points out the need to look for the cues that are guiding animal behavior. *Anthropomorphism*—ascribing human emotions, thoughts, or reasoning to other animals—should be minimized in the study of behavior.

Proximate and Ultimate Cause   *Proximate causes*, occurring in ecological time, explain behavior in terms of present cues and responses. Behavior is subject to natural selection and evolution. *Ultimate causes* of behavior relate to the evolutionary basis of a behavior.

Nature Versus Nurture   Patterns of behavior are, to a greater or lesser degree, under genetic control. The age-old controversy of nature versus nurture concerns how much of an organism's behavior is innate or instinctive and how much is learned. Much of vertebrate behavior appears to result from the complex interaction between physiology and external cues. Although behavior often can be modified by experience, and thus be phenotypically variable, it still has its basis in an organism's genes.

Innate Components of Behavior

Fixed Action Patterns   The work of Lorenz and Tinbergen provided evidence for the innate components of behavior and established the foundation of ethology. The *fixed action pattern*, or *FAP* (also called fixed motor pattern), was the fundamental concept to emerge from their work. An FAP is a highly stereotyped, innate behavior that, once begun, is carried through to completion. An FAP is triggered by a *sign stimulus*—some external sensory stimulus that, when perceived by the animal, "releases" the specific behavior.

The term *releaser* usually refers to sign signals that communicate between individuals of the same species.

The releaser for attack behavior in stickleback fish is the red belly of the intruder and may extend to any red colored object entering its territory. FAPs often are associated with interactions between parents and young, such as the begging and feeding behavior of many bird species. A human infant's smile and grasp are FAPs in response to visual or tactile stimuli.

The mechanistic aspect of FAPs is illustrated by the repeated stereotyped behavior seen in some animals, such as a digger wasp performing the drag and inspect subroutine over and over when the paralyzed cricket it placed outside its nest is experimentally moved.

The Nature of Sign Stimuli   The experiments of ethologists have shown that sign stimuli usually involve one or two simple characteristics of the object or organism that triggers the behavior. The pecking of a herring chick on its parent's beak also can be released by a spot swung horizontally at the end of a long object, corresponding to the red spot on the parent's moving beak.

An animal's sensitivity to certain stimuli is closely correlated to the sign stimuli to which it responds. Frogs are quite good at detecting movement, and the movement of objects releases the frog's feeding behavior.

Stronger responses often are elicited by exaggerated sign stimuli. The nestling with the widest-gaping beak will be most likely to be fed. The experimental use of *supernormal stimuli*, such as giant models of eggs, illustrates these behavioral tendencies.

The use of simple cues in the animal's environment to release pre-programmed behavior ensures behavioral responses requiring the least amount of processing and integrating of input.

Innate Releasing Mechanisms and Drive   *Innate releasing mechanisms* (IRM), consisting of networks of neurons or cells, were proposed by early ethologists as the link between stimuli and responses. Limited evidence for IRMs comes from the identification of a giant neuron in the nudibranch *Tritonia* that triggers escape behavior when artificially stimulated. In vertebrates in particular, the concepts of releasers and IRMs are consistent with the observation of behaviors but are not explanations of how these behaviors happen.

The idea of *drive* or motivation is used to explain the release of some behaviors. Feeding, reproduction, and other behaviors are not continuously performed. Stronger tendencies toward a behavior may be correlated with a physiological change, such as seen by the increase in stereotyped pecking when chickens are hungry. The concept of drive does not apply to all behavior, and the use of the concept of drives does not give any information about the mechanisms causing behaviors.

Learning and Behavior

Learning, the modification of behavior as a result of experience, can affect even innate components of behavior. *Habituation* refers to the loss of sensitivity to unimportant or repeated stimuli.

Practice may result in the more effective performance of innate behaviors. This improvement, however, may derive from the natural development and maturation of the animal. Birds prevented from flying until older are able, upon release, to fly without practice or learning.

Imprinting   The phenomenon of *imprinting* is a clear case of learning closely associated with innate behavior. Newly hatched ducks and geese instinctively follow the movement of an object, whether it be their mother, a ticking box, or Konrad Lorenz. Species recognition and reproductive behavior are tied to this early imprinting in ducks and geese. The return of salmon to their home stream to spawn is an example of olfactory imprinting.

A *critical period* is a limited time during which some type of imprinting may occur. Geese totally isolated during their first two days fail to imprint on anything. Adult birds appear to imprint on their young in the period following hatching. The length and timing of the critical period of sexual imprinting may vary. Male finches, after being placed only with members of another closely related species during their critical period for sexual identity, readily mated with females of the second species but reluctantly mated with their own species.

Learning and innate behavior may closely interact. A songbird learns to sing normally only when it can hear the song of its species. If a bird hears only the song of a related species during its critical period, it does not learn that song. Genetics determine what song a bird can sing, but it can learn that song only by hearing it from other birds.

Classical Conditioning   In *associative learning*, animals learn to associate one stimulus with another. *Classical conditioning*, described through the work with salivating dogs done by Pavlov in the 1900s, illustrates the association of an irrelevant stimulus with a fixed physiological response. The same type of conditioning may be associated with the sharpening of perceptions of release stimuli in an animal's natural environment.

Operant Conditioning   *Operant conditioning* refers to the trial-and-error learning through which an animal associates a behavior with a reward or punishment. Skinner's work with rats is the best known laboratory study of operant conditioning. The association of good or bad tastes with food items is probably a common form of operant conditioning in nature. Animals may also be able to learn by watching the behavior of others. The spread of milk-bottle pecking through the English tit population illustrates this type of learning.

Insight Insight learning, sometimes called reasoning, involves the ability to perform a behavior correctly in a novel situation. The stacking of boxes by a chimpanzee to reach a banana is a commonly cited example. Insight learning is most often observed in primates and some other mammals.

The capacity to learn confers a survival and reproductive advantage and can be acted upon by natural selection. The genetic base and internal mechanisms of learning are largely unknown. Some simple kinds of learning have been linked to biochemical or physiological changes. Most animal behaviors can be explained as relatively fixed patterns in response to simple stimuli, which may be modified by simple kinds of learning.

Behavioral Rhythms

Feeding, sleeping, reproducing, and migrating are all regularly repeated behaviors that show a temporal rhythm. Animals, as well as plants, show circadian rhythms of roughly 24-hours. To determine whether these rhythms are based on exogenous (external) cues or endogenous (internal) timers, researchers have placed animals into environments with no external cues and monitored their rhythmic behaviors. A biological clock appears to be involved, although exogenous cues are necessary to keep the behavior timed with the real world. The exogenous cue, sometimes called a *zeitgeber*, is usually light. Humans living in free-running conditions with no external time cues have a biological clock with a period of about 25 hours.

In many species, *circannual* behaviors, such as reproduction and hibernation, are based on physiological and hormonal changes that are linked with changes in day length. One of the few studies of the endogenous control of these long-term behaviors has shown that fat deposition in ground squirrels, associated with hibernation, occurs regularly, even in a constant environment.

The timing mechanism of the biological clock is unknown. Current hypotheses propose some sort of biochemical clock, perhaps based on regular molecular interactions.

Orientation and Navigation

Animals use a variety of cues to orient themselves in and navigate through their environments. A *taxis* is a more or less automatic movement directed toward or away from a stimulus. A *kinesis* is a change in activity rate in response to a stimulus. Although kinetic movements are randomly directed, they tend to maintain organisms in favorable environments as the animal's random movements slow down in response to favorable stimuli.

Long-distance migration requires sophisticated mechanisms of orientation. Some species of birds and other animals navigate by the heavens, using the sun or stars as directional cues. Internal timing mechanisms are needed to compensate for the continuous daily movement of these reference bodies. Many migrating organisms can adjust to the changing positions of the sun and stars both throughout the day and along the migratory route.

Many bird species are able to continue migrating under clouds or through fog, perhaps because of their ability to detect and orient to Earth's magnetic field. Experimentally manipulating the magnetic field changes the migratory orientation of some species.

Foraging Behavior

The large variety of animal feeding patterns is accompanied by an array of foraging behaviors. Feeding often involves relatively simple behavior, including fixed action patterns released by the sight of prey. Some animals are generalists, feeding on a large variety of items, whereas others are specialists with quite restricted diets. Specialists usually have morphological and behavioral adaptations that are specific for their food item and make them extremely efficient at foraging.

Generalists will often concentrate on an abundant food item, developing what we call a *search image* for the favored item. When that item becomes less abundant, the animal appears to change its search image and concentrate on a new item. Search images may allow for short-term specialization while maintaining the advantages of being a generalist. The concept of search image describes what appears to happen, but, once again, does little to explain the underlying mechanisms.

Behavioral ecologists study the foraging adaptations of animals, with the prediction of *optimal foraging*: that natural selection would favor animals that make efficient choices that maximize energy intake over expenditure. Studies, such as those on prey selection by bluegill sunfish, have indicated that animals can modify their foraging behavior in ways that tend to maximize overall energy intake. The ability appears to be innate, although experience and physical maturation are thought to increase foraging efficiency.

Competitive Social Interaction

Social behavior involves the interaction between two or more animals, usually of the same species. Mating behavior is mutually beneficial to the reproductive fitness of both individuals, but many types of interactions are competitive, especially when they involve conflict over limited resources.

Agonistic Behavior   *Agonistic behavior* involves a contest that determines which competitor gains access to resources such as food or mates. The encounter may include a test of strength or, more commonly, threat displays and *ritual* behavior, all of which serve to avoid actual physical conflicts. A submissive or appeasement display by one of the competitors inhibits further aggressive activity and indicates surrender. Natural selection apparently favors the ending of a conflict without a violent combat that would reduce the reproductive fitness of the winner as well as the loser.

Dominance Hierarchies   *Dominance hierarchies*, as illustrated by the "pecking order" in hens, establish which animals get first access to resources and prevent continual combats. In wolf packs, the top female controls mating of the others, in accordance with the abundance of food.

Territoriality   A *territory* is an area that is defended from other members of the same species. Territories typically are established for feeding, mating, rearing young, or a combination of these activities. Territory size varies with species, function, and resource abundance. A home range is the area in which an animal may roam, but a territory is the area that the animal defends. Agonistic behavior is used to establish and defend territories. Vocal displays, scent marks, and patrolling may be used to continually proclaim ownership. Usually only conspecifics are excluded from an animal's territory.

Both dominance systems and territories help to stabilize population density. They ensure that at least some individuals have a sufficient supply of a resource when the resource may be in limited supply.

Mating Behavior

Courtship   Most animals are programmed to view conspecifics as threatening competitors to be driven off. This aversion must be overcome for mating to be accomplished. Complex courtship interactions, unique to each species, assure that individuals are not a threat and are of the proper species, sex, and physiological mating condition. These complex rituals often consist of a series of fixed action patterns, each released in sequence by the reciprocal behavior of the individuals involved. Appeasement gestures are often part of the mating ritual.

Ritualized acts probably evolved from actions that had a more direct meaning. The behaviors of various species in the predaceous fly family Empididae show a progression from a male carrying a dead insect for the female to eat while he mates with her, to carrying a dead insect inside a silk balloon, to carrying simply an empty silk balloon.

Courtship behavior may also function as a basis for the female's selection of a specific mate whose "performance" may indicate his genetic fitness. Some male courtship displays show more of a connection with potential fitness than others do.

Mating Systems   Mating relationships vary a great deal among species. Many species have *promiscuous* mating, with no strong pair bonds forming. Longer-lasting relationships may be *monogamous* or *polygamous*. Polygamous relationships are most often polygynous (one male and many females), although a few are polyandrous.

According to Darwinian considerations, an individual is valuable to a member of the opposite sex only as a vehicle to help his or her genes get into subsequent generations. Reproductive behavior can be analyzed on the basis of maximizing fitness and may differ between the sexes based on their *sexual investment* in reproduction. Especially in vertebrates, females have a greater inherent investment in an offspring than do males, and it adds to the female's reproductive fitness to show discriminant selection of a mate. For males, sperm do not represent a large energy investment, and it may be reproductively advantageous to try to mate often and with a number of partners.

The needs of young offspring are reflected in the reproductive patterns of the parents. Newly hatched birds require more food than one parent may be able to supply. A male will ultimately leave more offspring, and thus increase his reproductive fitness, by helping to care for these young rather than by going off in search of more mates. With mammals, often only the female is needed as a source of food, and males are not monogamous.

Communication

Vision, hearing, and olfaction are common methods of communication between animals. *Pheromones* are chemical signals that often are involved in reproductive behavior, both to attract mates and to release specific courtship behaviors. The trailing behavior of ants involves scents released by scouts.

Honeybees have one of the most complex communication systems, using ritualized dances to communicate the location of food sources. These dances were first described by von Frisch. Round dances are used when the food source is relatively close to the hive, and regurgitated nectar provides a scent to direct other bees to the food. The location of more distant food is communicated by waggle dances, in which the speed and vertical orientation of the dance on the comb respectively indicate the distance from the hive and the direction to the food relative to the horizontal angle to the sun.

Altruistic Behavior

Altruistic behavior is difficult to explain in Darwinian terms of improving the reproductive fitness of the individual, especially in examples such as warning calls by Belding's ground squirrels that may increase the caller's risk of being killed, or the attack behavior of honeybees that results in the death of the stinging bee.

Natural selection can act on genes that produce altruistic behavior if the individual benefitting by an unselfish act also carries the altruistic genes. Cooperating individuals often are siblings or close relations, and thus are carrying many common genes. *Kin selection*, by which the inclusive fitness of related individuals is increased, suggests that altruistic behavior should be strongest between close relatives and less common as genetic relatedness decreases. Alarm calls usually are given by female ground squirrels. Males tend to disperse to other colonies after mating, while females are surrounded by offspring and other family relations. Thus, survival of a female's genes is enhanced by this altruistic behavior.

When altruistic behavior involves nonrelated animals, the explanation offered is *reciprocal altruism*; there is no immediate benefit for the altruistic individual, but some future benefit may occur when the helped animal may "return the favor." Reciprocal altruism often is used to explain altruism in humans.

Genetically maintained truly unselfish behavior, especially if it ended in the death of the altruistic animal, would be selected against. Altruistic behavior in animals must somehow enhance the survival of the individual's own genes. Many researchers maintain that seeming examples of kin selection in animals can be explained adequately as individual selection. Controversies over kin selection and altruistic behavior, particularly as they relate to human behavior, continue.

Animal Cognition

Because it is impossible to know what goes on inside the minds of animals, most researchers have taken a mechanistic approach, called *behaviorism*, to the study of animal behavior. Recently, Griffin and others have argued that conscious thinking is an inherent part of animal behavior. According to *cognitive ethology*, animal cognitive ability forms an evolutionary continuum through many animal groups and has adaptive advantages that have been acted on by natural selection.

Human Sociobiology

E. O. Wilson has elaborated the thesis that social behavior has an evolutionary basis and that behavioral characteristics are expressions of genes that have been acted on by natural selection. Much of Wilson's work dealt with the application of evolutionary theory to social behavior of insects, but he also speculated on the evolutionary basis of certain social behaviors of humans.

The sociobiology debate continues the nature-versus-nurture controversy. In the example of human incest, the "nurture" side would argue that the aversion to incest must not be innate or else the taboos and laws against it would be unnecessary. The avoidance of incest must be a learned behavior, based on the experience that individuals who break the taboo are more likely to have defective children. The "nature" side would argue that the occurrence of a behavior, such as the avoidance of incest, across many diverse cultures is evidence that it must have an innate component.

The Israeli kibbutzim provide evidence of the innate aversion to incest. Parents encourage their children to marry within the kibbutz, but marriages between these children, who were raised essentially as siblings, are rare.

Some sociobiologists, including Wilson, explain that cultural and genetic components of social behavior are linked together in a cycle of reinforcement. The development of cultural regulations of innate behaviors serves as an additional environmental factor in the natural selection of that behavior. According to this view, human behavior represents an integration of genes and culture.

The parameters of human social behavior may be set by genetics, but the environment undoubtedly shapes behavioral traits just as it influences the expression of physical traits. Genes involved with complex behavior probably have very broad "norms of reaction." An evolutionary interpretation of human social behavior must include environmental as well as genetic components.

STRUCTURE YOUR KNOWLEDGE

1. How does the nature vs. nurture controversy apply to behavior?

2. Distinguish the following types of learning: imprinting, classical conditioning, operant conditioning, and insight learning.

3. Why are many interactions between members of the same species agonistic? What phenomena reduce violent encounters?

4. What are the advantages and disadvantages of being a generalist or a specialist feeder?

5. How does the concept of Darwinian fitness apply to reproductive and altruistic behavior?

TEST YOUR KNOWLEDGE

MULTIPLE CHOICE: *Choose the one best answer.*

1. Behaviorism is the
   a. mechanistic study of the behavior of animals, focusing on stimulus and response.
   b. application of human emotions and thoughts to other animals.
   c. study of animal cognition.
   d. study of the origin and ultimate causes of animal behavior.

2. Proximate causes
   a. explain the evolutionary significance of a behavior.
   b. are immediate causes of behavior such as hunger or external cues.
   c. indicate that much of animal behavior is innate.
   d. are the focus of ethology.

3. A fixed action pattern
   a. shows that nature is the cause of animal behavior, not nurture.
   b. is an innate behavior, performed in its entirety in response to a sign stimulus.
   c. is released by an IRM, an innate releasing mechanism.
   d. is a stereotyped behavior that has been used to explain imprinting in young animals.

4. Supernormal stimuli
   a. are innate releasing mechanisms between individuals of the same species.
   b. may elicit stronger responses or FAPs.
   c. are illustrated by the removal of the cricket from near a digger wasp's nest.
   d. are illustrated by the bobbing of a stick with a red spot past herring chicks.

5. An animal's "drive" to perform a certain behavior
   a. may be related to its physiological condition.
   b. may be lessened through habituation.
   c. will be greater with supernormal stimuli.
   d. is correlated with the sensitivity of its receptors to sign stimuli.

6. Learning
   a. cannot modify innate components of behavior.
   b. is a type of imprinting.
   c. must involve the use of insight.
   d. is a change in behavior as a result of experience or practice.

7. A critical period
   a. is the time right after birth when sexual identity is developed.
   b. usually follows the receiving of a sign stimulus.
   c. is a limited time during which imprinting can occur.
   d. is the period during which birds can learn to fly.

8. In classical conditioning,
   a. an animal associates a behavior with a reward or punishment.
   b. an animal learns as a result of trial and error.
   c. an irrelevant stimulus can elicit a response because of its association with a normal stimulus.
   d. a bird can learn the song of a related species if it hears only that song.

9. Insight learning
   a. only occurs in human beings.
   b. involves the ability to perform successfully in a novel situation.
   c. was studied by Skinner in his use of Skinner boxes.
   d. All of the above are correct.

10. Regularly repeated behaviors that show a rough 24-hour rhythm are
    a. timed to the real world by the use of a biological clock.
    b. circadian rhythms.
    c. usually stopped in free-running conditions.
    d. a combination of learning and innate behavior.

11. Circannual behaviors
    a. are often linked to changes in day length.
    b. rely solely on endogenous cues.
    c. involve foraging, reproduction, and migration.
    d. do not occur in free-running conditions.

12. A kinesis
    a. is a randomly directed movement that is not caused by external stimuli.
    b. is a movement that is directed toward or away from a stimulus.
    c. is a change in activity rate in response to a stimulus.
    d. is illustrated by trout swimming upstream.

13. Which of the following is *not* true of long-distance migrations?
    a. Animals may use the stationary north star as a point of reference.
    b. Navigation using celestial bodies requires an internal clock to compensate for the daily movement of these objects.
    c. Birds that are able to continue migrating during fog and overcast weather may be sensing Earth's magnetic field.
    d. They are the result of operant learning.

14. A dominance hierarchy
    a. may be established by agonistic behavior.
    b. determines which animals get first access to resources.
    c. helps to avoid potential injury of competitors.
    d. All of the above are correct.

15. An animal's territory may

16. In a species in which females provide all the needed food and protection for the young,
    a. males are likely to be promiscuous.
    b. mating systems are likely to be monogamous.
    c. mating systems are likely to be polyandrous.
    d. males most likely will show sexual selection.

17. The ability of honeybees to fly directly to a food source, after having to wait several hours from the time of the waggle dance, indicates
    a. that the waggle dance provided directions relative only to the position of the hive.
    b. that bees have an internal clock that compensates for the movement of the sun during the elapsed time.
    c. that the bees must have been to that food source before.
    d. that bees are directed more by olfactory cues than by directional cues.

18. Sociobiology
    a. explains the evolutionary basis of behavioral characteristics within animal societies.
    b. applies evolutionary explanations to human social behaviors.
    c. studies the roles of culture and genetics in social behavior.
    d. does all of the above.

a. be larger than its home range.
b. change in size with changing resource density.
c. exclude both conspecifics and members of other species.
d. be all of the above.

CHAPTER 46
DIVERSE ENVIRONMENTS OF THE BIOSPHERE:
AN INTRODUCTION TO ECOLOGY

Suggested Answers to Structure Your Knowledge

1. Ecology is the study of the relationships of organisms to their abiotic and biotic environments, focusing on the abundance and distribution of species. Questions cover a range that includes the consideration of individuals and their physiological adaptations, the growth and regulation of populations, the interactions (such as predator-prey and competition) within communities, and the flow of energy and chemicals through ecosystems. Experimental manipulation of variables is done both in the laboratory and in the field, along with observational measurements of various physical factors and kinds and numbers of organisms. Mathematical models and computer simulations are useful for predicting the effects of variables, especially in situations where experimentation is impractical or impossible. Evolution, with the principles of natural selection and adaptation, is the key organizing theory.

2. Biomes include the eight major world communities, which are identified on the basis of abiotic and biotic similarities. Temperature and precipitation, which are related to latitude, altitude, and location (as in coast or interior of continent), are the factors that most influence the plant and animal life forms found in each community. Convergent evolution, as organisms adapt to similar environments, may account for the similarities in life forms in each biome.

Answers to Test Your Knowledge

Multiple Choice:

1.	b	5.	b	9.	d
2.	d	6.	c	10.	b
3.	c	7.	b	11.	c
4.	a	8.	a	12.	d

Matching:

1.	C	4.	E	7.	B
2.	H	5.	A	8.	F
3.	D	6.	G		

CHAPTER 47
POPULATION ECOLOGY

Suggested Answers to Structure Your Knowledge

1.

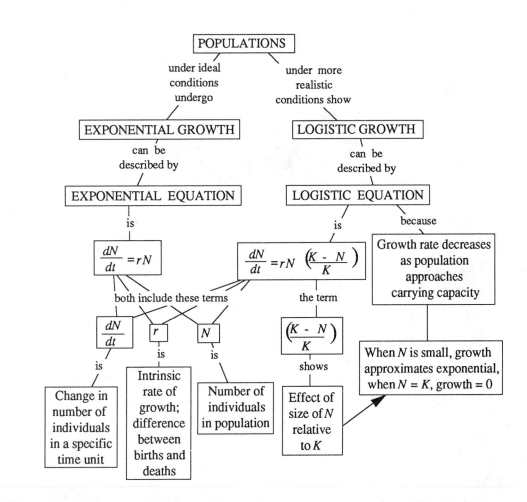

2. The age structure of a population impacts on population growth due to the differential reproductive capabilities of various age groups. The larger the proportion of reproductively active individuals, the higher the growth rate. The shorter the generation time (age at which reproduction first begins), the greater the population growth rate. The sex structure of a population may also determine growth rate, depending on the mating patterns of the population.

Answers to Test Your Knowledge

Matching:

1. H	3. B	5. A	7. J	9. E
2. G	4. C	6. L	8. K	10. D

Multiple Choice:

1. a	4. b	7. d	10. d	13. a
2. c	5. a	8. c	11. a	14. c
3. c	6. b	9. d	12. b	15. d

CHAPTER 48
COMMUNITIES

Suggested Answers to Structure Your Knowledge

1.

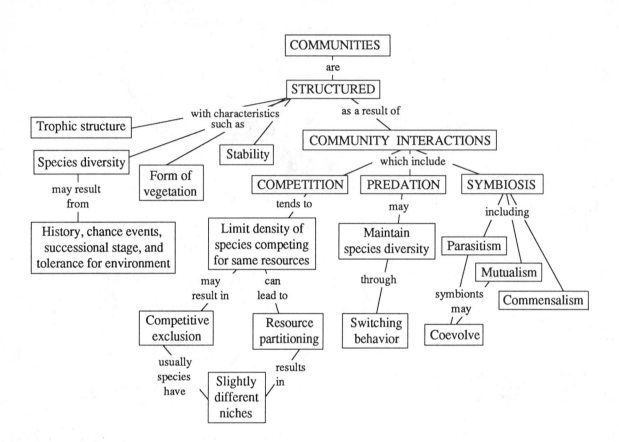

2. Succession is the gradual transition in species composition in a community, following a disturbance or creation of a new habitat. Each stage of vegetation alters the environment somewhat and may facilitate the transition to the next stage. Early pioneers in the habitat may also inhibit the establishment of other species, until more competitive species are able to become established. The change in species composition also may be due to the gradual success of the species that are most tolerant of the environment found in that habitat. Chance events such as dispersion and allogenic disturbances are important factors during succession.

There is evidence that even communities that appear to be in final successional stages are in continual nonequilibrium, with species composition and diversity still changing.

Answers to Test Your Knowledge

Multiple Choice:

1. c	5. b	9. a	13. c	17. d
2. d	6. a	10. c	14. d	18. b
3. a	7. a	11. a	15. a	19. a
4. d	8. b	12. b	16. b	20. c

CHAPTER 49
ECOSYSTEMS

Suggested Answers to Structure Your Knowledge

1.

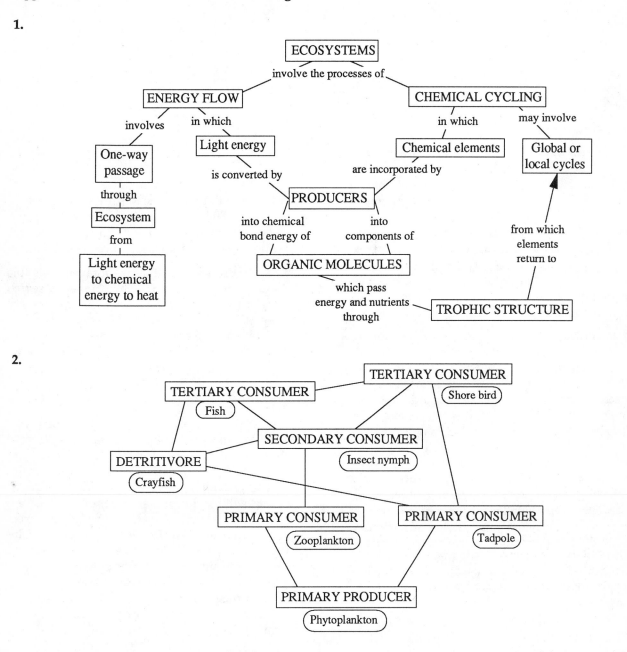

2.

3. a. Deforestation causes an increase in water runoff with an accompanying loss of soil and minerals. The cutting of tropical forests results in a loss of species diversity, a reduction in productivity of the area, and weather changes.

b. Dumping of wastes and runoff from agricultural lands has produced eutrophication in many lakes, killing off many fish and other organisms.

c. The introduction of toxic chemicals into the environment has resulted in their being incorporated into the food chain. As a result of biological magnification, these substances pose added threats to top-level consumers.

d. An increase in CO_2 levels in the atmosphere as a result of the combustion of fossil fuels and wood adds to the greenhouse effect. The resulting increase in temperature may have far-reaching effects on climate and sea level. Changes in CO_2 levels may also affect species composition of C_3 and C_4 plants in natural communities and in agricultural crops.

Answers to Test Your Knowledge

Multiple Choice:

1. b	5. d	9. a	13. d
2. a	6. d	10. c	14. a
3. c	7. b	11. d	15. a
4. c	8. c	12. c	16. d

CHAPTER 50
BEHAVIOR

Suggested Answers to Structure Your Knowledge

1. There is controversy over how much of animal behavior is innate and genetically programmed and how much is a product of experience and learning. Fixed action patterns are clearly preprogrammed behavior, and many seemingly complex animal behaviors can be isolated into a series of FAPs. In other cases, genetics may set the parameters for an organism's behavior, but experience can modify behavior and learning is clearly evident.

2. Imprinting is a type of learning that occurs during a critical time in an organism's development. The animal appears to be genetically programmed to identify with whatever species it is placed with at that time, to remember a smell as a guide to return "home" to spawn, or to follow whatever moves as "mother." Innate behavior and learning are tightly coupled in imprinting.

   The concept of classical conditioning was developed by Pavlov. As classical conditioning is used in a laboratory study, an animal learns to associate an unrelated stimulus with a stimulus that elicits a behavioral or physiological response. Later, the unrelated stimulus can cause the behavior. In nature, classical conditioning is the type of learning in which an organism adds to its collection of stimuli that may be associated with particular cues or releasers.

   Operant conditioning, as studied by Skinner, is learning in which an animal comes to associate a positive or negative reward with a particular behavior. In a laboratory or circus, it can be used to teach animals all sorts of "tricks." In nature, it is a way in which an organism can learn the good or bad tastes of food or similar lessons.

   Insight learning involves the ability of an animal to reason and to figure out an appropriate behavior in a novel situation.

3. Members of the same species occupy the same niche and are thus competitors for resources such as food, territory, and mates. In agonistic encounters, one animal may establish its right to these resources. Threat displays, appeasement gestures, dominance hierarchies, and territories all serve to reduce violent encounters between conspecifics.

4. A generalist is not as efficient a forager but is able to use more food items. The development of search images may improve the efficiency of generalists. A specialist often has anatomical and behavioral adaptations to feeding efficiently on one type of food item. Should this item become scarce, however, the specialist is at a disadvantage.

5. According to the concept of Darwinian fitness, an animal's behavior should help to increase its reproductive success. The best strategy for reproductive behavior will be determined on the basis of the probability of producing the most offspring to carry one's genes into the next generation. Females that have a high sexual investment in each offspring will maximize their fitness by a discriminate choice of a mate. A male's reproductive fitness may be maximized by frequent mating or by helping to rear young if two parents are needed for offspring to survive. Altruistic behavior has been explained on the basis of kin selection; animals will show altruistic behavior if their efforts will benefit related animals who may be carrying duplicates of their genes.

Answers to Test Your Knowledge

Multiple Choice:

1. a	6. d	11. a	16. a
2. b	7. c	12. c	17. b
3. b	8. c	13. d	18. d
4. b	9. b	14. d	
5. a	10. b	15. b	